# NATRIURETIC PEPTIDES IN HEALTH AND DISEASE

# CONTEMPORARY ENDOCRINOLOGY

P. Michael Conn, SERIES EDITOR

6. ***G Protein-Coupled Receptors and Disease,*** edited by *ALLEN M. SPIEGEL,* 1997
5. ***Natriuretic Peptides in Health and Disease,*** edited by *WILLIS K. SAMSON AND ELLIS R. LEVIN,* 1997
4. ***Endocrinology of Critical Diseases,*** edited by *K. PATRICK OBER,* 1997
3. ***Diseases of the Pituitary:*** *Diagnosis and Treatment,* edited by *MARGARET E. WIERMAN*, 1997
2. ***Diseases of the Thyroid,*** edited by *LEWIS E. BRAVERMAN,* 1997
1. ***Endocrinology of the Vasculature,*** edited by *JAMES R. SOWERS*, 1996

# Natriuretic Peptides in Health and Disease

*Edited by*

Willis K. Samson, PhD

*University of North Dakota, Grand Forks, ND*

*and*

Ellis R. Levin, MD

*Department of VA Medical Center, Long Beach, CA*

Humana Press

Totowa, New Jersey

Softcover reprint of the hardcover 1st edition 1997

999 Riverview Drive, Suite 208
Totowa, New Jersey 07512

For additional copies, pricing for bulk purchases, and/or information about other Humana titles, contact Humana at the above address or at any of the following numbers: Tel: 201-256-1699; Fax: 201-256-8341; E-mail: humana@mindspring.com or visit our website at http://humanapress.com



This publication is printed on acid-free paper. ∞
ANSI Z39.48-1984 (American National Standards Institute)
Permanence of Paper for Printed Library Materials.



Natriuretic peptides in health and disease/edited by Willis K. Samson and Ellis R. Levin.
p. cm.—(Contemporary endocrinology; 5)
Includes index.
ISBN-13: 978-1-4612-8441-3 e-ISBN-13: 978-1-4612-3960-4
DOI: 10.1007/978-1-4612-3960-4

1. Atrial natriuretic peptides—Physiological effect. 2. Atrial natriuretic peptides—Pathophysiology. I. Samson, Willis Kendrick, 1947– . II. Levin, Ellis R. III. Series: Contemporary endocrinology (Totowa, NJ); 5.
[DNLM: 1. Atrial Natriuretic Factor—physiology. QU 68 N279 1997]
QP572.A82N38 1997
612.4'05—dc21
DNLM/DLC
for Library of Congress 97-18996
CIP

# PREFACE

Beginning with the initial observation of natriuretic activity of cardiac extracts, the field of cardiovascular endocrinology has experienced tremendous growth in recent years. The discovery of the natriuretic peptides not only provided insight into the hormonal mechanisms controlling fluid and electrolyte homeostasis, but also introduced the concept of paracrine control of vascular function and further established the importance of the coordinated peripheral and central actions of vasoactive factors. The field has provided the scientific and medical communities with numerous important findings. Studies examining the mechanisms of actions of these peptides led to discoveries in cell signaling (the structure of the particulate guanylyl cyclase receptor), cell communication (the paracrine interactions of the endothelium and vascular smooth muscle), cell growth regulation (antimitogenic actions of the peptides), and gene regulation (physical factors and gene transcription). Secretion rates of these peptides are now proposed to be markers of disease process and therapeutic strategies for the treatment of heart failure have incorporated these peptides.

It has been several years since a comprehensive review of the field has been published and we have therefore assembled in *Natriuretic Peptides in Health and Disease* up-to-date reviews of the literature on molecular and cellular biology of the natriuretic peptides. Additionally, the developing relationship of the cardiovascular hormones with the immune and central nervous systems is addressed. Although no one monograph can hope completely to summarize the wealth of information available on these potent hormones, we have attempted here to detail their most widely studied aspects and to introduce novel aspects of their biochemistry, physiology, and pathology. Finally, although the field continues to move forward during production of such a monograph as this, attempts have been made to present as current a status report as possible. To be sure, more will soon be known about the importance of these peptides and their numerous, diverse effects in experimental animals and humans. It is the promise of such new developments that motivates this review of the field.

***Willis K. Samson***

# CONTENTS

# CONTRIBUTORS

TOSHIHIKO BAN, *Department of Medical and Clinical Sciences, Kyoto University College of Medicine, Kyoto, Japan*

JOHN C. BURNETT, JR., *Cardiorenal Laboratories, Department of Medicine, Mayo Foundation, Rochester, MN*

TAE-HWA CHUN, *Department of Medical and Clinical Sciences, Kyoto University College of Medicine, Kyoto, Japan*

KENTARO DOI, *Department of Medical and Clinical Sciences, Kyoto University College of Medicine, Kyoto, Japan*

ERIC A. ESPINER, *Department of Endocrinology, Christchurch Hospital, Christchurch, New Zealand*

ALASTAIR V. FERGUSON, *Department of Physiology, Queen's University School of Medicine, Kingston, Canada*

T. GEOFFREY FLYNN, *Department of Biochemistry, Queen's University School of Medicine, Kingston, Canada*

WOLF-GEORG FORSSMANN, *Lower Saxony Institute for Peptide Research, Hannover, Germany*

DAVID C. FOSTER, *Department of Pharmacology, University of Texas Southwestern Medical Center, Dallas, TX*

RONALD H. FREEMAN, *Department of Physiology, University of Missouri, Columbia, MO*

DAVID L. GARBERS, *Department of Pharmacology, University of Texas Southwestern Medical Center, Dallas, TX*

DAVID G. GARDNER, *Metabolic Research Unit, Department of Medicine, University of California at San Francisco, CA*

CHRISTOPHER C. GLEMBOTSKI, *Department of Biology, San Diego State University, San Diego, CA*

J. AARON GRANTHAM, *Cardiorenal Laboratories, Department of Medicine, Mayo Foundation, Rochester, MN*

TOSHIO IGAKI, *Department of Medical and Clinical Sciences, Kyoto University College of Medicine, Kyoto, Japan*

TADASHI IKEDA, *Department of Medical and Clinical Sciences, Kyoto University College of Medicine, Kyoto, Japan*

MAYUMI INOUE, *Department of Medical and Clinical Sciences, Kyoto University College of Medicine, Kyoto, Japan*

HIROSHI ITOH, *Department of Medical and Clinical Sciences, Kyoto University College of Medicine, Kyoto, Japan*

GORO KATSUURA, *Department of Medical and Clinical Sciences, Kyoto University College of Medicine, Kyoto, Japan*

YASATO KOMATSU, *Department of Medical and Clinical Sciences, Kyoto University College of Medicine, Kyoto, Japan*

BRANKA KOVACIC-MILIVOJEVIC, *Metabolic Research Unit, Department of Medicine, University of California at San Francisco, CA*

Ellis R. Levin, *Department of Medicine and Pharmacology, University of California at Irvine; Department of Veterans Affairs Medical Center, Long Beach, CA*
John A. Lewicki, *Scios/Nova, Mountain View, CA*
David G. Lowe, *Genentech, South San Francisco, CA*
Ken Masatsugu, *Department of Medical and Clinical Sciences, Kyoto University College of Medicine, Kyoto, Japan*
Katsuhiko Matsuda, *Department of Medical and Clinical Sciences, Kyoto University College of Medicine, Kyoto, Japan*
Yuzuru Matsuda, *Tokyo Research Laboratories, Kyowa Hakko Kogyo Co., Tokyo, Japan*
Markus Meyer, *Lower Saxony Institute for Peptide Research, Hannover, Germany*
Tonya C. Murphy, *Department of Physiology, University of North Dakota School of Medicine, Grand Forks, ND*
Kazuwa Nakao, *Department of Medical and Clinical Sciences, Kyoto University College of Medicine, Kyoto, Japan*
Yoshihiro Ogawa, *Department of Medical and Clinical Sciences, Kyoto University College of Medicine, Kyoto, Japan*
Kailash N. Pandey, *Department of Biochemistry and Molecular Biology, Medical College of Georgia, Augusta, GA*
Andrew A. Protter, *Scios/Nova, Mountain View, CA*
Zachary Resch, *Department of Physiology, University of North Dakota School of Medicine, Grand Forks, ND*
Willis K. Samson, *Department of Physiology, University of North Dakota School of Medicine, Grand Forks, ND*
Gotaro Shirakami, *Department of Medical and Clinical Sciences, Kyoto University College of Medicine, Kyoto, Japan*
Shinichi Suga, *Department of Medical and Clinical Sciences, Kyoto University College of Medicine, Kyoto, Japan*
Naohisa Tamura, *Department of Medical and Clinical Sciences, Kyoto University College of Medicine, Kyoto, Japan*
George J. Trachte, *Professor of Pharmnacology, University of Minnesota, Duluth, MN*
Makiko Ueda, *Department of Medical and Clinical Sciences, Kyoto University College of Medicine, Kyoto, Japan*
Daniel Villarreal, *Division of Cardiology, Department of Internal Medicine, University of Missouri, Columbia, MO*
Angelika M. Vollmar, *Institute of Pharmacology, Toxicology, and Pharmacy, University of Munich, Germany*
Barbara J. Wedel, *Department of Pharmacology, University of Texas Southwestern Medical Center, Dallas, TX*
Jianming Wu, *Metabolic Research Unit, Department of Medicine, University of California at San Francisco, CA*
Jun Yamashita, *Department of Medical and Clinical Sciences, Kyoto University College of Medicine, Kyoto, Japan*

# 1 The Biochemistry of Atrial Natriuretic Peptides

*T. Geoffrey Flynn*

*CONTENTS*



## INTRODUCTION

It is now some 15 yr since de Bold and Sonnenberg and colleagues discovered that the injection of atrial extracts into rats gave rise to a profound diuresis, natriuresis, and hypotension *(1)*. Within a very short time following this discovery it was shown that the agent responsible for this effect was a 28-amino acid, disulfide-bonded peptide that was given a variety of names by different investigators but is now known as atrial natriuretic peptide (ANP) *(2)*. The discovery of ANP and its biological effects heralded a new era of research on the control and maintenance of blood pressure and body volume, as well as on the specific role played by ANP. Because the discovery of ANP coincided with the development of new and improved techniques in protein sequencing and with rapid advances in recombinant DNA technology, the structure of the ANP precursor-pro-ANP and of its gene was accomplished very quickly. At the same time a large amount of information was being gathered on the physiology and pharmacology of this unique peptide. The pace of research was so rapid that reviews were being written within 2–3 yr of the discovery *(2,3)*.

Toward the end of the 1980s, additional peptides related to ANP were discovered. These peptides, named B-type natriuretic peptide (BNP) (4) and C-type natriuretic peptide (CNP) *(5)*, respectively, were structurally similar to ANP and their presence in a large number of species signified that ANP was part of a natriuretic peptide family. The natriuretic peptides exert their biological actions through three receptors *(6)*, two of which are membrane-bound guanylyl cyclases (GC-A and -B). The third serves as a clearance receptor (C-receptor). Interactions of the natriuretic peptides with their receptors are complex and not completely understood.

The biochemistry and molecular biology of the natriuretic peptides has helped us to understand the basic biology of the system and, more importantly, has given us the tools

From: *Contemporary Endocrinology: Natriuretic Peptides in Health and Disease*
Edited by: W. K. Samson and E. R. Levin Humana Press Inc., Totowa, NJ

**A**

```
              ┌─────────── S ─────────── S ───────────┐
S-L-R-R-S-S-C-F-G-G-R-I-D-R-I-G-A-Q-S-G-L-G-C-N-S-F-R-Y
1                                                     28
(99)                                               (126)
```

**B**

```
              ┌─────────── S ─────────── S ───────────┐
S-L-R-R-S-S-C-F-G-G-R-M-D-R-I-G-A-Q-S-G-L-G-C-N-S-F-R-Y
1                                                     28
(99)                                               (126)
```

**Fig. 1.** Amino acid sequence of rat ANP **(A)** *(7)* and human ANP **(B)** *(12)*. The numbers in parenthesis indicate the position occupied by ANP in pro-ANP.

to unravel the role played by these peptides in the homeostatic control of blood pressure and extracellular volume.

## ATRIAL NATRIURETIC PEPTIDE

Atrial natriuretic peptide was first purified to homogeneity from rat atria by Flynn et al. *(7)* in 1983. These workers also produced the first amino-acid sequence of the molecule (Fig. 1). Several investigators were involved initially in the purification of ANP and a variety of purification procedures were used *(8–12)*. Many of these involve an acid extraction as the key step followed by a preliminary purification on octadecylsilylsilica (Sep-Pak [Waters, Mississauga, Ontario, Canada] columns). Subsequent purification to homogeneity is achieved by high performance liquid chromatography (HPLC). Figure 2 shows the HPLC elusion profile obtained by Flynn et al. *(13)* and the legend to Fig. 2 gives the details of the purification method.

The purification obtained by Flynn et al. *(13)* gave one of the first indications that ANP was part of a larger precursor. Amino acid sequencing of peptides obtained from the peaks designated III and IV in Fig. 2 yielded amino-end extensions of ANP. (Peak II gave a much different sequence, as described below). The structure of the precursor derived

**Fig. 2.** *(opposite)* Gel-permeation chromatography of atrial extracts **(A)** and the first purification step by HPLC or atrial natriuretic peptides I **(B)**, II **(C)**, III **(D)**, and IV **(E)**. One thousand frozen rat atria were ground together with solid $CO_2$ chips and homogenized (Polytron, Brinkman, Mississauga, Ontario, Canada) in 10 vol (by weight) in 1*M* acetic acid, 1*N* HCl, and 1% NaCl. After centrifugation of the extracts, the supernatants (20-mL portions) were passed through wetted Sep-Pak (Waters) cartridges. Each cartridge was washed with 20 mL of 0.1% trifluoroacetic acid (TFA) and eluted material was combined, freeze-dried, dissolved in 5 mL of 1*M* acetic acid and 1% NaCl, and fractionated on a BioGel (Bio-Rad, Mississauga, Ontario, Canada) P-10 column (2.5 × 40 cm) equilibrated with the same solution. Fraction pools I, II, III, and IV (A) were processed further by HPLC by pumping them directly into two serially connected Vydac Separations Group (Mississauga, Ontario, Canada) $C_{18}$ columns (10 × 250 mm) that had been equilibrated with 12% acetonitrile in 0.1% TFA. Elution was carried out at 3 mL/min with acetonitrile gradients (12–44%) containing 0.1% TFA. These gradients were achieved by varying the input of the organic pump (percent of B, B–E), which delivered an 80% (weight by volume) acetonitrile solution in 0.1% TFA. Elution times for atrial natriuretic peptides I, II, III, and IV were determined by the rat bioassay *(1)* and are indicated as CI (B), CII (C), CIII (D), and CIV (E), respectively. Atrial natriuretic peptides I and II were isolated

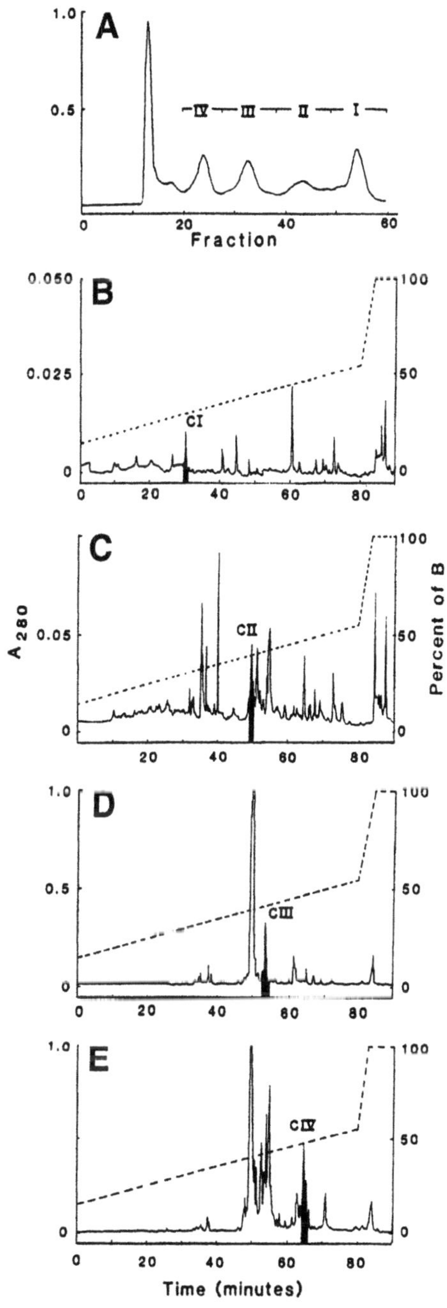


as described in *(7)* and by further purification in similar acetonitrile gradients containing 0.13% heptafluorobutyric acid as counterion. Atrial natriuretic peptides III and IV were purified further by high-performance, cation-exchange chromatography (Spherogel TSK, 4× 300 mm; Altex, Beckman, Mississauga, Ontario, Canada) with a gradient (over 60 min at 1 mL/min) of 0.010–1*M* ammonium formate buffer (pH 6.5) containing 100% acetonitrile. All atrial natriuretic peptides went through a final purification step performed in a Vydac $C_{18}$ column (4.6 × 250 mm), eluted at 1.5 mL/min with acetonitrile gradients containing 0.1% TFA. Legend and figure are reproduced with permission from Flynn TG, Davies PL, Kennedy BP, de Bold ML, de Bold AJ. Alignment of rat cardionatrin sequences with the preprocardionatrin sequence from complementary DNA. Science 1985; 228(4697): 323–325. Copyright 1985, American Association for the Advancement of Science.

```
                                                    50
GAGAGAGAAACCAGAGAGTGAGCCGAGACAGCAAACATCAGATCGTGCCCCGACCCACGCCAGC

                                          100
ATG GGC TCC TTC TCC ATC ACC AAG GGC TTC TTC CTC TTC CTG GCC TTT TGG CTC CCA GGC CAT ATT GGA GCA
Met Gly Ser Phe Ser Ile Thr Lys Gly Phe Phe Leu Phe Leu Ala Phe Trp Leu Pro Gly His Ile Gly Ala
-24             -20                                     -10                                  -1
C IV            150                                                                 200
AAT CCC GTA TAC AGT GCG GTG TCC AAC ACA GAT CTG ATG GAT TTC AAG AAC CTG CTA GAC CAC CTG GAG GAG
∇ Asn Pro Val Tyr Ser Ala Val Ser Asn Thr Asp Leu Met Asp Phe Lys Asn Leu Leu Asp His Leu Glu Glu
1                                                                                   20
                                                    250
AAG ATG CCG GTA GAA GAT GAG GTC ATG CCT CCG CAG GCC CTG AGC GAG CAG ACC GAT GAA GCG GGG GCG GCA
Lys Met Pro Val Glu Asp Glu Val Met Pro Pro Gln Ala Leu Ser Glu Gln Thr Asp Glu Ala Gly Ala Ala
                                                            40
C III                   300                                                                 350
CTT AGC TCC CTC TCT GAG GTG CCT CCC TGG ACT GGG GAA GTC AAC CCG TCT CAG AGA GAT GGA GGT GCT CTC
Leu Ser Ser Leu Ser Glu Val Pro Pro Trp Thr Gly Glu Val Asn Pro Ser Gln Arg Asp Gly Gly Ala Leu
                                            60
                                                            400
GGG CGC GGC CCC TGG GAC CCC TCC GAT AGA TCT GCC CTC TTG AAA AGC AAA CTG AGG GCT CTG CTC GCT GGC
Gly Arg Gly Pro Trp Asp Pro Ser Asp Arg Ser Ala Leu Leu Lys Ser Lys Leu Arg Ala Leu Leu Ala Gly
                            80
        C I                 450
CCT CGG AGC CTG CGA AGA TCA AGC TGC TTC GGG GGT AGG ATT GAC AGG ATT GGA GCC CAG AGC GGA CTA GGC
Pro Arg Ser Leu Arg Arg Ser Ser Cys Phe Gly Gly Arg Ile Asp Arg Ile Gly Ala Gln Ser Gly Leu Gly
        99  100                                                                             120
500                                                     550
TGC AAC AGC TTC CGG TAC CGA AGA TAA CAGCCAAATCTGCTCGAGCAGATCGCAAAAGATCCCAAGCCCTTGCGGTGTGTCACACA
Cys Asn Ser Phe Arg Tyr Arg Arg
                    126
                600                                                     650
GCTTGGTCGCATTGCCACTGAGAGGTGGTGAATACCCTCCTGGAGCTGCAGCTTCCTGTCTTCATCTATCACGATCATGTTAAGTGTAGATGAG

                    700                                                     750
TGGTTTAGTGAGGCCTTACCTCTCCCACTCTGCATATTAAGGTAGATCCTCACCCCTTTCAGAAAGCAGTTGGAAAAAAATAAATCCGAATAAAC

TTCAGCACCACGGAC
```

**Fig. 3.** Nucleotide and amino acid sequence of rat prepro-ANP and isolated cleavage products *(13)*. Nucleotides are numbered from the 5'-end, whereas the amino acid sequence is numbered starting with the first residue of prepro-ANP (indicated by ∇). The amino acid sequences of peptides isolated by HPLC separation (*see* Fig. 2) are indicated by C I, C III, and C IV. C I is the sequence of ANP, whereas C III and C IV are underlined to indicated the residues sequenced by Edman protein sequencing. Legend and figure are reproduced with permission from Flynn TG, Davies PL, Kennedy BP, de Bold ML, de Bold AJ. Alignment of rat cardionatrin sequences with the preprocardinatrin sequence from complementary DNA. Science 1985; 228(4697): 323–325. Copyright 1985, American Association for the Advancement of Science.

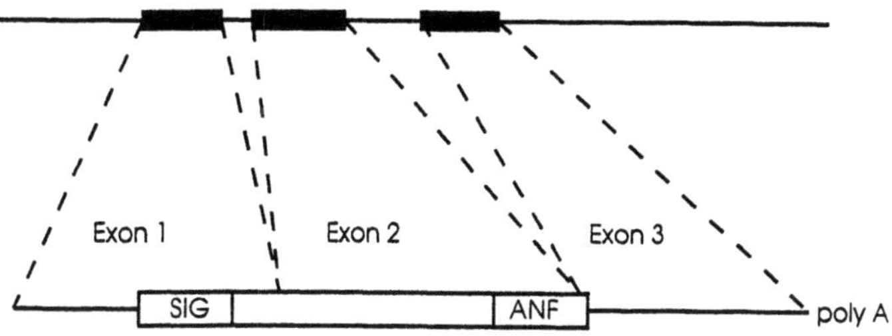


**Fig. 4.** Schematic diagram of the gene and cDNA structure for rat ANP *(18)*.

from the cDNA sequence (Fig. 3) was obtained from a number of different species by several laboratories *(13–15)*. The structure of the prohormone revealed what was previously expected, i.e., that the ANP peptide formed the C-terminal 28 amino acids of the precursor (ANP refers to the 28-amino-acid peptide). It is sometimes written ANP(1–28), and often written ANP(99–126), utilizing the numbering system of pro-ANP. Despite a very high degree of sequence identity in several parts of the amino-terminal segment, until very recently it was believed (indeed there was evidence from several laboratories to show), that only the C-terminal portion had biological activity. Several publications suggest now that peptides derived from the amino-terminal end have potent natriuretic, diuretic, and vasoactive activity (for review *see* ref. *16*). These results are mainly from a single laboratory and work from other laboratories does not support these results *(17)*. This aspect of the ANP story is discussed below.

Elucidation of the cDNA sequence paved the way for the determination of the sequence and structure of the gene for ANP. The structure of the gene for human ANP, shown in Fig. 4, is typical of the gene from other species *(15,18)*. There are three exons separated by two introns. Exon 1 codes for the signal sequence and 16 amino acids at the 5' end. Exon 2 codes for the majority of the remainder of the prohormone, whereas exon 3 codes only for the three C-terminal amino acids together with Arg-Arg in the rat and human *(19)*. The di-Arg sequence is presumably the C-terminal cleavage site.

The biochemical actions of ANP are completely dependent on its structural integrity. Reduction of the disulfide bond abolishes all biological activity *(11)*, indicating that the ring formed by the disulfide bond is critical—presumably for receptor recognition. Removal of the amino terminal residues does not appear to attenuate biological activity very much—although there is some effect if all amino-terminal residues are removed *(20)*. As far as the C-terminal end is concerned, removal of the Phe-Arg-Tyr peptide causes a marked reduction in effect on both natriuresis and blood pressure *(20–22)*, and causes a distinct reduction in receptor binding *(23)*. The importance of the C-terminal Phe-Arg was actually shown during the testing of truncated ANP analogs. The incorporation of these C-terminal amino acids, either within or outside the disulfide ring of truncated peptides, converted these peptides from relatively inactive compounds into full agonists *(24)*. Replacing amino acids within the disulfide-bridge by D-amino acids does not substantially affect the binding of these analogs for the clearance or C-receptor, but has a marked effect on binding to the guanylyl cyclase-linked receptors, underscoring the fact that structural and conformational integrity of the ANP molecule is essential for biological activity *(25)*.

As far as three-dimensional structure is concerned, there have been no X-ray crystallographic studies, and the general consensus of several NMR studies is that ANP has no

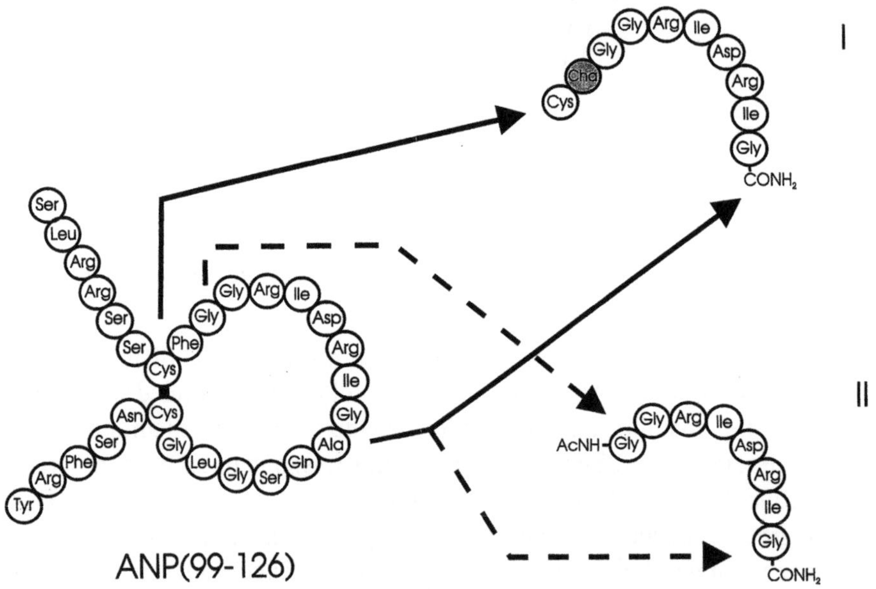


**Fig. 5.** Linear peptides derived from ANP(99–126). The figure is adapted from figures in refs. *(44,100)*.

defined tertiary structure *(26–30)*. It is very likely, therefore, that recognition by the guanylyl cyclase-linked receptors is based on amino-acid sequence coupled with induction of the correct ANP conformation to fit the receptor binding site. The best study that has been done to date in this regard is that of Fairbrother et al. *(31)*, who developed a solution configuration of an ANP heptapeptide variant *(32)* that was less flexible than previous NMR structures of ANP. Their data are compatible with a number of structure-function studies *(33–37)* that suggest this structure is likely to be similar to the receptor-bound structure. Recognition by the C-receptor is less discriminate and it is clear that a wide variety of linear and cyclic analogs are capable of binding to this receptor.

A major thrust for truncating and modifying the structure of ANP has been largely from a therapeutic standpoint. Circulating ANP has a very short half-life (0.5–2 mini *[38–41]*); therefore, a reduced-size molecule that retains full activity and has a greater half-life would be useful as a therapeutic agent that interacts with a receptor. Several laboratories have reported the synthesis and activity of a wide range of ANP analogs—both linear and nonlinear. A number of linear peptides representing different fragments of ANP have been shown to bind to the noncyclase-coupled ANP receptor *(42,43)*. These peptides, representing 7–10 amino acids of residues 106–115 of the native ANP(99–126) structure, discriminate between the two types of receptors since they retain nanomolar activity for the nonguanylyl cyclase-linked or C-receptor, but do not recognize the guanylyl cyclase-linked receptors. A linear peptide in this category is shown as peptide II in Fig. 5. One linear peptide, however, does activate particulate guanylyl cyclase, but it does so with a much reduced potency (10–100-fold) than does ANP *(44)*. This decapeptide [(cyclohexylalanine 106)ANP(105–114)NH2], peptide I in Fig. 5, is the shortest linear peptide that has agonist activity for the particulate guanylyl cyclase receptors. From a structure–activity perspective, it is interesting that the replacement of Phe 106 by cyclohexylalanine (Cha) greatly enhances receptor binding and activity in this linear

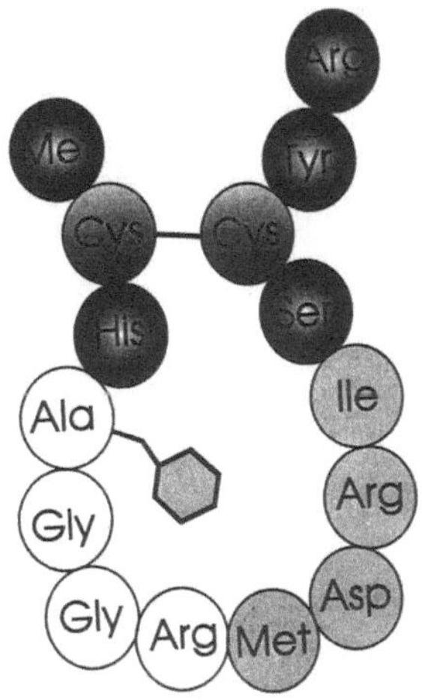


**Fig. 6.** Reduced-size (mini-ANP) cyclic ANP derived from ANP(99–126). This molecule developed by Li et al. *(45)* retains high binding affinity for receptors and biopotency.

peptide compared with other similar peptides (Fig. 5). This also appears to be the case for cyclic analogs in which the most potent have a cyclohexylalanine in place of Phe at position 106 *(24,44)*. Interestingly, one of the most potent of the Cha-containing peptides also contains the C-terminal Phe-Arg in its structure—and it does not appear to matter whether this dipeptide is within or outside the disulfide ring *(24)* (Fig. 6).

The main conclusion from the studies with linear peptides is that a sequence of seven amino acids in the cyclic core (Gly108–Gly114) form a structural element that is responsible for specific recognition of the nonguanylyl cyclase-coupled C-receptor. This linear heptapeptide does not bind to the GC-linked receptors unless it contains additional amino acids in the form of a N-terminal Cys and replacement of Phe 106 by cyclohexylalanine. This does not make the peptide cyclic—although the Cys allows the possibility of dimerization—but clearly the added hydrophobicity of the cyclohcxyl ring in some way enables this peptide to bind and activate the GC-linked receptors.

The cyclohexylalanine residue has also proved useful in recent studies on reducing the size of ANP to a smaller cyclic structure while retaining high biological potency *(45)*. Using an elegant and systematic approach, scientists identified a discontinuous functional epitope for receptor binding and activation that was placed onto a smaller disulfide ring created by repositioning the ANP-native disulfide bond (Fig. 6). The result was a peptide (mini-ANP) that bound to the GC-A receptor and stimulated cGMP production only about seven times less than the native peptide. This is a remarkable achievement. It remains to be seen if this mini-ANP is effective in vivo.

## B-TYPE NATRIURETIC PEPTIDE (BNP)

B-type natriuretic peptide (BNP) was originally isolated from pig brain as either a 26- or 32-amino acid peptide by Matsuo and his group *(4,46)*. The highest concentration is found in the heart so it is also a cardiac peptide *(47–51)* and the name brain natriuretic peptide is clearly a misnomer. In contrast to ANP, BNP is a ventricular hormone *(52,53)* and its levels are increased in diseases that affect the ventricle, e.g., congestive heart failure *(54)* and myocardial infarction *(55)*. Around the time of the discovery of pig BNP,

```
A-type Natriuretic Peptide

    Rat ANP                 SLRRSS-CFGGRIDRIGAQSGLGCNSFR-Y
    Human ANP               SLRRSS-CFGGRMDRIGAQSGLGCNSFR-Y
    Eel ANP                  SKSSSPCFGGKLDRIGSYSGLGCNS-RK
    Bullfrog ANP                SSDCFGSRIDRIGAQSGMGC-G-RRF

    Eel VNP                  KSFNS-CFGTRMDRIGSWSGLGCNSL-KNGTKKKIFGN

B-type Natriuretic Peptide

    Rat BNP                 S-K-MAHSSSCFGQKIDRIGAVSRLGCDGLRLF
    Human BNP               SPK-MVQGSGCFGRKMDRISSSSGLGCKVLRRH
    Fowl BNP                    MMRDSGCFGRRIDRIGSLSGMGCNGSRKN
    Porcine BNP             SPKTM-RDSGCFGRRLDRIGSLSGLGCNVLRRY
    Canine BNP              SPK-MMHKSGCFGRRLDRIGSLSGLGCNVLRKY
    Bovine ASIF              PK-MMRDSGCFGRRLDRIGSLSGLGCNVLRRY

C-type Natriuretic Peptide

    Rat CNP                 GLSKGCFGLKLDRIGSMSGLGC
    Human CNP               GLSKGCFGLKLDRIGSMSGLGC
    Porcine CNP             GLSKGCFGLKLDRIGSMSGLGC
    Mouse CNP               GLSKGCFGLKLDRIGSMSGLGC
    Killifish brain CNP     GWNRGCFGLKLDRIGSMSGLGC
    Eel CNP                 GWNRGCFGLKLDRIGSLSGLGC

D-type Natriuretic Peptide
    Green Mamba DNP         EVKYDPCFGHKIDRINHVSNLGCPSLRDPRPNAPSTSA
```

**Fig. 7.** Sequence alignment of members of the natriuretic family. Regions of identity within a family are indicated by boxes. Sequences were as follows: **A-type natriuretic peptides:** rat ANP *(7)*, human ANP *(12)*, eel ANP *(101)*, bullfrog ANP *(102)*, eel VNP *(103)*; **B-type natriuretic peptides:** rat BNP *(56)*, human BNP *(47)*, fowl BNP *(104)*, porcine BNP *(4)*, canine BNP *(105)*, bovine ASIF *(106)*; **C-type natriuretic peptides:** Rat CNP *(107)*, human CNP *(108)*, porcine CNP *(5)*, mouse CNP *(86)*, killifish CNP *(109)*, eel CNP *(110)*; and **D-type natriuretic peptides:** green mamba DNP *(111)*. DNP may represent a new category. It is present in the venom of the green mamba *(Dendroaspis angusticeps)* as a 38-residue peptide. It recognizes GC-A and C-receptor, but not the CNP-specific GC-B receptor.

Flynn and coworkers isolated a 45-amino acid peptide with a similar amino-acid sequence from rat hearts, which they termed iso-ANP *(56)*. This peptide is clearly the rat form of BNP *(57)*, but a single sequence difference (or a sequence error) did not gain it the same recognition as pig BNP. It may be of interest to note that the peptide giving rise to peak II in Fig. 2 was actually BNP. Even though this substance exhibited natriuretic and diuretic activity, the sequence of its amino terminus was so unlike that of ANP that it was put aside in favor of research on ANP.

B-type natriuretic peptide has been purified and sequenced from a number of different species. The similarity of the overall BNP sequence to ANP is, of course, quite striking

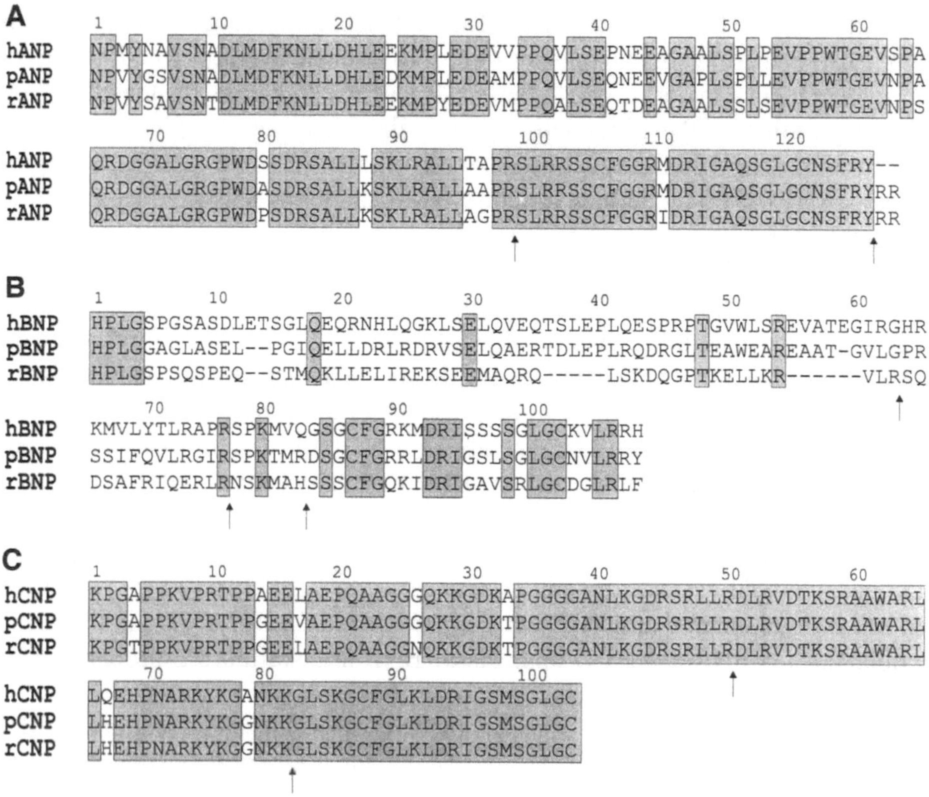


**Fig. 8.** Amino acid sequence alignment of human, pig, and rat natriuretic peptide precursors. Shaded boxes indicate conservation of identical amino acid sequences. **(A)** Human *(15)*, pig *(112)*, and rat *(13)* pro-ANPs. **(B)** Human *(61)*, pig *(59)*, and rat *(60)* pro-BNPs. **(C)** Human *(108)*, pig *(113)*, and rat *(107)* pro-CNPs. Arrows in (A) indicate cleavage sites for the processing of the mature hormone ANP(99–126). (B) arrows indicate the cleavage sites for processing of BNP-45 (pig and rat), BNP-32 (pig, rat, and human), and BNP-26 (pig and rat); the mature forms of the hormone. Arrows in (C) denote the cleavage sites for CNP-53 and CNP-22, which are believed to be the active forms of the hormone.

(Fig. 7). BNPs contain a disulfide bridge and several of the intraring residues are identical or homologous to those of ANP. What is quite obvious and different, though, about sequence comparisons among the BNPs is the degree of sequence difference (Fig. 7). Apart from certain residues enclosed by the disulfide ring, there is very little sequence identity. BNPs, in fact, may differ by as much as 50% in amino acid sequence *(58)*. This is also true for sequence comparisons of the prohormones (Fig. 8). Like ANP, BNP is synthesized as a precursor-pre-pro-BNP and the cDNA for the precursor has been cloned and sequenced from pig *(46,59)*, rat *(60)*, and human *(61)* sources. In each case the peptides form the C-terminal end of the pro-BNP structure and, like ANP, incorporate a disulfide bridge. The posttranslational processing of BNP precursors appears to be quite different to that occurring for ANP, at least in rat and humans, where it seems that processing of BNP precursors occurs within the heart, in contrast to ANP, in which proteolytic cleavage occurs at or during secretion *(61,62)*. The net result, however, is essentially the same—the biologically active peptide is cleaved from the C-terminus of the precursor protein. The structure of BNP yields little clue regarding any differential function from ANP. The peptide is as potent as ANP in pharmacological doses in invoking hypotension, natriuresis, and diuresis *(4,63)*. BNP binds to receptors but it does not appear to interact in a major way with the GC-B receptor *(6,64,65)*. On the other hand, there is a report that BNP preferentially activates the GC-A receptor *(66)*.

So far as three-dimensional structure is concerned, there is some controversy regarding whether BNP has a well-defined structure. NMR studies of BNP in aqueous solution show that the peptide appears to be flexible with the possibility of some β-type structure in the C-terminus *(67,68)*. Similar conclusions were reached for both BNP and ANP using circular dichroism (CD) *(69)*. On the other hand, NMR of BNP in dimethyl sulfoxide suggest that the peptide has a well-defined structure *(70)*. The reason for this discrepancy is not clear, but it probably indicates that the secondary structure of peptides is very sensitive to the nature of the solvent, the presence of different counterions, and pH. Like ANP, however, it may well be that BNP has no defined conformation in a bulk solution, like water, and that it only adopts its biologically active conformation when bound to its receptor. This idea is, in fact, supported for ANP by the results of infrared and CD spectroscopic studies in the presence of membrane-simulating reagents, such as anionic phospholipid vesicles or sodium dodecyl sulfate (SDS) micelles, which suggested much more ordered structures for the peptide *(26,27,29,71,72)*. Similar results were obtained for BNP using CD analysis of BNP in hexafluoro-2-propanol-water solutions *(69)*.

Little work has been done on the necessity of the various parts of the sequence of BNP for activity. It is presumed that the disulfide-bonded ring is essential, but the variation in sequence of BNPs from species to species makes it difficult to assess the importance of various residues, particularly those in the N- and C- termini. In one study it was shown that deletion of the C-terminus of BNP had very little effect compared to a similar deletion in ANP *(73)*. What was apparent from this study, also, is that the ring structures of both BNP and CNP are more important for receptor recognition than is the ring structure of ANP. From the work that has gone on so far with all of the natriuretic peptides, it would appear that the precise conformation of any of the natriuretic peptides is so governed by interaction with the appropriate receptor that the only meaningful picture of the conformation of these peptides will come from X-ray crystallography of receptor-peptide complexes.

The structure of the gene for BNP has been determined in both mouse and rat *(74–76)*. It is very similar to the gene for ANP in that it is composed of three exons and two introns, and analysis of the deduced amino-acid sequence shows that BNP is the C-terminal portion of the expressed protein. A notable difference in the genes for ANP and BNP is the presence of several copies of a motif (ATTTA) in the 3'-untranslated region that is implicated in mRNA stability *(77)*. Analysis of the genomic sequences of human and canine BNP genes *(78)* reveal that the regulatory element that appears to control tissue-specific expression of human ANP (–400 to –333 of human ANP *[79]*) is not present in the 5'-flanking region of the BNP gene. This suggests that the two genes may be regulated by separate mechanisms. However, Steinhelper et al. *(74)* have shown that the genes for ANP and BNP are within 20 kb of each other on chromosome 4 of the mouse, suggesting that they may be closely linked.

What is the biological function of BNP? The peptide is present in both atria and ventricles, and in such diseases as congestive heart failure and myocardial infarction, there is greater augmentation of gene expression in the ventricle than atria. Also, in disease states that cause hypertrophy, there is greater induction and more rapid turnover of BNP mRNA and a much greater increase in circulating hormone levels over basal levels compared to ANP. For these reasons, it has been suggested that BNP may act as an emergency molecule against ventricular overload and that BNP is more in the nature of a pathophysiological than physiological hormone *(80)*. If this were the case, it might be expected that when ANP is deficient or lacking, BNP synthesis would be increased and compensate for the lack of ANP. To date, however, this has not been shown to be the case. In "knockout" mice that have a targeted ANP gene disruption *(81)*, we have not detected any increase in either BNP mRNA or in cardiac or plasma BNP (Flynn TG, Pang SC, unpublished results). Transgenic mice that overexpress the BNP gene show large elevations in plasma BNP, and these mice manifest the same features, e.g., hypotension and increased plasma cGMP, as do transgenic ANP mice *(75)*. This shows that BNP functions through the same mechanisms as does ANP, but it does not tell us anything about the basic role of BNP.

## C-TYPE NATRIURETIC PEPTIDE (CNP)

CNP is the most highly conserved of all the natriuretic peptides. The active hormone, usually the C-terminal 22 amino acids of the prohormone, is virtually identical in all species (Fig. 7) and the remainder of its precursor is 90% identical among species (Fig. 8). The CNP was originally isolated from pig brain *(5)* and it is distributed throughout the brain in concentrations that are at least one order of magnitude higher than those of ANP and BNP *(82)*, suggesting that CNP is a major peptide of the central nervous system (CNS). There is evidence, however, that CNP is present in the endothelial cells of blood vessels and may act as an endothelium-derived autocrine or paracrine regulator-mediating vasorelaxation *(83)*. Recently, CNP has been detected in seminal vesicles *(84)* and also in ovaries *(85)* and it is becoming clear that CNP plays a much different role than ANP or BNP.

The difference between CNP and the other natriuretic peptides is also evident from its interaction with the natriuretic peptide receptors. Both ANP and BNP bind to the GC-A receptor with relatively high affinity and both can effectively stimulate GC-A to cause an increase in intracellular cGMP, with BNP being about tenfold less potent *(65)*. In contrast, CNP does not bind to GC-A even at very high concentrations. The CNP only binds with high affinity to GC-B and it is the only natriuretic peptide to stimulate GC-B to produce significant amounts of cGMP. Both the gene for CNP and GC-B receptor are widely expressed in vascular walls *(83)* and throughout the CNS, and this, along with the fact that CNP is not normally found in the circulation, supports the view that CNP is synthesized and acts locally in the peripheral and CNS.

The gene for CNP is similarly constructed to those of BNP and CNP. It consists of at least two exons and one intron, and the expressed gene product is a preprohormone of 126 amino acids, the C-terminal 22 amino acids of which consist of the active peptide *(86)*. The 5'-flanking region of mouse CNP contains an array of *cis*-acting elements and a dinucleotide CA repeat (microsatellite) that are not present in the 5'-flanking regions of either the ANP or BNP genes. Also, taking advantage of microsatellite polymorphisms among recombinant strains of mice, Ogawa et al. *(86)* have mapped the gene for CNP to chromosome 1. Both ANP and BNP genes map to chromosome 4 and therefore CNP appears to be quite distinct in many aspects to the other natriuretic peptides.

The distinctiveness of CNP is also evident from its structure. Clearly, the molecular action of all of the natriuretic peptides is dependent on the presence of a disulfide ring enclosing certain residues critical for function. In terms of mechanism of action and receptor interaction it is difficult to define a role for residues outside the ring. Elimination of N-terminal residues of ANP does not seem to greatly impair function, but truncation of C-terminal residues has a much greater effect. The CNP, however, has no C-terminal tail, yet it appears to bind to GC-B and elicit a cGMP response. It also appears to be able to bind to the C-receptor, as do virtually all ANP-like peptides. Clearly, the recognition of various natriuretic peptides by the different receptors depends very critically on the conformation of the peptide and on the nature and sequence of certain amino acids. Li et al. *(45)* have gone a long way to defining the functional epitopes and structural requirements for the binding of ANP to the GC-A receptor, but little, if any, work has been done yet on why some residues seem to be of critical importance in some natriuretic peptides yet can be dispensed with in others (either artificially or naturally) and leave the molecule with biological potency.

## URODILATIN

Urodilatin is the name given to an ANP homolog that appears only in the urine. It was first purified by Forssmann and colleagues *(87)* from human urine and its amino acid sequence reveals that it is an N-terminal extension of ANP. It in fact comprises residues 95–126 of pro ANP. Urodilatin is presumed to be a kidney-derived peptide *(88)* since it is not present in the circulation *(89)* and there is evidence for the synthesis of pro-ANP in distal cortical nephrons *(90)*. The biological actions of urodilatin closely resemble those of ANP *(91)*, and it is believed in some quarters that urodilatin rather than ANP regulates renal sodium excretion *(92)*.

It is possible that urodilatin is a renal hormone and acts in a paracrine fashion in the kidney. The only difference between ANP and urodilatin is an additional four amino acids at the N-terminus, which raises the question regarding whether urodilatin is functionally different from ANP because of this amino-terminal extension. The NMR studies show that urodilatin, like ANP and BNP, has no defined structure in solution *(93)* and the biological effects of urodilatin closely resemble those of ANP. One has to conclude, therefore, that urodilatin interacts with receptors in much the same way as ANP and that the additional four amino acids make little difference to its mode of action in comparison with ANP.

## PRO-ANP AMINO-TERMINAL PEPTIDES

There is evidence that peptides derived from the N-terminus of pro-ANP are biologically active. Vesely and his coworkers *(16)* have obtained evidence that in addition to ANP(1–98), three other peptides derived from the N-terminus of pro-ANP circulate in the blood. These peptides, ANP(1–30), ANP(31–67), and ANP(79–88), are designated long-acting sodium stimulator, vessel dilator, and kaliuretic stimulator, respectively, according to their reported functions *(94)*. They have much longer plasma half-lives than ANP and circulate at higher concentrations *(95–97)*. It is clear that these peptides do not act through the ANP receptor system *(17)* and there is controversy regarding whether they have specific binding sites *(17,98)*. Much is puzzling about their mode of action. A major criticism is that studies previously carried out with these peptides utilized nonhomologous species and peptides, i.e., human peptide in rat. In our own laboratory using rat ANP(11–30) and (31–67), as well as rats and rat tissues, we found no evidence that either ANP(1–30) or ANP(31–67) had any biological activity. The peptides did not displace radioactive ANP from binding sites on plasma membranes and did not appear to have distinct binding sites in either lung or kidney *(17)*. Another puzzling feature about these peptides is that they only appear to work in the intact animal. If the abdomen is opened, say, for placement of a renal catheter, then the effect of these peptides is markedly attenuated. In fact, in the rat their natriuretic effects may disappear completely without an intact abdomen *(16)*.

The implications of the presence of these N-terminally derived peptides seems to lie in their proposed long-term effects, in contrast to the acute effects of ANP in responding to volume changes *(94)*. Much more work needs to be done on these potentially important peptides by other workers in the field so that their exact role in volume homeostasis can be defined. An interesting and potentially important finding regarding these N-terminal peptides is that the plasma concentration of ANP(31–67) increases in early heart failure *(96)* even in otherwise asymptomatic people. This can, therefore, be used as a distinguishing feature. Indeed, Burnett and his group have used an RIA to ANP(1–25) to detect N-terminal ANPs and to use this as a marker for symptomless left ventricular dysfunction, i.e., to distinguish early signs of heart failure in healthy patients with no signs of congestive heart failure *(99)*. None of these studies, of course, address the mechanism of action or physiological importance of these N-terminal pro-ANP peptides. Why their concentrations are raised under some circumstances is an intriguing question and merits further investigation.

## CONCLUSIONS AND PERSPECTIVE

The mechanism of action of the atrial natriuretic peptides in relation to their structure and their physiology is not completely understood. A great deal of work has gone into trying to define the essential and minimal functional entities that allow the molecule to elicit a biological response. To some extent, this has been successful as far as ANP is concerned and a reasonable picture has emerged of those features that are essential for receptor binding. A precise picture of how any of the natriuretic peptides bind to receptors will not be forthcoming until crystallography of a receptor–ligand complex is obtained. Even then there will be a unilateral view. This whole area of structural analysis is complicated by the fact that there are at least three receptors and at least three major peptides, each of which has a sufficiently different structure to suggest that their interactions with each receptor will be more than subtly different. A further, limiting complication is that the natriuretic peptide receptors are all membrane-bound and expression of the extracellular, heavily glycosylated, ligand-binding domains has proved difficult.

The solution of these problems and the presentation of the three-dimensional structure of a receptor–natriuretic complex will, of course, be a landmark event and will pave the way for the development of nonpeptide natriuretic peptide analogs. The natriuretic peptide field has come a long way since Adolfo de Bold's pioneering work in the early 1980s. There is no doubt that there is much more exciting work to come.

## ACKNOWLEDGMENT

The author thanks David Hyndman for his considerable help in the preparation of this chapter. Research on natriuretic peptides in the author's laboratory is funded by the Medical Research Council of Canada.

## REFERENCES


1. de Bold AJ, Borenstein HB, Veress AT, Sonnenberg H. A rapid and potent natriuretic response to intravenous injection of atrial myocardial extract in rats. Life Sci 1981;28:89–94.
2. Flynn TG, Davies PL. The biochemistry and molecular biology of atrial natriuretic factor. Biochem J 1985;232:313–321.
3. Cantin M, Genest J. The heart and the atrial natriuretic factor. Endocr Rev 1985;6:107–127.
4. Sudoh T, Kangawa K, Minamino N, Matsuo H. A new natriuretic peptide in porcine brain. Nature 1988;332:78–81.
5. Sudoh T, Minamino N, Kanagawa K, Hisayauki M. C-type natriuretic peptide (CNP): a new member of natriuretic peptide family identified in porcine brain. Biochem Biophys Res Commun 1990;168: 863–870.
6. Koller KJ, Goeddel DV. Molecular biology of the natriuretic peptides and their receptors. Circulation 1992;86:1081–1088.
7. Flynn TG, de Bold ML, de Bold AJ. The amino acid sequence of an atrial peptide with potent diuretic and natriuretic properties. Biochem Biophys Res Commun 1983;117:859–865.
8. Cahill PA, Hassid A. ANF-C-receptor-mediated inhibition of aortic smooth muscle cell proliferation and thymidine kinase activity. Am J Physiol Regul Integr Comp Physiol 1994;266: R194–R203.
9. Currie MG, Geller DM, Cole BR, Siegel NR, Fok KF, Adams SP, Eubanks SR, Galluppi GR, Needleman P. Purification and sequence analysis of bioactive atrial peptides (atriopeptins). Science 1984;223(4631):67–69.
10. Atlas SA, Kleinert HD, Camargo MJ, Januezewicz A, Sealey JE, Laragh JH, Schilling JW, Lewicki JA, Johnson LK, Maack T. Purification, sequencing and synthesis of natriuretic and vasoactive rat atrial peptide. Nature 1984;309(5970):717–719.
11. Misono KS, Fukumi H. Grammer RT, Inagami T. Rat atrial natriuretic factor: complete amino acid sequence and disulfide linkage essential for biological activity. Biochem Biophys Res Commun 1984;119:524–529.
12. Kangawa K, Matsuo H. Purification and complete amino acid sequence of a-human atrial natriuretic polypeptide (a-hANP). Biochem Biophys Res Commun 1984;118:131–139.
13. Flynn TG, Davies PL, Kennedy BP, de Bold ML, de Bold AJ. Alignment of rat cardionatrin sequences with the preprocardionatrin sequence from complementary DNA. Science 1985;228(4697): 323–325.
14. Seidman CE, Duby AD, Choi E, Graham RM, Haber E, Homcy C, Smith JA, Seidman JG. The structure of rat preproatrial natriuretic factor as defined by a complementary DNA clone. Science 1984;225(4659):324–326.
15. Oikawa S, Imai M, Ueno A, Tanaka S, Noguchi T, Nakazato H, Kangawa K, Fukuda A, Matsuo H. Cloning and sequence analysis of cDNA encoding a precursor for human atrial natriuretic polypeptide. Nature 1984;309(5970):724–726.
16. Vesely DL. Atrial natriuretic hormones originating from the N-terminus of the atrial natriuretic factor prohormone. (Review). Clin Exp Pharm Phys 1995;22:108–114.
17. Weir ML, Honrath U, Flynn TG, Sonnenberg H. Lack of biologic activity or specific binding of amino-terminal pro-ANP segments in the rat. Regul Pept 1994;53:111–122.
18. Argentin S, Nemer M, Drouin J, Scott GK, Kennedy BP, Davies PL. The gene for rat atrial natriuretic factor. J Biol Chem 1985;260:4568–4571.
19. Lewicki JA, Greenberg B. Yamanaka M, Vlasuk G, Brewer M, Gardner D, Baxter J, Johnson LK, Fiddes JC. Cloning, sequence analysis, and processing of the rat and human atrial natriuretic peptide precursors. Fed Proc 1986;45:2086–2090.
20. Thibault G, Garcia R, Carrier F, Seidah NG, Lazure C, Chrétien M, Cantin M, Genest J. Structure–activity relationships of atrial natriuretic factor (ANF). I. Natriuretic activity and relaxation of intestinal smooth muscle. Biochem Biophys Res Commun 1984;125:938–946.
21. Sugiyama M, Fukumi H, Grammer RT, Misono KS, Yabe Y, Morisawa Y, Inagami T. Synthesis of atrial natriuretic peptides and studies on structural factors in tissue specificity. Biochem Biophys Res Commun 1984;123:338–344.
22. Garcia R, Thibault G, Seidah NG, Lazure C, Cantin M, Genest J, Chretien M. Structure–activity relationships of atrial natriuretic factor (ANF). II. Effect of chain-length modifications on vascular reactivity. Biochem Biophys Res Commun 1985;126:178–184.

23. De Lean A, Thibault G, Seidah NG, Lazure C, Gutkowska J, Chretien M, Genest J, Cantin M. Structure–activity relationships of atrial natriuretic factor (ANF). III. Correlation of receptor affinity with relative potency on aldosterone production in zone glomerulosa cells. Biochem Biophys Res Commun 1985;132:360–367.
24. Holleman WH, Budzik GP, Devine EM, Pollock DM, Opgenorth TJ, Thomas AM, Von Geldern TW, Rockway TW. Truncated atrial natriuretic factor analogs retain full agonist activity. Can J Physiol Pharmacol 1991;69:1622–1627.
25. Scarborough RM, McEnroe GA, Arfsten A, Kang L-L, Schwark K, Lewicki JA. D-amino acid-substituted atrial natriuretic peptide analogs reveal novel receptor recognition requirements. J Biol Chem 1988;263:16,818–16,822.
26. Kobayashi Y, Ohkubo T, Kyogoku Y, Koyama S, Kobayashi M, Go N. The conformation of a-human atrial natriuretic polypeptide in solution. J Biochem 1988;104:322–325.
27. Fesik SW, Holleman WH, Perun TJ. Two-dimensional 1H NMR studies of rat atrial natriuretic factor (1–23). Biochem Biophys Res Commun 1985;131:517–523.
28. Gampe RT Jr, Connolly PJ, Rockway T, Fesik SW. Two-dimensional NMR studies of [Pro-10] atrial natriuretic factor [7–23]. Biopolymers 1988;27:313–321.
29. Olejniczak ET, Gampe RT Jr, Rockway TW, Fesik SW. NMR study of the solution conformation of rat atrial natriuretic factor 7–23 in sodium dodecyl sulfate micelles. Biochemistry 1988;27: 7124–7131.
30. Theriault Y, Boulanger Y, Weber PL, Reid BR. Two-dimensional 1H-NMR investigation of the water conformation of the atrial natriuretic factor (ANF 101–126). Biopolymers 1987;26: 1075–1086.
31. Fairbrother WJ, McDowell RS, Cunningham BC. Solution conformation of an atrial natriuretic peptide variant selective for the type A receptor. Biochemistry 1994;33:8897–8904.
32. Cunningham BC, Lowe DG, Li B, Bennett BD, Wells JA. Production of an atrial natriuretic peptide variant that is specific for type A receptor. EMBO J 1994;13:2508–2515.
33. Nutt RF, Veber DF. Chemical synthesis and structure–activity relations for ANF analogues. Endocrinol Metab Clin North Am 1987;16:19–41.
34. Fok KF, Tjoeng FS, Houbion JA, Spear KL, Nugent ST, Eubanks SR, Zupec ME, Olins GM, Blehm DJ, Adams SP, Deber CM, Hruby VJ, Kopple KD, eds. Peptides: Structure and Function. Pierce Chemical Co., Rockford, IL, 1985, pp. 953–956.
35. Brady SF, Ciccarone TM, Williams TM, Veber DF, Nutt RF. Rivier JE, Marshall GR, eds. Peptides: Chemistry, Structure and Biology. ESCOM, Leiden, The Netherlands, 1990, pp. 598–599.
36. Nutt RF, Brady SF, Lyle TA, Ciccarone TM, Colton CD, Paleveda WJ, Williams TM, Smith GM, Winquist RJ, Veber DF, Tam JP, Kaiser ET, eds. Synthetic Peptides: Approaches to Biological Problems. Liss, New York, 1989, pp. 267–279.
37. Von Geldern TW, Rockway TW, Davidsen SK, Budzik GP, Bush EN, Chu-Moyer MY, Devine EM Jr., Holleman WH, Johnson MC, Lucas SD, et al. Small atrial natriuretic peptide analogues: design, synthesis, and structural requirements for guanylate cyclase activation. J Med Chem 1992;35: 808–816.
38. Katsube N, Schwartz D, Needleman P. Atriopeptin turnover: quantitative relationship between in vivo changes in plasma levels and atrial content. J Pharmacol Exp Ther 1986;239:474–479.
39. Murthy KK, Thibault G, Schiffrin EL, Garcia R, Chartier L, Gutkowska J, Genest J, Cantin M. Disappearance of atrial natriuretic factor from circulation in the rat. Peptides 1986;7:241–246.
40. Nakao K, Sugawara A, Morli N, Sakamoto M, Yamada T, Itoh H, Shiono S, Saito Y, Nishimura K, Ban T, et al. The pharmacokinetics of alpha-human atrial natriuretic polypeptide in healthy subjects. Eur J Clin Pharmacol 1986;31:101–103.
41. Yandle TG, Richards AM, Nicholls MG, Cuneo R, Espiner EA, Livesey JH. Metabolic clearance rate and plasma half life of alpha-human atrial natriuretic peptide in man. Life Sci 1986;38: 1827–1833.
42. Olins GM, Patton DR, Bovy PR, Mehta PP. A linear analog of atrial natriuretic peptide (ANP) discriminates guanylate cyclase-coupled ANP receptors from non-coupled receptors. J Biol Chem 1988;263:10,989–10,993.
43. Scarborough RM, Schenk DB, McEnroe GA, Aristen A, Kang LL, Schwartz K, Lewicki JA. Truncated atrial natriuretic peptide analogs. Comparison between receptor binding and stimulation of cyclic GMP accumulation in cultured vascular smooth muscle cells. J Biol Chem 1986;261: 12,960–12,964.

44. Bovy PR, O'Neal JM, Olins GM, Patton DR, Mehta PP, McMahon KG, Palomo M, Schuh J, Blehm D. A synthetic linear decapeptide binds to the atrial natriuretic peptide receptors and demonstrates cyclase activation and vasorelaxant activity. J Biol Chem 1989;264:20,309–20,313.
45. Li B, Tom JYK, Oare D, Yen R, Fairbrother WJ, Wells JA, Cunningham BC. Minimization of a polypeptide hormone. Science 1995;270:1657–1660.
46. Maekawa K, Sudoh T, Furusawa M, Minamino N, Kangawa K, Ohkubo H, Nakanishi S, Matsuo H. Cloning and sequence analysis of cDNA encoding a precursor for porcine brain natriuretic peptide. Biochem Biophys Res Commun 1988;157:410–416.
47. Kambayashi Y, Nakao K, Mukoyama M, Saito Y, Ogawa Y, Shiono S, Inouye K, Yoshida N, Imura H. Isolation and sequence determination of human brain natriuretic peptide in human atrium. FEBS Lett 1990;259:341–345.
48. Hino J, Tateyama H, Minamino N, Kangawa K, Matsuo H. Isolation and identification of human brain natriuretic peptides in cardiac atrium. Biochem Biophys Res Commun 1990;167:693–700.
49. Minamino N, Aburaya M, Ueda S, Kangawa K, Matsuo H. The presence of brain natriuretic peptide of 12,000 daltons in porcine heart. Biochem Biophys Res Commun 1988;155:740–746.
50. Saito Y, Nakao K, Itoh H, Yamada T, Mukoyama M, Arai H, Hosoda K, Shirakami G, Suga S, Minamino N, Kangawa K, Matsuo H, Imura H. Brain natriuretic peptide is a novel cardiac hormone. Biochem Biophys Res Commun 1989;158:360–368.
51. Mukoyama M, Nakao K, Hosoda K, Suga S, Saito Y, Ogawa Y, Shirakami G, Jougasaki M, Obata K, Yasue H, Kambayashi Y, Inouye K, Imura H. Brain natriuretic peptide as a novel cardiac hormone in humans. Evidence for an exquisite dual natriuretic peptide system, atrial natriuretic peptide and brain natriuretic peptide. J Clin Invest 1991;87:1402–1412.
52. Ogawa Y, Nakao K, Mukoyama M, Hosoda K, Shirakami G, Arai H, Saito Y, Suga S, Jougasaki M, Imura H. Natriuretic peptides as cardiac hormones in normotensive and spontaneously hypertensive rats: The ventricle is a major site of synthesis and secretion of brain natriuretic peptide. Circ Res 1991;69:491–500.
53. Hasegawa K, Fujiwara H, Doyama K, Miyamae M, Fujiwara T, Suga S, Mukoyama M, Nakao K, Imura H, Sasayama S. Ventricular expression of brain natriuretic peptide in hypertrophic cardiomyopathy. Circulation 1993;88:372–380.
54. Mukoyama M, Nakao K, Saito Y, Ogawa Y, Hosoda K, Suga S, Shirakami G, Jougasaki M, Imura H. Increased human brain natriuretic peptide in congestive heart failure. N Engl J Med 1990;323:757,758.
55. Morita E, Yasue H, Yoshimura M, Ogawa H, Jougasaki M, Matsumura T, Mukoyama M, Nakao K. Increased plasma levels of brain natriuretic peptide in patients with acute myocardial infarction. Circulation 1993;88:82–91.
56. Flynn TG, Brar A, Tremblay L, Sarda I, Lyons C, Jennings DB. Isolation and characterization of ISO-rANP, a new natriuretic peptide from rat atria. Biochem Biophys Res Commun 1989;161: 830–837.
57. Kambayashi Y, Nakao Y, Itoh H. Hosoda K, Saito Y, Yamada T, Mukoyama M, Arai H, Shirakami G, Suga S, Ogawa Y, Jougasaki M, Minamino N, Kangawa K, Matsuo H, Inouye K, Imura H. Isolation and sequence determination of rat cardiac natriuretic peptide. Biochem Biophys Res Commun 1989;163:233–240.
58. Inagami T. Atrial natriuretic factor. J Biol Chem 1989;264:3043–3046.
59. Porter JG, Arfaten A, Palisi T. Scarborough RM, Lewicki JA, Seilhamer JJ. Cloning of a cDNA encoding porcine brain natriuretic peptide. J Biol Chem 1989;264:6689–6692.
60. Kojima M, Minamino N, Kangawa K, Matsuo H. Cloning and sequence analysis of cDNA encoding a precursor for rat brain natriuretic peptide. Biochem Biophys Res Commun 1989;159:1420–1426.
61. Sudoh T, Maekawa K, Kojima M, Minamino N, Kangawa K, Matsuo H. Cloning and sequence analysis of cDNA encoding a precursor for human brain natriuretic peptide. Biochem Biophys Res Commun 1989;159:1427–1434.
62. Ogawa Y, Nakao K, Mukoyama M, Shirakami G, Itoh H, Hosoda K, Saito Y, Arai H, Suga S, Jougasaki M, Yamada T, Kambayashi Y, Inouye K, Imura H. Rat brain natriuretic peptide—Tissue distribution and molecular form. Endocrinology 1990;126:2225–2227.
63. Jennings DB, Flynn TG. Cardiovascular and renal effects of iso-rANP, a second natriuretic peptide from rat atria. Can J Physiol Pharmacol 1989;67:1372–1379.
64. Koller KJ, Lowe DG, Bennett GL, Minamino N, Kangawa K, Matsuo H, Goeddel DV. Selective activation of the B natriuretic peptide receptor by C-type natriuretic peptide (CNP). Science 1991;252:120–123.

65. Koller KJ, Lowe DG, Minamino N, Matsuo H, Kanagawa K, Goeddel DV. Differential activation of the human natriuretic peptide receptor guanylate kineses. In: Imura H. Matsuo H. Masaki T, eds. *Peptide Regulation of Cardiovascular Function.* Takeda Scilence Foundation, Osaka, 1991, pp. 91–99.
66. Iida T, Hirata Y, Takemura N, Togashi K, Nakagawa S, Marumo F. Brain natriuretic peptide is cosecreted with atrial natriuretic peptide from porcine cardiocytes. FEBS Lett 1990;260:98–100.
67. Craik D, Munro S, Nielsen K, Shehan P, Tregear G, Wade J. The conformation of porcine-brain natriuretic peptide by two-dimensional NMR spectroscopy. Eur J Biochem 1991;201:183–190.
68. Nielsen KJ, Barron PR, Craik DJ, Lambert PF, Wade JD. Two-dimensional nuclear magnetic resonance study of brain natriuretic peptide in aqueous solution. Biochem Int 1991;25:199–210.
69. Mimeault M, De Lean A, Lafleur M, Bonenfant D, Fournier A. Evaluation of conformational and binding characteristics of various natriuretic peptides and related analogs. Biochemistry 1995;34: 955–964.
70. Inooka H, Kikuchi T, Endo S, Ishibashi Y, Wakimasu M, Mizuta E. Conformation in solution of porcine brain natriuretic peptide determined by combined use of nuclear magnetic resonance and distance geometry. Eur J Biochem 1990;193:127–134.
71. Surewicz WK, Mantsch HH, Stahl GL, Epand RM. Infrared spectroscopic evidence of conformational transitions of an atrial natriuretic peptide. Proc Natl Acad Sci USA 1987;84:7028–7030.
72. Epand RM, Stahl GL. Interaction of atriopeptin III with lipids and detergents. Int J Pept Protein Res 1987;29:238–243.
73. Shimekake Y, Kawabata T, Nakamura M, Nagata K. The role of the C-terminal region of rat brain natriuretic peptide in receptor selectivity. FEBS Lett 1992;309:185–189.
74. Steinhelper ME. Structure, expression, and genomic mapping of the mouse natriuretic peptide type-B gene. Circ Res 1993;72:984–992.
75. Ogawa Y, Itoh H, Tamura N, Suga S-l, Yoshimasa T, Uehira M, Matsuda S, Shiono S, Nishimoto H, Nakao K. Molecular cloning of the complementary DNA and gene that encode mouse brain natriuretic peptide and generation of transgenic mice that overexpress the brain natriuretic peptide gene. J Clin Invest 1994;93:1911–1921.
76. Roy RN, Flynn TG. Organization of the gene for *iso*-rANP, a rat B-type natriuretic peptide. Biochem Biophys Res Commun 1990;171:416–423.
77. Shaw G, Kamen R. A conserved AU sequence from the 3' untranslated region of GM-CSF mRNA mediates selective mRNA degradation. Cell 1986;46:659–667.
78. Seilhamer JJ, Arfsten A, Miller JA, Lundquist P, Scarborough RM, Lewicki JA, Porter JG. Human and canine gene homologs of porcine brain natriuretic peptide. Biochem Biophys Res Commun 1989;165:650–658.
79. LaPointe MC, Wu J, Greenberg B, Gardner DG. Upstream sequences confer atrial-specific expression on the human atrial natriuretic factor gene. J Biol Chem 1988;263:9075–9078.
80. Ogawa Y, Itoh H, Nakao K. Molecular biology and biochemistry of natriuretic peptide family. [Review]. Clin Exp Pharmacol Physiol 1995;22:49–53.
81. John SW, Krege JH, Oliver PM, Hagaman JR, Hodgin JB, Pang SC, Flynn TG, Smithies O. Genetic decreases in atrial natriuretic peptide and salt-sensitive hypertension [published erratum appears in Science 1995 Mar 24;267(5205):1753]. Science 1995;267:679–681.
82. Komatsu Y, Nakao K, Suga S, Ogawa Y, Mukoyama M, Arai H, Shirakami G, Hosoda K, Nakagawa O, Hama N, Kishimoto I, Imura H. C-type natriuretic peptide (CNP) in rats and humans. Endocrinology 1991;129:1104–1106.
83. Komatsu Y, Nakao K, Itoh H, Suga S-I, Ogawa Y, Imura H. Vascular natriuretic peptide. Lancet 1992;340:622.
84. Chrisman TD, Schulz S, Potter LR, Garbers DL. Seminal plasma factors that cause large elevations in cellular cyclic GMP are C-type natriuretic peptides. J Biol Chem 1993;268:3698–3703.
85. Huang H, Acuff CG, Steinhelper ME. Am J Physiol 1996;271:H1565–1575.
86. Ogawa Y, Itoh H, Yoshitake Y, Inoue M, Yoshimasa T, Serikawa T, Nakao K. Molecular cloning and chromosomal assignment of the mouse C-type natriuretic peptide (CNP) gene (Nppc): comparison with the human CNP gene *(NPPC)*. Genomics 1994;24:383–387.
87. Schulz-Knappe P, Forssmann K, Herbst F, Hock D, Pipkorn R, Forssmann WG. Isolation and structural analysis of "urodilatin," a new peptide of the cardiodilatin-(ANP)-family, extracted from human urine. Klin Wochenschr 1988;66:752–759.
88. Abassi ZA, Golomb E, Klein H, Keiser HR. Urodilatin: a natriuretic peptide of renal origin. Cardiovasc Drug Rev 1992;10:199–210.

89. Saxenhofer H, Raselli A, Weidmann P, Forssmann W-G, Bub A, Ferrari P, Shaw SG. Urodilatin, a natriuretic factor from kidneys, can modify renal and cardiovascular function in men. Am J Physiol Renal, Fluid Electrolyte Physiol 1990;259:F832–F838.
90. Greenwald JE, Needleman P, Wilkins MR, Schreiner GF. Renal synthesis of atriopeptin-like protein in physiology and pathophysiology. Am J Physiol Renal, Fluid Electrolyte Physiol 1991;260: F602–F607.
91. Goetz KL. Renal natriuretic peptide (urodilatin?) and atriopeptin: evolving concepts. Am J Physiol Renal Fluid Electrolyte Physiol 1991;261:F921–F932.
92. Goetz K, Drummer C, Zhu JL, Leadley R, Fiedler F, Gerzer R. Evidence that urodilatin, rather than ANP, regulates renal sodium excretion. J Am Soc Nephrol 1990;1:867–874.
93. Weber J, Rosch P, Adermann K, Forssmann WG, Wokaun A. $^1$H-NMR studies of the natriuretic peptide urodilatin: sequence-specific resonance assignment. Biochem Biophys Acta Protein Struct Mol Enzymol 1994;1207:231–235.
94. Vesely DL, Douglass MA, Dietz JR, Gower WR, Jr., McCormick MT, Rodriguez-Paz G, Schocken DD. Three peptides from the atrial natriuretic factor prohormone amino terminus lower blood pressure and produce diuresis, natriuresis, and/or kaliuresis in humans. Circulation 1994;90:1129–1140.
95. Winters CJ, Sallman AL, Meadows J, Rico DM, Vesely DL. Two new hormones: prohormone atrial natriuretic peptides 1–30 and 31–67 circulate in man. Biochem Biophys Res Commun 1988;150: 231–236.
96. Winters CJ, Sallman AL, Baker BJ, Meadows J, Rico DM, Vesely DL. The N-terminus and a 4,000-MW peptide from the midportion of the N-terminus of the atrial natriuretic factor prohormone each circulate in humans and increase in congestive heart failure. Circulation 1989;80:438–449.
97. Ackerman BH, Wyeth RP, Vesely DL, Ngo WL, Bissett JK, Winters CJ, Sallman AL. Pharmacokinetic characterization of the postdistribution phase of prohormone atrial natriuretic peptides amino acids 1–98, 31–67, and atrial natriuretic factor during and after rapid right ventricular pacing in dogs. J Clin Pharmacol 1992;32:415–421.
98. Vesely DL, Cornett LE, MacLeod SL, Nash AA, Norris JS. Specific binding sites for prohormone atrial natriuretic peptides 1–30, 31–67 and 99–126. Peptides 1990;11:193–197.
99. Lerman A, Gibbons RJ, Rodeheffer RJ, Bailey KR, McKinley LJ, Heublein DM, Burnett JC Jr. Circulating N-terminal atrial natriuretic peptide as a marker for symptomless left-ventricular dysfunction. Lancet 1993;341:1105–1109.
100. Page E, Upshaw-Earley J, Goings GE, Hanck DA.1 Effect of external $Ca^{2+}$ concentration on stretch-augmented natriuretic peptide secretion by rat atria. Am J Physiol Cell Physiol 1991;260: C756–C762.
101. Takei Y, Takahashi A, Watanabe TX, Nakajima K, Sakakibara S. Amino acid sequence and relative biological activity of eel atrial natriuretic peptide. Biochem Biophys Res Commun 1989;164:537–543.
102. Sakab J, Kangawa K, Matsuo H. Identification of new atrial natriuretic peptides in frog heart. Biochem Biophys Res Commun 1988;155:1338–1345.
103. Takei Y, Takahashi A, Watanabe TX, Nakajima K, Sakakibara S. A novel natriuretic peptide isolated from eel cardiac ventricles. FEBS Lett 1991;282:317–320.
104. Miyata A, Minamino N, Kangawa K, Matsuo H. Identification of a 29-amino acid natriuretic peptide in chicken heart. Biochem Biophys Res Commun 1988;155:1330–1337.
105. Seilhamer JJ, Arfsten A, Miller JA, Lundquist P, Scarborough RM, Lewicki JA, Porter JG. Human and canine gene homologs of porcine brain natriuretic peptide. Biochem Biophys Res Commun 1989;165:650–658.
106. Nguyen TT, Lazure C, Babinski K, Chretien M, Ong H, De Lean A. Aldosterone secretion inhibitory factor: a novel neuropeptide in bovine chromaffin cells. Endocrinology 1989;124:1591–1593.
107. Kojima M, Minamino N, Kangawa K, Matsuo H. Cloning and sequence analysis of a cDNA encoding a precursor for rat c-type natriuretic peptide (CNP). FEBS Letters 1990;276:209–213.
108. Tawaragi Y, Fuchimura K, Tanaka S, Minamino N, Kangawa K, Matsuo H. Gene and precursor structures of human C-type natriuretic peptide. Biochem Biophys Res Commun 1991;175:645–651.
109. Price DA, Doble KE, Lee TD, Galli SM, Dunn BM, Parten B, Evans DH. The sequencing, synthesis, and biological actions of an ANP-like peptide isolated from the brain of the killifish *Fundulus heteroclitus*. Biol Bull 1990;178:279–285.
110. Takei Y, Takahashi A, Watanabe TX, Nakajima K, Sakakibara S, Takao T, Shimonishi Y. Amino acid sequence and relative biological activity of a natriuretic peptide isolated from eel brain. Biochem Biophys Res Commun 1990;170:883–891.

111. Schweitz H, Vigne P, Moinier D, Frelin C, Lazdunski M. A new member of the natriuretic peptide family is present in the venom of the green mamba *(Dendroaspis angusticeps)*. J Biol Chem 1992;267:13,928–13,932.
112. Mägert H-J, Appelhans H, Gassen H-G, Forssmann W-G. Nucleotide sequence of a porcine prepro atrial natriuretic peptide (AN P) cDNA. Nucleic Acids Res 1990;8:6704.
113. Tawaragi Y, Fuchimura K, Nakazato H, Tanaka S, Minamino N, Kangawa K, Matsuo H. Gene and precursor structure of porcine c-type natriuretic peptide. Biochem Biophys Res Commun 1990;172:627–632.

# 2 The Guanylyl Cyclase-A Receptor

*David C. Foster, David L. Garbers, and Barbara J. Wedel*

*CONTENTS*



## INTRODUCTION

There are at least seven plasma membrane forms of guanylyl cyclase found in mammals, one of which serves as the receptor for atrial natriuretic peptide (ANP) and probably B-type natriuretic peptide (BNP). This isoform, known as the guanylyl cyclase-A (GC-A; NPR-A) receptor, is similar in overall topology to that of all other known membrane forms of guanylyl cyclases in that it contains an extracellular ligand-binding domain, a single transmembrane region, and intracellular protein kinase homology (KHD) and cyclase catalytic domains *(1,2)*.

The physiological roles of GC-A have been principally suggested based on the effects seen with ANP or BNP, the proposed ligands for GC-A. Examples of the proposed signaling functions of GC-A include the relaxation of smooth muscle, induction of natriuresis and diuresis in the kidney, inhibition of aldosterone production in the adrenal gland, inhibition of water drinking through the hypothalamus, and decreased cardiac output *(3)*.

There have been many reviews on the natriuretic peptides and on guanylyl cyclases, and therefore in this chapter, we will concentrate on the most recent work that addresses both mechanisms of regulation and the physiological roles of the GC-A receptor.

## CELLULAR DISTRIBUTION OF GC-A

GC-A appears to be expressed in many different areas of the body, including the kidney, smooth muscle vasculature, and the adrenal gland, tissues thought to be primary

From: *Contemporary Endocrinology: Natriuretic Peptides in Health and Disease*
Edited by: W. K. Samson and E. R. Levin Humana Press Inc., Totowa, NJ

targets of ANP action. However, the receptor is also found in the intestine, eye, heart, testis, olfactory mucosa, brain, and other tissues *(3)*. Since ANP and BNP are also produced in many regions of the body other than the heart, the receptor is conceivably regulated through bloodborne signals as well as peptides produced locally.

## REGULATION OF GC-A

### *Oligomerization State*

Various laboratories have attempted to address the state of GC-A in the presence or absence of ligand. Although none of the studies clearly define whether GC-A exists as a dimer, trimer, or tetramer, the conclusion reached by each laboratory has been that the receptor exists as an oligomer in the presence or absence of ANP *(4–7)*. Whether GC-A could fit a model proposed by Vaandrager et al. *(8)* for the guanylyl cyclase-C receptor is not known. In that model, although GC-C polypeptide chains are suggested to be associated with each other independent of ligand, the heat-stable enterotoxin is proposed to cause an internal, functional dimerization, possibly then accounting for ligand-induced activation.

The site(s) of dimerization are not clearly defined, but an amphipathic region lies between the protein-kinase homology and cyclase catalytic domains (*see* Fig. 1). Chinkers *(9)*, through the use of the yeast two-hybrid system as well as deletion mutations, has presented evidence to support this region as a dimerization domain. All of the membrane guanylyl cyclases contain an amphipathic region within this part of the molecule; furthermore, the soluble forms of guanylyl cyclase also contain an amphipathic region to the amino-terminal side of the cyclase catalytic domain *(2)*. When a GC-A construct is synthesized that results in a protein containing the catalytic domain, the amphipathic region, and a small segment of the protein kinase homology domain, dimerization occurs and guanylyl cyclase activity is evident *(10–12)*, supportive of the concept that functional dimerization normally occurs within this region of the cyclase independent of the protein-kinase homology domain, the transmembrane region, or the extracellular domain. Wedel and colleagues *(13)* have also shown that it is possible to delete a considerable portion of the amino terminus of either the α or β subunits of the soluble cyclase, leaving the amphipathic region and the cyclase catalytic domain, and catalytic activity is retained, indicating that dimerization still occurred with these mutant subunits.

### *Extracellular Ligand Binding Domain*

There are a limited number of studies on the extracellular domain of GC-A; however, some inferences with respect to this domain also can be drawn from studies on the natriuretic peptide clearance (NPR-C) receptor. Two studies on the NPR-C receptor have identified disulfide linkages and sites of glycosylation *(14,15)*. Of the six cysteines present in NPR-C, the two nearest the plasma membrane, which lie at positions 428 and 431 in human NPR-C, do not form an intramolecular disulfide bond; they instead likely form intermolecular disulfide bonds responsible for covalent dimerization (the NPR-C

**Fig. 1.** *(opposite)* Helical wheels of the amphipathic region of GC-A and of the $\beta_1$ and $\alpha_1$ subunits of soluble guanylyl cyclase. Both GC-A and the soluble guanylyl cyclase continue to dimerize when regions amino terminal to this amphipathic region are deleted from the proteins *(4–7)*. The

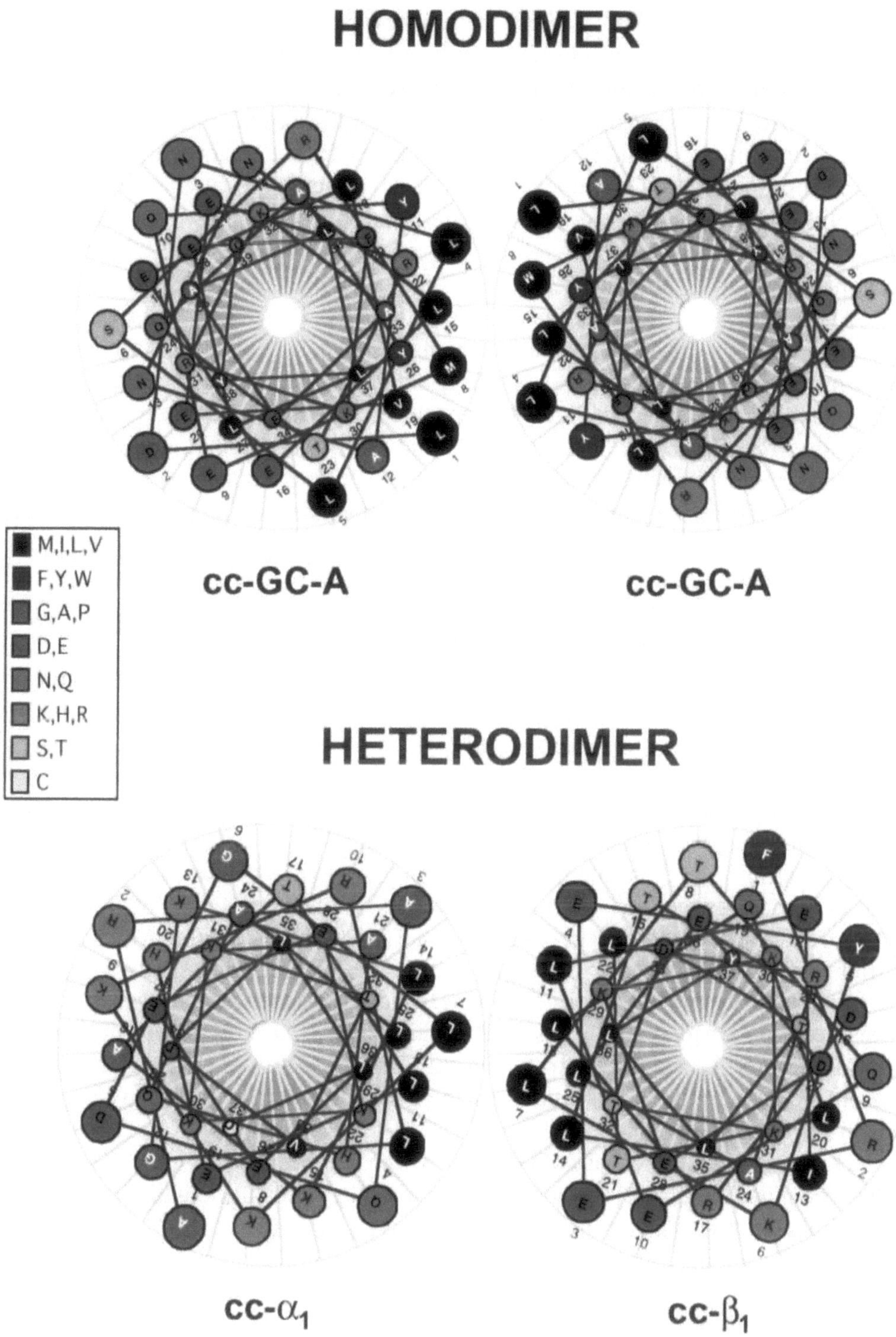


region lies just to the amino-terminal side of the cyclase catalytic domain. Shown are the helical wheel structures of amino acids 809–847 of GC-A (cc-GC-A), 418–454 of $\alpha_1$ (cc-$\alpha_1$), and 359–395 of $\beta_1$ (cc-$\beta_1$). The hydrophobic interface between the respective helices is formed mostly by leucines and methionines.

receptor only contains approx 37 amino acids within the intracellular region). Human and rat GC-A also contain six cysteine residues; however, the two most proximal to the plasma membrane are nine amino acids apart and do not align with NPR-C. The cysteine residues that form the two intramolecular disulfide loops align between GC-A and NPR-C and are within the amino-terminal regions of the extracellular ligand binding domain. Therefore, it is likely that these same residues also form intramolecular disulfide bonds in GC-A. Of amino acids which have been implicated in ANP binding to the NPR-C receptor, the various sites are within or near the two intramolecular disulfide loops *(14,16,17)*.

The number of ANP-binding sites per GC-A dimer has been recently explored in detail by Rondeau et al. *(18)*. Although early studies suggested one ANP bound per monomer *(19–21)*, the recent work suggests one ANP bound per dimer *(18)*. This is intriguing since early work on NPR-C suggested that a soluble, monomeric form of NPR-C retained high-affinity ANP binding, and it was therefore concluded that NPR-C did not require dimerization for ANP binding *(22)*. Since GC-A and NPR-C are homologous within the extracellular region, it seems unlikely that GC-A and NPR-C would show differences in stoichiometry. Therefore, in lieu of technical problems, the binding of one ANP to the dimeric receptor possibly negates binding of a second ANP, a problem not faced by the monomeric species.

### *Adenine Nucleotides/Protein Kinase Homology Domain*

The addition of ANP alone only marginally activates GC-A. However, the inclusion of ATP or an ATP analog markedly increases cyclase activity in the presence of ANP *(23–25)*. Whether ATP is a physiological regulator of GC-A has not been carefully examined, however. Since nonhydrolyzable analogs of ATP also activate, it has been suggested that protein kinase or ATPase activity is not required for transduction of the ANP-binding signal to activation of the cyclase catalytic domain. Whether enzyme activity is intrinsic to the protein-kinase homology domain has not been vigorously tested, yet protein-kinase activity seems unlikely given the replacement of an invariant Asp (also suggested to serve as the catalytic base, *[26]*) in all active protein kinases, with a different amino acid in every guanylyl cyclase yet cloned. The Asp is a part of a consensus sequence of HXD (most often HRD) and is replaced with the sequence HGX (most often HGN) in guanylyl cyclases; in domain 2 of the JAK kinases, and in a number of protein kinase-like sequences found in plants, the same HGX sequence is found. The apparent pressure for the HGX sequence, in these apparently inactive protein kinase-like domains, is not understood.

The kinase homology domain of GC-A also seems to be a negative regulator of the cyclase catalytic domain. Deletion of the kinase homology domain yields an enzyme with elevated basal activity that is no longer responsive to the activators ANP/ATP *(27)*. Certain point mutations within the kinase homology domain also disrupt signaling. These mutations do not appear to affect basal activity or ANP binding, but they do perturb ANP/ATP signaling *(28)*.

Early work suggested that associated ATP binding proteins were required for stimulation of GC-A *(24)*. This was primarily because of the loss of ANP/ATP stimulation during purification of the enzyme *(19,20)*. The recent work of Wong et al. *(29)* demonstrates that purified GC-A can retain ANP/ATP responsiveness; however, it remains unknown if associated proteins play a significant role in GC-A signaling.

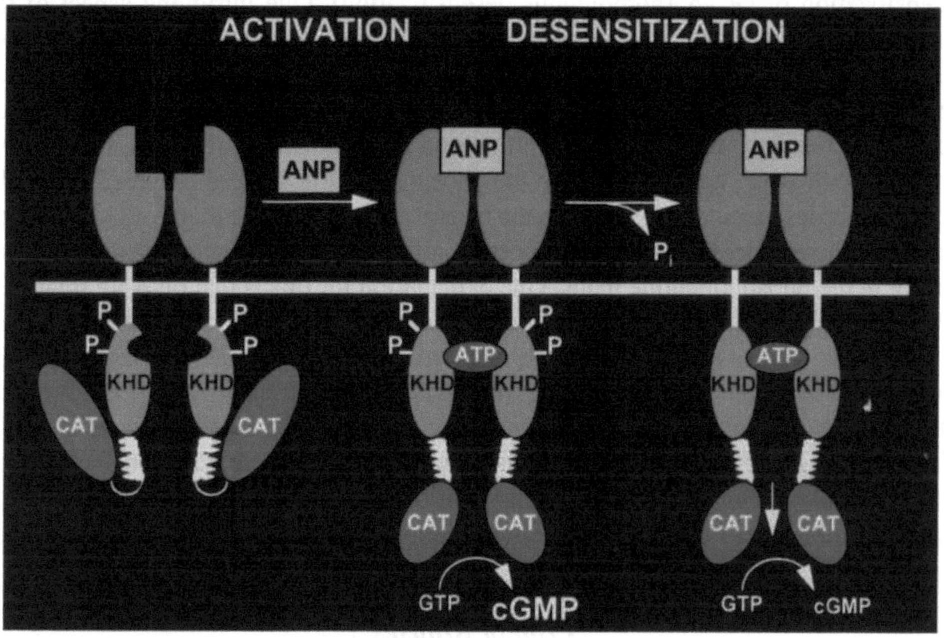


**Fig. 2.** A model for GC-A. The guanylyl cyclase-A receptor contains an extracellular ligand-binding domain that binds ANP or BNP. The receptor exists in at least a homodimeric state, but may exist as a trimer or tetramer in the absence of ligand. Ligand could induce an internal dimerization as suggested for GC-C by Vaandrager et al. *(8)*. The two proposed cysteine loops that are most amino terminal appear necessary for ANP binding, and recent studies suggest a stoichiometry of 1 mol ANP bound per dimeric GC-A *(18)*. GC-A exists in a phosphorylated state in the absence of ligand (the mol phosphate/mol enzyme is not known); in the liganded state, the receptor is dephosphorylated. The dephosphorylation correlates with a desensitization of the receptor to its ligand. Whether a protein phosphatase recognizes only the ligand receptor complex or whether a protein phosphatase is activated in response to GC-A signaling is not yet known. An adenine nucleotide (apparently ATP) is required for transduction of the ligand-binding signal to activation of the cyclase catalytic domain. Although ATP is suggested as the physiological regulator, a determination of whether another endogenous nucleoside/nucleotide would be more effective has not been tested. An amphipathic region distal to the protein kinase homology domain appears to represent a dimerization domain. The putative catalytic domain fails to demonstrate activity as a monomer, but whether this is because GTP shares binding sites across both subunits or the subunit association stabilizes a catalytic site intrinsic to each subunit is not known.

A model of activation of GC-A is emerging. ANP binding to the extracellular domain and ATP binding, perhaps to the kinase homology domain, leads to a conformational change to relieve kinase-homology domain inhibition of the catalytic domain. This results in a marked increase in cyclase catalytic activity. In this model, GC-A would contain ligand binding, regulatory, and effector domains within the same protein (Fig. 2).

### *Phosphorylation State*

The sea urchin sperm guanylyl cyclase was the first cyclase shown to be highly phosphorylated in the unliganded state *(30,31)*; binding of ligand resulted in dephosphorylation of the cyclase and coincident desensitization. Later work of Potter and Garbers *(32,33)* demonstrated that GC-A was also phosphorylated in the absence of ANP. Addition of ANP resulted in a loss of phosphate (principally phosphoserine) and coincident desensitization of GC-A (homologous desensitization). One intriguing aspect of the correlation of phosphorylation state with desensitization was that phorbol esters caused desensitization as well (heterologous desensitization). Although the initial conclusion from other studies *(34,35)* was that phorbol esters activated protein kinase C (PKC), which then resulted in the phosphorylation and desensitization of GC-A, the phorbol esters, as with ANP, actually reduced the phosphorylation state of GC-A *(33)*. Unlike homologous desensitization, in which the tryptic phosphopeptide maps of GC-A do not change (although dephosphorylation occurs), after phorbol-ester treatment, there is a change in the phosphopeptide maps. One explanation for this is that on ANP treatment of GC-A there is a population of GC-A that is completely dephosphorylated, whereas the remaining molecules are not. However, on phorbol-ester treatment, dephosphorylation of only some of the phosphorylation sites occurs, leading to desensitization *(33)*. Whether this decrease in phosphorylation is caused by activation of a GC-A phosphatase or inhibition of a GC-A protein kinase is not known. Although there is a strong correlation between dephosphorylation and desensitization of GC-A, major questions remain regarding this aspect of regulation, since rephosphorylation of GC-A has not been accomplished.

### *Protein Kinases*

The protein kinase(s) that are responsible for regulation of GC-A have not been identified. Although consensus phosphorylation sites for several known protein kinases are found in GC-A, it is not known if any of them regulate the receptor.

### *Protein Phosphatases*

The only report on a potential protein phosphatase that may associate with GC-A comes from the laboratory of Chinkers *(36)*. The protein phosphatase initially was identified by the use of the yeast two-hybrid method and subsequently was shown to bind to GC-A in in vitro binding assays. However, whether the protein phosphatase normally associates with GC-A is not yet known. The protein phosphatase is predicted to be 55,000 in molecular weight and contains a consensus Ser/Thr phosphatase catalytic domain at its carboxyl tail. Its amino terminus contains four tetratricopeptide repeats similar to that found in some immunophilins and is implicated in binding to heat-shock protein 90 *(37)*. Although its subcellular localization is primarily nuclear *(38)*, the physiological role(s) of this phosphatase remains unknown.

### *Associated Proteins*

The work of Wong et al. *(29)* demonstrated that ANP/ATP could elevate GC-A cyclase activity in the absence of protein cofactors. However, these studies did not address the question of whether activity could be modulated by associated factors, or whether associated proteins might be involved in cell signaling through pathways not involving cyclic GMP.

### *Loss of Function/Dominant-Negative Mutations*

As discussed above, all of the cyclases appear to require at least dimerization in order to demonstrate cyclase activity. Whether this is necessary in order to form a shared GTP-binding site or to stabilize a GTP-binding site on each polypeptide chain is not known. Various mutations in different guanylyl cyclase isoforms have resulted in the

Table 1
**Overview of Guanylyl Cyclase Mutations**

| *Guanylyl cyclase* | *Mutation* | *Phenotype* | *Reference* |
|---|---|---|---|
| Complete loss of function mutations | | | |
| GC-A | Asp-893 (catalytic domain) | Dominant-negative mutant | *(12)* |
| GC-A | Deletion of the cytoplasmic domain | Dominant-negative subunits | *(6)* |
| GC-C | Asp-834 (catalytic domain) | Dominant-negative mutant | *(50)* |
| $\alpha_1$ | Asp-513/Asp529 (catalytic domain) | Dominant-negative subunits | *(39)* |
| Partial loss of function mutations | | | |
| GC-A | N-terminal deletion | Catalytically active | *(10)* |
| GC-A | Deletion of most of the cytoplasmic domain | Loss of dimerization | *(5)* |
| GC-A | Point mutations/deletions in the kinase homology domain | Loss of ANP/ATP sensitivity | *(51)* |
| GC-A | Point mutations within the kinase homology domain | Loss of enzymatic activity | *(28)* |
| GC-A | Deletion of a possible dimerization region | Loss of dimer formation | (9) |
| GC-A and GC-B | Partial and complete deletion of the kinase homology domain | Loss of ANP/ATP sensitivity | *(52)* |
| GC-B | Point mutations within the extracellular domain | Loss of CNP binding and CNP-stimulated activity | *(53)* *(53)* *(53)* |
| $\beta_1$ | His-105 (heme binding domain?) | Loss of NO sensitivity | *(54)* |
| $\beta_1$ | $NH_2$-terminal deletions of $\beta_1$ | Loss of NO sensitivity | *(13)* |
| Gain of function mutations | | | |
| GC-A | Kinase-homology domain deletion | Constitutive activity | *(27)* |
| GC-C | Kinase-homology domain deletion | Constitutive activity | *(50)* |
| GC-A | Glu-974A (catalytic domain) | Hyperactive mutant | *(55)* |

formation of proteins that can still associate with a wild-type form and in so doing lead to a loss of function (*see* Table 1). These mutants are called dominant-negative mutants. Mutation of highly conserved aspartates cause complete loss of function and dominant-negative proteins not only in the membrane GC-A and GC-C receptors (E893 or E834, respectively), but also in the $\alpha_1$ subunit of soluble guanylyl cyclase (E513 and E52). A splice variant of the $\alpha_2$ subunit has also been described as a dominant-negative mutant *(56)*.

Mutations that may lead to the most information and to development of reagents to define cyclase function may be those that lead to a partial loss of function (e.g., retention of basal activity but loss of ligand sensitivity). Mutations that have resulted in a loss of ligand sensitivity include certain point mutations as well as deletion mutations of regions that appear to serve regulatory functions. A loss in ligand responsiveness has been described for deletion mutants as well as point mutations within the protein kinase homology domain of GC-A, GC-B, and GC-C. A point mutation within the N-terminus of the $\beta_1$ subunit of soluble cyclase or deletion mutants of N-terminal regions of $\alpha_1$ and $\beta_1$ result in a loss of NO sensitivity.

The potential selectivity of loss of function mutations for a respective guanylyl cyclase isoform was addressed by Yuen et al. *(39)*. In their work, the α subunit mutations discussed above were introduced into rat insulinoma cells, which also contain very high levels of the guanylyl cyclase-C receptor. The dominant-negative α subunits markedly inhibited elevations of cyclic GMP induced by sodium nitroprusside, an activator of soluble guanylyl cyclase, but failed to affect elevations caused by a heat-stable enterotoxin, the ligand for GC-C.

### *Gain of Function Mutations*

The first apparent gain of function (elevated cyclase activity) mutation was produced when the protein-kinase homology domain was deleted from GC-A (*see* Table 1). Although enzyme activity was higher than that of the wild-type receptor giving a gain of function, the GC-A also became unreceptive to ANP/ATP. It was proposed that the protein-kinase domain normally served as a negative regulatory element and that its removal therefore relieved a constraint on the catalytic domain. In this model, ANP/ATP would activate GC-A by relief of this inhibition.

Various models can be proposed regarding the mechanisms of activation of the catalytic domain of GC-A. Models include a derepression of protein-kinase domain inhibition of the cyclase catalytic domain as discussed above, an internal "functional" dimerization (*see* Vaandrager et al. *[8]*), and a direct conformational change within the catalytic domain itself. With any of these models it is conceivable that site-directed mutations could mimic the ligand effect, and thereby yield an activated form of GC-A.

We, therefore, mutated various residues highly conserved or invariant in the guanylyl cyclases to search for a point mutation that would mimic the ligand-activated state of GC-A *(55)*. Mutation of a glutamate residue that is conserved not only in all guanylyl cyclases but also in the cytoplasmic domains of eukaryotic adenylyl cyclases resulted in an enzyme with a dramatically increased catalytic activity that was basically ANP/ATP insensitive. That this mutation also causes activation in a deletion mutant that only comprises a putative dimerization region and the catalytic domain suggests that this mutation does not cause relief of kinase-domain inhibition of catalytic activity, but results in an activated state of the catalytic domain.

## DISCOVERY OF THE PHYSIOLOGICAL ROLES OF GC-A

### *Specific Inhibitors*

A high-mol-wt polysaccharide called HS 142–1 has been used in various laboratories as an inhibitor of guanylyl cyclase-A *(40,41)*; it appears to act as a competitor of ANP. This compound also effectively inhibits GC-B *(42)*. Although not specific for GC-A, the compound has yielded effects consistent with GC-A inhibition in various model systems, including the inhibition of ANP-induced natriuresis and diuresis *(43)*. Although the specificity of the effects of HS 142–1 on GC-A and GC-B has not been vigorously tested (e.g., determination if these receptors are the only membrane receptors to which HS 142–1 binds), the effects seen in various tissues, such as the kidney and adrenal gland, are consistent with binding of the antagonist exclusively to the natriuretic peptide receptors.

An antibody has been produced to GC-B that appears to block CNP binding *(44)*. However, an antibody that competes with ANP for binding to GC-A has not been reported.

### *Specific Agonists*

Previously Maack and colleagues *(45)* synthesized rather specific agonists for the natriuretic peptide clearance receptor (NPR-C). Recently, through the use of random mutagenesis and phage display, a modified ANP containing an arginine instead of a glycine at position 16 has yielded a peptide that now competes with [$^{125}$I] with an $IC_{50}$ of 85 p*M* for GC-A but an $IC_{50}$ of 17,800 p*M* for NPR-C *(46)*. Therefore, the peptide is highly selective for GC-A compared to NPR-C. These relatively receptor-specific agonists may clearly prove powerful reagents to define function.

### *Dominant-Negative Subunits*

Although the dominant negative mutations of guanylyl cyclases discussed above have the potential of being powerful reagents to block a specific guanylyl cyclase pathway, they have not yet been used for such purposes with GC-A. Whether the expression level of a dominant-negative variant would be high enough when driven by a tissue-specific promotor in a transgenic animal has not yet been tested.

### *Antisense RNA*

There have been no reports on the use of antisense RNA to block the expression of the GC-A receptor.

### *Gene Disruption*

A powerful method by which to identify a gene that accounts for a human genetic disease is the disruption of that gene in an animal model. Additionally, it becomes possible to define the functions of a gene product with such an approach, although this must be done with great caution.

Prior to the disruption of the GC-A gene, the ANP gene was disrupted in the mouse *(47)*. The phenotype was that of a salt-sensitive form of hypertension, and although BNP (a potential second ligand for GC-A) was presumably still present, and although the granules from the atrium of the heart were apparently also eliminated by disruption of the ANP gene, there was an expectation that disruption of the GC-A gene would lead to some form of salt-sensitive hypertension. However, this was not the case. Animals lacking GC-A had an elevated blood pressure independent of dietary salt *(48)*. Therefore, the kidney appeared to satisfactorily handle the dietary salt load, but to reset the blood pressure at the abnormally high level. It is worth noting that the salt sensitivity has only been tested in younger (3–4-mo-old) animals so far and therefore it remains possible that with age, salt sensitivity will become evident. Nevertheless, how do these kidneys of the young animals handle the salt load in the absence of the apparent ANP receptor?

One possible explanation for the phenotype would be that ANP acts on the kidney through a receptor other than GC-A. ANP, for example, is known to bind with high affinity to the NPR-C receptor and to bind with low affinity to GC-B *(2)*. To test this possibility, ANP was infused into mice lacking GC-A. Although ANP caused a diuresis and natriuresis in wild-type mice, it failed to have any demonstrable effect in the null mice *(49)*. Therefore, ANP appears to act through GC-A in the kidney to cause a diuresis and natriuresis.

This observation raised the possibility that the granules of the atrium of the heart contain other natriuretic principles that act through receptors other than GC-A. Therefore, an iv administration of an isooncotic solution was used to cause volume expansion. That volume expansion occurred in these studies was ascertained by measurement of ANP concentrations in plasma. In wild-type or null mice, ANP concentrations were markedly elevated by the infusion of the solution. Thus, a volume expansion apparently occurred. In wild-type mice, the volume expansion caused a marked diuresis and natriuresis, but in the GC-A-null mice there were no apparent effects. Thus, in the anesthetized mouse, communication between the heart and the kidney appears to occur exclusively through the GC-A receptor in the acute setting.

## CONCLUSIONS

The guanylyl cyclase A-receptor, found in many cell types throughout the body, appears to be an important regulator of blood pressure even under chronic conditions. Furthermore, in the acute setting, at least in the anesthetized animal, it appears to be the lone receptor for natriuretic/diuretic factors released from the heart. The phenotype of a salt-resistant hypertension is similar to that of about 50% of humans who have elevated blood pressure, however, it is not yet known if defects in the GC-A receptor gene account for some forms of hypertension in the human.

## ACKNOWLEDGMENTS

This work was supported in part by the Cell and Molecular Biology Training Grant NIH 5-T32-GM08203 (DCF) and by the Deutsche Forschungsgesellschaft (BJW).

## REFERENCES

1. Garbers DL, Lowe DG. Guanylyl cyclase receptors. J Biol Chem 1994;269:30,741–30,744.
2. Garbers DL. Guanylyl cyclase receptors and their endocrine, paracrine and autocrine ligands. Cell 1992;71:1–4.
3. Drewett JG, Garbers DL. The family of guanylyl cyclase receptors: their ligands and functions. Endocrine Reviews 1994;15:135–162.
4. Iwata T, Uchida-Mizuno K, Katafuchi T, Ito T, Hagiwara H, Hirose S. Bifunctional atrial natriuretic peptide receptor (type A) exists as a disulfide-linked tetramer in plasma membranes of bovine adrenal cortex. J Biochem 1991;110:35–39.
5. Lowe DG. Human natriuretic peptide receptor-A guanylyl cyclase is self-associated prior to hormone binding. Biochemistry 1992;31:10,421–10,425.
6. Chinkers M, Wilson EM. Ligand-independent oligomerization of natriuretic peptide receptors: Identification of heteromeric receptors and a dominant negative mutant. J Biol Chem 1992;267:18,589–18,597.
7. Yuen, PST, Takada, M, Thompson, DK, Garbers, D, Masaki T, ed. Production of dominant negative mutations of guanylyl cyclase. Endothelium-Derviced Factors and Vascular Functions. Elsevier Science BV, Amsterdam, 1994; pp. 215–210.
8. Vaandrager AB, van der Wiel E, de Jonge HR. Heat-stable enterotoxin activation of immunopurified guanylyl cyclase C. Modulation by adenine nucleotides. J Biol Chem 1993;268:19,598–19,603.

9. Wilson EM, Chinkers M. Identification of sequences mediating guanylyl cyclase dimerization. Biochemistry 1995;34:4696–4701.
10. Thorpe DS, Morkin E. The carboxyl region contains the catalytic domain of the membrane form of guanylate cyclase. J Biol Chem 1990;265:14,717–14,720.
11. Thorpe DS, Niu S, Morkin E. The guanylyl cyclase core of an atrial natriuretic peptide receptor: enzymatic properties and basis for cooperativity. Biochem Biophys Res Commun 1996;218:670–673.
12. Thompson DK, Garbers DL. Dominant negative mutations of the guanylyl cyclase-A receptor: extracellular domain deletion and catalytic domain point mutations. J Biol Chem 1995;270:425–430.
13. Wedel B, Harteneck C, Foerster J, Friebe A, Schultz G, Koesling D. Functional domains of soluble guanylyl cyclase. J Biol Chem 1995;270:24,871–24,875.
14. Stults JT, O'Connell KL, Garcia C, Wong S, Engel AM, Garbers DL, Lowe DG. The disulfide linkages and glycosylation sites of the human natriuretic peptide receptor-C homodimer. Biochemistry 1994;33:11,372–11,381.
15. Itakura M, Iwashina M, Mizuno T, Ito T, Hagiwara H, Hirose S. Mutational analysis of disulfide bridges in the type C atrial natriuretic peptide receptor. J Biol Chem 1994;269:8314–8318.
16. Iwashina M, Mizuno T, Hirose S, Ito T, Hagiwara H. His145-Trp146 residues and the disulfide-linked loops in atrial natriuretic peptide receptor are critical for the ligand-binding activity. J Biochem 1994;115:563–567.
17. Engel AM, Schoenfeld JR, Lowe DG. A single residue determines the distinct pharmacology of rat and human natriuretic peptide receptor-C. J Biol Chem 1994;269:17,005–17,008.
18. Rondeau JJ, McNicoll N, Gagnon J, Bouchard N, Ong H, De Lean A. Stoichiometry of the atrial natriuretic factor-R1 receptor complex in the bovine zona glomerulosa. Biochemistry 1995;34: 2130–2136.
19. Kuno T, Andresen JW, Kamisaki Y, Waldman SA, Chang LY, Saheki S, Leitman DC, Nakane M, Murad F. Co-purification of an atrial natriuretic factor receptor and particulate guanylate cyclase from rat lung. J Biol Chem 1986;261:5817–5823.
20. Takayanagi R, Inagami T, Snajdar RM, Imada T, Tamura M, Misono KS. Two distinct forms of receptors for atrial natriuretic factor in bovine adrenocortical cells: Purification, ligand binding, and peptide mapping. J Biol Chem 1987;262:12,104–12,113.
21. Meloche S, McNicoll N, Liu B, Ong H, De Léan A. Atrial natriuretic factor $R_1$ receptor from bovine adrenal glomerulosa: Purification, characterization, and modulation by amiloride. Biochemistry 1988;27:8151–8158.
22. Fuller F, Porter JG, Artsten AE, Miller J, Schilling JW, Scarborough RM, Lewicki JA, Schenk DB. Atrial natriuretic peptide clearance receptor: Complete sequence and functional expression of cDNA clones. J Biol Chem 1988;263:9395–9401.
23. Kurose H, Inagami T, Ui M. Participation of adenosine 5'-triphosphate in the activation of membrane-bound guanylate cyclase by the atrial natriuretic factor. FEBS Letts 219:375–379.
24. Chang C, Kohse KP, Chang B, Hirata M, Jiang B, Douglas JE, Murad F. Characterization of ATP-stimulated guanylate cyclase activation in rat lung membranes. Biochim Biophys Acta 1990;1052:159–165.
25. Chinkers M, Singh S, Garbers DL. Adenine nucleotides are required for activation of rat atrial natriuretic peptide receptor/guanylyl cyclase expressed in a baculovirus system. J Biol Chem 1991;266: 4088–4093.
26. Taylor SS, Knighton DR, Zheng J, Ten Eyck LF, Sowadski JM. Structural framework for the protein kinase family [Review]. Ann Rev Cell Biol 1992;8:429–462.
27. Chinkers M, Garbers DL. The protein kinase domain of the ANP receptor is required for signaling. Science 1989;245:1392–1394.
28. Koller KJ, Lipari MT, Goeddel DV. Proper glycosylation and phosphorylation of the type A natriuretic peptide receptor are required for hormone-stimulated guanylyl cyclase activity. J Biol Chem 1993;268:5997–6003.
29. Wong SKF, Ma C, Foster DC, Chen A, Garbers DL. The guanylyl cyclase-A receptor transduces an atrial natriuretic peptide/ATP activation signal in the absence of other proteins. J Biol Chem 1995;270:30,818–30,822.
30. Suzuki N, Shimomura H, Radany EW, Ramarao CS, Ward GE, Bentley JK, Garbers DL. A peptide associated with eggs causes a mobility shift in a major plasma membrane protein of spermatozoa. J Biol Chem 1984;259:14,874–14,879.
31. Ward GE, Garbers DL, Vacquier VD. Effects of extracellular egg factors on sperm guanylate cyclase. Science 1985;227:768–770.

32. Potter LR, Garbers DL. Dephosphorylation of the guanylyl cyclase-A receptor causes desensitization. J Biol Chem 1992;267:14,531–14,534.
33. Potter LR, Garbers DL. Phorbol ester-dependent desensitization of the atrial natriuretic peptide receptor is mediated by dephosphorylation. J Biol Chem 1994;269:14,636–14,642.
34. Ballerman BJ, Marala RB, Sharma RK. Characterization and regulation by protein kinase C of renal glomerular atrial natriuretic peptide receptor-coupled guanylate cyclase. Biochem Biophys Res Commun 1988;157:755–761.
35. Duda T, Sharma RK. Regulation of guanylate cyclase activity by atrial natriuretic factor and protein kinase C. Mol Cell Biochem 1990;93:179–184.
36. Chinkers M. Targeting of a distinctive protein-serine phosphatase to the protein kinase-like domain of the atrial natriuretic peptide receptor. Proc Natl Acad Sci USA 1994;91:11,075–11,079.
37. Owensgrillo JK, Czar MJ, Hutchison KA, Hoffman K, Perdew GH, Pratt, WB. A model of protein targeting mediated by immunophilins and other proteins that bind to HSP90 via tetratricopeptide repeat domains. J Biol Chem 1996;271:13,468–13,475.
38. Chen MX, McPartlin AE, Brown L, Chen YH, Barker HM, Cohen PT. A novel human protein serine/threonine phosphatase, which possesses four tetratricopeptide repeat motifs and localizes to the nucleus. EMBO J 1994;13:4278–4290.
39. Yuen PST, Doolittle LK, Garbers DL. Dominant negative mutants of nitric oxide-sensitive guanylyl cyclase. J Biol Chem 1994;269:791–793.
40. Sano T, Morishita Y, Yamada K, Matsuda Y. Effects of HS-142-1, a novel non-peptide ANP antagonist, on diuresis and natiuresis induced by acute volume expansion in anesthetized rats. Biochem Biophys Res Commun 1992;182:824–829.
41. Imura R, Sano T, Goto J, Yamada K, Matsuda Y. Inhibition by HS-142-1, a novel nonpeptide atrial natriuretic peptide antagonist of microbial origin, of atrial natriuretic peptide-induced relaxation of isolated rabbit aorta through the blockade of guanylyl cyclase-linked receptors. Mol Pharmacol 1992;42:982–990.
42. Ohyama Y, Miyamoto K, Morishita Y, Matsuda Y, Saito Y, Minamino N, Kangawa K, Matsuo H. Stable expression of natriuretic peptide receptors: Effects of HS-142-1, a non-peptide ANP antagonist. Biochem Biophys Res Commun 1992;189:336–342.
43. Stevens TL, Wei CM, Aahrus LL, Heublein DM, Kinoshita M, Matsuda Y, Burnett JC Jr. Modulation of exogenous and endogenous atrial natriuretic peptide by a receptor inhibitor. Hypertension 1994;23:613–618.
44. Drewett JG, Fendly BM, Garbers DL, Lowe DG. Natriuretic peptide receptor-B (guanylyl cyclase-B) mediates C-type natriuretic peptide relaxation of precontracted rat aorta. J Biol Chem 1995;270: 4668–4674.
45. Maack T, Suzuki M, Almeida FA, Nussenzveig D, Scarborough M, McEnroe GA, Lewicki JA. Physiological role of silent receptors of atrial natriuretic factor. Science 1987;238:675–678.
46. Cunningham BC, Lowe DG, Li B, Bennett BD, Wells JA. Production of an atrial natriuretic peptide variant that is specific for type A receptor. EMBO J 1994;13:2508–2515.
47. John SW, Krege JH, Oliver PM, Hagaman JR, Hodgin JB, Pang SC, Flynn TG, Smithies O. Genetic decreases in atrial natriuretic peptide and salt-sensitive hypertension [published erratum appears in Science 1995 Mar 24;267(5205):1753]. Science 1995;267:679–681.
48. Lopez MJ, Wong SKF, Kishimito I, Dubois S, Mach V, Friesen J, Garbers DL, Beuve A. Salt-resistant hypertension in mice lacking the guanylyl cyclase-A receptor gene. Nature 1995;378:65–68.
49. Kishimoto I, Dubois SK, Garbers DL. The heart communicates with the kidney exclusively through the guanylyl cyclase-A receptor: Acute handling of sodium and water in response to volume expansion. Proc Natl Acad Sci of the USA 1996;93:6215–6219.
50. Rudner XL, Mandal KK, de Sauvage FJ, Kindman LA, Almenoff JS. Regulation of cell signaling by the cytoplasmic domains of the heat- stable enterotoxin receptor: identification of autoinhibitory and activating motifs. Proc Natl Acad Sci USA 1995;9295:5169–5173.
51. Goraczniak RM, Duda T, Sharma RK. A structural motif that defines the ATP-regulatory module of guanylate cyclase in atrial natriuretic factor signalling. Biochem J 1992;282:533–537.
52. Koller KJ, de Sauvage FJ, Lowe DG, Goeddel DV. Conservation of the kinaselike regulatory domain is essential for activation of the natriuretic peptide receptor guanylyl cyclases. Mol Cell Biol 1992;12:2581–90.
53. Duda T, Goraczniak RM, Sharma RK. Site-directed mutational analysis of a membrane guanylate cyclase cDNA reveals the atrial natriuretic factor signaling site. Proc Natl Acad Sci USA 1991;88:7882–7886.

54. Wedel B, Humbert P, Harteneck C, Foerster J, Malkewitz J, Bohme E, Schultz G, Koesling D. Mutation of His-105 in the beta 1 subunit yields a nitric oxide-insensitive form of soluble guanylyl cyclase. Proc Natl Acad Sci USA 1994;91:2592–2596.
55. Wedel BJ, Foster DC, Miller DE, Garbers DL. A mutation of the atrial natriuretic peptide (guanylyl cyclase-A) receptor results in a constitutively hyperactive enzyme. Proc Natl Acad Sci USA 1997; 94: 459–462.
56. Behrends S, Harteneck C, Schultz G, Koesling D. A variant of the alpha 2 subunit of soluble guanylyl cyclase contains an insert homologous to a region within adenylyl cyclases and functions as a dominant negative protein. J Biol Chem 1995; 270: 21,109–21,113.

# 3 The Guanylyl Cyclase-B Receptor

*David G. Lowe*

*CONTENTS*



## MOLECULAR BIOLOGY OF GC-B

### *cDNA Cloning*

The cDNA cloning of GC-A and GC-B, the first two mammalian-membrane guanylyl cyclases to be identified, was accomplished in 1989 *(1–4)* through the convergence of two unrelated areas of investigation. The first area involved studies on the molecular basis of fertilization in sea urchins. This research resulted in the identification in 1981 of the egg peptide speract, which played a role in sperm activation *(5)*. Subsequent studies identified sperm-activating peptides from the eggs of several species of sea urchin. Work on the identification and purification of the receptor for one of these peptides, resact, demonstrated that this sperm receptor was associated with membrane guanylyl cyclase activity (reviewed in ref. *6*).

The second independent line of research began in 1981 with the identification of a natriuretic and diuretic activity in a heart atrial extract *(7)* that was later identified as atrial natriuretic peptide (reviewed in ref. *8*). Intense efforts on the identification of an ANP receptor revealed two classes of binding site, one of which was associated with unregulated guanylyl cyclase activity; however, purification of the receptors was hampered by the low abundance of the binding site(s) *(9)*. Successful purification and cDNA cloning of the sea urchin resact receptor, a putative membrane guanylyl cyclase *(10)*, provided an alternative route to test the hypothesis that the ANP receptor and membrane guanylyl cyclase were identical. The strong evolutionary conservation of structural motifs among many classes of enzymes suggested that it might be possible to use a cDNA fragment encoding the cytoplasmic domain of the sea urchin guanylyl cyclase to isolate mammalian homologs. Low stringency hybridization of the sea urchin cDNA probe to a human

From: *Contemporary Endocrinology: Natriuretic Peptides in Health and Disease*
Edited by: W. K. Samson and E. R. Levin Humana Press Inc., Totowa, NJ

Table 1
The Many Names of GC-B

| |
|---|
| Guanylyl cyclase-B, GC-B |
| Natriuretic peptide receptor-B, NPR-B |
| NP-B receptor |
| ANPR-B |
| Atrial natriuretic factor receptor-$1_c$, $ANFR_{1c}$ |
| ANP-B receptor |
| NPR2 |
| C-Type natriuretic peptide receptor guanylyl cyclase, CNP-RGC |
| ANP-RGC(B) |

kidney cDNA library resulted in the identification of homologous human cDNAs. The protein encoded by the cloned cDNAs was identified as the ANP-receptor guanylyl cyclase and subsequently termed guanylyl cyclase-A (GC-A) or natriuretic peptide receptor-A (NPR-A) *(1,2)*. The cloning of the first mammalian homolog of the sea urchin resact receptor immediately set off a search for additional guanylyl cyclases by the same low stringency hybridization strategy. Fragments of the mammalian GC-A cDNA were used for low-stringency screening of a human placenta cDNA library and a rat brain cDNA library to identify the second mammalian membrane guanylyl cyclase, GC-B *(3,4)*. The recent cloning of bovine GC-B *(11)* supports the idea that this receptor is highly conserved across mammals. There is 98% amino acid identity over the 1047-amino-acid precursor sequence in pairwise comparisons among human, rat, and bovine GC-B. A variety of names have been used to designate this receptor (Table 1), with the most commonly used being GC-B and NPR-B. To avoid confusion, the different names of this receptor should be kept in mind by those planning to read the primary literature. In this review, the term GC-B is used.

## *Sequence Homologies*

### Extracellular Domain

Removal of the 16-amino-acid signal sequence *(12)* generates a 1031-amino-acid protein with a predicted $M_r$ of 115,530 for human GC-B. The amino-terminal sequence from GC-B extracellular domain (ECD) expressed in Chinese hamster ovary cells *(12)* showed processing by signal peptidase was occurring six residues upstream of the originally predicted amino terminus *(3)*. The ECD of GC-B is predicted to be 439 amino acids, with seven consensus N-linked glycosylation sites, and six cysteine residues *(3)*. There is no direct evidence for glycosylation of GC-B; however, in direct analogy to what has been observed with GC-A *(13)*, Western blots of recombinant GC-B suggest that the enzyme is modified by carbohydrate addition, with an immunoreactive protein larger than the core polypeptide (Lowe DG, unpublished data).

Based on sequence homology in the ECD, GC-B is related to only two other cell surface receptors, GC-A, the ANP receptor guanylyl cyclase, and the natriuretic peptide receptor-C (NPR-C), also known as the clearance receptor or C-receptor. This latter receptor lacks the large cytoplasmic domain characteristic of the guanylyl cyclases. Alignment of the ECD amino-acid sequences of human GC-B, GC-A, and NPR-C *(2,3,14)* is shown in Fig. 1. These proteins are 30–43% identical, or 39–53% similar based on

```
GC-B      1 VRPPGARNLTLAVVLPEHNLSYAWAWPRVGPAVALAVEAL-GRA-----L
GC-A      1 ------GNLTVAVVLPLANTSYPWSWARVGPAVELALAQVKARPD---LL
NPR-C     1 -EALPPQKIEVLVLLPQDD-SYLFSLTRVRPAIEYALRSVEGNGTGRRLL

GC-B     45 PV-DLRFVSSELE---GACSE--YLAPLSAVDLKLYHDPDLLLGPGCVYP
GC-A     42 PGWTVRTVLGSSENALGVCSD--TAAPLAAVDLKWEHNPAVFLGPGCVYA
NPR-C    49 PP-GTRFQVAYED---SDCGNRALFSLVDRVAAARGAKPDLILGPVCEYA

GC-B     89 AASVARFASHWRLPLLTAGAVASGFSAKNDHYRTLVRTGPSAPKLGEFVV
GC-A     90 AAPVGRFTAHWRVPLLTAGAPALGFGVK-DEYALTTRAGPSYAKLGDFVA
NPR-C    95 AAPVARLASHWDLPMLSAGALAAGFQHKDSEYSHLTRVAPAYAKMGEMML

GC-B    139 TLHGHFNWTARAALLYLDARTDDRPHYFTIEGVFEALQGS-NLSVQHQVY
GC-A    139 ALHRRLGWERQALMLYAYRPGDEEHCFFLVEGLFMRVRDRLNITVDHLEF
NPR-C   145 ALFRHHHW-SRAALVYSDDKL-ERNCYFTLEGVHEVFQEE-GLHTSIYSF

GC-B    188 AREPGGP-EQATHFIRANGRIVYICGPLEMLHEILLQAQRENLTNGDYVF
GC-A    189 AEDDLSHYTRLLRTMPRKGRVIYICSSPDAFRTLMLLALEAGLCGEDYVF
NPR-C   192 DETKDLDLEDIVRNIQASERVVIMCASSDTIRSIMLVAHRHGMTSGDYAF

GC-B    237 FYLDVFGESLRAGPTRATGRPWQDNRTREQAQALREAFQTVLVITYREPP
GC-A    239 FHLDIFGQSLQGGQGPAPRRPWE--RGDGQDVSARQAFQAAKIITYKDPD
NPR-C   242 FNIELFNSSSYGDG------SWK--RGDKHDFEAKQAYSSLQTVTLLRTV

GC-B    287 NPEYQEFQNRLLIRAREDFGVELGPSLMNLIAGCFYDGILLYAEVLNETI
GC-A    287 NPEYLEFLKQLKHLAYEQFNFTMEDGLVNTIPASFHDGLLLYIQAVTETL
NPR-C   284 KPEFEKFSMEVKS-SVEKQGLNMED-YVNMFVEGFHDAILLYVLALHEVL
             *                         *
GC-B    337 QEGGTREDGLRIVEKMQGRRYHGVTGLVVMDKNNDRETDFVLWAMGDLDS
GC-A    337 AHGGTVTDGENITQRMWNRSFQGVTGYLKIDSSGDRETDFSLWDM-DPEN
NPR-C   332 RAGYSKKDGGKIIQQTWNRTFEGIAGQVSIDANGDRYGDFSVIAMTDVEA

GC-B    387 GDFQPAAHYSGAEKQ-IWWTGRPIPWVKGAPPSDNPPCAFDLDDPSCDKT
GC-A    386 GAFRVVLNYNGTSQELVAVSGRKLNWPLGYPPPDIPKCGFDNEDPACNQD
NPR-C   382 GTQEVIGDYFGKEGRFEMRPNVKYPWGPLKLRIDENRIVEHTNSSPCKSC

GC-B    436 PLST
GC-A    436 HLSTLE
NPR-C   432 GLEES
```

**Fig. 1.** Alignment of human GC-B, GC-A, and NPR-C amino acid sequences for the mature extracellular domain. The first residue of each receptor is based on N-terminal amino acid sequencing of secreted receptor-IgG fusion proteins *(12)*. Cysteine residues are indicated by shading, and residues implicated in ligand binding are marked by an asterisk.

conservative amino-acid substitutions, with NPR-C the more distantly related receptor. The sequence homology among these receptors is reflected in their related function of binding to one or more of the three natriuretic peptides; ANP, BNP, and CNP (reviewed in ref. *8*).

Sequence homology among the natriuretic peptide receptors is likely to extend to a similar tertiary structure or protein fold. This may in part be evident in the conservation of cysteine residues. In NPR-C the first four cysteine residues, out of a total of five or six *(15)*, are paired sequentially to form two disulfide-linked loops *(16,17)*. Residues forming the first loop are conserved in GC-B, C59, and C85, suggesting a common structural motif in this region (Fig. 1). For the second pair of cysteines, the first is missing in GC-B, replaced by H164, and the second is present as a conserved residue at C211. This residue may in fact pair with GC-B residue C320, a cysteine residue not found in GC-A or NPR-C. The location of the divergent Cys residue in GC-B has been conserved in eel GC-B *(18)*, suggesting that this difference with the other receptors was fixed early in evolution. The role of, or disulfide bond pattern of, the two juxtamembrane cysteine residues in GC-B, C423, and C432, has not been defined.

#### Cytoplasmic Domain

The cytoplasmic domain of GC-B is closely related to GC-A, and can be subdivided into three structurally distinct subdomains *(19,20)*. The membrane proximal region of approx 300 residues is referred to as the kinase homology domain, or kinase-like domain, based on its identity with the large family of protein kinases. This is a regulatory domain for GC-A and GC-B that functions similarly in both receptors (*see* Signal Transduction). Following the kinase homology domain is a conserved, 43-amino-acid region that is necessary and sufficient to mediate dimerization of the receptor *(20)*. The functional identity of this region between GC-A and GC-B is illustrated by the formation of heteromeric complexes between these two receptors *(21)*.

For the 240-residue C-terminal cyclase catalytic domain, there is homology to other membrane guanylyl cyclases, soluble guanylyl cyclases, and adenylyl cyclases *(19)*. This homology indicates a conserved protein fold common among this class of enzyme regardless of the mechanism of enzyme regulation.

### *Chromosomal Localization*

Assignment of the chromosomal location of a newly identified gene may lead to a correlation between gene location and the map position for a human genetic disease. Using mouse–human and hamster–human somatic cell hybrids, and PCR to distinguish the human GC-B gene from the mouse gene by length polymorphism, human GC-B was initially mapped to the interval 9p11→p22. *In situ* metaphase chromosome hybridization provided a finer localization to 9p12→p21 *(22)*. So far, no disease genes have been identified in this interval (*see* "Online Mendelian Inheritance in Man," www3.ncbi.nlm.nih.gov/omim/, locus name ANPRB).

## PHARMACOLOGY OF GC-B

### *Natriuretic Peptide Specificity*

Expression of the cloned GC-B cDNA originally showed that this membrane guanylyl cyclase was more sensitive to BNP than ANP, suggesting that there might be selectivity in the natriuretic peptide responses in target tissues *(3,4)*. Porcine BNP and rat BNP had $ED_{50}$s of approx 6 μ*M* for the stimulation of cGMP production on intact cells, and ANP was much less potent. In comparison, the $ED_{50}$ for ANP and BNP on GC-A was 3 and 25 n*M*, respectively. This major difference in the absolute potency of these natriuretic peptides suggested at the time that BNP was probably not the physiological ligand for GC-B.

The purification in 1990 of a third natriuretic peptide, called C-type natriuretic peptide, or CNP *(23)*, was soon followed by the demonstration that CNP was a potent agonist of transiently expressed GC-B *(24)*. Subsequent studies on a stable cell line expressing rat GC-B gave an $ED_{50}$ of 2.2 n*M* for CNP stimulation of cGMP production *(25)*, comparable to the $ED_{50}$ of 12 n*M* for CNP on transiently expressed bovine GC-B *(11)*. ANP had an $EC_{50} > 10$ μ*M* in these experiments, suggesting that at physiological concentrations ANP does not signal through GC-B.

The first affinity measurements for CNP on GC-B were made on binding to intact cells expressing a GC-B cytoplasmic domain truncation *(24)*. This deleted version of the receptor is expressed at high levels, and may be expected to show the intrinsic affinity of the cell surface ligand-binding domain independent of influences of the cytoplasmic domain (*see* Signal Transduction). In these experiments, a Ki of 30 p*M* was determined, which is comparable to the CNP Kd of 12 p*M* for binding to a truncated bovine GC-B in a membrane preparation *(11)*. Binding to a GC-B ECD linked to IgGγ1 Fc, a disulfide-linked dimer of two ECDs joined by the hinge of the IgG, showed a similar high-affinity binding with Kd = 7 p*M (12)*. Recently reported Kis for CNP binding to full-length GC-B show two classes of binding sites at 0.2 and 1 n*M* for high-affinity and low-affinity states, respectively. The low-affinity state is ATP-dependent *(26)*, in analogy to what has been observed for GC-A *(27)* (*see* Signal Transduction).

Many studies have employed the ligand specificity of CNP for GC-B, and of ANP for GC-A, to determine the relative abundance of these receptors through stimulation of cGMP production. This is a valid approach given the very low relative potency of CNP on GC-A and of ANP on GC-B *(4,28,29)*. However, care must be taken in comparing CNP $ED_{50}$ values among tissues or cell lines since there may be intrinsic variability resulting from differences in receptor expression level, or from coexpression of receptor guanylyl cyclases along with NPR-C. Reconstitution of receptor expression in a model system with cloned GC-A and NPR-C has shown that NPR-C can cause a shift in the concentration response curve to higher $ED_{50}$ values for ANP-stimulated cGMP production *(30)*. A similar result has been obtained with GC-B and NPR-C for CNP stimulation of cGMP production (Lowe DG, unpublished observations). Receptor coexpression, therefore, affords one possible explanation for the variable potency of CNP in stimulating cGMP production. Reported values for the CNP $ED_{50}$ on cells ranges from 4 to 100 n*M*, independent of maximal stimulation, i.e., receptor expression level *(28,31)*. Although coexpression with NPR-C could account for this, there may be trivial explanations, such as differences in CNP handling and quantification between labs.

### *Determinants of Receptor Selectivity*

Mutational analysis of the GC-B extracellular domain has been carried out by Sharma and coworkers. These investigators have focused on two amino acid residues in GC-B, E338, and V364 (E332 and V358 by their nomenclature; Fig. 1). Mutations in the analogous positions in GC-A were originally identified in a variant GC-A cDNA clone *(32)* that most likely represents a cloning artifact. The GC-A protein encoded by this mutated cDNA clone was inactive for ligand binding, and correction of the two residues back to the original sequence, H338 and L364 (Fig. 1) restored ligand binding. Extending these GC-A mutation studies to GC-B, these investigators found that mutation of E332 by charge reversal to His or Lys abolished CNP binding *(33)*. Caution is required in the interpretation of these data since the incorporation of a charge reversal mutation may disrupt ECD structure, rendering it incapable of CNP binding. Mutation of V364 suggested no role for this residue in CNP signal transduction *(33)*, however the mutation V364L actually resulted in some ANP stimulation of GC-B *(34)*. The maximal efficacy for ANP stimulation of cGMP production on the mutant receptor was approx 50% that seen with CNP, suggesting that the receptor mutation changed ANP from a weak agonist to a partial but more potent agonist. Comprehensive studies on the GC-B ligand binding site and ECD structure–function remain to be accomplished.

### *GC-B Antibodies as Pharmacological Tools*

In addition to a native ligand, receptor-selective agonists or antagonists are useful to exploring the physiological significance of a particular receptor. These types of molecules are typically powerful tools in pharmacology; however, no potent peptide or nonpeptide antagonists or receptor-selective agonists (CNP binds to the NPR-C receptor) of GC-B have been described. Experiments in two labs have, however, identified antagonistic monoclonal antibodies (MAbs) against GC-B *(25,35)*. Both of these groups showed that the antibody blocked CNP binding to GC-B, and Drewett and coworkers used the MAb 3G12 to demonstrate that CNP targets GC-B in the relaxation of precontracted aortic rings *(25)*. This experiment illustrates the utility of an antagonistic MAb in proving that a specific receptor (GC-B) is present and mediates a particular response to agonist (CNP).

## SIGNAL TRANSDUCTION

Membrane guanylyl cyclases are a pastiche of several structurally distinct functional domains or protein structural modules that act in a concerted fashion to transduce a ligand binding event in the extracellular domain to activation of the cytoplasmic catalytic domain. Deletion studies on GC-A have demonstrated that the C-terminal catalytic domain has intrinsic cyclase activity. Removal of the kinase homology domain resulted in a constitutively activated receptor, indicating that this region mediates basal inhibition of cyclase activity *(36)*. Furthermore, guanylyl cyclase activation for both GC-A and GC-B has been shown to require ATP, or a nonhydrolyzable ATP analog, as a cofactor for signal transduction, and this requirement is mediated by the kinase homology domain *(36–38)*.

The kinase homology domains of GC-A and GC-B are functionally interchangeable *(39)*, suggesting that observations on the role of this domain in either system are potentially valid for the other receptor as well. This conclusion is supported by the work of Sharma and colleagues, who have shown a similarity between the two receptors on the ATP requirements for GC-B signal transduction *(38)*. The sequence similarity of the kinase homology domain to the family of protein kinases extends to a select group of residues and structural motifs that are invariant or highly conserved among protein kinases *(40)*. Among these are a glycine-rich motif at the amino terminal portion of kinases that serves as part of a sharp turn between the first two strands of an antiparallel β-sheet. In the crystal structure of protein kinase A, this region is in contact with the phosphates of MgATP and serves as a phosphate anchor *(41)*. Mutation of this region ($Gly^{505}$-x-x-x-$Gly^{509}$) in GC-B with the substitution Gly505 Ala (residue 499 by Duda et al.) inhibited signal transduction and induction of a low-affinity state for CNP *(26,42)*. This result may be explained by steric interference from the Ala side chain, causing inhibition of the MgATP phosphate anchor function of this region and inhibiting ATP binding.

The induction of an ATP-dependent low-affinity state, seen as a fivefold reduction in affinity for CNP *(26)*, may be a general phenomena for GC-A and GC-B. Maack and colleagues *(43)* have described a rapid release of ANP bound to GC-A at 37°C. This is probably related to the finding that ATP is required for the induction of a low-affinity state for ANP binding to GC-A in membrane preparations *(27)*. These experiments suggest a model in which bound ligand triggers ATP binding to the kinase homology domain. Consequently, MgATP-dependent conformational changes in the KHD (in direct analogy to MgATP-dependent conformational changes in protein kinases) results in a release of the KHD-inhibitory effect on the cyclase domain, and generates, in an unde-

Table 2
Localization of GC-B Expression

| *Species* | *Method* | *Tissue* | *Reference* |
|---|---|---|---|
| Rat | PCR | Whole brain, cerebral cortex, cerebellum, brain stem, whole kidney, renal cortex, renal medulla, renal papilla, adrenal, aorta, liver, lung, pituitary, spleen, testis | *45* |
| | | Kidney glomerulus, vasa recta, arcuate artery, distal nephron, cortical, outer medullary, inner medullary, tubules, cardiac fibroblasts | *47,51,83* |
| | | Eye: retina, choroid, ciliary process | *84* |
| | | Xiphisternum, chondrocytes, bone marrow | *77,78* |
| | RNase protection | Aorta > inferior vena cava | *85* |
| | Northern | Lung ≈ uterus/oviduct > brain > adrenal, kidney > small intestine | *86* |
| | | Cultured primary astrocytes | |
| | *In situ* hybridization | Pituitary (posterior > anterior, intermediate), hypothalamus (throughout) | *73* |
| | In vitro ligand autoradiography | Pituitary pars intermedia > anterior ≈ neural lobe, area prostrema > subformical organ > olfactory bulb > cerebellum > adrenal cortex, kidney cortex ≈ kidney medulla | *87* |
| | | Uterus: endometrium, myometrum | *72* |
| | Immunohistochemistry | Glomerulus, cortical collecting tubule, inner medullary collecting duct | *88* |
| Cattle | RNase protection | Adrenal cortex, adrenal medulla, renal cortex, inner medulla, outer medulla | *89* |
| | Northern | Trachea | *71* |
| Pig | Northern | Uterus | *71* |
| Monkey | In situ hybridization | Adrenal medulla, pituitary (anterior, intermediate, posterior), cerebellum, skin | *44* |
| Human | PCR | Thoracic aorta | *66* |
| | Northern | Uterus | *90* |
| | CNP binding | Kidney mesangial cells | *63* |
| | Cloning | Retina | *38* |

fined way, a transition in the extracellular domain to a lower affinity ligand-binding state, causing release of bound hormone. The mechanism by which these events are coordinated across the membrane remains to be elucidated, but probably is dependent on the aggregation state of the receptor and the quaternary interactions among subunit domains *(27)*.

## GC-B EXPRESSION

### *Regional Localization*

A variety of techniques have been used to establish the presence of GC-B in numerous tissues (Table 2). These methods have different levels of sensitivity, resolution, and caveats associated with interpreting the data. In general, abundant sites of receptor expression may correspond to major target tissues for hormone action. This appears to be the case, for example, with the ANP receptor GC-A, which is detected by *in situ* hybridization in the kidney glomerulus, adrenal zona glomerulosa, and pituitary *(44)*. Each of these sites of expression corresponds to a major site of action for ANP as a circulating hormone and is consistent with the known biological actions of ANP. However, it is also true that GC-A is expressed in tissues in which it is not detected by *in situ* hybridization but may be biologically relevant. For GC-B and its ligand CNP, we do not have a central theme underlying the biological significance of the system, so it is more difficult to judge the potential relevance of receptor localization and the magnitude of receptor expression. The most meaningful conclusions can be made for those tissues in which a combination of gene expression and functional data (ligand pharmacology and cGMP stimulation) are available. Table 2 summarizes experiments from many labs on the localization of GC-B expression, with the most extensive work having been done in the rat. The GC-B is clearly very widely expressed, with one study reporting detection of GC-B mRNA by reverse transcription-polymerase chain reaction (RT-PCR) in all of 15 rat tissues analyzed *(45)*. Whereas this result reinforces the notion of widespread GC-B expression, caution must be exercised in the interpretation of RT-PCR assays. This technique is so sensitive that essentially "illegitimate transcription" can be detected, corresponding to less than one transcript per cell *(46)*. With isolated rat myocytes in culture, GC-B expression has been documented by RT-PCR, but stimulation of cGMP production by CNP was negative *(47)*. A number of posttranslational mechanisms could account for the absence of signal transduction, so this result does not necessarily mean that the receptor protein is not synthesized. The observations of these workers reinforce the need for determining meaningful expression both by mRNA detection and functional signal transduction with the appropriate pharmacology. Detection of ligand binding need not necessarily be detectable for there to be measurable cGMP production. However, the biological relevance of receptor expression and second messenger synthesis may remain unknown in the absence of a measurable biological or phenotypic response.

Brain localization studies by *in situ* hybridization and in vitro ligand autoradiography (Table 2) suggest a role for this receptor in mediating responses to CNP as a neurotransmitter or neuromodulator. The source of CNP and the physiological role of the hormone in the brain remain to be determined. However, the sites of GC-B expression in the hypothalamus and pituitary (Table 2) suggest a role in neuroendocrine regulation (*see* Central and Peripheral Nervous System).

A number of studies have described GC-B expression in the adrenal and kidney using complimentary techniques, all of which indicate that there is functional expression of the receptor. Again, the physiological role is not well defined and the responses may derive from locally produced CNP or from fluctuations in circulating levels of hormone. In cultured bovine adrenal medullary cells, CNP is an agonist for stimulation of catecholamine production, an effect inhibited by endothelin-1 and angiotensin II *(48)*. This is consistent with the general theme of natriuretic peptides as mutual antagonists to the renin/angiotensin, endothelin, and ACTH systems in a dynamic yin/yang relationship (*see* Perspective).

Conflicting reports suggest that CNP may be weakly diuretic and natriuretic in rats *(23)*, whereas in dogs it may be antinatriuretic *(49)*. There is also data to suggest that, despite evidence of GC-B mRNA and perhaps protein expression in the kidney glomerulus (Table 2), CNP is not effective at stimulating cGMP production in isolated glomeruli

*(50)*. Concentrations of $10^{-7}M$ are required to stimulate cGMP synthesis in the glomerulus *(51)*, calling into question the physiological relevance of this signaling, despite evidence for CNP mRNA and protein expression in the kidney *(51–53)*.

### *Regulation of Expression*

Data on the regulation of GC-B expression is limited. Although the functional relevance of renal GC-B is open to questions (*see* Regional Localization), increased GC-B mRNA has been reported in sheep kidney in response to sodium depletion *(54)*. Using Northern blot analysis, receptor mRNA levels were examined 5–6 d post high- or low-salt diet. Only in the renal cortex, and not the medulla, of the salt-depleted animals was a difference observed, with a twofold increase in mRNA expression. No differences were found for GC-A or NPR-C in the cortex or medulla of either experimental group *(54)*.

In a detailed study of NPR mRNA levels using quantitative RT-PCR, changes were measured during the development of cardiac hypertrophy in the aortavenocaval fistula rat *(55)*. Whereas NPR-C expression was lost, GC-A and GC-B were both increased, with up to a 70-fold increase in GC-B in the right ventricle at 28 d postsurgery *(55)*. The role of increased GC-B in this model of cardiac hypertrophy could relate to in vitro effects of CNP in attenuating proliferation of smooth muscle cells and cardiac fibroblasts *(56,57)*. Correlation between changes in mRNA and protein need to be determined in this system.

The third area in which modulation of GC-B expression has been studied is in arterial smooth muscle cells. In normal arteries smooth muscle cells contract in response to chemical and mechanical stimuli, and are not actively biosynthetic. In this normal physiological context, the cells are referred to as having a "contractile phenotype." Both the mRNA and protein for GC-B and GC-A are expressed and may be involved in mediating the direct smooth-muscle-relaxant activities of CNP and ANP *(25,58)*. When smooth-muscle cells are placed in culture, a rapid change occurs in the phenotype of the cells with decreased contractile activity together with proliferation and increased protein synthesis, the so-called "synthetic phenotype." It is during this transition to the synthetic phenotype that there is a very pronounced increase in GC-B and NPR-C mRNA expression together with a decrease in GC-A mRNA *(58)*. A more recent study has reported up to a 1000-fold increase in GC-B mRNA together with a 10-fold increase in cGMP stimulation by CNP in comparison to a minor or insignificant decrease in GC-A mRNA and activity *(59)*. These studies were extended to an in vivo study of smooth-muscle proliferation in a rabbit central-ear artery compression model *(60)*. Although there was extensive mitosis of smooth-muscle cells in response to the injury in this animal model, no increase in GC-B was detected by CNP stimulation of cGMP production, or by *in situ* ligand autoradiography to measure binding sites. In marked contrast, experiments in an air-drying injury model with the rat common carotid artery showed a marked increase in CNP-responsive cGMP production in injured arteries *(61)*. Several possibilities could account for the different results of these two groups. Clearly, more detailed studies with physiologically relevant models of smooth-muscle proliferation are required to establish the nature of the response and biological relevance of changes in GC-B expression. A recent report on the suppression of GC-B mRNA by TGF-$\beta_1$ hints at a potentially complicated pattern of regulated vascular expression and GC-B responsiveness *(59)* (*see* Vascular Natriuretic Peptide Hypothesis).

# BIOLOGICAL ROLES OF GC-B

## *Vascular Natriuretic Peptide Hypothesis*

The most interesting hypothesis proposed to date on the potential significance of this pathway is the idea that CNP and GC-B constitute a natriuretic peptide/receptor system localized to vessel walls. According to this model, regulated expression of CNP derived from endothelial cells targets GC-B on adjacent smooth-muscle cells *(62)*. This hypothesis has been supported by data on the regulation of CNP expression from cultured-endothelial cells, demonstrating that vasoactive substances, such as TGF-β, IL-1α, or to a lesser degree ANP or BNP, stimulate synthesis and release of CNP *(62–65)*. The CNP has been measured in the endothelial scrapings from bovine aorta *(9)*, providing in vivo support for the cell culture experiments and suggesting that CNP from endothelial cells may act in an autocrine or paracrine manner. In addition, GC-B mRNA expression has been detected in human aorta *(66)*. One action of CNP derived from endothelial cells may be as a direct vasodilator involved in the regulation of vascular tone through targeting of GC-B on smooth muscle cells. This is supported by the demonstration of GC-B-dependent relaxation of rat aorta by CNP *(25)* (*see* Signal Transduction). As such, CNP may play a role as an endothelial-derived relaxing factor, targeting GC-B expressed in arteries and veins (Table 2). Experiments with cultured smooth muscle cells have shown an inhibitory effect of TGF-$\beta_1$ on GC-B expression, hinting at a potentially complex system of regulated ligand release from endothelial cells and modulated receptor expression in smooth muscle *(59)*. Because of experimental material requirements, the studies described above have been done with samples from large arteries. These conduction vessels are not involved in the regulation of vascular resistance, hence these data are only supportive of a working hypothesis in which CNP modulates blood pressure. Experiments designed to examine the role of GC-B in the tone of resistance arterioles will be required to productively extend the vascular natriuretic peptide hypothesis.

Data on the expression of GC-A and GC-B in human saphenous vein and mammary artery, obtained by measuring the magnitude of cGMP responses, shows that GC-B is more prominently expressed in venous preparations *(67)*. These data are consistent with the results of Wei et al. *(68)*, who showed that CNP is more potent than ANP at relaxing isolated canine veins, with the reverse situation occurring in arteries. Measurements of CNP levels in patients with septic shock showed up to a 10-fold increase in plasma levels, suggesting that CNP may play a role in clinical hypotension by decreasing venous return *(69)*.

In addition to its role as a paracrine EDRF, CNP may also have a role in autocrine regulation of endothelial cells by inhibition of endothelin production *(70)*. The net effect would be both direct and indirect promotion of relaxation of vascular smooth muscle. Drewett and Garbers *(9)* have also proposed that endothelial-derived CNP could inhibit norepinephrine release in the vasculature. This proposed sympathoinhibitory function of GC-B would be consistent with its tentative role in vascular regulation.

Another aspect to the potential vascular role of CNP is in the regulation of vascular structure, as compared to a role in vascular reactivity. In the rat carotid artery, air-dried injury model, CNP infusion attenuates up to 70% of the neointimal formation *(61)*. These data are consistent with the observations that CNP is a potent inhibitor of DNA synthesis in cultured vascular smooth-muscle cells *(56)*, and has modest effects on growth inhibition of cultured cardiac fibroblasts *(57)*. Establishing the role of GC-B in normal vascular physiology and pathology is going to require the development of receptor-selective antagonists. Additional understanding may also develop through the application of gene-targeting techniques to generate homozygous null mutations in the GC-B gene.

### Reproductive System

A potential role for GC-B in reproductive biology is suggested by localization studies (Table 2). In the rat, Northern blot hybridizations show the most abundant sites of GC-B mRNA to be the lung and uterus/oviduct *(71)*. The abundance of GC-B in the uterus may differ among species, with lower amounts of GC-B mRNA in porcine and bovine uterus *(71)*. By ligand autoradiography, GC-B was identified predominately in the rat endometrium and myometrium *(72)*. Northern blotting and a cGMP stimulation have also shown functional GC-B expression in human uterus and placenta during pregnancy. Whereas the role in pregnancy is uncertain, an earlier role for GC-B in fertilization is suggested by the huge amounts of CNP found in porcine seminal plasma *(71)*. The physiological significance of this source of CNP is obscure, but may be targeted to epithelial GC-B, where cellular secretion, proliferation, or differentiation could be affected.

### Central and Peripheral Nervous System

Although GC-B is localized in diverse regions of the brain (Table 2), the main focus of research has been on the hypothalamic/pituitary/adrenal axis. Detailed *in situ* hybridization studies suggest that GC-B is the predominant membrane guanylyl cyclase in the hypothalamus *(73)*, and additional studies have shown GC-B expression in the pituitary and adrenal (Table 2). Whereas CNP and ANP may mediate similar responses in peripheral target cells, for example in the vasculature, several lines of evidence indicate opposing actions for these peptides centrally. This may be caused by entirely different cell populations expressing GC-B, GC-A, and/or NPR-C.

Using toxin-conjugated ligands, Samson and coworkers *(74)* have provided evidence that GC-B mediates CNP stimulation of prolactin secretion in the hypothalamus, whereas a different population of cells mediate ANP inhibition of prolactin secretion. Opposing actions of CNP and ANP have also been suggested for the central control of water intake, with CNP stimulating, and ANP inhibiting drinking. Whereas GC-B is probably signaling the CNP response, for ANP the clearance receptor NPR-C could be involved *(75)*. In this case NPR-C may be sequestering CNP from GC-B (*see* Natriuretic Peptide Specificity), an explanation also suggested in experiments examining the inhibitory effect of hypothalamic GC-B on mediating CNP inhibition of leutinizing hormone secretion *(76)*.

In the pituitary there is no evidence as yet on what actions of CNP may be directly mediated by GC-B. However, in the adrenal, there is evidence that GC-B mediates CNP stimulation of catecholamine secretion in antagonism to endothelin, angiotensin II, and salt *(48)*. This is in direct contrast to the proposal that GC-B could be playing a role in the inhibition of vascular norepinephrine release *(9)* (*see* Vascular Natriuretic Peptide Hypotheses). Clearly the role of GC-B in mediating neurotransmission or neuromodulation is dependent on the specific context of GC-B expression, combined with factors controlling CNP production, and integration of other regulatory signals. Further investigation will be required to develop a unifying hypothesis on the role of GC-B in the central and peripheral nervous systems.

### *Bone and Cartilage Metabolism*

Two studies have indicated a role for GC-B in the modulation of bone metabolism and cartilage formation. In mouse bone-marrow cultures, GC-B mediates CNP stimulation of osteoclast bone resorption *(77)*. In these cultures, I,25-$(OH)_2D_3$ increased CNP production, either directly or indirectly, and stimulated osteoclast resorption of bone, suggesting that there may be a local autocrine or paracrine CNP/GC-B system regulating bone resorption.

Screening of rat tissues for GC-B activity revealed that the highest levels of activity were in cartilage *(78)*. When exposed to CNP, chondrocytes are growth inhibited by up to 50%. Similar to endothelial cells, TGF-$\beta_1$ stimulates CNP production in these cells, suggesting that GC-B mediates a TGF-β-dependent autocrine pathway of growth regulation *(78)*.

### *Perspective*

For the GC-B/CNP system, specific responses will depend not only on the regulation of GC-B expression and mechanisms governing exposure to CNP, but also on what specific intracellular targets are present for the second messenger cGMP. A variety of proteins responsive to cGMP have been described, including ion channels, phosphodiesterases, and protein kinases *(79)*. The potential for complex regulation is apparent in the combinatorial expression of these different cGMP effectors, combined with interactions with other signaling pathways. Despite this potential complexity, there are some general themes that appear to be emerging from the studies of CNP and effects mediated by GC-B in tissues and cultured cells. The peripheral actions of CNP, whether in the adrenal or the vasculature, appear to be in antagonism either directly or indirectly with vasoactive hormones, such as endothelin I, angiotensin II, and ACTH. This is also true for ANP, and extends to the direct and indirect actions of ANP in the kidney in antagonism to substances involved in fluid and electrolyte homeostasis *(80,81)*. These observations suggest a dynamic balance between the activity of CNP/GC-B and ANP/GC-A, with vasoactive hormones in a yin/yang relationship. This physiological antagonism refers to a dynamic balance between the opposing actions of these hormone systems and is not to be confused with the cGMP/cAMP yin/yang hypothesis of second-messenger signal transduction proposed 20 yr ago *(82)*. Although there is some evidence in the periphery for this physiological yin/yang, this concept may not necessarily extend to the nervous system where there appear to be some opposing actions of ANP and CNP.

An interesting relationship may be found in the expression of GC-B in mesenchymal-derived cells that are related by function and ontogeny. The GC-B is expressed prominently in fibroblast cell lines *(71)*, and is found in fibroblast/mesenchymal-derived smooth muscle cells *(25,28)* and chondrocytes *(78)*. Two additional cell types that are developmentally downstream of the fibroblast/mesenchymal precursor are adipocytes and osteoblasts. The regulatory activities of GC-B in smooth muscle and chondrocytes therefore makes it tempting to speculate that adipocytes and osteoblasts will also be responsive to GC-B signaling in the modulation of their specialized functions. Although this idea does not present a unifying model of the role of GC-B in mesenchyme-derived cells, the role of many of these cell types in connective-tissue formation suggests a potential role for GC-B in modulating this activity, perhaps in balance with TGF-β and/or other vasocative molecules.

# REFERENCES


1. Chinkers M, Garbers DL, Chang M-S, Lowe DG, Chin H, Goeddel DV, Schulz S. A membrane form of guanylate cyclase is an atrial natriuretic peptide receptor. Nature 1989;338:78–83.
2. Lowe DG, Chang MS, Hellmiss R, Chen E, Singh S, Garbers DL, Goeddel DV. Human atrial natriuretic peptide receptor defines a new paradigm for second messenger signal transduction. EMBO J 1989;8:1377–1384.
3. Chang MS, Lowe DG, Lewis M, Hellmiss R, Chen E, Goeddel DV. Differential activation by atrial and brain natriuretic peptides of two different receptor guanylate cyclases. Nature 1989;341:68–72.
4. Schulz S, Singh S, Bellet RA, Singh G, Tubb DJ, Chin H, Garbers DL. The primary structure of a plasma membrane guanylate cyclase demonstrates diversity within this new receptor family. Cell 1989;59:1155–1162.
5. Hansbrough JR, Garbers DL. Speract: purification and characterization of a peptide associated with eggs that activates spermatozoa. J Biol Chem 1981;256:1447–1452.
6. Garbers DL. Molecular basis of fertilization. Annu Rev Biochem 1989;58:719–742.
7. deBold AJ, Borenstein HB, Veress AT, Sonnenberg H. A rapid and potent natriuretic response to intravenous injection of atrial myocardial extract in rats. Life Sci 1981;28:89–94.
8. Espiner EA, Richards AM, Yandle TG, Nicholls MG. Natriuretic hormones. Endocrinol Metab Clin North Am 1995;24:481–509.
9. Drewett JG, Garbers DL. The family of guanylyl cyclase receptors and their ligands. Endocrine Rev 1994;15:135–162.
10. Singh S, Lowe DG, Thorpe DS, Rodriguez H, Kuang WJ, Dangott LJ, Chinkers M, Goeddel DV, Garbers DL. Membrane guanylate cyclase is a cell-surface receptor with homology to protein kinases. Nature 1988;334:708–712.
11. Fenrick R, Babinski K, McNicoll N, Therrien M, Drouin J, DeLean A. Cloning and functional expression of the bovine natriuretic peptide receptor-B (Natriuretic factor $R_{1c}$ subtype). Mol Cell Biochem 1994;137:173–182.
12. Bennett BD, Bennett GL, Vitangcol RV, Jewett JRS, Burnier J, Henzel W, Lowe DG. Extracellular domain-IgG fusion proteins for three human natriuretic peptide receptors. J Biol Chem 1991;266:23,060–23,067.
13. Lowe DG, Fendly BM. Human natriuretic peptide receptor-A guanylyl cyclase. Hormone cross-linking and antibody reactivity distinguish receptor glycoforms. J Biol Chem 1992;267:21,691–21,697.
14. Lowe DG, Camerato TR, Goeddel DV. cDNA sequence of the human atrial natriuretic peptide clearance receptor. Nucleic Acids Res 1990;18:3412.
15. Saheki T, Mizuno T, Iwata T, Saito Y, Nagasawa T, Mizuno KU, Ito F, Ito T, Hagiwara H, Hirose S. Structure of the bovine atrial natriuretic peptide receptor (type C) gene. J Biol Chem 1991;266:11,122–11,125.
16. Itakura M, Iwashina M, Mizumo T, Ito T, Hagiwara H, Hirose S. Mutational analysis of disulfide bridges in the type C atrial natriuretic peptide receptor. J Biol Chem 1994;269:8314–8318.
17. Stults JT, O'Connell KL, Garcia C, Wong S, Engel AM, Garbers DL, Lowe DG. The disulfide linkages and glycosylation sites of the human natriuretic peptide receptor-C homodimer. Biochem 1994;33:11,372–11,381.
18. Katafuchi T, Takashima A, Kashiwagi M, Hagiwara H, Takei Y, Hirose S. Cloning and expression of eel natriuretic-peptide receptor B and comparison with its mammalian counterparts. Eur J Biochem 1994;333:835–842.
19. Garbers DL, Lowe DG. Guanylyl cyclase receptors. J Biol Chem 1994;269:30,741–30,744.
20. Wilson EM, Chinkers M. Identification of sequences mediating guanylyl cyclase dimerization. Biochemistry 1995;34:4696–4701.
21. Chinkers M, Wilson EM. Ligand-independent oligomerization of natriuretic peptide receptors. Identification of heteromeric receptors and a dominant negative mutant. J Biol Chem 1992;267:18,589–18,597.
22. Lowe DG, Klisak I, Sparkes RS, Mohandas T, Goeddel DV. Chromosomal distribution of three members of the human natriuretic peptide receptor/guanylyl cyclase gene family. Genomics 1990;8:304–312.
23. Sudoh T, Minamino N, Kangawa K, Matsuo H. C-type natriuretic peptide (CNP): a new member of natriuretic peptide family identified in porcine brain. Biochem Biophys Res Commun 1990;168:863–870.
24. Koller KJ, Lowe DG, Bennett GL, Minamino N, Kangawa K, Matsuo H, Goeddel DV. Selective activation of the B natriuretic peptide receptor by C-type natriuretic peptide (CNP). Science 1991;252:120–123.

25. Drewett JG, Fendly BM, Garbers DL, Lowe DG. Natriuretic peptide receptor-B (guanylyl cyclase-B) mediates C-type natriuretic peptide relaxation of precontracted rat aorta. J Biol Chem 1995;270:4668–4674.
26. Duda T, Sharma RK. ATP modulation of the ligand binding and signal transduction activities of the type C natriuretic peptide receptor guanylate cyclase. Mol Cell Biochem 1995;152:179–183.
27. Jewett JRS, Koller KJ, Goeddel DV, Lowe DG. Hormonal induction of low affinity receptor guanylyl cyclase. EMBO J 1993;12:769–777.
28. Suga S, Nakao K, Hosoda K, Mukoyama M, Ogawa Y, Shirakami G, Arai H, Saito Y, Kambayashi Y, Inouye K, Imura H. Receptor selectivity of natriuretic peptide family, atrial natriuretic peptide, brain natriuretic peptide, and C-type natriuretic peptide. Endocrinology 1992;130:229–239.
29. Schoenfeld JR, Sehl P, Quan C, Burnier JP, Lowe DG. Agonist selectivity for three species of natriuretic peptide receptor-A. Mol Pharm 1995;47:172–180.
30. Cunningham BC, Lowe DG, Li B, Bennett BD, Wells JA. Production of an atrial natriuretic peptide variant that is specific for type A receptors. EMBO J 1994;13:2508–2515.
31. Féthière J, Graihle R, Larose L, Babinski K, Ong H, De Léan A. Distribution and regulation of natriuretic factor-$R_{1c}$ receptor subtypes in mammalian cell lines. Mol Cell Biochem 1993;124:11–16.
32. Duda T, Goraczniak R, Sharma R. Site-directed mutational analysis of a membrane guanylate cyclase cDNA reveals the atrial natriuretic factor signalling site. Proc Natl Acad Sci USA 1991;88:7882–7886.
33. Duda T, Goraczniak RM, Sharma RK. Glutamic acid-332 residue of the type C natriuretic peptide receptor guanylate cyclase is important for signaling. Biochemistry 1994;33:7430–7433.
34. Duda T, Goraczniak RM, Sharma RK. Single amino acid residue-linked signaling shifts in the transduction activities of atrial and type C natriuretic factor receptor guanylate. Biochem Biophys Res Commun 1995;212:1046–1053.
35. Kitano K, Fukuda Y, Nagahira K, Nasu T, Izumi R, Kawashima K, Nakanishi T. Production and characterization of monoclonal antibodies against human natriuretic peptide receptor-A or -B. Immunol Letts 1995;47:215–222.
36. Chinkers M, Garbers DL. The protein kinase domain of the ANP receptor is required for signalling. Science 1989;245:1392–1394.
37. Chinkers M, Singh S, Garbers DL. Adenine nucleotides are required for activation of rat atrial natriuretic peptide receptor/guanylyl cyclase expressed in a baculovirus system. Biochemistry 1991;266:4088–4093.
38. Duda T, Goraczniak RM, Sitaramayya A, Sharma RK. Cloning and expression of an ATP-regulated human retina C-type natriuretic factor receptor guanylate cyclase. Biochemistry 1993;32:1391–1395.
39. Koller KJ, deSauvage FJ, Lowe DG, Goeddel DV. Conservation of the kinase-like regulatory domain is essential for activation of the natriuretic peptide receptor guanylyl cyclases. Mol Cell Biol 1992;12:2581–2590.
40. Taylor SS, Knighton DR, Zheng J, Sowadski JM, Gibbs CS, Zoller MJ. A template for the protein kinase family. Trends Biochem Sci 1993;18:84–89.
41. Knighton DR, Zheng J, Eyck LFT, Ashford VA, Xuong NH, Taylor SS, Sowadski JM. Crystal structure of the catalytic subunit of cyclic adenosine monophospate-dependent protein kinase. Science 1991;253:407–414.
42. Duda T, Goraczniak RM, Sharma RK. The glycine residue of ATP regulatory module in receptor guanylate cyclases that is essential in natriuretic factor signaling. FEBS Lett 1993;335:309–314.
43. Koh GY, Nussenzveig DR, Okolicany J, Price DA, Maack T. Dynamics of atrial natriuretic factor-guanylate cyclase receptors and receptor-ligand complexes in cultured glomerular mesangial and renomedullary interstitial cells. J Biol Chem 1992;267:11,987–11,994.
44. Wilcox JN, Augustine A, Goeddel DV, Lowe DG. Differential regional expression of three natriuretic peptide receptor genes within primate tissues. Mol Cell Biol 1991;11:3454–3462.
45. Tallerico-Melnyk T, Yip CC, Watt VM. Widespread co-localization of mRNAS encoding the guanylate cyclase-coupled natriuretic peptide receptors in rat tissues. Biochem Biophys Res Commun 1992;189:610–616.
46. Chelly J, Concordet J-P, Kaplan J-C, Kahn A. Illegitimate transcription: transcription of any gene in any cell type. Proc Natl Acad Sci USA 1989;86:2617–2621.
47. Lin X, Hänze J, Heese F, Sodmann R, Lang RE. Gene expression of natriuretic peptide receptors in myocardial cells. Circ Res 1995;77:750–758.
48. Tsutsui M, Yanagihara N, Minami K, Kobayashi H, Nakashima Y, Kuroiwa A, Izumi F. C-type natriuretic peptide stimulates catecholamine synthesis through the accumulation of cyclic GMP in cultured bovine adrenal medullary cells. J Pharmacol Exper Ther 1994;268:584–589.

49. Stingo AJ, Clavell AL, Aarhus LL, Burnett JC. Cardiovascular and renal actions of C-type natriuretic peptide. Am J Physiol 1992;262:H308–H312.
50. Furuya M, Takehisa M, Minamitake Y, Kitajima Y, Hayashi Y, Ohnuma N, Ishihara T, Minamino N, Kangawa K, Matsuo H. Novel natriuretic peptide, CNP, potently stimulates cyclic GMP production in rat cultured vascular smooth muscle cells. Biochem Biophys Res Commun 1990;170:201–208.
51. Terada Y, Tomita K, Nonoguchi H, Yang T, Marumo F. PCR localization of C-type natriuretic peptide and B-type receptor mRNAs in rat nephron segments. Am J Physiol 1994;267:F215–F222.
52. Suzuki E, Hirata Y, Hayakawa H, Omata M, Kojima M, Kangawa K, Minamino N, Matsuo H. Evidence for C-type natriuretic peptide production in the rat kidney. Biochem Biophys Res Commun 1993;192:532–538.
53. Dean AD, Vehaskari VM, Greenwald JE. Synthesis and localization of C-type natriuretic peptide in mammalian kidney. Am J Physiol 1994;266:F491–F496.
54. Fraenkel MB, Aldred GP, McDougall JG. Sodium status affects GC-B natriuretic peptide receptor mRNA levels, but not GC-A or C receptor mRNA levels, in the sheep kidney. Clin Sci 1994;86:517–522.
55. Brown LA, Nunez DJ, Wilkins MR. Differential regulation of natriuretic peptide receptor messenger RNAs during the development of cardiac hypertrophy in the rat. J Clin Invest 1993;92:2702–2712.
56. Porter JG, Catalano R, McEnroe G, Lewicki JA, Protter AA. C-type natriuretic peptide inhibits growth factor-dependent DNA synthesis in smooth muscle cells. Am J Physiol 1992;263:C1001–C1006.
57. Cao L, Zlock DW, Gardner DG. Differential regulation of natriuretic peptide receptor activity in vascular cells. Hypertension 1994;24:329–338.
58. Suga S-I, Nakao K, Kishimoto I, Hosoda K, Mukoyama M, Arai H, Shirakami G, Ogawa Y, Komatsu Y, Nakagawa O, Hama N, Imura H. Phenotype-related alteration in expression of natriuretic peptide receptors in aortic smooth muscle cells. Circ Res 1992;71:34–39.
59. Fujio N, Gossard F, Bayard F, Tremblay J. Regulation of natriuretic peptide receptor A and B expression by transforming growth factor-$\beta_1$ in cultured aortic smooth muscle cells. Hypertension 1994;23:908–913.
60. Brown J, Chen Q. Regional expression of natriuretic peptide receptors during the formation of arterial neointima in the rabbit. Circ Res 1995;77:906–918.
61. Furuya M, Aisaka K, Miyazaki T, Honbou N, Kawashima K, Ohno T, Tanaka S, Minamino N, Kangawa K, Matsuo H. C-type natriuretic peptide inhibits intimal thickening after vascular injury. Biochem Biophys Res Commun 1993;193:248–253.
62. Suga S, Nakao K, Itoh H, Komatsu Y, Ogawa Y, Hama N, Imura H. Endothelial production of C-type natriuretic peptide and its marked augmentation by transforming growth factor-β. J Clin Invest 1992;90:1145–1190.
63. Suga S-I, Itoh H, Komatsu Y, Ogawa Y, Hama N, Yoshimasa T, Nakao K. Cytokine-induced C-type natriuretic peptide (CNP) secretion from vascular endothelial cells-evidence for CNP as a novel autocrine/paracrine regulator from endothelial cells. Endocrinology 1993;133:3038–3041.
64. Stingo AJ, Clavell AL, Heublein DM, Wei C-M, Pittelkow MR, Burnett JJC. Presence of C-type natriuretic peptide in cultured human endothelial cells and plasma. Am J Physiol 1992;263:H1318–H1321.
65. Nazario B, Hu R-M, Pedram A, Prins B, Levin ER. Atrial and brain natriuretic peptides stimulate the production and secretion of C-type natriuretic peptide from bovine aortic endothelial cells. J Clin Invest 1995;95:1151–1157.
66. Komatsu Y, Nakao K, Itoh H, Suga S-I, Ogawa Y, Imura H. Vascular natriuretic peptide. Lancet 1992;340:622.
67. Zhang L, Castresana M, McDonald M, Johnson J, Newman W. Response of human artery, vein, and cultured smooth muscle cells to atrial and C-type natriuretic peptides. Crit Care Med 1996;24:306–310.
68. Wei C-M, Aarhus LL, Miller VM, Burnett J John C. Action of C-type natriuretic peptide in isolated canine arteries and veins. Am J Physiol 1993;264:H71–H73.
69. Hama N, Itoh H, Shirakami G, Suga S-I, Komatsu Y, Yoshimasa T, Tanaka I, Mori K, Nakao K. Detection of C-type natriuretic peptide in human circulation and marked increase of plasma CNP level in septic shock patients. Biochem Biophys Res Commun 1994;198:1177–1182.
70. Kohno M, Horio T, Yokokawa K, Kurihara N, Takeda T. C-type natriuretic peptide inhibits thrombin- and angiotensin II-stimulated endothelin release via cyclic guanosine 3',5'-monophosphate. Hypertension 1992;19:320–325.
71. Chrisman TD, Schulz S, Potters LR, Garbers DL. Seminal plasma factors that cause large elevations in cellular cyclic GMP are C-type natruiretic peptides. J Biol Chem 1993;268:3698–3703.

72. Dos Reis AM, Fujio N, Dam T-V, Mukaddam-Daher S, Jankowski M, Tremblay J, Gutkowska J. Characterization and distribution of natriuretic peptide receptors in the rat uterus. Endocrinology 1995;136:4247–4253.
73. Langub JMC, Dolgas CM, Watson JRE, Herman JP. The C-type natriuretic peptide receptor is the predominant natriuretic peptide receptor mRNA expressed in rat hypothalamus. J Neuroendocrin 1995;7:305–309.
74. Samson WK, Huang F-LS, Fulton RJ. Opposing neuroendocrine actions of the natriuretic peptides: C-type and A-type natriuretic peptides do not interact with the same hypothalamic cells controlling prolactin secretion. J Neuroendocrinol 1995;7:759–763.
75. Samson W, Skala K, Huang F. CNP-22 stimulates, rather than inhibits, water drinking in the rat: evidence for a unique biological action of the C-type natriuretic peptides. Brain Res 1991;568:285–288.
76. Samson W, Skala K, Huang F, Fulton R. C-type natriuretic peptide mediates the hypothalamic actions of the natriuretic peptides to inhibit luteinizing hormone secretion. Endocrinology 1993;132:504–509.
77. Holliday LS, Dean AD, Greenwald JE, Gluck SL. C-type natriuretic peptide increases bone resorption in 1,25-dihydroxyvitamin $D_3$-stimulated mouse bone marrow cultures. J Biol Chem 1995;270:18,983–18,989.
78. Hagiwara H, Sakaguchi H, Lodhi KM, Suda K, Hirose S. Subtype switching of natriuretic peptide receptors in rat chondrocytes during *in vitro* culture. J Biochem 1994;116:606–609.
79. Lincoln TM, Cornwell TL. Intracellular cyclic GMP receptor proteins. FASEB J 1993;7:328–338.
80. Brenner BM, Ballermann BJ, Gunning ME, Zeidel ML. Diverse biological actions of atrial natriuretic peptide. Physiol Rev 1990;70:665–699.
81. Ruskoaho H. Atrial natriuretic peptide: synthesis, release, and metabolism. Pharmacol Rev 1993;44: 479–602.
82. Goldberg ND, Haddox MK, Nicol SE, Glass DB, Sanford CH, Kuehl JFA, Estensen R. Biologic regulation through opposing influences of cyclic GMP and cyclic AMP: the yin yang hypothesis. Adv Cyclic Nucleotide Res 1975;5:307–330.
83. Lohe A, Yeh I, Hyver T, Pratt R, Jamison R. Natriuretic peptide B receptor and C-type natriuretic peptide in the rat kidney. J Am Soc Nephrol 1995;6:1552–1558.
84. Fernandez-Durango R, Nunez DJR, Brown MJ. Messenger RNAs encoding the natriuretic peptides and their receptors are expressed in the eye. Exp Eye Res 1995;61:723–729.
85. Yoshimoto T, Naruse M, Naruse K, Shionoya K, Tanabe A, Tanaka M, Hagiwara H, Hirose S, Muraki T, Demura H. Differential gene expression of vascular natriuretic peptide receptor subtype in artery and vein. Biochem Biophys Res Commun 1995;216:535–539.
86. Deschepper CF, Picard S. Effects of C-type natriuretic peptide on rat astrocytes: regional differences and characterization of receptors. J Neurochem 1994;62:1974–1982.
87. Konrad EM, Thibault G, Schiffrin EL. Autoradiographic visualization of the natriuretic peptide receptor-B in rat tissues. Regulatory Peptides 1992;39:177–189.
88. Ritter D, Dean AD, Gluck SL, Greenwald JE. Natriuretic peptide receptors A and B have different cellular distributions in rat kidney. Kidney Int 1995;48:1758–1766.
89. Yamamoto T, Feng L, Mizuno T, Hirose S, Kawasaki K, Yaoita E, Kihara I, Wilson CB. Expression of mRNA for natriuretic peptide receptor subtypes in bovine kidney. Am J Physiol 1994;267:F318–F324.
90. Itoh H, Sagawa N, Hasegawa M, Nanno H, Kobayashi F, Ihara Y, Mori T, Komatsu Y, Suga S-I, Yoshimasa T, Itoh H, Nakao K. Expression of biologically active receptors for natriuretic peptides in the human uterus during pregnancy. Biochem Biophys Res Commun 1994;203:602–607.

# 4 Molecular Determinants of Natriuretic Peptide Clearance Receptor Function

*John A. Lewicki and Andrew A. Protter*

**CONTENTS**



## INTRODUCTION

The natriuretic peptides, atrial natriuretic peptide (ANP), B-type natriuretic peptide (BNP), and C-type natriuretic peptide (CNP), elicit their hemodynamic, renal, and neuroendocrine actions via a series of specific receptors on target tissues. To date, three natriuretic peptide receptors have been elucidated and characterized in detail. Two of these receptors are membrane-associated guanylyl cyclases, termed GC-A and GC-B. These receptors, which are reviewed in detail Chapters 3, 4, and 6–8, mediate many of the direct biological actions of these peptides. The natriuretic peptide clearance (NP-C) receptor comprises the third member of this receptor family. As its name implies, the NP-C receptor is a unique protein that plays a central role in mediating the metabolic clearance of the natriuretic peptides from the circulation. The NP-C receptor thus serves to regulate plasma concentrations of the natriuretic peptides. Recent data also suggest that this receptor may be responsible for some direct actions of the natriuretic peptides, although the physiological significance of these events is unclear.

Figure 1 is a schematic representation of the three known natriuretic peptide receptors. The NP-C receptor exhibits some structural homology with GC-A and GC-B. The extracellular ligand-binding domains of all three receptors are approx 35% homologous with conservation of several key cysteine residues involved in disulfide bonding. Furthermore, each receptor has a single transmembrane domain. Despite this homology, all three receptors exhibit unique ligand specificities (Fig. 1). Within the cytoplasmic region, GC-A and GC-B contain approx 566 amino acids that can be subdivided into kinase homology

From: *Contemporary Endocrinology: Natriuretic Peptides in Health and Disease*
Edited by: W. K. Samson and E. R. Levin Humana Press Inc., Totowa, NJ

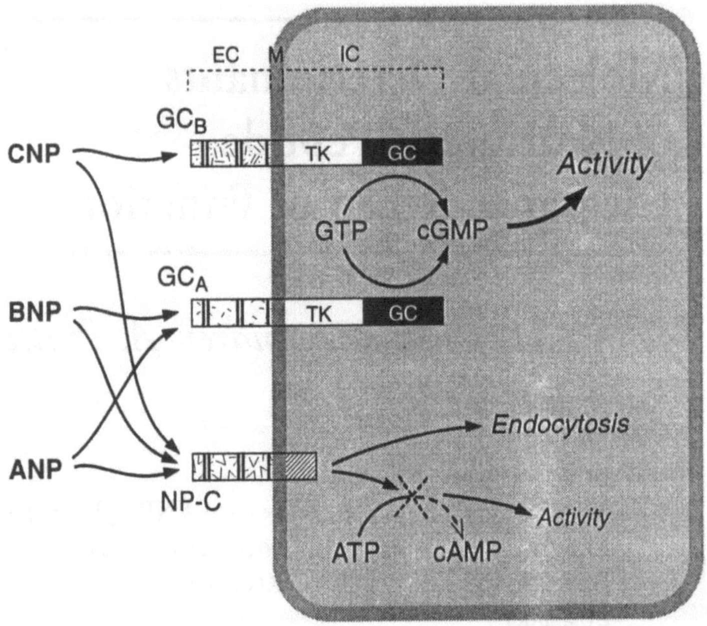


**Fig. 1.** Pharmacological specificity of the natriuretic peptides for the biological (GC-A and GC-B) and NP-C receptors. Each of the three receptors have a single transmembrane domain (M) and exhibit amino acid sequence homology in their extracellular ligand-binding domains (EC). The short (37 amino acids) intracellular domain of the NP-C receptor is contrasted with the large intracellular domains of the GC-A (566 amino acids) and GC-B (568 amino acids) receptors, which have a thymidine kinase-like region (TK) and a guanylyl cyclase region (GC) contained within this domain. The coupling of GC-A and GC-B to the stimulation of cyclic GMP formation is noted. In addition, the functions of the NP-C receptor to internalize and clear the natriuretic peptides and inhibit adenylate cyclase activity are summarized.

and guanylyl cyclase catalytic domains. Conversely, the NP-C receptor contains a short cytoplasmic tail of only 37 amino acids. Short cytoplasmic domains are observed in many proteins that function in the clearance and transport of proteins and macromolecules (e.g., low density lipoprotein, asialoglycoprotein, and transferrin receptors). This structural feature is consistent with the central role of the NP-C receptor in the metabolic clearance and regulation of plasma concentrations of the natriuretic peptides.

The purpose of this chapter is to provide insight into the molecular determinants of NP-C receptor function. In particular, attention is devoted to our knowledge about structural features of the NP-C receptor, the regulation of its expression, localization, and turnover, its novel ligand specificity, and primary role in mediating the metabolic clearance of the natriuretic peptides. More generally, we intend to highlight the unique and specialized role of the NP-C receptor in regulating the complex biological actions of the natriuretic peptide family.

## MOLECULAR CHARACTERIZATION OF THE NP-C RECEPTOR

### *Molecular Biology of the NP-C Receptor*

#### Cloning and Structural Analysis

Initial studies characterizing the NP-C receptor indicated that the protein is a homodimer of two 60 kDa subunits *(1)*. The purification of the NP-C receptor from bovine aortic smooth muscle cells by ANP-sepharose affinity chromatography *(1)* led to the isolation and characterization of bovine *(2)*, human *(3,4)*, and rat *(5)* cDNA clones. The NP-C receptor protein is well conserved through evolution with the rat and bovine amino acid sequences showing 95% identity to the human sequence.

The gene for the bovine NP-C receptor has been isolated and characterized by sequence analysis *(6)*. The gene comprises eight exons distributed over approx 85 kilobases of genome. The transcription start site has been mapped and potential TATA box and AP-2 binding sites have been identified in the promoter region; however, no structure/function analysis of the promoter region has been performed. A comparison of the exon/intron organizations of the bovine NP-C gene and the rat GC-A receptor gene indicates striking similarity, suggesting the two genes evolved from a common ancestral gene.

Analysis of the amino acid sequence deduced from the human DNA clones (*see* Fig. 2A) reveals a 541-amino-acid protein with three distinct regions: an 18-amino-acid, amino-terminal signal sequence; a 27-amino-acid, glycine-rich spacer region; and a region of 496 amino acids whose amino terminus is homologous to the amino acid sequence obtained from amino-terminal sequence analysis of the isolated bovine receptor. The function of the glycine-rich spacer region is unknown. As noted previously, the mature receptor protein can be represented as a tripartite structure with a single hydrophobic (membrane-spanning) domain separating a large, amino-terminal extracellular ligand-binding domain and a short (37 amino acids) cytoplasmic tail (*see* Fig. 2B).

Several structural features have been reported that afford high-affinity ligand binding to the extracellular domain of the NP-C receptor. In vitro mutagenesis and recombinant expression techniques have been deployed to generate truncated variants of the NP-C receptor that lack the transmembrane and cytoplasmic domains *(7,8)*. These truncated, soluble receptors bind native forms of the natriuretic peptides and peptide analogs with similar rank-order kinetics as observed with the membrane-associated holoreceptor, thus demonstrating that the extracellular domain alone is fully capable of associating with ligands. In vitro mutagenesis and mass spectroscopy techniques have been used to define the disulfide linkage pattern of the bovine and human NP-C receptors *(9,10)*. These studies have revealed that four of the five cysteines found within the extracellular domain form intrachain disulfide bonds (cysteine$^{108}$-cysteine$^{136}$ and cysteine$^{213}$-cysteine$^{261}$). The final cysteine residue at position 473 appears to form an interchain bond, and thus mediates receptor dimerization *(11)*. Specific amino acid residues important for ligand binding have been identified by several groups. Replacing residues histidine$^{149}$-tryptophan$^{150}$ (amino acids conserved among the NP-C, GC-A, and GC-B receptors) with leucine$^{149}$-leucine$^{150}$ markedly reduces the binding affinity of ANP *(12)*. In addition, amino acid substitutions at residues 233 (rat: alanine, human: isoleucine) and 250 (rat: tyrosine, human: asparagine) modulate ligand binding and specificity *(5,13)*. This result suggests that these two residues form part of the ligand-binding pocket of the NP-C receptor. Finally, a transcription splice variant of the NP-C receptor has been described in bovine *(14)* and human clones *(3,4)* that results in a cysteine residue being substituted for serine$^{476}$ glycine$^{477}$ in the mature human sequence. Analysis of expression of the mRNAs encoding the two forms of the NP-C receptor in various bovine tissues by polymerase chain reaction (PCR) amplification have shown that this variant is a relatively minor species *(14)*. Recombinant expression of the two bovine NP-C receptor proteins demonstrate the two proteins are identical with regard to homodimer formation, affinities for ANP, BNP, and CNP, and kinetics of internalization *(14)*.

**A**

Signal Sequence / Glycine-rich Spacer

```
1    M P S L L V L T F S P C V L L G W A L L A G G T G G G G V G G G G G
36   A G I G G G R Q E R E A L P P Q K I E V L V L L P Q D D S Y L F S L T
71   R V R P A I E Y A L R S V E G N G T G R R L L P P G T R F Q V A Y E D
106  S D C G N R A L F S L V D R V A A A R G A K P D L I L G P V C E Y A A
141  A P V A R L A S H W D L P M L S A G A L A A G F Q H K D S E Y S H L T
176  R V A P A Y A K M G E M M L A L F R H H H W S R A A L V Y S D D K L E
211  R N C Y F T L E G V H E V F Q E E G L H T S I Y S F D E T K D L D L E
246  D I V R N I Q A S E R V V I M C A S S D T I R S I M L V A H R H G M T
281  S G D Y A F F N I E L F N S S S Y G D G S W K R G D K H D F E A K Q A
316  Y S S L Q T V T L L R T V K P E F E K F S M E V K S S V E K Q G L N M
351  E D Y V N M F V E G F H D A I L L Y V L A L H E V L R A G Y S K K D G
386  G K I I Q Q T W N R T F E G I A G Q V S I D A N G D R Y G D F S V I A
421  M T D V E A G T Q E V I G D Y F G K E G R F E M R P N V K Y P W G P L
456  K L R I D E N R I V E H T N S S P C K S S G G L E E S A V T G I V V G
491  A L L G A G L L M A F Y F F R K K Y R I T I E R R T Q Q E E S N L G K
526  H R E L R E D S I R S H F S V A
```

**Fig. 2. (A)** Primary amino acid sequence of the human NP-C receptor. The signal peptide region, glycine-rich spacer region, and the mature receptor protein are indicated. The transmembrane domain is underlined. Boxed residues indicate amino acids in common among the NP-C, GC-A, and GC-B receptors. The amino acids are numbered with residue #1 being the methionine translation initiation codon.

## Sites of Expression

NP-C receptors have been localized to a wide variety of tissues, including the kidney, adrenal, heart, vasculature, lung, brain, and intestine. The receptor is also expressed in numerous cultured cell types derived from these tissues. Numerous disparities between laboratories concerning the relative abundance of NP-C receptors in cells and tissues likely reflect differences and variations in experimental conditions. Nevertheless, in many settings, the NP-C receptor is clearly expressed as the most abundant natriuretic-peptide-receptor subtype.

The widespread distribution and strategic localization of NP-C receptors are consistent with their central role in sequestering and clearing the natriuretic peptides. Accordingly, the following section summarizes our understanding of the distribution of NP-C receptors in tissues that are major targets of natriuretic peptide action.

**Renal.** Using equilibrium displacement studies in an isolated perfused rat kidney preparation, Maack et al. *(15)* have shown that NP-C receptors comprise the vast majority (>95%) of natriuretic peptide receptors in the rat kidney cortex. The NP-C receptors are also present in the renal papilla; however, in this region they represent fewer of the total ANP binding sites. Additional ligand-binding studies have shown expression of the NP-C receptor in whole glomeruli *(16,17)*, in cultured mesangial cells *(18,19)*, and in glomeru-

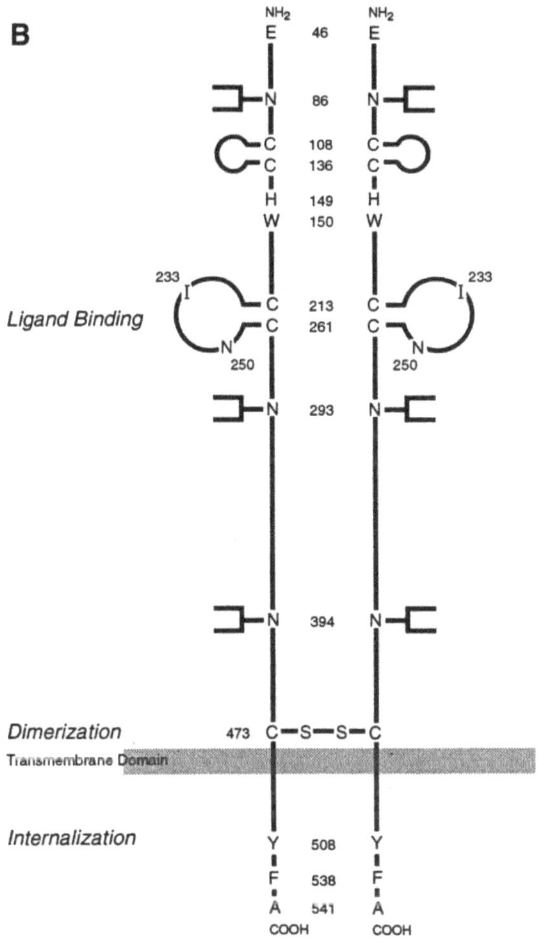


**Fig. 2. (B)** *(continued)* Schematic representation of the mature clearance receptor homodimer with important structural features. The signal sequence and glycine rich spacer region are not included (*see* A). Highlighted residues include N-linked glycosylation sites, cysteine-mediated loop structures, cysteine residue mediating homodimer formation (cysteine$_{473}$), residues involved in ligand binding (histidine$_{149}$, tryptophan$_{150}$, isoleucine$_{233}$, asparagine$_{250}$), and residues involved in internalization (cysteine$_{473}$, tyrosine$_{508}$, phenylalanine$_{538}$). The amino acid numbers correspond to the sequence shown in (A).

lar epithelial cells *(19)*. These functional binding studies are confirmed by Northern blot analysis that demonstrates abundant NP-C receptor mRNA expression in bovine kidney cortex and papillae *(2)* and in human whole kidney tissue *(3)*. In confirmation of the above studies, quantitative ribonuclease protection assays have revealed that the NP-C receptor mRNA is remarkably abundant in the glomerulus and relatively rare in the inner medulla *(20)*. More refined *in situ* hybridization and immunohistochemical techniques have localized NP-C receptor mRNA to visceral and parietal epithelial cells of the glomerulus. However, NP-C receptors are notably absent from epithelial structures of the inner medulla *(16)*. In vitro autoradiography using the NP-C receptor-specific ligand C-ANP *(4–23)* (*see* Fig. 3) has shown that the NP-C receptor is also expressed on intrarenal arteries *(21,22)*. The presence of NP-C receptors in glomerular and vascular structures

desQ,S,G,L,G, ANP(4-23)-$NH_2$
(C-ANP(4-23))

ANP(8-15)-$NH_2$

2-Naphthoxyacetyl-Isonipecotyl
ANP(11-15)-$NH_2$
(Nap ANP(11-15))

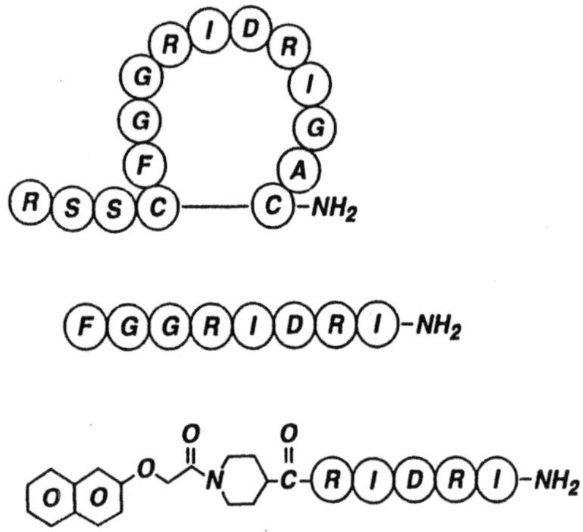


**Fig. 3.** Primary structures of NP-C receptor-specific ligands.

of the kidney is consistent with the clearance function of this receptor, because these structures receive a large proportion of the total cardiac output.

**Adrenal.** The NP-C receptors have been identified in discrete cell clusters of the adrenal cortex *(23)*. There is some controversy whether these receptors are expressed on aldosterone-secreting cells of the zona glomerulosa, but if present, their density appears lower than that of GC-A receptors *(24)*. The NP-C receptors have also been detected in the adrenal medulla and adrenal chromaffin cell, although their physiological role in these cells is not understood *(23,25)*.

**Cardiac.** The NP-C receptor is expressed prominently in cardiac tissue, including heart myocardial, endothelial, and fibroblast cells. Using *in situ* hybridization on postmortem primate tissue, Wilcox and coworkers found expression to be highest in endocardial endothelial cells of all four chambers of the heart *(23)*. They also found expression throughout the myocardium of the right atrium. This led to the speculation that a majority of ANP secreted by the atria is sequestered by NP-C receptors before reaching the circulation. A subsequent study has demonstrated expression in all chambers of the heart, although levels were higher in the atria *(26)*. In this same report, significant NP-C receptor mRNA was detected from ventricular tissue taken from a cystic-fibrosis patient following heart-lung transplantation. The NP-C receptor mRNA has been measured during the development of cardiac hypertrophy in rats *(27)*. Whereas NP-C receptor mRNA was present in all four cardiac chambers in normal animals, there was a gradual, yet striking, loss of transcripts during the development of hypertrophy. In marked contrast, the mRNAs for the GC-A and GC-B receptors increased during this time. Thus, the phenotypic changes associated with cardiac hypertrophy in the rat are associated with a downregulation of the NP-C receptor. The NP-C receptor mRNA has been detected in neonatal cardiac fibroblasts *(28)*; however, expression in adult cardiac fibroblasts and myocytes is not well understood.

**Vascular.** The NP-C receptors are widely expressed in vascular tissues. Northern blot analysis of NP-C receptor mRNA in rat tissues has demonstrated significant expression in numerous large vessels, including the thoracic aorta, superior vena cava, and the main pulmonary artery *(29)*. Remarkably high densities of NP-C receptors have been reported (up to $5.5 \times 10^6$ receptors/cell) on cultured bovine aortic smooth muscle cells *(30–33)* with somewhat lower levels measured on endothelial cells (0.1–2 × $10^5$ receptors/cell) *(30,34)*. This observation is paralleled by the finding that NP-C receptor mRNA levels are greater in aortic smooth muscle cells than in endothelial cells *(2)*.

The high density of NP-C receptors measured in cultured vascular smooth-muscle cells does not appear to recapitulate NP-C receptor expression in intact vascular tissue. Suga et al. have shown that although cultured rat aortic smooth muscle cells express high levels of NP-C receptor binding sites and mRNA, receptor expression is reduced markedly in aortic media tissue derived from the same animals *(35)*. By contrast, although cultured cells express low levels of GC-A and GC-B receptors, tissue derived from aortic media tissue contains significantly higher levels of these receptors. Thus, the process of culturing the vascular smooth muscle cells may result in a significant phenotypic shift involving altered expression of the natriuretic peptide receptors.

**Pulmonary.** As with many other peptide hormones, the lung contributes to the metabolic clearance of the natriuretic peptides *(36–38)*. Accordingly, NP-C receptor binding sites and NP-C receptor mRNA have been shown to be abundant in lung tissue *(29,34,39–41)*. Infusion of NP-C receptor-selective ligands has been reported to reduce the pulmonary binding and clearance of ANP in rats and mice *(42,43)*, although no effect was observed in dogs *(44)*. This disparity suggests the role of the NP-C receptor in the pulmonary clearance of the natriuretic peptides may vary as a function of physiological status and experimental conditions. For example, it has been reported *(29)* that hypoxia significantly downregulates expression of the NP-C receptor in the lung. The lung is a rich source of peptidases (including neutral endopeptidase 24.11) that also contribute to natriuretic peptide metabolism *(43,45–47)*.

### Regulation of NP-C Receptor Gene Expression

Considering the proposed role of the NP-C receptor in regulating plasma concentrations of the natriuretic peptides, one would anticipate that the expression and resultant density of the receptor would change as a function of the physiological status of the organism. The demonstrations that the NP-C receptor is downregulated during cardiac hypertrophy and hypoxia, or upregulated in the setting of dehydration *(17)*, are consistent with this expectation. Nevertheless, at the molecular level, the mechanisms regulating NP-C receptor expression are poorly understood. Surprisingly, no information has been published to date regarding the regulation of the NP-C gene promoter.

One factor that appears to regulate the expression of the NP-C receptor is ANP itself. Incubation of cultured vascular smooth muscle or endothelial cells with ANP reduces NP-C receptor density *(48–51)*. The ability of ANP to downregulate NP-C expression was enhanced by concomitant preincubation with a phosphodiesterase inhibitor and mimicked by the addition of sodium nitroprusside (which activates soluble guanylate cyclase) or 8-bromo cyclic GMP (cGMP). The effect was not observed on incubation of cells with three selective NP-C receptor ligands. These results suggest that ANP-mediated downregulation of the NP-C receptor occurs via the activation of the GC-A receptor, which is associated with an accumulation of intracellular cGMP. Interestingly, ANP itself does not appear to regulate the expression of GC-A.

Several other hormones and growth factors have been reported to affect NP-C receptor expression. For instance, serum deprivation of cultured rat mesangial cells was reported to double the density of NP-C receptors and markedly increase steady-state levels of NP-C receptor mRNA *(52)*. Upregulation of the NP-C receptor induced by serum deprivation was inhibited by treatment with platelet-derived growth factor or phorbol acetate, suggesting a role for protein kinase C in this regulatory process. Indeed, protein kinase C appears to mediate phorbol acetate-induced downregulation of the NP-C receptor in rat brain astroglial cell cultures *(53)*. Pretreatment with angiotensin II has also been reported to cause a time- and dose-dependent decrease (70%) in ANP binding sites in cultured vascular smooth-muscle cells without affecting ANP-induced cGMP accumulation *(48)*. This study suggests that the renin angiotensin system may influence NP-C receptor expression. Finally, norepinephrine and isoproterenol have been shown to downregulate NP-C receptor expression in cultured rat vascular smooth-muscle cells without affecting the expression of GC-A or GC-B *(54)*. Isoproterenol decreases the steady-state NP-C receptor mRNA levels as well as the rate of NP-C receptor gene transcription. The effects can be blocked by a selective β2-adrenergic antagonist but not a β1-, β2-, or β1-antagonist. The NP-C receptor downregulation is also induced by treatment with 8-bromo cyclic AMP (cAMP) and forskolin, suggesting a cAMP-mediated effect.

The observations that NP-C receptor expression is influenced by growth factors, neurotransmitters, and hormones, including ANP itself, is consistent with a key role for the receptor in regulating cardiovascular homeostasis. Future investigations stand to shed more light on the molecular mechanisms regulating NP-C receptor expression.

## IDENTIFICATION OF NP-C RECEPTOR-SPECIFIC LIGANDS

As depicted in Fig. 1, the NP-C receptor binds all three native natriuretic peptides with high affinity. The peptides ANP, BNP, and CNP exhibit binding constants for this site of $10^{-9}$–$10^{-11}M$, depending on the receptor source and assay conditions. The human NP-C receptor exhibits a rank order of binding affinity of ANP>CNP>BNP *(34)*. The binding affinity of BNP for the human NP-C receptor is about one order of magnitude lower than that of ANP. The biological consequences of this modestly reduced affinity of the human NP-C receptor for BNP are unclear because the NP-C receptor has been shown to mediate the metabolic clearance of all three natriuretic peptides (*55*; Protter, in preparation).

The structural requirements for ligand binding to the NP-C receptor are much less stringent than those required for binding to the GC-A and GC-B receptors. Indeed, our early work demonstrated that the NP-C receptor binds natriuretic peptide analogs with high affinity even when significant exocyclic truncations, endocyclic amino-acid deletions, and conformational constraints are introduced within the natriuretic peptide ring structure *(33,56)*. The ligand C-ANP (4–23) (Fig. 3), which is missing the complete, five-amino-acid carboxyl terminal amino-acid tail and five amino acids from within the disulfide-bridged ring of ANP, was the initial NP-C receptor-specific ligand constructed *(15)*. The C-ANP (4–23) was used as a probe molecule to help elucidate the clearance function of the NP-C receptor (*see* Biologic Role of the NP-C Receptor). Subsequently, C-ANP (4–23) has been deployed by many laboratories to explore the biological roles of NP-C receptors. It should be noted that although this ligand is relatively specific for the NP-C receptor, it does activate the GC-A receptor at concentrations approaching the high nanomolar range. This raises the concern that some actions ascribed to the NP-C receptor may actually result from C-ANP (4–23) activation of GC-A, or other yet unknown natriuretic peptide receptors. In any case, investigators should exercise caution when using this probe to assess NP-C receptor function.

Although C-ANP (4–23) served as a prototype of a specific ligand for the NP-C receptor, subsequent work has revealed that the minimal linear sequence necessary for binding to this site comprises the 8–15 sequence of the native ANP molecule, $Phe^{8}$, $Gly^{9}$, $Gly^{10}$, $Arg^{11}$, $Ile^{12}$, $Asp^{13}$, $Arg^{14}$, $Ile^{15}$ (Fig. 3). This sequence is contained within the amino-terminal region of the 17-membered, disulfide-bridged ring of ANP. Indeed, this linear, eight-amino-acid peptide exhibits an affinity for the NP-C receptor within an order of magnitude of that of ANP. The corresponding linear sequences of BNP ($Phe^{11}$, $Gly^{12}$, $Arg^{13}$, $Lys^{14}$, $Met^{15}$, $Asp^{16}$, $Arg^{17}$, $Ile^{18}$) and CNP ($Phe^{7}$, $Gly^{8}$, $Leu^{9}$, $Lys^{10}$, $Leu^{11}$, $Asp^{12}$, $Arg^{13}$, $Ile^{14}$) also bind with high affinity to the clearance receptor.

The identification of a linear, eight-amino fragment of the endogenous natriuretic peptides that confers high affinity binding to the NP-C receptor is representative of a unique structure-activity relationship. Most peptide hormones require structural contributions from distant regions of the peptide backbone to induce local conformations that bind effectively to their receptors. For instance, the high-affinity binding of ANP and BNP to the GC-A receptor requires an intact, 17-membered, disulfide-bridged ring along with contributions from exocyclic amino- and carboxyl-terminal amino acids. The relatively indiscriminant structural requirements of NP-C receptors for the natriuretic peptides may be a result of the specialized clearance function of these receptors. By recognizing a linear octapeptide fragment of these molecules, NP-C receptors are capable of clearing native forms of ANP, BNP, and CNP as well as metabolites of these peptides. At the same time, this site specifically binds only the natriuretic peptides and does not interact with any other known peptide hormones.

Subsequent structure-activity studies of NP-C receptor ligands have focused on modifications of the linear octapeptide sequence (ANP *[8–15]*) toward the goal of generating small molecule peptidomimetics. Indeed, the $Phe^{8}$, $Gly^{9}$, $Gly^{10}$ tripeptide has been effectively substituted by a number of nonpeptidic moieties. Whereas some substitutions in this region of the molecule result in a loss in receptor affinity, other substitutions maintain optimal receptor binding kinetics. This has allowed us to conclude that the key receptor-binding determinants of the NP-C receptor are represented by an amino-terminal hydrophobic moiety spaced a certain distance from the core sequence $Arg^{11}$, $Ile^{12}$, $Asp^{13}$, $Arg^{14}$, $Ile^{15}$ of ANP. One such representative compound, *N*-2-naphthoxyacetyl-*N*-isonipecotyl ANP (11–15)-$NH_2$ (Nap ANP [11–15]; Fig. 3), is a high-affinity, selective ligand for the NP-C receptor *(57)*. It is unclear whether attempts have been made to further modify the structure of these small NP-C receptor ligands. To date, no small molecule, nonpeptidic ligands of the NP-C receptor have been reported.

## BIOLOGIC ROLE OF THE NP-C RECEPTOR

### *Role in the Metabolic Clearance of the Natriuretic Peptides*

#### Function in In Vivo Clearance

The clearance function of the NP-C receptor was originally elucidated in studies deploying the selective ligand C-ANP (4–23) *(15)*. It was observed that although C-ANP (4–23) displaced a large percentage of $^{125}I$-ANP binding to isolated tissue preparations, like the kidney, it did not elicit any discernible biological actions in these same isolated tissue preparations. Nevertheless, infusion of C-ANP (4–23) into rats led to a decrease in blood pressure and an increase in urine volume and sodium excretion. This apparent discrepancy between the in vitro and in vivo actions of the ligand was reconciled by the finding that C-ANP (4–23) infusion led to a threefold increase in circulating levels of endogenous ANP. A subsequent pharmacokinetic analysis revealed that C-ANF (4–23) infusion reduced the volume of distribution and metabolic clearance rate of $^{125}$I-ANP to one-third and one-fourth of control levels, respectively, thus explaining the observed increase in circulating levels of ANP. The infusion of C-ANP (4–23) also reduced the appearance of hydrolytic products of ANP in the plasma, consistent with the hypothesis that NP-C receptors mediate the metabolism of ANP *(58)*. Thus, by saturating NP-C receptors, C-ANP (4–23) confines ANP to the plasma space and reduces its metabolic clearance. Based on these studies, it has been proposed that NP-C receptors act as a hormonal buffer system to impede large and inappropriate fluctuations in plasma levels of ANP.

Subsequent studies have confirmed that the in vivo blockade of NP-C receptors by C-ANP (4–23), or NapANP *(11–15)*, increases circulating levels of ANP and is accompanied generally by hemodynamic, renal, and neuroendocrine effects *(59–62)*. An example of one study is highlighted in Fig. 4. In this study, C-ANP (4–23) was infused into anesthetized dogs at doses ranging from 0.1 to 3.0 µg/kg/min. As shown, C-ANP (4–23) infusion led to a dose-dependent increase in circulating levels of ANP (Fig. 4A). Concomitant with the enhancement of circulating ANP levels, C-ANP (4–23) infusion promoted a dose-dependent increase in urinary sodium excretion (Fig. 4B) and a marked suppression of plasma aldosterone levels (Fig. 4C). In addition to its effects on ANP clearance, the NP-C receptor also mediates the metabolic clearance of BNP and CNP *(55)*. It is unclear how effects on circulating concentrations of these peptides contribute to the in vivo biologic actions of NP-C receptor blockers in this and related studies. Nevertheless, this study shows that small changes in circulating natriuretic peptide levels elicited by NP-C receptor blockade have profound biological consequences. The fact that several laboratories have failed to note significant renal or hemodynamic effects induced by acute or chronic administration of NP-C receptor ligands suggests that the biological actions of these compounds may depend on the physiological status of the organism *(63,64)*.

It is now apparent that the NP-C receptor represents one of two distinct clearance pathways for removal of circulating natriuretic peptides. The enzyme neutral endopeptidase 24.11 (NEP) also contributes to the metabolism of this peptide family. The relative contribution of these two molecular pathways to the metabolic clearance of the natriuretic peptides is unclear, but probably varies according to the pathophysiological status of the organism. It has been suggested that the NP-C receptor pool is sufficiently large to rapidly sequester circulating natriuretic peptides under normal physiological conditions. Indeed, the observed pharmacokinetic effects of NP-C receptor-selective ligands suggest a dominating role for the NP-C receptor in normal animals. Nevertheless, coinfusion of NP-C receptor ligands and NEP inhibitors into normal animals leads to a marked synergistic effect in decreasing the metabolic clearance rate of ANP and enhancing hemodynamic and renal responses in rats and dogs *(65–67)*. Under conditions in which ambient plasma levels of the natriuretic peptides are high, NEP may contribute even more significantly to the clearance of the peptides *(68)*. Thus, NEP inhibitors have been shown to increase plasma levels of ANP when these levels are elevated by infusion of the peptide, or in pathophysiological conditions, like congestive heart failure.

The overall contribution of the NP-C receptor in mediating the metabolic clearance of the natriuretic peptides in pathophysiological states has not been intensively investigated. As noted previously, cardiac expression of the gene encoding the NP-C receptor has been shown to be reduced to undetectable levels in a model of AV fistula-induced congestive heart failure, suggesting a reduced role for the receptor in this condition *(27)*.

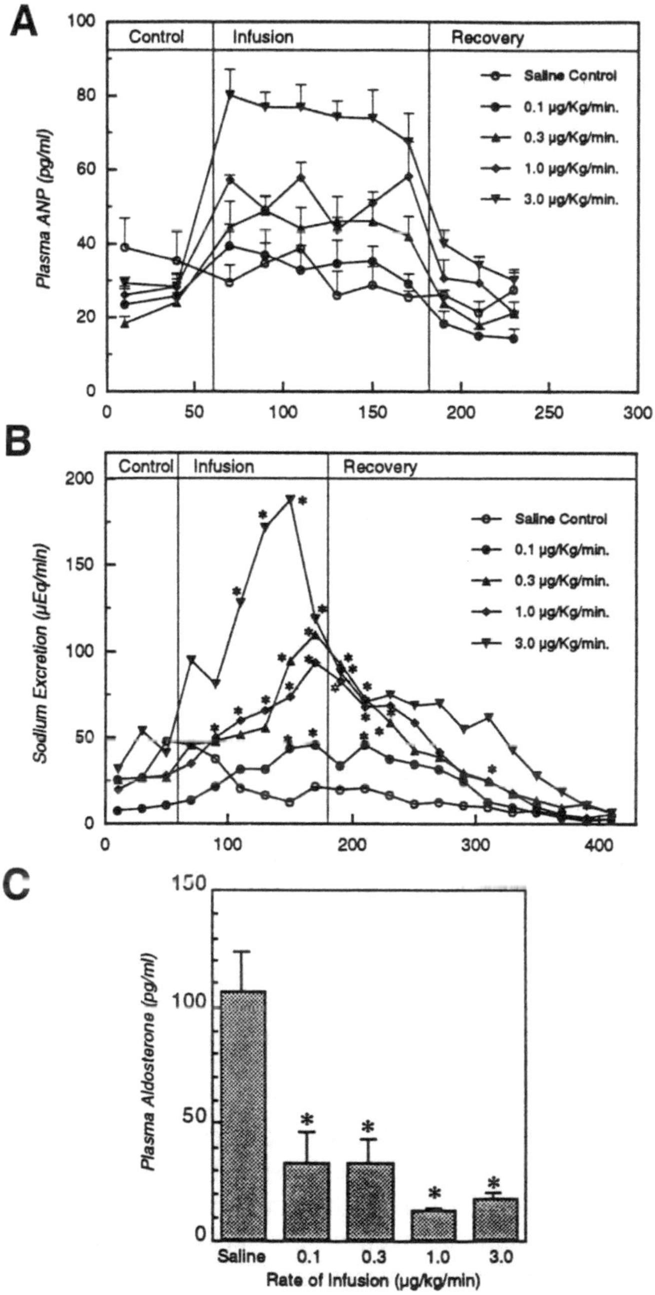


**Fig. 4.** Renal and hormonal effects of the clearance receptor-specific ligand, des[$Gln_{116}$, $Ser_{117}$,$Gly_{118}$,$Leu_{119}$,$Gly_{120}$]rANP(102–126)$NH_2$, termed C-ANP (4–23), in conscious dogs. C-ANF (4–23) at 0.1, 0.3, 1.0, and 3.0 μg/kg/minute or saline was infused for 2 h and plasma ANP **(A)**, urine sodium excretion rate **(B)**, and plasma aldosterone **(C)** was determined. C-ANF (4–23) increases plasma ANP ($p < 0.01$ compared with control), increases urine sodium excretion rate ($p < 0.01$ compared with control), and decreases plasma aldosterone ($p < 0.05$ compared with control) *(see* ref. *94*).

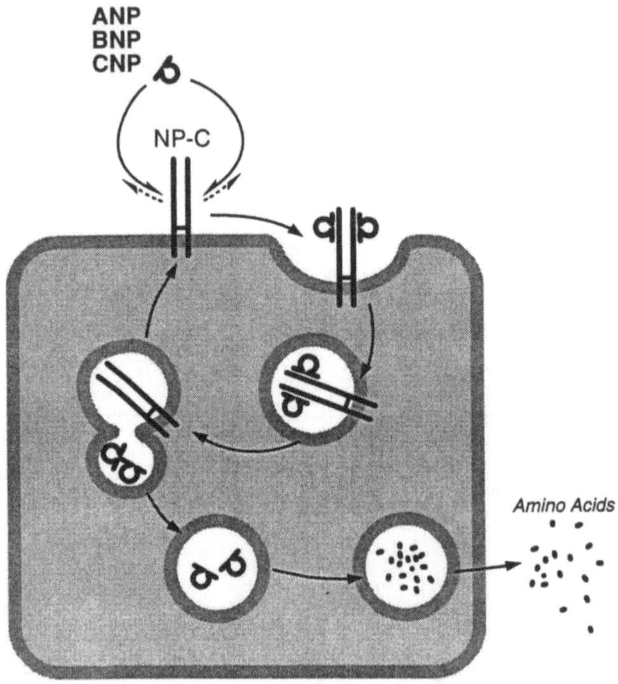


**Fig. 5.** Natriuretic peptide metabolism and cycling of the NP-C receptor. Cell surface NP-C receptors bind tightly ANP, BNP, or CNP, cluster in membrane coated pits, and internalize the bound ligand. The ligand dissociates from the receptor in endocytic vesicles and becomes associated with lysosomes, where it is hydrolyzed and ultimately released from the cell as free amino acids. The NP-C receptor recycles to the cell surface. Although the NP-C receptor can bind ligand as a monomer, the homodimer, which internalizes in a more efficient manner, is the predominant form.

Conversely, infusion of C-ANP (4–23) into animals with experimental heart failure induced by rapid ventricular pacing elicits profound pharmacokinetic and biological responses, consistent with a regulatory role for the NP-C receptor *(44,69)*.

### Cellular Mechanisms of the Clearance Function

The cellular mechanisms of the NP-C receptor clearance function have been elucidated largely in studies using cultured vascular smooth-muscle cells, vascular endothelial cells, and genetically modified Chinese hamster ovary cells *(11,70–72)*. These studies have revealed common features between NP-C receptor-mediated ligand clearance and mechanisms used by other clearance receptors of proteins and macromolecules, including low-density lipoprotein, asiayloglycoprotein, and hyaluronic-acid receptors *(73,74)*. These features include NP-C receptor internalization, lysosomal hydrolysis of bound ligand, and recycling of NP-C receptors to the cell surface (*see* Fig. 5; *11,70*).

NP-C receptor internalization is a constitutive process that is not stimulated by ligand binding to the receptor. The NP-C receptors are internalized at a rate of 5–7%/min, although this rate may vary with the cell type. This internalization rate is comparable to that exhibited by other clearance receptors. The NP-C receptor internalization is disrupted by hyperosmotic sucrose, which causes clathrin disassembly, suggesting that receptor endocytosis occurs via clathrin-coated pits *(11,75)*. The fact that the rate of ANP dissociation from NP-C receptors at 37°C is much slower that the rate of NP-C receptor internalization, ensures that all bound ligand is efficiently transported into the cell where it is then degraded.

The intracellular hydrolysis of receptor-associated ligand occurs on targeting of the ligand to an intracellular acidic, lysosomal-like compartment. Accordingly, intracellular hydrolysis of receptor-associated ligands can be disrupted by treatment of cells with the lysosomotropic agents, $NH_4Cl$, chloroquine, and monensin *(70,72)*. Following ligand degradation, NP-C receptors reappear on the cell-membrane surface. The reappearance of receptor is not impeded by treatment of cells with inhibitors of protein synthesis, thus implying that receptors are recycled in their native form from intracellular compartments to the cell surface.

The NP-C receptor lacks known cytoplasmic-tail internalization motifs (e.g., NPXY, YXXZ, where X is any amino acid and Z is a large hydrophobic amino acid) for coated-pit internalization that are common to other known receptors that undergo rapid internalization and recycling. Nevertheless, a recent study has revealed that deletion of the entire 37-amino-acid tail of the receptor markedly abrogates the rate of receptor internalization *(11)*. Receptor mutagenesis studies have revealed that replacement of the single tyrosine residue in this cytoplasmic domain at position 508 with an alanine residue reduces net endocytosis and hydrolysis of bound ANP by approx 50% (*see* Fig. 2). Similar results were observed in a double mutant in which $Tyr^{508}$ and $Phe^{538}$ were replaced by alanine residues. Finally, replacement of a extracellular-domain cysteine at position 473 with an alanine reduced the rate of endocytosis by approx 50%. Since this cysteine residue appears to contribute to NP-C receptor dimerization, this result implies that receptor dimerization is another determinant of receptor internalization *(11)*.

The process of internalization, lysosomal hydrolysis of bound ligand, and receptor recycling ensures that NP-C receptors are always available to sequester and clear the natriuretic peptides from the plasma pool. This phenomenon, along with the large density, high affinity, and strategic location of NP-C receptors, is consistent with a pivotal role of this protein in mediating the metabolic clearance of the natriuretic peptides.

### *Role in Signal Transduction*

The widespread tissue distribution of the NP-C receptor raises the possibility of other functions unrelated to peptide clearance. Indeed, in recent years a number of laboratories have reported that the NP-C receptor elicits direct biological signals in various cells and tissues, including effects on cAMP and inositol phosphate. This topic is addressed in other chapters of this monograph, and as a result, is summarized in an abbreviated form here.

Several groups have reported that NP-C receptor-selective analogs (e.g., C-ANF[4–23]) inhibit adenylate cyclase activity in membrane and cell preparations from platelets, anterior pituitary, aorta, brain striatum, adrenal cortex, heart, and pheochromocytoma cells *(76–79)*. These effects are sensitive to pertussis toxin, suggesting the involvement of a Gi regulatory protein. Indeed, the recent observation that the intact 37-amino-acid cytoplasmic tail of the NP-C receptor itself inhibits rat heart particulate adenylate cyclase in a pertussis toxin-sensitive manner confirms these earlier observations *(79)*. Taken together, these results strongly implicate the coupling of the NP-C receptor to the inhibition of at least certain isoforms of adenylate cyclase.

The ANP-induced phospholipase C activation and phosphoinositol hydrolysis via a Gi regulatory protein has also been noted in bovine vascular smooth-muscle cells *(80)* and rat inner-medullary collecting-tubule cells *(81)*. These effects on phospholipase C activation did not correlate with increases in intracellular cGMP accumulation. These results suggest that under some conditions the NP-C receptor may be coupled to the phospholipase C-phosphoinositol pathway.

Potential biological correlates of NP-C receptor-mediated signaling have been investigated. It has been reported that the NP-C receptor mediates the antiproliferative effects of ANP on astroglial cells *(82)*, neonatal rat cardiac fibroblasts *(28)*, and rat-aortic smooth-muscle cells *(83,84)*, although conflicting results have been reported by others *(85)*. Additional studies suggest that certain neuromodulatory effects of ANP on pheochromocytoma cells *(78)* and isolated vas deferens *(86)* are mediated by the NP-C receptor through a Gi-regulatory protein pathway linked to the inhibition of adenylate cyclase. Finally, it has been reported that the inhibitory effects of ANP on endothelin production in aortic endothelial cells can be mimicked with C-ANP (4–23), suggesting that the NP-C receptor is involved in this activity *(87)*. To date, no direct NP-C receptor-mediated effects of the natriuretic peptides have been shown to occur in vivo. As a result, the physiological significance of the above observations is unclear.

These NP-C receptor-mediated biological effects are intriguing and deserve attention. However, they largely ignore several important issues. These include the possibility that some actions attributed to NP-C receptor may be mediated by noncyclic GMP-coupled actions of GC-A receptors, or by other, yet undescribed natriuretic peptide receptors. Furthermore, because ANP has been shown to increase cGMP-stimulated phosphodiesterase activity *(88–90)*, some of the effects of ANP on cAMP metabolism may be mediated by the GC-A receptor. To date, many studies have failed to demonstrate a coherent pharmacological profile correlating NP-C receptor binding of natriuretic peptide analogs with a biological signal. Finally, the ultimate proof of a biological role of the NP-C receptor awaits the demonstration of direct biological effects in a reconstituted receptor system. In summary, whereas it is apparent that the NP-C receptors do not mediate many of the hallmark actions of ANP and BNP on target tissues (e.g., vasodilation, diuresis, and natriuresis) since these have been shown to occur via GC-A receptors *(91–93)*, the compelling possibility that there are other important NP-C receptor-mediated actions of the natriuretic peptides must be examined in further detail.

## SUMMARY

The NP-C receptor plays a central role in modulating the actions of the natriuretic peptides by regulating plasma and local tissue concentrations of these peptide hormones. The high level of NP-C receptor expression in tissues like the kidney, lung, and vasculature ensures that the receptor has complete access to the total circulating pool of the natriuretic peptides. Furthermore, its substantial density, high affinity, and continuous route of internalization and rapid recycling ensures a constant supply of unoccupied binding sites to meet the demands of a clearance receptor. It is particularly noteworthy that the NP-C receptor recognizes its ligands in a most unusual manner. Structure-activity studies have revealed that the NP-C receptor binds only a continuous eight-amino-acid linear fragment of the natriuretic peptides that themselves range in length from 23 to 32 amino acids. Interestingly, the receptor is able to associate with the same corresponding eight-amino-acid fragment of ANP, BNP, and CNP, yet is not known to interact with any other proteins or peptides. This ligand recognition property of the NP-C receptor is unprecedented within the family of peptide hormone receptors.

The receptor-mediated mechanism for the metabolic clearance of the natriuretic peptides is unique. Virtually all other peptide hormones involved in the regulation of cardiovascular homeostasis are inactivated and removed from plasma and tissues stores by the proteolytic actions of relatively nonselective proteinases. As a result, it is intriguing to speculate why such a specific and specialized clearance mechanism has evolved for the natriuretic peptides. At the very least, this implies that the NP-C receptor fine tunes plasma and tissue concentrations of the natriuretic peptides as a novel biological regulatory mechanism. The finding that ANP and other factors regulate NP-C receptor gene expression is consistent with the novel regulatory function of this receptor. Perhaps more importantly, these and other findings summarized in this chapter suggest that we do not yet appreciate the overall significance of the NP-C receptor.

Future research will shed further light on the function of the NP-C receptor. In particular, we can anticipate that we will gain an enhanced understanding of the manner in which the expression of the NP-C receptor gene is regulated. Additional studies of possible direct biological events mediated by these receptors will surely bolster our understanding of the physiological role of this peptide hormone family. Finally, we can anticipate that the future design of novel small-molecule ligands of the NP-C receptor will prompt investigation into whether the modulation of NP-C receptor function can be exploited for therapeutic purposes.

## REFERENCES

1. Schenk DB, Phelps MN, Porter JG, Fuller F, Cordell B, Lewicki JA. Purification and subunit composition of atrial natriuretic peptide receptor. Proc Natl Acad Sci USA 1987;84:1521–1525.
2. Fuller F, Porter JG, Arfsten AE, Miller J, Schilling JW, Scarborough RM, Lewicki JA, Schenk DB. Atrial natriuretic peptide clearance receptor. Complete sequence and functional expression of cDNA clones. J Biol Chem 1988;263:9395–9401.
3. Porter JG, Arfsten A, Fuller F, Miller JA, Gregory LC, Lewicki JA. Isolation and functional expression of the human atrial natriuretic peptide clearance receptor cDNA. Biochem Biophys Res Commun 1990;171:796–803.
4. Lowe DG, Camerato TR, Goeddel DV. cDNA sequence of the human atrial natriuretic peptide clearance receptor. Nucleic Acids Res 1990;18:3412.
5. Engel AM, Schoenfeld JR, Lowe DG. A single residue determines the distinct pharmacology of rat and human natriuretic peptide receptor-C. J Biol Chem 1994;269:17,005–17,008.
6. Saheki T, Mizuno T, Iwata T, Saito Y, Nagasawa T, Mizuno KU, Ito F, Ito T, Hagiwara H, Hirose S. Structure of the bovine atrial natriuretic peptide receptor (type C) gene. J Biol Chem 1991;266:11,122–11,125.
7. Porter JG, Scarborough RM, Wang Y, Schenk, D, McEnroe GA, Kang LL, Lewicki JA. Recombinant expression of a secreted form of the atrial natriuretic peptide clearance receptor. J Biol Chem 1989;264:14,179–14,184.
8. Bennett BD, Bennett GL, Vitangcol RV, Jewett JR, Burnier J, Henzel W, Lowe DG. Extracellular domain-IgG fusion proteins for three human natriuretic peptide receptors. Hormone pharmacology and application to solid phase screening of synthetic peptide antisera. J Biol Chem 1991;266:23,060–23,067.
9. Itakura M, Iwashina M, Mizuno T, Ito T, Hagiwara H, Hirose S. Mutational analysis of disulfide bridges in the type C atrial natriuretic peptide receptor. J Biol Chem 1994;269:8314–8318.
10. Stults JT, O'Connell KL, Garcia C, Wong S, Engel AM, Garbers DL, Lowe DG. The disulfide linkages and glycosylation sites of the human natriuretic peptide receptor-C homodimer. Biochemistry 1994;33:11,372–11,381.
11. Cohen D, Koh GY, Nikonova LN, Porter JG, Maack T. Molecular determinants of the clearance function of type C receptors of natriuretic peptides. J Biol Chem 1996;271:9863–9869.
12. Iwashina M, Mizuno T, Hirose S, Ito T, Hagiwara H. His145-Trp146 residues and the disulfide-linked loops in atrial natriuretic peptide receptor are critical for the ligand-binding activity. J Biochem 1994;115:563–567.

13. Engel AM, Lowe DG. Characterization of the hormone binding site of natriuretic peptide receptor-C. FEBS Lett 1995;360:169–172.
14. Mizuno T, Iwashina M, Itakura M, Hagiwara H, Hirose S. A variant form of the type C atrial natriuretic peptide receptor generated by alternative RNA splicing. J Biol Chem 1993;268:5162–5167.
15. Maack T, Suzuki M, Almeida FA, Nussenzveig D, Scarborough RM, McEnroe GA, Lewicki JA. Physiological role of silent receptors of atrial natriuretic factor. Science 1987;238:675–678.
16. Martin ER, Lewicki JA, Scarborough RM, Ballermann BJ. Expression and regulation of ANP receptor subtypes in rat renal glomeruli and papillae. Am J Physiol 1989;257:F649–657.
17. Kollenda MC, Vollmar AM, McEnroe GA, Gerbes AL. Dehydration increases the density of C receptors for ANF on rat glomerular membranes. Am J Physiol 1990;258:R1084–1088.
18. Koh GY, Nussenzveig DR, Okolicany J, Price DA, Maack T. Dynamics of atrial natriuretic factor-guanylate cyclase receptors and receptor-ligand complexes in cultured glomerular mesangial and renomedullary interstitial cells. J Biol Chem 1992;267:11,987–11,994.
19. Chansel D, Pham P, Nivez MP, Ardaillou R. Characterization of atrial natriuretic factor receptors in human glomerular epithelial and mesangial cells. Am J Physiol 1990;259:F619–627.
20. Yamamoto T, Feng L, Mizuno T, Hirose S, Kawasaki K, Yaoita E, Kihara I, Wilson CB. Expression of mRNA for natriuretic peptide receptor subtypes in bovine kidney. Am J Physiol 1994;267:F318–324.
21. Bea ML, Dussaule JC, Bens M, Ardaillou R. Characterization of ANF receptors in cultured renal cortical vascular smooth muscle cells. Am J Physiol 1991;260:C424–432.
22. Brown J, Salas SP, Singleton A, Polak JM, Dollery CT. Autoradiographic localization of atrial natriuretic peptide receptor subtypes in rat kidney. Am J Physiol 1990;259:F26–39.
23. Wilcox JN, Augustine A, Goeddel DV, Lowe DG. Differential regional expression of three natriuretic peptide receptor genes within primate tissues. Mol Cell Biol 1991;11:3454–3462.
24. Meloche S, Ong H, Cantin M, De Lean A. Affinity cross-linking of atrial natriuretic factor to its receptor in bovine adrenal zona glomerulosa. J Biol Chem 1986;261:1525–1528.
25. Heisler S, Morrier E. Bovine adrenal medullary cells contain functional atrial natriuretic peptide receptors. Biochem Biophys Res Commun 1988;150:781–787.
26. Nunez DJ, Dickson MC, Brown MJ. Natriuretic peptide receptor mRNAs in the rat and human heart. J Clin Invest 1992;90:1966–1971.
27. Brown LA, Nunez DJ, Wilkins MR. Differential regulation of natriuretic peptide receptor messenger RNAs during the development of cardiac hypertrophy in the rat. J Clin Invest 1993;92:2702–2712.
28. Cao L, Gardner DG. Natriuretic peptides inhibit DNA synthesis in cardiac fibroblasts. Hypertension 1995;25:227–234.
29. Li H, Oparil S, Meng QC, Elton TS, Chen YF. Selective downregulation of ANP-clearance-receptor gene expression in lung of rats adapted to hypoxia. Am J Physiol 1995;268:L328–335.
30. Schenk DB, Johnson LK, Schwartz K, Sista H, Scarborough RM, Lewicki JA. Distinct atrial natriuretic factor receptor sites on cultured bovine aortic smooth muscle and endothelial cells. Biochem Biophys Res Commun 1985;127:433–442.
31. Leitman DC, Andresen JW, Catalano RM, Waldman SA, Tuan JJ, Murad F. Atrial natriuretic peptide binding, cross-linking, and stimulation of cyclic GMP accumulation and particulate guanylate cyclase activity in cultured cells. J Biol Chem 1988;263:3720–3728.
32. Pandey KN, Pavlou SN, Inagami T. Identification and characterization of three distinct atrial natriuretic factor receptors. Evidence for tissue-specific heterogeneity of receptor subtypes in vascular smooth muscle, kidney tubular epithelium, and Leydig tumor cells by ligand binding, photoaffinity labeling, and tryptic proteolysis. J Biol Chem 1988;263:13,406–13,413.
33. Scarborough RM, Schenk DB, McEnroe GA, Arfsten A, Kang LL, Schwartz K, Lewicki JA. Truncated atrial natriuretic peptide analogs. Comparison between receptor binding and stimulation of cyclic GMP accumulation in cultured vascular smooth muscle cells. J Biol Chem 1986;261:12,960–12,964.
34. Suga S, Nakao K, Hosoda K, Mukoyama M, Ogawa Y, Shirakami G, Arai H, Saito Y, Kambayashi Y, Inouye K, et al. Receptor selectivity of natriuretic peptide family, atrial natriuretic peptide, brain natriuretic peptide, and C-type natriuretic peptide. Endocrinology 1992;130:229–239.
35. Suga S, Nakao K, Kishimoto I, Hosoda K, Mukoyama M, Arai H, Shirakami G, Ogawa Y, Komatsu Y, Nakagawa O, et al. Phenotype-related alteration in expression of natriuretic peptide receptors in aortic smooth muscle cells. Circ Res 1992;71:34–39.
36. Perrella MA, Margulies KB, Wei CM, Aarhus LL, Heublein DM, Burnett J Jr. Pulmonary and urinary clearance of atrial natriuretic factor in acute congestive heart failure in dogs. J Clin Invest 1991;87:1649–1655.

37. Hollister AS, Rodeheffer RJ, White FJ, Potts JR, Imada T, Inagami T. Clearance of atrial natriuretic factor by lung, liver, and kidney in human subjects and the dog. J Clin Invest 1989;83:623–628.
38. Masuda T, Shichiri M, Marumo F. Degradation of atrial natriuretic peptide in dogs. Acta Endocrinol 1989;120:170–174.
39. von Schroeder HP, Nishimura E, McIntosh CH, Buchan AM, Wilson N, Ledsome JR. Autoradiographic localization of binding sites for atrial natriuretic factor. Can J Physiol Pharmacol 1985;63:1373–1377.
40. Shimonaka M, Saheki T, Hagiwara H, Ishido M, Nogi A, Fujita T, Wakita K, Inada Y, Kondo J, Hirose S. Purification of atrial natriuretic peptide receptor from bovine lung. Evidence for a disulfide-linked subunit structure. J Biol Chem 1987;262:5510–5514.
41. Uchida K, Mizuno T, Shimonaka M, Sugiura N, Nara K, Ling N, Hagiwara H, Hirose S. Purification and properties of active atrial-natriuretic-peptide receptor (type C) from bovine lung. Biochem J 1989;263:671–678.
42. Klinger JR, Moalli R, Warburton RR, Wrenn DS, Hill NS. C-receptor ligand blocks pulmonary clearance of atrial natriuretic peptide in isolated rat lungs. Proc Soc Exp Biol Med 1992;201:154–158.
43. Souque A, Gros C, Schwartz JC. Binding of [125I]atrial natriuretic factor to mouse lung membranes in vivo: characterization and effects of peptidase inhibitors. J Pharmacol Exp Ther 1992;260:1373–1378.
44. Brandt RR, Redfield MM, Aarhus LL, Lewicki JA, Burnett J Jr. Clearance receptor-mediated control of atrial natriuretic factor in experimental congestive heart failure. Am J Physiol 1994;266:R936–943.
45. Lindberg BF, Bengtsson HI, Lundin S, Andersson KE. Degradation and inactivation of rat atrial natriuretic peptide 1–28 by neutral endopeptidase-24.11 in rat pulmonary membranes. Regul Pept 1992;42:85–96.
46. Numan NA, Gillespie MN, Altiere RJ. Pulmonary clearance of atrial natriuretic peptides. Pulm Pharmacol 1990;3:25–28.
47. Zhao L, Hughes JM, Winter RJ. Effects of natriuretic peptides and neutral endopeptidase 24.11 inhibition in isolated perfused rat lung. Am Rev Respir Dis 1992;146:1198–1201.
48. Chabrier PE, Roubert P, Lonchampt MO, Plas P, Braquet P. Regulation of atrial natriuretic factor receptors by angiotensin II in rat vascular smooth muscle cells. J Biol Chem 1988;263:13,199–13,202.
49. Cahill PA, Redmond EM, Keenan AK. Vascular atrial natriuretic factor receptor subtypes are not independently regulated by atrial peptides. J Biol Chem 1990;265:21,896–21,906.
50. Kato J, Lanier-Smith KL, Currie MG. Cyclic GMP down-regulates atrial natriuretic peptide receptors on cultured vascular endothelial cells. J Biol Chem 1991;266:14,681–14,685.
51. Zhang LM, Tao H, Newman WH. Regulation of atrial natriuretic peptide receptors in vascular smooth muscle cells: role of cGMP. Am J Physiol 1993;264:H1753–1759.
52. Paul RV, Wackym PS, Budisavljevic M, Everett E, Norris JS. Regulation of atrial natriuretic peptide clearance receptors in mesangial cells by growth factors. J Biol Chem 1993;268:18,205–18,212.
53. Paulding WR, and Sumners C. Protein kinase C modulates natriuretic peptide receptors in astroglial cultures from rat brain. Am J Physiol 1996;270:C740–747.
54. Kishimoto I, Yoshimasa T, Suga S, Ogawa Y, Komatsu Y, Nakagawa O, Itoh H, Nakao K. Natriuretic peptide clearance receptor is transcriptionally down-regulated by beta 2-adrenergic stimulation in vascular smooth muscle cells. J Biol Chem 1994;269:28,300–28,308.
55. Brandt RR, Heublein DM, Aarhus LL, Lewicki JA, Burnett J Jr. Role of natriuretic peptide clearance receptor in in vivo control of C-type natriuretic peptide. Am J Physiol 1995;269:H326–331.
56. Scarborough RM, McEnroe GA, Arfsten A, Kang LL, Schwartz K, Lewicki JA. D-amino acid-substituted atrial natriuretic peptide analogs reveal novel receptor recognition requirements. J Biol Chem 1988;263:16,818–16,822.
57. McEnroe GA, Scarborough RM, Gregory LC, Okolicany J, Porter G, Maack T, Lewicki JA. Synthesis and biological properties of specific ANP C-receptor ligands. In: Peptide Regulation of Cardiovascular Function. Takeda Science Foundation, 1991, pp. 65–77.
58. Almeida FA, Suzuki M, Scarborough RM, Lewicki JA, Maack T. Clearance function of type C receptors of atrial natriuretic factor in rats. Am J Physiol 1989;256:R469–475.
59. Abassi ZA, Tate J, Hunsberger S, Klein H, Trachewsky D, Keiser HR. Pharmacokinetics of ANF and urodilatin during cANF receptor blockade and neutral endopeptidase inhibition. Am J Physiol 1992;263:E870–876.
60. Chiu PJ, Tetzloff G, Romano MT, Foster CJ, Sybertz EJ. Influence of C-ANF receptor and neutral endopeptidase on pharmacokinetics of ANF in rats. Am J Physiol 1991;260:R208–216.
61. Castro R, Ervin MG, Leake RD, Ross MG, Fisher DA. Effect of a ring-deleted atrial natriuretic factor analogue on ovine fetal renal and cardiovascular function. Pediatr Res 1991;29:342–346.

62. Okolicany J, McEnroe GA, Gregory LC, Lewicki JA, Maack T. Effects of small C-ANF receptor ligands on plasma levels of atrial natriuretic factor, blood pressure, and renal function in the rat. Can J Physiol Pharmacol 1991;69:1561–1566.
63. Kukkonen P, Vuolteenaho O, Ruskoaho H. Basal and volume expansion-stimulated plasma atrial natriuretic peptide concentrations and hemodynamics in conscious rats: effects of SCH 39.370, an endopeptidase inhibitor, and C-ANF-(4–23), a clearance receptor ligand. Endocrinology 1992;130:755–765.
64. Seymour AA, Norman JA, Asaad MM, Fennell SA, Abboa-Offei B, Little DK, Kratunis VJ, Delaney NG, Hunt JT, Di Donato G. Possible regulation of atrial natriuretic factor by neutral endopeptidase 24.11 and clearance receptors. J Pharmacol Exp Ther 1991;256:1002–1009.
65. Okolicany J, McEnroe GA, Koh GY, Lewicki JA, Maack T. Clearance receptor and neutral endopeptidase-mediated metabolism of atrial natriuretic factor. Am J Physiol 1992;263:F546–553.
66. Wegner M, Stasch JP, Hirthdietrich C, Dressel J, Voges KP, Kazda S. Interaction of a neutral endopeptidase inhibitor with an ANP-C receptor ligand in anesthetized dogs. Clin Exp Hypertension 1995;17:861–876.
67. Koepke JP, Tyler LD, Trapani AJ, Bovy PR, Spear KL, Olins GM, Blaine EH. Interaction of non-guanylate cyclase-linked atriopeptin receptor ligand and endopeptidase inhibitor in conscious rats. J Pharmacol Exp Ther 1989;249:172–176.
68. Hashimoto Y, Nakao K, Hama N, Imura H, Mori S, Yamaguchi M, Yasuhara M, Hori R. Clearance mechanisms of atrial and brain natriuretic peptides in rats. Pharm Res 1994;11:60–64.
69. Perrella MA, Aarhus LL, Heublein DM, Lewicki JA, Burnett J Jr. Biologic role of atrial natriuretic factor clearance receptor in congestive heart failure. Am J Physiol 1993;265:H401–408.
70. Nussenzveig DR, Lewicki JA, Maack T. Cellular mechanisms of the clearance function of type C receptors of atrial natriuretic factor. J Biol Chem 1990;265:20,952–20,958.
71. Johnson GR, Arik L, Pitts BJ, Foster CJ. Rapid receptor-mediated catabolism of 125I-atrial natriuretic factor by vascular endothelial cells. Biochem J 1990;268:771–776.
72. Murthy KK, Thibault G, Cantin M. Binding and intracellular degradation of atrial natriuretic factor by cultured vascular smooth muscle cells. Mol Cell Endocrinol 1989;67:195–206.
73. Hobbs HH, Russell DW, Brown MS, Goldstein JL. The LDL receptor locus in familial hypercholesterolemia: mutational analysis of a membrane protein. Annu Rev Genet 1990;24:133–170.
74. Bridges K, Harford J, Ashwell G, Klausner RD. Fate of receptor and ligand during endocytosis of asialoglycoproteins by isolated hepatocytes. Proc Natl Acad Sci USA 1982;79:350–354.
75. Fuhrer C, Geffen I, Spiess M. Endocytosis of the ASGP receptor H1 is reduced by mutation of tyrosine-5 but still occurs via coated pits. J Cell Biol 1991;114:423–431.
76. Anand-Srivastava MB, Sairam MR, Cantin M. Ring-deleted analogs of atrial natriuretic factor inhibit adenylate cyclase/cAMP system. Possible coupling of clearance atrial natriuretic factor receptors to adenylate cyclase/cAMP signal transduction system. J Biol Chem 1990;265:8566–8572.
77. Anand-Srivastava MB, Gutkowska J, Cantin M. The presence of atrial-natriuretic-factor receptors of ANF-R2 subtype in rat platelets. Coupling to adenylate cyclase/cyclic AMP signal-transduction system. Biochem J 1991;278:211–217.
78. Drewett JG, Ziegler RJ, Trachte GJ. Neuromodulatory effects of atrial natriuretic peptides correlate with an inhibition of adenylate cyclase but not an activation of guanylate cyclase. J Pharmacol Exp Ther 1992;260:689–696.
79. Anand-Srivastava MB, Sehl PD, Lowe DA. Cytoplasmic domain of natriuretic peptide receptor-C inhibits adenylyl cyclase: involvement of a pertussis toxin-sensitive G protein. J Biol Chem 1996;271:19,324–19,329.
80. Hirata M, Chang CH, Murad F. Stimulatory effects of atrial natriuretic factor on phosphoinositide hydrolysis in cultured bovine aortic smooth muscle cells. Biochem Biophys Acta 1989;1010:346–351.
81. Berl T, Mansour J, Teitelbaum I. ANP stimulates phospholipase C in cultured RIMCT cells: roles of protein kinases and G protein. Am J Physiol 1991;260:F590–595.
82. Levin ER, Frank HJ. Natriuretic peptides inhibit rat astroglial proliferation: mediation by C receptor. Am J Physiol 1991;261:R453–457.
83. Cahill PA, Hassid A. Clearance receptor-binding atrial natriuretic peptides inhibit mitogenesis and proliferation of rat aortic smooth muscle cells. Biochem Biophys Res Commun 1991;179:1606–1613.
84. Cahill PA, Hassid A. ANF-C-receptor-mediated inhibition of aortic smooth muscle cell proliferation and thymidine kinase activity. Am J Physiol 1994;266:R194–203.
85. Porter JG, Catalano R, McEnroe G, Lewicki JA, Protter AA. C-type natriuretic peptide inhibits growth factor-dependent DNA synthesis in smooth muscle cells. Am J Physiol 1992;263:C1001–1006.

86. Johnson BG, Trachte GJ, Drewett JG. Neuromodulatory effect of the atrial natriuretic factor clearance receptor binding peptide, cANF(4–23)-NH2 in rabbit isolated vasa deferentia. J Pharmacol Exp Ther 1991;257:720–726.
87. Hu RM, Levin ER, Pedram A, Frank HJ. Atrial natriuretic peptide inhibits the production and secretion of endothelin from cultured endothelial cells. Mediation through the C receptor. J Biol Chem 1992;267:17,384–17,389.
88. Kishi Y, Ashikaga T, Watanabe R, Numano F. Atrial natriuretic peptide reduces cyclic AMP by activating cyclic GMP-stimulated phosphodiesterase in vascular endothelial cells. J Cardiovasc Pharmacol 1994;24:351–357.
89. MacFarland RT, Zelus BD, Beavo JA. High concentrations of a cGMP-stimulated phosphodiesterase mediate ANP-induced decreases in cAMP and steroidogenesis in adrenal glomerulosa cells. J Biol Chem 1991;266:136–142.
90. Whalin ME, Scammell JG, Strada SJ, Thompson WJ. Phosphodiesterase II, the cGMP-activatable cyclic nucleotide phosphodiesterase, regulates cyclic AMP metabolism in PC12 cells. Mol Pharmacol 1991;39:711–717.
91. Kishimoto I, Dubois SK, Garbers DL. The heart communicates with the kidney exclusively through the guanylyl cyclase-A receptor: acute handling of sodium and water in response to volume expansion. Proc Natl Acad Sci USA 1996;93:6215–6219.
92. Sano T, Morishita Y, Yamada K, Matsuda Y. Effects of HS-142-1, a novel non-peptide ANP antagonist, on diuresis and natriuresis induced by acute volume expansion in anesthetized rats. Biochem Biophys Res Commun 1992;182:824–829.
93. Morishita Y, Sano T, Kase H, Yamada K, Inagami T, Matsuda Y. HS-142-1, a novel nonpeptide atrial natriuretic peptide (ANP) antagonist, blocks ANP-induced renal responses through a specific interaction with guanylyl cyclase-linked receptors. Eur J Pharmacol 1992;225:203–207.
94. Gregory LC, Scarborough RM, Metzler CH, McEnroe GA, Maack T, Lewicki JA. Effects of acute infusion of a C-ANP receptor specific compound in conscious dogs. J Cell Biochem Suppl 1988;12A:23.

# 5 Cellular and Molecular Aspects of the A-Type Natriuretic Peptide

*David G. Gardner, Jianming Wu, and Branka Kovacic-Milivojevic*

***Contents***



## INTRODUCTION

Physiologists have long suspected that the cardiac atria play an important role in the regulation of intravascular volume and blood pressure. The first suggestive evidence that these structures might function as endocrinc organs came from the studies of Kisch *(1)* that demonstrated the presence of membrane-bound secretory granules within atrial myocytes, a finding that was confirmed in the studies of Jamison and Palade *(2)* several years later. In 1981 DeBold et al. *(3)* demonstrated that extracts from atrial, but not ventricular, tissue effected a pronounced natriuresis in test animals. The active component of these extracts, the natriuretic factor or natriuretic peptides, as they are now known, immediately attracted the interest of physiologists, pharmacologists, cell and molecular biologists, and clinicians, reflecting their broad-based physiological effects, novel signal-transduction mechanisms, and potential therapeutic utility. This chapter will focus on the cell and molecular biology of the atrial, or A-type, natriuretic peptide. Where appropriate, efforts will be made to contrast and/or integrate these findings with those obtained for other members of the natriuretic peptide family.

From: *Contemporary Endocrinology: Natriuretic Peptides in Health and Disease*
Edited by: W. K. Samson and E. R. Levin Humana Press Inc., Totowa, NJ

## STRUCTURE OF THE ATRIAL NATRIURETIC PEPTIDE GENE

The identification, subsequent purification, and sequencing of the atrial natriuretic peptide permitted the generation of oligonucleotide probes that were used to isolate cDNA clones for the rat *(4–9)* and human *(6,10,11)* ANP coding sequence. Subsequently, cDNA sequences were isolated from a number of other species, including the rabbit *(12)* and dog *(12)*. Each of these cDNAs is approx 1 kilobase long and encodes a precursor proANP molecule that varies from 126 to 128 amino acids in length depending on the species (Fig. 1). The mature, fully processed form of ANP is 28 amino acids in length and is positioned at the extreme carboxy terminus of the proANP precursor.

The genomic or chromosomal counterparts of the cDNAs for rat *(13)*, human *(14–17)*, mouse *(15)*, and bovine *(18)* proANP have been isolated and characterized. They have a simple structural organization with three exons separated by two introns. In the rat, the first exon harbors the 5'-untranslated region of the coding region and the first 16 amino acids of preproANP, including the signal peptide required for translocating the nascent protein across the membrane of the endoplasmic reticulum. The second exon contains most of the proANP structural sequence, including that for ANP itself, whereas the third exon encodes the tripeptide Tyr-Arg-Arg preceding a translation stop codon that leads into the 3'-untranslated region of the mRNA. The Arg-Arg peptide is not present in the human or dog sequence and exists in the rat sequence for only a short time following translation. Thus, it does not appear to be critical for the function of ANP or for the processing of its precursor. Its functional role in the rat molecule, if one exists, is unknown at this time.

Genomic sequences for proANP, as well as those for the other natriuretic peptide precursors, are present as single copies in the mammalian genome. In the human the proANP gene is located on chromosome 1 *(19,20)*, in the mouse on chromosome 4 *(20,21)*, and in the rat on chromosome 5 *(22–25)*. Several informative restriction fragment-length polymorphisms have been identified in the vicinity of the human locus but, to date, none have been linked convincingly to clinical disease *(26–28)*. Interestingly, the gene encoding brain natriuretic peptide (BNP) is located on the same chromosome as ANP (i.e., chromosome 4 in the mouse and 1 in humans) *(29,30)*, suggesting that an ancestral duplication event gave rise to the pair. It also raises the possibility that regulatory factors (e.g., changes in chromatin conformation) operating over a distance could alter expression of these genes in a concerted fashion.

## TISSUE-SPECIFIC EXPRESSION OF THE ANP GENE

### *Tissue Distribution of Expression*

Expression of the ANP gene is overwhelmingly concentrated in the atria of the heart. The ANP mRNA levels in atrial myocardium are on the order of 30–50-fold higher than those seen in most extra-atrial tissues *(31)*. The adult cardiac ventricle expresses low levels of ANP transcript under basal conditions; however, expression is increased dramatically in certain developmental (e.g., late fetal and early neonatal periods) *(32–38)* and pathophysiological (e.g., following hemodynamic overload) *(39–42)* paradigms. Expression of the ANP gene has also been documented in a number of extra-atrial tissues, including the central nervous system (CNS) *(43–45)*, aortic arch *(46)*, kidney *(47–50)*, gonadotropes of the anterior pituitary gland *(31,47,51)*, thymus *(52,53)*, lung *(31,54)*, and macrophage *(55,56)*. Such expression has been documented

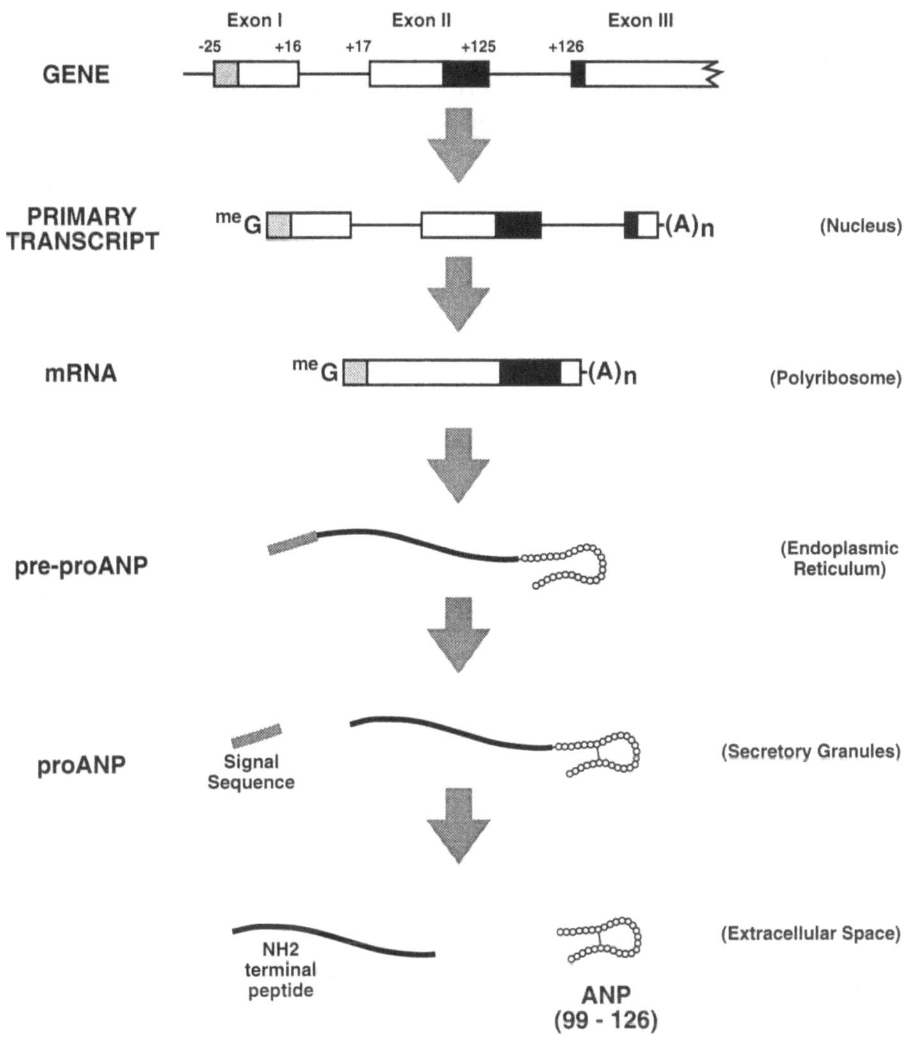


**Fig. 1.** Flow of genetic information involved in ANP synthesis. The ANP gene contains three exons (thick bar) separated by two introns (thin line). Numbers describe the positions of amino acids at the individual exon borders relative to the first amino acid encoded in proANP. The stippled bar indicates the signal peptide of the preproANP sequence; the solid bar identifies the mature ANP peptide. The ANP primary transcript is assumed to be capped and polyadenylated in the nucleus prior to intron processing, although this has not been demonstrated formally. The position of ANP in the proANP molecule is indicated by the open circles at the carboxy terminus of the latter molecule. ProANP is the dominant storage form in the secretory granule. Processing is thought to occur at the time of, or shortly after secretion, releasing both mature ANP and the amino-terminal peptide into the extracellular space.

by different combinations of mRNA measurements, *in situ* hybridization, radioimmunoassay, and immunocytochemistry.

In the CNS, expression is concentrated in areas of the diencephalon and brainstem *(43–45)* thought to be important in the regulation of fluid and electrolyte balance and blood pressure (e.g., anteroventral third ventricular region) and administration of exogenous ANP has been shown to have pronounced effects on neuronal firing rates in these regions *(57–59)*. This translates into effects on water consumption, vasopressin secretion, salt appetite, blood pressure, urine output, and sympathetic outflow in the whole animal *(60)*. The ANP transcripts have been identified in the adventitia of the aortic arch *(46)* where the peptide may play a role in modulating autonomic reflex activity *(61–63)*.

Low levels of expression are also seen in the kidney *(48,49)*, presumably arising from the distal cortical nephron. Interestingly, renal expression of proANP is accompanied by a unique processing event (*see* Storage and Processing of ANP) that generates an amino-terminally extended form of ANP called urodilatin or renal natriuretic peptide *(64,65)*. Like ANP, this peptide binds to the A-type natriuretic peptide receptor and triggers increased guanylyl cyclase activity. This peptide does not circulate at appreciable levels in plasma, but it is present in urine, suggesting that it may function as a paracrine regulator of tubular cell function further downstream in the nephron. Such a model is supported by the findings of Sonnenberg et al. *(66)*, who have demonstrated efficacy of natriuretic peptides operating at the luminal surface of the medullary collecting duct.

### *Identification of Molecular Factors Involved in Tissue-Specific Expression*

Considerable effort has been devoted to the identification of tissue-specific regulatory elements on the promoter of the ANP gene. This has been approached in vitro using transient transfection of neonatal cardiac myocytes with ANP promoter-driven reporter constructs and, in vivo, using transgenic mice. For the in vitro experiments, varying lengths of 5' flanking sequence (5'FS) from the human or rat ANP promoters have been linked to coding sequence for bacterial chloramphenicol acetyl transferase (CAT), firefly luciferase (LUC), or human growth hormone (GH). These reporters are not expressed to a significant degree in cardiac myocytes. Thus, by inference, expression of the reporter is dependent on regulatory sequences present in the contiguous ANP promoter fragment. Introduction of mutations into these promoters has permitted investigators to identify specific elements responsible for regulating transcriptional activity.

Using the rat-ANP promoter, Seidman et al. showed that constructs harboring approx 3000 base pairs of 5' FS linked to a CAT reporter were selectively expressed in primary cultures of neonatal rat atrial (vs ventricular) myocytes *(67)*. Subsequent studies from the same group indicated that this atrial selectivity was maintained as the 5' terminus was truncated to –609, although the overall level of expression was reduced *(68)*. Knowlton et al. *(69)* made similar observations using the ventricular myocyte model. In their hands, deletion of rANP 5' FS from –3003 to –638 reduced cardiac-specific expression by approx 60%.

LaPointe et al. *(70)* showed that a construct harboring as little as 410 bp of the human (h) ANP 5' FS (i.e., –410 hANP CAT) retained atrial-specific expression, although as noted with the rat gene, extending the FS in the 5' direction clearly increased the level of expression *(70–72)*. A more proximal element was localized to a segment of DNA lying between –400 and –332 relative to the CAP site, based on deletion analysis and the ability of this segment to confer atrial-specific expression on a neutral thymidine kinase gene promoter *(71)*. Additional studies showed that this DNA fragment associated with a protein found in cardiocyte nuclear extracts with a high degree of sequence specificity *(71,72)*. Analysis of specific DNA–protein contact points indicated that the critical element is bordered by two GC-rich segments, arranged in an inverted palindrome-like array. These GC-rich segments may have the capacity to stabilize helical structure in the region and thereby facilitate critical DNA–protein interactions.

McBride et al. *(73)* identified a pair of *cis*-acting elements, which they termed myocyte-specific elements, or MSEs, between –370 and –190 in the rANP promoter. These elements appeared to be responsible for at least a portion of this promoter's cardiac specificity. Interestingly, the MSEs were also found to be targets for negative regulation by *c-jun* and *c-fos*. Subsequent studies *(74)*, carried out by the same group demonstrated that DNA sequence positioned between –700 and –136 relative to the CAP site was capable of activating a core promoter structure in cardiac myocytes. Moreover, deletion of this sequence from the native rANP promoter resulted in a dramatic reduction in activity. These same deletion mutants were highly active in nonmyocytes (i.e., non-ANP-expressing fibroblasts) in culture. Of interest, proximal deletions in the hANP promoter, leaving 100 bp or less of 5' FS attached to the CAT reporter, also resulted in a significant increase in reporter expression in cardiac fibroblast cultures *(72)*. This suggests that elements involved in stimulating ANP promoter activity in the cardiac myocyte may also be responsible for regulating (in a negative sense) transcriptional activity in the nonmyocyte population.

The enhancer described by Argentin et al. *(74)* (i.e., positions –700 to –136 in the rANP gene) harbors at least four individual elements, as defined by DNase footprinting analysis, including an AT-rich segment and a CArG element. The latter, which is defined by the sequence $CC(A/T)_6GG$, is present in many muscle-specific gene promoters *(75–77)* and forms the core of the serum response element (SRE) in the *c-fos* promoter *(78)*. The CArG element in the rANP promoter appears to be important for both tissue- and stage-specific expression of this gene in cultured rat cardiac myocytes. The presence of the element conferred higher levels of expression in ventricular myocytes vs nonmyocytes and in 4- vs 1-d-old cultures *(74)*. Argentin et al. showed that the CArG element associated with a cardiocyte nuclear protein complex that did not recognize a similar CArG motif present in the SRE from the *c-fos* promoter, leading them to infer that SRF is not the protein responsible for establishing the primary DNA contacts in the complex. This finding remains controversial. Sprenkle et al. have demonstrated that SRF itself binds to elements present in the proximal promoter of the rANP gene (centered at –114 and –406) and that these interactions are important for transcriptional activity *(79)*. A functional CArG motif, bearing a superficial resemblance to that present in the *c-fos* gene, has also been identified in the hANP promoter *(80)*.

The proximal portion of the rANP promoter also contains a GATA motif that could prove important for this gene's expression. Overexpression of the relatively cardiac-specific GATA-4 isoform resulted in a modest increase in the expression of a cotransfected rANP-luciferase reporter in either Hela or P19 cells *(81)*. GATA-4 is even more effective in activating brain natriuretic peptide (BNP) promoter activity *(81,82)*. A schematic summary of the functional regulatory elements present in the ANP gene promoter is presented in Fig. 2.

Tissue-specific expression of ANP-reporter constructs has also been investigated in vivo using transgenic animals. Field *(83)* linked approx 500 bp of 5' FS from the hANP gene to coding sequence for the SV40 T antigen and introduced the chimeric construct into transgenic mice. Animals bearing the transgene expressed T antigen immunoreactivity in the atrial, but not adjacent ventricular, myocardium. In addition, the atria developed hyperplastic nodules that eventually compromised the anatomical and functional

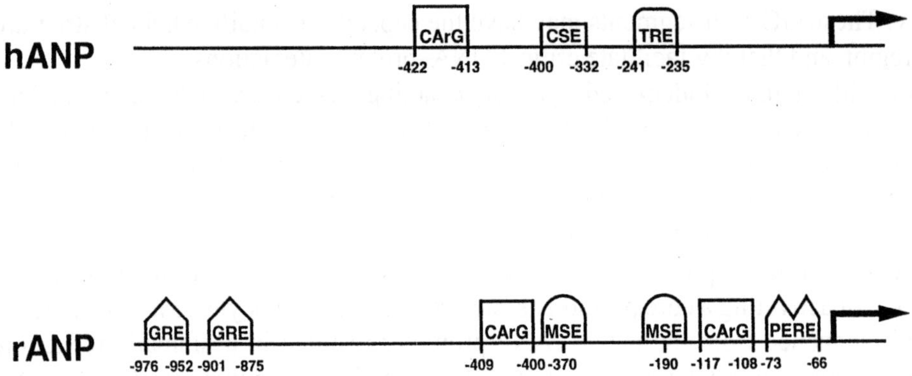


**Fig. 2.** Schematic representation of the functional elements identified in the rat and human ANP promoters. Each has been shown to contribute to basal or agonist-stimulated rat (r) or human (h) ANP-promoter activity in transiently transfected cardiac myocytes. CArG box *(79,80)*; GRE, glucocorticoid response element *(94)*; MSE, cardiac myocyte-specific element *(73)*; CSE, cardiocyte-specific element *(71)*; TRE, TPA response element *(206)*; PERE, phenylephrine response element *(79,109)*. Numbers delineate the approximate position of each element with regard to the start site of transcription for the individual promoter. *See text* for details.

integrity of the heart. Far fewer of these nodules were found in ventricular myocardium. Cells derived from these tumors can be placed in culture *(84)* and retain the phenotype of the atrial myocyte *(85,86)*. A similar approach was adopted by Seidman et al. *(87)*, who introduced a transgene bearing 2400 bp of 5' FS from the rANP gene linked to a CAT reporter into mice. Expression was seen in adult atrium and neonatal ventricle but not in adult ventricular myocardium. In fact, CAT levels in adult atrial extracts were more than three orders of magnitude higher than in ventricular extracts from the same animals. Interestingly, low level expression was also seen in the hypothalamus of these animals, suggesting that the genetic information required for neural specification is present in the reporter construct. More recently, Knowlton et al. *(88)* showed that transgenes harboring either 3003 or 638 bp of rANP 5' FS linked to a luciferase reporter were expressed in both atrial and neonatal ventricular myocardium, but not in adult ventricle. These studies demonstrate a pattern of transgene expression that parallels that for the endogenous gene, suggesting that all the information required for both tissue- and chamber-specific expression is present in the 5' FS included in the relevant test construct.

## REGULATION OF THE A-TYPE NATRIURETIC PEPTIDE GENE

The ANP gene is regulated by a variety of neurotransmitters, growth factors, cytokines, hormones, and physical factors. In a number of cases, specific signaling pathways linking these various ligands to enhanced ANP secretion and/or ANP gene expression have been identified.

### *Regulation by Nuclear Receptor Ligands*

#### Glucocorticoids

A number of hormonal ligands for the nuclear receptors (i.e., glucocorticoid, thyroid, vitamin D, and retinoid receptors) have been shown to regulate ANP secretion and/or gene expression. Glucocorticoids activate ANP gene expression both in vivo *(89,90)* and in vitro *(91–94)*. This is reflected through increases in secretion *(90–94)*, steady-state mRNA levels *(89–91,93,94)*, and transcription *(91,94)* of the ANP gene. Argentin et al. *(94)* identified a DNA segment within the rANP gene promoter (–1029 to –697) that appears to account for its glucocorticoid sensitivity. This region harbors two glucocorticoid-response elements (GREs) that are homologous to GREs identified in other systems *(95)*. Functional analysis of 5' deletion mutants confirmed that these elements conferred glucocorticoid sensitivity on the rANP promoter. A combination of gel-mobility shift analyses and DNase protection showed that the glucocorticoid receptor associated with these elements with high affinity and specificity.

Interestingly, glucocorticoids also regulate the secretion and expression of ANP in cultured hypothalamic neurons *(96,97)*. In the case of the neonatal rat neurons *(97)*, the effect is inhibitory, whereas in the adult it is stimulatory *(96)*, implying both tissue and developmental specificity of the steroid effect.

#### Thyroid Hormones

Thyroid hormone (TH) increases ANP gene expression and secretion of the protein product, both in vivo *(98,99)* and in vitro *(98,100)*. The regulatory element(s) responsible for signalling TH-dependent activity to the ANP promoter remain undefined.

#### Mineralocorticoids

Mineralocorticoids have also been shown to increase levels of ANP mRNA in vivo *(90,101–103)*. In this case, the increase in ANP gene expression and secretion appears to be linked to the intravascular volume-expanding properties of these agents. In fact, elevations in circulating ANP may contribute to the "escape phenomenon" observed in animals treated for extended periods with mineralocorticoids *(101–103)*. Preliminary evidence suggests little in the way of direct effects of mineralocorticoids on ANP gene expression in the heart *(91)*; however, this has not been explored in great detail.

#### Vitamin D and Retinoids

1,25 $(OH)_2$ vitamin $D_3$ and retinoic acid have been shown to suppress secretion of ANP, reduce steady-state ANP mRNA levels, and inhibit transcriptional activity of the ANP promoter in cultured atrial myocytes *(104,105)*. When the two agents are employed together, they display either additive or synergistic activity. In ventricular myocytes, these agents suppress many of the phenotypic markers associated with cardiac hypertrophy, including ANP gene expression *(106,107)*. The vitamin D effects are shared by oxacalcitriol *(104,107)*, a nonhypercalcemic analog of 1,25 $(OH)_2$ vitamin $D_3$. Retinoids seem to promote their antihypertrophic effects through occupancy of either the retinoic acid (RAR) or retinoid X (RXR) receptor. The relative importance of these two retinoid receptor systems (i.e., RAR vs RXR) in mediating this activity remains controversial *(106,107)*.

### *Regulation by Surface Receptor Ligands*

A number of different proteins, peptides, and neurotransmitters act through receptors on the cell surface to regulate ANP secretion and/or gene expression. These include the $\alpha_1$-adrenergic agonists (e.g., phenylephrine) *(69,108,109)*, endothelin-1 (operating through the A-type endothelin receptor) *(110–113)*, angiotensin *(114–117)*, thrombin *(118)*, fibroblast growth factor *(119,120)*, and transforming growth factor β *(119)*. A number of signaling pathways have been invoked to account for these effects, including activation of protein kinase C *(113,121,122)*, mitogen-activated protein kinase *(123–125)*, and calcium/calmodulin kinase *(108)*.

Phenylephrine has been shown to increase protein kinase C activity in cultured ventricular myocytes *(126)* and phorbol ester (TPA), a potent activator of protein kinase C, is a potent secretagog for ANP *(127)*. The TPA also increases steady-state levels of the ANP gene transcript *(121,128)* and activates a cotransfected hANP-CAT reporter *(129)*. In addition, transient overexpression of PKC $\alpha$ or $\beta$ in neonatal ventricular myocytes results in a significant increase in expression of a cotransfected –638 rANP luciferase reporter *(130)*.

LaMorte et al. *(131)* showed that stimulation of ANP gene expression by phenylephrine requires the $G\alpha_{q,11}$ transducer protein. A constitutively active (i.e., independent of liganded receptor) $G\alpha_{q,11}$ stimulated ANP gene expression, whereas microinjection of an antibody directed against the transducer protein inhibited the phenylephrine response.

Thornburn et al. *(132)* employed a dominant-negative mutant of the protooncogene ras to show that at least a portion of phenylephrine's ANP-inducing properties requires an intact ras-signaling pathway. Hunter et al. *(133)* recently demonstrated that overexpression of a constitutively activated ras protein, targeted to the ventricular myocardium by placing it under control of a myosin light-chain (MLV-2v) promoter, produced a hypertrophic ventricular myocardium with diastolic dysfunction in animals bearing two allelic copies of the transgene. Activation of the hypertrophic phenotype in these animals included significant increases in ANP gene expression. Noteworthy in the studies of LaMorte et al. *(131)*, described above, overexpression of the dominant-negative ras mutant did not reverse the increase in ANP gene expression triggered by the constitutively active $G\alpha_{q11}$ mutant. When assembled together, these findings suggest that phenylephrine activates at least two independent signaling pathways leading to increased ANP gene expression and the other phenotypic changes associated with cardiac hypertrophy; one that is ras-dependent and a second, presumably operating through $G\alpha_{q11}$, that is ras-independent. Either or both of these pathways may employ PKC and/or MAPK as distal effectors in the cascade.

Recent studies have shown that phenylephrine and endothelin *(123,134,135)* increase MAPK activity in ventriculocyte cultures. Thornburn et al. *(123)* suggested that activation of MAPK by these hypertrophic agents results in the increased transcriptional activity, although not the cytoskeletal reorganization that accompanies hypertrophy.

Gillespie-Brown et al. *(125)* implicated both mitogen-activated protein kinase (MAPK) and MAPK kinase (MAPKK or MEK) as effectors in phenylephrine-signaled transcriptional activity in the cardiac myocyte. In their studies, phenylephrine-stimulated ANP expression was inhibited by a cotransfected, dominant-negative MEK I mutant. Cotransfection of this mutant together with a dominant-negative MAPK mutant amplified this inhibition. Conversely, a constitutively active MEK I construct doubled ANP promoter activity, whereas the combination of constitutively active MEK I and MAPK increased promoter activity by approximately fivefold. Collectively, these studies lend support to the model implicating the ras/MEK/MAPK cascade in the development of the hypertrophic phenotype. However, more recently Post et al. *(136)* presented evidence that seemingly dissociates activation of MAPK from the development of the hypertrophic phenotype. At present, the linkage must be regarded as unproven, but certainly worthy of further investigation.

Ardati and Nemer *(109)* carried out studies in an effort to localize the phenylephrine response element (i.e., PERE) in the rANP gene promoter. They identified a GC-rich, SP-1 like sequence between –73 and –66 relative to the transcription start (i.e., CAP) site that was both necessary and sufficient to confer $\alpha_1$-adrenergic sensitivity on the core transcriptional machinery. This PERE formed a phenylephrine-inducible complex with soluble proteins extracted from cardiocyte nuclei. The complex was not perturbed by antisera directed against SP-1, leading the authors to conclude that the nuclear protein in the complex was distinct from SP-1 and possibly unique to myocardial cells.

A more recent study by Sprenkle et al. *(79)* showed that either the PERE (located at position –69) or a proximal SRE (located at position –114) was capable of conferring phenylephrine sensitivity on the core rANP promoter (i.e., a construct harboring 134 bp of rANP 5' FS). In this context, mutation of either element eliminated the response to phenylephrine; however, in the context of a larger regulatory region (i.e., –638 rANP LUC), mutation of either element in isolation had little effect on the response to phenylephrine. Mutation of a second SRE, positioned further upstream (i.e., centered at –406), was effective in truncating the phenylephrine induction when paired with mutations of either the proximal SRE or the PERE, as was the combined mutation of the proximal SRE and the PERE. Both rANP SREs associated with a nuclear protein present in ventriculocyte nuclei. Competition analysis and challenge with anti-SRF antisera suggested that the protein was indistinguishable from the serum response factor. The aggregate data argue that agonist- and, presumably, hypertrophy-driven ANP gene expression are controlled by a collection of regulatory elements rather than a single "master" element. This collection of elements presumably serves to anchor the protein complex that controls the gene's transcriptional activity.

The majority of the agonists that trigger ANP secretion and/or gene expression through interaction with cell surface receptors require calcium in the medium *(108,137,138)*. Elevations in extracellular calcium promote both ANP secretion and ANP gene expression *(121)*. The calcium ionophore, A23187, has only modest effects on steady-state levels of the ANP mRNA in ventriculocytes; however, when employed together with TPA, it promotes a robust increase in ANP gene expression *(121)*. It should be emphasized, however, that the dependence of ANP gene expression on calcium remains a controversial issue, with several groups suggesting that higher levels of extracellular calcium *(139–141)*, and in one instance intracellular calcium *(140)*, inhibit ANP secretion and gene expression. One of these studies has implicated a potential calcium regulatory element (CaRE), which is shared between ANP and a number of other calcium-regulated genes *(139)*. The functional activity of this element in the ANP promoter remains to be demonstrated.

Both cyclooxygenase *(142,143)* and lipoxygenase *(144)* products of arachidonic acid metabolism have been shown to stimulate ANP secretion and/or gene expression in cultured cardiac myocytes, as have a variety of growth factors, including FGF and TGF$\alpha$ *(119)*. More recently, a cytokine *(145,146)* named cardiotropin and a protein growth factor called myotrophin *(147)*, each of which are found in the hypertrophied myocardium, have been shown to increase ANP gene expression. Other factors known to regulate ANP secretion and/or gene expression directly include calcitonin gene-related peptide *(148,149)*, $\alpha$-adrenergic agonists (presumably through their capacity to increase intracellular calcium levels) *(138)*, ouabain *(150,151)*, and nitric oxide *(152)*. The relative importance of these factors, either as intermediates for more conventional activators (e.g., angiotensin, endothelin, and so on) or as primary regulators of ANP gene expression, remains unknown.

### *Regulation by Physical Stimuli*

Whereas the biochemical stimuli described above are clearly potent in promoting ANP synthesis and secretion, it is generally agreed that the primary stimulus for ANP production in the intact animal is mechanical stress, or tension, delivered across the myocardial wall. Stretch, either directly applied in vitro *(153–155)* or inferred from cardiac dimensions and intracavitary pressures in vivo *(39,42,156,157)*, has been shown to be a potent stimulus for ANP gene expression. Isolated perfused hearts *(156,158)*, isolated atria *(155,159,160)*, and myocardial cells on distensible culture surfaces *(153,154)* have each been shown to increase ANP release in response to mechanical strain. In some instances, strain acts synergistically with biochemical secretagogs to amplify secretory activity *(110,160,161)*. Stretch-dependent increments in steady-state levels of the ANP transcript have also been demonstrated in cultured atrial and ventricular myocytes *(153,154)*; however, increased perfusion pressures in the isolated perfused heart *(162)* failed to increase ANP mRNA levels despite a robust effect on brain natriuretic peptide (BNP) transcripts. This discrepancy may reflect, in part, the differences in the half-lives of these two transcripts (ANP: 12–24 h vs BNP: <4 h). Such differences might preclude detection of modest changes in transcript levels that would be expected to occur over the short experimental interval employed in the these studies.

Several studies have attempted to identify regulatory elements capable of conveying stretch sensitivity to responsive genes. In the *c-fos* gene, stretch-dependent activation of transcription seems to require an intact serum response element (SRE) *(124)*. Identification of specific stretch-dependent elements in the ANP gene promoter has proven more elusive. Sadoshima et al. *(153)* failed to demonstrate activation of a rat ANP-linked reporter following application of linear stretch to ventricular myocyte cultures despite evidence for activation of the endogenous gene. Rockman et al. *(157)* used an in vivo aortic banding model to overload the left ventricle of the hANP Tag transgene-bearing mice of Field *(83)*. Despite a robust increase in expression of the endogenous ANP gene in the ventricle of the coarcted mice, there was no effect on transgene expression. Employing a similar system, Knowlton et al. *(88)* showed that neither a –3003 rANP luciferase transgene nor the –2400 rANP CAT transgene constructed by Seidman et al. *(87)* were activated in the ventricles of banded mice, despite appropriate expression of the transgene in the adult atrial myocardium and in the early neonatal ventricle. Taken as a whole, these studies imply that the signals responsible for driving ventricular ANP gene expression in the developmental program (i.e., fetal and neonatal heart) are not sufficient for initiating load-dependent reactivation of the promoter in the adult animal. They also suggest that such elements, if they exist, lie outside the 5' FS included in the test constructs (i.e., further upstream in the 5' FS, within intragenic sequences or in the 3' FS). Alternatively, stretch-dependent ANP gene expression may operate through mechanisms independent of transcription.

The signaling mechanisms linking stretch to enhanced ANP gene expression have not been well defined. Stretch-dependent hypertrophy activates both angiotensin II *(115)* and endothelin *(163)* accumulation in the cardiac myocyte, suggesting that these peptides may represent paracrine mediators of stretch-dependent activity. Stretch has been shown to increase PKC, MAPK, tyrosine kinases, p21ras, S6 kinases (pp90RSK), phospholipase C, phospholipase D, and probably the phospholipase A2 and P450 pathways. Gadolinium, an ionic inhibitor of stretch-dependent ion channels in the membrane, has been shown to inhibit stretch-dependent ANP secretion in one case *(164)* but not in another *(165)*.

Electrical depolarization of either atrial strips *(166–168)* or cultured myocytes *(169,170)* also stimulates ANP secretion and/or gene expression. Rate-dependent activation of ANP gene expression in the cultured myocyte studies *(169,170)* was dependent on intracellular calcium and calcium/calmodulin kinase activity, but not protein kinase C activity. Hypoxia has been shown to increase ANP release from isolated hearts in vitro *(171,172)*. The signaling mechanism underlying the induction remains poorly understood.

## STORAGE AND PROCESSING OF ANP

The ANP gene product in the rat is synthesized as a nascent, 152-amino-acid, precursor molecule called preproANP *(173–175)*. Removal of the amino-terminal signal peptide, through proteolytic cleavage between $Ala_{-1}$ and $Asn_{+1}$ (amino acids 26 and 27 relative to the amino terminus) leads to generation of the 126-amino-acid precursor molecule proANP. This precursor form is stored in the dense core granules of the atrial myocytes *(176)*. Such granules are generally lacking in ventricular myocytes except under conditions associated with cardiac hypertrophy in which a limited number are seen to accrue *(39,42)*. Differential accumulation of granules in atrial vs ventricular myocytes led to the early speculation that secretion occurred through a regulated secretory pathway (i.e., secretion from a preformed storage pool) in the former vs a constitutive secretory pathway (i.e., direct secretion of newly synthesized hormone) in the latter *(176)*. Subsequent studies *(177)* have demonstrated that ventricular cells possess the capacity to employ the regulated secretory pathway, despite the paucity of secretory granules, rendering this distinction questionable.

Atrial granules are rich in calcium and hydrogen ions and maintain an anion gradient *(178)*. Calcium binding proteins termed annexins are known to be present in the granules *(179)*, implying that intragranular calcium exists primarily in a chelated form. As noted above, granules contain predominantly unprocessed precursor (i.e., proANP) *(174–176)*. Brain natriuretic peptide (BNP), on the other hand, although synthesized in the same atrial myocytes and, to some degree, packaged in the same secretory granules *(180)*, is stored as both γBNP (i.e., proBNP) and its carboxy-terminal, 45-amino-acid peptide (i.e., mature BNP) *(181)*. This differential processing may reflect the presence of two different processing activities. Alternatively, folding of the precursors may be highly idiosyncratic, precluding access to the processing site in one (i.e., proANP) while permitting cleavage of the other (i.e., proBNP).

The actual location of the proANP processing event remains controversial, although there is general agreement that cleavage occurs at the time of or shortly following the secretory event *(174)*. Under normal conditions, the majority of irANP leaves the heart in the fully processed form (an exception exists in congestive heart failure in which a modest amount of proANP may find its way into the peripheral circulation) *(182)*. In the isolated perfused heart, mature, fully processed ANP is typically the dominant form present in the cardiac effluent. Ito et al. *(183)* showed that removing calcium from or adding the protease inhibitor aprotinin to the perfusion medium resulted in a predominance of proANP over ANP in the effluent. Furthermore, when they infused proANP into these hearts, they found that a significant fraction of the precursor was converted into mature ANP. This process, once again, was reversed by inclusion of aprotinin in the perfusion medium. These findings suggest an extracellular locus for the proANP processing event. This activity could operate at the surface of the secretory cell or a neighboring nonmyocardial cell or, alternatively, in contiguous extracellular matrix. The same group has, in fact, provided evidence that cardiac mesenchymal cells (predominantly fibroblasts) possess the ability to cleave proANP at the appropriate processing site *(184)*. Dube et al. *(185)* on the other hand, were unsuccessful in demonstrating proANP processing by adult cardiac fibroblast cultures. Corthorn et al. found no evidence for proANP processing activity in granular fractions from atrial myocytes *(186)*, leading the authors to conclude that processing occurs distal to the point of granule release. It remains possible, however, that conditions for isolation and/or assay of the processing activity were not optimal in these studies, leaving the question of an intragranular processing enzyme unresolved. Sei et al. *(187)*, using neonatal atrial myocyte cultures, failed to demonstrate significant conversion of extracellular proANP to ANP. They also found the kinetics of immunoreactive (ir)ANP appearance in the extracellular media to closely parallel depletion of intracellular proANP pools, leading them to conclude that the processing event occurred cosecretionally within the atrial granule.

Cultured neonatal cardiac myocytes (atrial or ventricular) release proANP rather than ANP following stimulation with secretagog *(173)*. Adult myocytes, in culture, generate more of the mature peptide *(185)*; however, processing still falls short of completion (approx 67% of irANP is in the processed form). Treatment of neonatal atriocytes with glucocorticoid *(93)* increases the processed ANP fraction. Of interest, Greenwald et al. showed that dexamethasone treatment of mesenchymal cell cultures increased their proANP processing capability *(184)*, supporting their contention that processing occurs through a regulatable extracellular mechanism.

ProANP is processed in hypothalamic neurons to yield two smaller ANP peptides encoding 24 and 25 amino acids, respectively, and little of the 28-amino-acid peptide produced in the heart *(188)*. The neuronal cleavage sites (between $Arg_{102}$ and $Ser_{103}$ and between $Arg_{101}$ and $Arg_{102}$) are bordered by basic amino acids, suggesting that a different processing machinery is operative in the neuron vs cardiac myocyte. Shields et al. *(189)* expressed a cDNA encoding the proANP molecule in cultured AtT-20 pituitary tumor cells. They found that irANP was stored and secreted as the 24-amino-acid peptide ANP (103–126) (reflecting cleavage between $Arg_{102}$ and $Ser_{103}$), and the amino-terminal fragment proANP (1–97). These cells also proved capable of cleaving the precursor at the sole paired basic residues in proANP ($Arg_{102}$ and $Arg_{103}$), generating the 25-amino-acid peptide. Thus, proANP processing in the AtT-20 cell resembles that which takes place in the hypothalamic neuron.

A second variant processing event takes place in the kidney, where cleavage between $Leu_{94}$ and $Thr_{95}$ generates a 32-amino-acid peptide (i.e., ANP with a four-amino-acid N-terminal extension) called urodilatin or renal natriuretic factor *(64,65)*. This peptide does not circulate at appreciable levels in plasma but is present in renal tubular fluid, where it may play an important paracrine role in sodium handling by distal nephron segments *(66)*.

The enzyme responsible for processing proANP to ANP (1–28) in the heart has yet to be identified with certainty, although several candidates have been proposed. Processing activity that cleaves proANP in the correct position *(190,191)* has been identified and partially purified from serum; however, given its dependence on an activated clotting mechanism, this activity seems unlikely to play a major role in proANP processing in vivo. Imada and colleagues *(192)* identified an activity, which they termed atrioactivase, in the plasma membrane fraction of cardiac myocytes. This activity appeared to be capable of generating bona fide ANP (1–28). Wypij and Harris identified a second activity in the granular fraction of the same cells *(193)*. These two activities differ in a number of important biochemical and biophysical properties and are unlikely to arise from the same protein. Thus, the identity of the true processing enzyme remains unknown.

The only posttranslational modification of proANP identified to date is Ser/Thr phosphorylation by the cAMP-dependent protein kinase (PKA). Initial studies conflicted regarding whether phosphorylation of the mature ANP peptide was present in vivo. Rittenhouse et al. found significant, albeit low-level, phosphorylation of ANP in rat atria exposed to $^{32}$P-orthophosphate in vitro *(194)*. This phosphorylation took place at $Ser_{104}$, the same site employed by PKA for phosphorylation in vitro *(195)*. Treatment of atria with cAMP-elevating agents resulted in a 5–10-fold increase in the specific radioactivity of the purified proANP molecule. Bloch et al. *(196)*, on the other hand, found that the bulk of phosphorylation in vivo took place in the amino terminus of proANP with little incorporation of $^{32}$P into those residues present in mature ANP. This finding was confirmed by Wildey et al. *(197)*.

The functional importance of ANP phosphorylation remains to be demonstrated. Rittenhouse et al. *(195)* have noted an increase in the peptide's capacity to stimulate $Na^+$/$K^+$/$Cl^-$ cotransport in cultured vascular smooth muscle cells (VSMC) following PKA-dependent phosphorylation of the peptide in vitro. Others have found that phosphorylation reduces ANP's biological activity in VSMC *(198,199)* and alters binding of the peptide to its receptor *(199)*. Whether sufficient levels of phosphorylated ANP are generated in vivo to exert significant physiological effects is uncertain. Noteworthy, Kubler et al. *(200)* have identified a cAMP-dependent ecto-protein kinase (i.e., localized in the extracellular space) that is capable of phosphorylating human ANP (99–126) with a $K_m$ of 3–5 $\mu M$ for the peptide substrate. This enzymatic activity was increased following treatment with the cAMP-promoting agonist isoproterenol. Although it is too early to assess the physiological importance of this enzymatic activity, it represents a novel mechanism for modification of ANP and one that, at least theoretically, could be accessible to therapeutic manipulation.

## A-TYPE NATRIURETIC PEPTIDE AND CARDIAC HYPERTROPHY

Mature cardiac myocytes are terminally differentiated and, therefore, respond to hemodynamic overload with an increase in cell size (hypertrophy) rather than cell number (hyperplasia) *(201,202)*. Whereas there is no in vitro model that exactly replicates the response to hypertrophy seen in vivo, cultured neonatal ventricular myocytes respond to a number of different growth-promoting perturbations (e.g., $\alpha_1$-adrenergic agonists, endothelin, angiotensin, or linear stretch) with biochemical and morphological changes that closely parallel those seen with hypertrophy in the whole animal *(201)*. Several of the biochemical effectors have, in fact, been linked to hypertrophy in vivo *(163,203–205)*.

Hypertrophic stimuli have been shown to stimulate the expression of several classes of genes in these cells. The immediate-early gene (IEG) group is the first to be activated. This group includes a number of protooncogene transcription factors (e.g., *c-jun*, *c-fos*, and *c-myc*) that function in the cell nucleus. Expression of these genes increases within a manner of minutes following application of the stimulus and persists for varying periods of time thereafter. This is followed by augmented expression of a second group of genes, called the embryonic repertoire or fetal gene program *(201)*. This group includes ANP, β-myosin heavy chain, and α skeletal actin. The ANP gene expression is activated early and reproducibly following application of the hypertrophic stimulus, making it one of the most reliable markers of hypertrophy. As a group, these genes are generally expressed in the late fetal and early neonatal period. Expression decays in the weeks to months following birth and remains low in the adult myocardium unless reactivated by hemodynamic overload or one of the other hypertrophic stimuli alluded to above. The third group contains the constitutively expressed structural genes of the sarcomere (e.g., α cardiac actin and myosin light-chain-2) that provide the protein substrate for cellular hypertrophy.

It appears that the sequential activation of the IEG family and the embryonic repertoire may be linked not only temporally but mechanistically as well. Overexpression of *c-jun* and low levels of *c-fos*, for example, increase transcription from an hANP promoter in transiently transfected cardiac myocyte cultures *(206)*. This effect requires the presence of a consensus TPA response element (TRE, a cognate DNA recognition element that associates with hetero- or homodimeric complexes of proteins from the *jun/fos* family and, by virtue of this association, triggers increased transcriptional activity from contiguous promoters) located between –241 and –235 relative to the transcription start site. Transcription of the α skeletal actin gene, another member of the embryonic repertoire, is similarly activated by cotransfection of *c-jun* and *c-fos (207)*.

However, the system is more complex than it would appear on first inspection. Overexpression of higher levels of *c-fos* result in a suppression, rather than an activation, of the hANP promoter, and this suppression traffics through an element or elements located between –210 and the transcription start site. Moreover, McBride et al. reported that overexpression of either *c-fos* or *c-jun* resulted in a reduction in the expression of a cotransfected rANP CAT reporter *(73)*. These latter, seemingly discrepant findings *(73,129,206)* remain unexplained.

Other members of the extended *jun/fos* family display different activities at the level of the ANP gene promoter. Fra 1, for example, consistently increases hANP promoter activity even when expressed at high levels *(208)*, a property that is distinct from the biphasic activity seen with *c-fos (206)*. Overexpression of *jun B* leads to a modest activation of a cotransfected hANP CAT reporter *(209)*. Interestingly, in ventricular myocytes, *jun B* synergizes with *c-jun* in driving the activity of this promoter, whereas in atrial myocytes it suppresses *c-jun*-dependent transcription *(209)*. The latter activity is in keeping with *jun B's* behavior in other systems *(210–212)*; however, the former is quite unique and suggests that the interactions of the individual *jun/fos* family members with the hANP promoter do not conform to a single simplistic model. Their highly indiosyncratic behavior may add increased flexibility in the response of this promoter to a variety of physiological perturbations.

## PHYSIOLOGICAL CONSEQUENCES OF OVER- OR UNDEREXPRESSION OF THE ANP GENE IN VIVO

Transgenic animals have been employed to address a number of different questions regarding ANP molecular biology and physiology. We have alluded to their use in studies of ANP gene regulation above. Transgenic animals have also been used to overexpress ANP in the intact animal. Steinhelper et al. *(213)* placed the mouse ANP coding sequence under the control of the transthyretin promoter and introduced the chimeric construct into transgenic mice. The latter is a relatively strong promoter that is selectively expressed in hepatocytes of the liver. Mice harboring this transgene display very high levels of circulating immunoreactive ANP, in this case primarily proANP since hepatocytes lack both the regulated secretory pathway as well as the processing mechanism needed to generate mature ANP. In addition, they demonstrated a significant reduction in mean arterial pressure, presumably reflecting enhanced ANP bioactivity at the level of target tissues.

The latter finding suggests that at least a portion of the secreted proANP is processed to ANP independent of the conventional processing pathway and, perhaps more importantly, that sustained elevations in immunoreactive ANP in plasma lead to significant perturbations of cardiovascular homeostasis that are not easily overridden by opposing counterregulatory mechanisms (e.g., activation of the renin-angiotensin system).

The reverse experiment (i.e., reduction in endogenous ANP activity) has been carried out using the recently developed gene-knockout technology. John et al. *(214)* inactivated a single allele encoding the ANP gene locus, then employed conventional breeding techniques to generate animals homozygous for inactivation of both ANP alleles. The homozygotes had little or no circulating ANP in their plasma and were hypertensive on standard laboratory rat chow. Heterozygotes appeared phenotypically normal on the standard diet but became very hypertensive on a high salt (8% NaCl) diet. These studies suggest an important role for endogenous ANP as a natural defense against salt-dependent expansion of intravascular volume and elevated blood pressure. Of interest, Lopez et al. *(215)* recently carried out a similar experiment by deleting the A-type natriuretic peptide receptor (ANPR), ostensibly the cognate receptor for ANP. Animals homozygous for this deletion were hypertensive but not salt sensitive. The discrepant findings with the ANP vs ANPR knockouts remain unexplained but may imply the presence of an independent natriuretic peptide system *(215)*.

It will be of great interest to examine the phenotypes of crosses between these animals (i.e., ANP and ANPR knockouts) and transgenics manipulated at other gene loci involved in blood pressure regulation (e.g., amplification of the renin or angiotensinogen locus). Such crosses may generate complex phenotypes that more closely approximate those seen in human essential hypertension.

## ACKNOWLEDGMENTS

This work was supported by HL 45637 and HL 35753 from the NIH.

## REFERENCES

1. Kisch B. Electron microscopy of the atrium of the heart. I. Guinea pig. Exp Med Surg 1956;14:99–112.
2. Jamison JD, Palade GE. Specific granules in atrial muscle cells. J Cell Biol 1964;23:151–172.
3. de Bold AJ, Borenstein HB, Veress AT, Sonnenberg H. A rapid and potent natriuretic response to intravenous injection of atrial myocardial extract in rats. Life Sci 1981;28:89–94.
4. Maki M, Takayanagi R, Misono KS, Pandey KN, Tibbetts C, Inagami T. Structure of rat atrial natriuretic factor precursor deduced from cDNA sequence. Nature 1984;309:722–724.
5. Yamanaka M, Greenberg B, Johnson L, Seilhamer J, Brewer M, Friedemann T, Miller J, Atlas S, Laragh J, Lewicki J, et al. Cloning and sequence analysis of the cDNA for the rat atrial natriuretic factor precursor. Nature 1984;309:719–722.
6. Zivin RA, Condra JH, Dixon RA, Seidah NG, Chretien M, Nemer M, Chamberland M, Drouin J. Molecular cloning and characterization of DNA sequences encoding rat and human atrial natriuretic factors. Proc Natl Acad Sci USA 1984;81:6325–6329.
7. Flynn TG, Davies PL, Kennedy BP, de Bold ML, de Bold AJ. Alignment of rat cardionatrin sequences with the preprocardionatrin sequence from complementary DNA. Science 1985;228:323–325.
8. Kangawa K, Tawaragi Y, Oikawa S, Mizuno A, Sakuragawa Y, Nakazato H, Fukuda A, Minamino N, Matsuo H. Identification of rat gamma atrial natriuretic polypeptide and characterization of the cDNA encoding its precursor. Nature 1984;312:152–155.
9. Seidman CE, Duby AD, Choi E, Graham RM, Haber E, Homcy C, Smith JA, Seidman JG. The structure of rat preproatrial natriuretic factor as defined by a complementary DNA clone. Science 1984;225:324–326.

10. Nakayama K, Ohkubo H, Hirose T, Inayama S, Nakanishi S. mRNA sequence for human cardiodilatin-atrial natriuretic factor precursor and regulation of precursor mRNA in rat atria. Nature 1984;310:699–701.
11. Oikawa S, Imai M, Ueno A, Tanaka S, Noguchi T, Nakazato H, Kangawa K, Fukuda A, Matsuo H. Cloning and sequence analysis of cDNA encoding a precursor for human atrial natriuretic polypeptide. Nature 1984;309:724–726.
12. Oikawa S, Imai M, Inuzuka C, Tawaragi Y, Nakazato H, Matsuo H. Structure of dog and rabbit precursors of atrial natriuretic polypeptides deduced from nucleotide sequence of cloned cDNA. Biochem Biophys Res Commun 1985;132:892–899.
13. Argentin S, Nemer M, Drouin J, Scott GK, Kennedy BP, Davies PL. The gene for rat atrial natriuretic factor. J Biol Chem 1985;260:4568–4571.
14. Maki M, Parmentier M, Inagami T. Cloning of genomic DNA for human atrial natriuretic factor. Biochem Biophys Res Commun 1984;125:797–802.
15. Seidman CE, Bloch KD, Klein KA, Smith JA, Seidman JG. Nucleotide sequences of the human and mouse atrial natriuretic factor genes. Science 1984;226:1206–1209.
16. Greenberg BD, Bencen GH, Seilhamer JJ, Lewicki JA, Fiddes JC. Nucleotide sequence of the gene encoding human atrial natriuretic factor precursor. Nature 1984;312:656–658.
17. Nemer M, Chamberland M, Sirois D, Argentin S, Drouin J, Dixon RA, Zivin RA, Condra JH. Gene structure of human cardiac hormone precursor, pronatriodilatin. Nature 1984;312:654–656.
18. Vlasuk GP, Miller J, Bencen GH, Lewicki JA. Structure and analysis of the bovine atrial natriuretic peptide precursor gene. Biochem Biophys Res Commun 1986;136:396–403.
19. Dracopoli NC, Stanger BZ, Ito CY, Call KM, Lincoln SE, Lander ES, Housman DE. A genetic linkage map of 27 loci from PND to FY on the short arm of human chromosome I. Am J Hum Genet 1988;43:462–470.
20. Yang-Feng TL, Floyd-Smith G, Nemer M, Drouin J, Francke U. The pronatriodilatin gene is located on the distal short arm of human chromosome 1 and on mouse chromosome 4. Am J Hum Genet 1985;37:1117–1128.
21. Mullins JJ, Zeng Q, Gross KW. Mapping of the mouse atrial natriuretic factor gene. Evidence for tight linkage to the Fv-1 locus. Hypertension 1987;9:518–521.
22. Kobayashi E, Tachibana M, Ikadai H, Kunieda T. A genetic linkage map of rat chromosome 5 reveals extensive linkage conservation with mouse chromosome 4. Mamm Genome 1994;5:222–224.
23. Deng AY, Dene H, Pravenec M, Rapp JP. Genetic mapping of two new blood pressure quantitative trait loci in the rat by genotyping endothelin system genes. J Clin Invest 1994;93:2701–2709.
24. Serikawa T, Kuramoto T, Hilbert P, Mori M, Yamada J, Dubay CJ, Lindpainter K, Ganten D, Guenet JL, Lathrop GM, et al. Rat gene mapping using PCR-analyzed microsatellites. Genetics 1992;131:701–721.
25. Szpirer C, Riviere M, Szpirer J, Genet M, Dreze P, Islam MQ, Levan G. Assignment of 12 loci to rat chromosome 5: evidence that this chromosome is homologous to mouse chromosome 4 and to human chromosomes 9 and 1 (1p arm). Genomics 1990;6:679–684.
26. Nemer M, Sirois D, Drouin J. TaqI polymorphism at the 3' end of the human pronatriodilatin gene (hPND). Nucleic Acids Res 1986;14:8697.
27. Frossard PM, Coleman RT. Human atrial natriuretic peptides (ANP) gene locus: BglI RFLP. Nucleic Acids Res 1986;14:9223.
28. Nemer M, Sirois D, Drouin J. XhoI polymorphism at the human pronatriodilatin (hPND) gene locus. Nucleic Acids Res 1986;14:8696.
29. Steinhelper ME. Structure, expression, and genomic mapping of the mouse natriuretic peptide type-B gene. Circ Res 1993;72:984–992.
30. Arden KC, Viars CS, Weiss S, Argentin S, Nemer M. Localization of the human B-type natriuretic peptide precursor (NPPB) gene to chromosome 1p36. Genomics 1995;26:385–389.
31. Gardner DG, Deschepper CF, Ganong WF, Hane S, Fiddes J, Baxter JD, Lewicki J. Extra-atrial expression of the gene for atrial natriuretic factor. Proc Natl Acad Sci U S A 1986;83:6697–6701.
32. Zeller R, Bloch KD, Williams BS, Arceci RJ, Seidman CE. Localized expression of the atrial natriuretic factor gene during cardiac embryogenesis. Genes Dev 1987;1:693–698.
33. Kikuchi K, Nakao K, Hayashi K, Morii N, Sugawara A, Sakamoto M, Imura H, Mikawa H. Ontogeny of atrial natriuretic polypeptide in the human heart. Acta Endocrinol (Copenh) 1987;115:211–217.
34. Wei YF, Rodi CP, Day ML, Wiegand RC, Needleman LD, Cole BR, Needleman P. Developmental changes in the rat atriopeptin hormonal system. J Clin Invest 1987;79:1325–1329.
35. Wu JP, Deschepper CF, Gardner DG. Perinatal expression of the atrial natriuretic factor gene in rat cardiac tissue. Am J Physiol 1988;255:E388–396.

36. Claycomb WC. Atrial-natriuretic-factor mRNA is developmentally regulated in heart ventricles and actively expressed in cultured ventricular cardiac muscle cells of rat and human. Biochem J 1988;255:617–620.
37. Gardner DG, Hedges BK, Wu J, LaPointe MC, Deschepper CF. Expression of the atrial natriuretic peptide gene in human fetal heart. J Clin Endocrinol Metab 1989;69:729–737.
38. Camreon VA, Aitken GD, Ellmers LJ, Kennedy MA, Espiner EA. The sites of gene expression of atrial, brain, and C-type natriuretic peptides in mouse fetal development: temporal changes in embryos and placenta. Endocrinology 1996;137:817–824.
39. Day ML, Schwartz D, Wiegand RC, Stockman PT, Brunnert SR, Tolunay HE, Currie MG, Standaert DG, Needleman P. Ventricular atriopeptin. Unmasking of messenger RNA and peptide synthesis by hypertrophy or dexamethasone. Hypertension 1987;9:485–491.
40. Lee RT, Bloch KD, Pfeffer JM, Pfeffer MA, Neer EJ, Seidman CE. Atrial natriuretic factor gene expression in ventricles of rats with spontaneous biventricular hypertrophy. J Clin Invest 1988;81:431–434.
41. Saito Y, Nakao K, Arai H, Nishimura K, Okumura K, Obata K, Takemura G, Fujiwara H, Sugawara A, Yamada T, et al. Augmented expression of atrial natriuretic polypeptide gene in ventricle of human failing heart. J Clin Invest 1989;83:298–305.
42. Gu J, D'Andrea M, Seethapathy M. Atrial natriuretic peptide and its messenger ribonucleic acid in overloaded and overload-released ventricles of rat. Endocrinology 1989;125:2066–2074.
43. Saper CB, Standaert DG, Currie MG, Schwartz D, Geller DM, Needleman P. Atriopeptin-immunoreactive neurons in the brain: presence in cardiovascular regulatory areas. Science 1985;227:1047–1049.
44. Gardner DG, Vlasuk GP, Baxter JD, Fiddes JC, Lewicki JA. Identification of atrial natriuretic factor gene transcripts in the central nervous system of the rat. Proc Natl Acad Sci USA 1987;84:2175–2179.
45. Gundlach AL, Knobe KE. Distribution of preproatrial natriuretic peptide mRNA in rat brain detected by in situ hybridization of DNA oligonucleotides: enrichment in hypothalamic and limbic regions. J Neurochem 1992;59:758–761.
46. Gardner DG, Deschepper CF, Baxter JD. The gene for the atrial natriuretic factor is expressed in the aortic arch. Hypertension 1987;9:103–106.
47. McKenzie JC, Tanaka I, Misono KS, Inagami T. Immunocytochemical localization of atrial natriuretic factor in the kidney, adrenal medulla, pituitary, and atrium of rat. J Histochem Cytochem 1985;33:828–832.
48. Golomb E, Friedman E, Abassi ZA, Trachewsky D, Keiser HR. Rare renal transcription of the atrial natriuretic factor gene in rats. Demonstration by polymerase chain reaction. Am J Hypertens 1993;6:867–872.
49. Greenwald JE, Ritter D, Tetens E, Rotwein PS. Renal expression of the gene for atrial natriuretic factor. Am J Physiol 1992;263:F974–978.
50. Ritter D, Chao J, Needleman P, Tetens E, Greenwald JE. Localization, synthetic regulation, and biology of renal atriopeptin-like prohormone. Am J Physiol 1992;263:F503–509.
51. Morel G, Chabot JG, Gossard F, Heisler S. Is atrial natriuretic peptide synthesized and internalized by gonadotrophs? Endocrinology 1989;124:1703–1710.
52. Vollmar AM, Schulz R. Atrial natriuretic peptide is synthesized in the human thymus. Endocrinology 1990;126:2277–2280.
53. Vollmar AM, Wolf R, Schulz R. Co-expression of the natriuretic peptides (ANP, BNP, CNP) and their receptors in normal and acutely involuted rat thymus. J Neuroimmunol 1995;57:117–127.
54. Gutkowska J, Nemer M, Sole MJ, Drouin J, Sirois P. Lung is an important source of atrial natriuretic factor in experimental cardiomyopathy. J Clin Invest 1989;83:1500–1504.
55. Throsby M, Lee D, Huang WQ, Yang ZY, Copolov DL, Lim AT. Evidence for atrial natriuretic peptide-(5–28) production by macrophages of the rat spleen: an immunochemical and nonradioactive in situ hybridization approach. Endocrinology 1991;129:991–1000.
56. Vollmar AM, Schulz R. Expression and differential regulation of natriuretic peptides in mouse macrophages. J Clin Invest 1995;95:2442–2450.
57. Wong M, Samson WK, Dudley CA, Moss RL. Direct, neuronal action of atrial natriuretic factor in the rat brain. Neuroendocrinology 1986;44:49–53.
58. Standaert DG, Cechetto DF, Needleman P, Saper CB. Inhibition of the firing of vasopressin neurons by atriopeptin. Nature 1987;329:151–153.
59. Ermirio R, Ruggeri P, Cogo CE, Molinari C, Calaresu FR. Neuronal and cardiovascular responses to ANF microinjected into the solitary nucleus. Am J Physiol 1989;256:R577–582.
60. Imura H, Nakao K, Itoh H. The natriuretic peptide system in the brain: implications in the central control of cardiovascular and neuroendocrine functions. Front Neuroendocrinol 1992;13:217–249.

61. Volpe M, Cuocolo A, Vecchione F, Mele AF, Condorelli M, Trimarco B. Vagal mediation of the effects of atrial natriuretic factor on blood pressure and arterial baroreflexes in the rabbit. Circ Res 1987;60:747–755.
62. Thoren P, Mark AL, Morgan DA, O'Neill TP, Needleman P, Brody MJ. Activation of vagal depressor reflexes by atriopeptins inhibits renal sympathetic nerve activity. Am J Physiol 1986;251:H1252–1259.
63. Schultz HD, Gardner DG, Deschepper CF, Coleridge HM, Coleridge JC. Vagal C-fiber blockade abolishes sympathetic inhibition by atrial natriuretic factor. Am J Physiol 1988;255:R6–13.
64. Schulz-Knappe P, Forssmann K, Herbst F, Hock D, Pipkorn R, Forssmann WG. Isolation and structural analysis of "urodilatin," a new peptide of the cardiodilatin-(ANP)-family, extracted from human urine. Klin Wochenschr 1988;66:752–759.
65. Valentin JP, Humphreys MH. Urodilatin: a paracrine renal natriuretic peptide. Semin Nephrol 1993;13:61–70.
66. Sonnenberg H, Honrath U, Wilson DR. In vivo microperfusion of inner medullary collecting duct in rats: effect of amiloride and ANF. Am J Physiol 1990;259:F222–226.
67. Seidman CE, Wong DW, Jarcho JA, Bloch KD, Seidman JG. Cis-acting sequences that modulate atrial natriuretic factor gene expression. Proc Natl Acad Sci USA 1988;85:4104–4108.
68. Rosenzweig A, Halazonetis TD, Seidman JG, Seidman CE Proximal regulatory domains of rat atrial natriuretic factor gene. Circulation 1991;84:1256–1265.
69. Knowlton KU, Baracchini E, Ross RS, Harris AN, Henderson SA, Evans SM, Glembotski CC, Chien KR. Co-regulation of the atrial natriuretic factor and cardiac myosin light chain-2 genes during alpha-adrenergic stimulation of neonatal rat ventricular cells. Identification of cis sequences within an embryonic and a constitutive contractile protein gene which mediate inducible expression. J Biol Chem 1991;266:7759–7768.
70. LaPointe MC, Wu JP, Greenberg B, Gardner DG. Upstream sequences confer atrial-specific expression on the human atrial natriuretic factor gene. J Biol Chem 1988;263:9075–9078.
71. Wu J, LaPointe MC, West BL, Gardner DG. Tissue-specific determinants of human atrial natriuretic factor gene expression in cardiac tissue. J Biol Chem 1989;264:6472–6479.
72. Wu JP, Kovacic-Milivojevic B, Lapointe MC, Nakamura K, Gardner DG. Cis-active determinants of cardiac-specific expression in the human atrial natriuretic peptide gene. Mol Endocrinol 1991;5:1311–1322.
73. McBride K, Robitaille L, Tremblay S, Argentin S, Nemer M. fos/jun repression of cardiac-specific transcription in quiescent and growth-stimulated myocytes is targeted at a tissue-specific cis element. Mol Cell Biol 1993;13:600–612.
74. Argentin S, Ardati A, Tremblay S, Lihrmann I, Robitaille L, Drouin J, Nemer M. Developmental stage-specific regulation of atrial natriuretic factor gene transcription in cardiac cells. Mol Cell Biol 1994;14:777–790.
75. Sartorelli V, Webster KA, Kedes L. Muscle-specific expression of the cardiac alpha-actin gene requires MyoD1, CArG-box binding factor, and Sp1. Genes Dev 1990;4:1811–1822.
76. Parker TG, Chow KL, Schwartz RJ, Schneider MD. Positive and negative control of the skeletal alpha-actin promoter in cardiac muscle. A proximal serum response element is sufficient for induction by basic fibroblast growth factor (FGF) but not for inhibition by acidic FGF. J Biol Chem 1992;267:3343–3350.
77. MacLellan WR, Lee TC, Schwartz RJ, Schneider MD. Transforming growth factor-beta response elements of the skeletal alpha-actin gene. Combinatorial action of serum response factor, YY1, and the SV40 enhancer-binding protein, TEF-1. J Biol Chem 1994;269:16,754–16,760.
78. Treisman R. Journey to the surface of the cell: Fos regulation and the SRE. EMBO J 1995;14:4905–4913.
79. Sprenkle AB, Murray SF, Glembotski CC. Involvement of multiple cis elements in basal- and alpha-adrenergic agonist-inducible atrial natriuretic factor transcription. Roles for serum response elements and an SP-1-like element. Circ Res 1995;77:1060–1069.
80. Kovacic-Milivojevic B, Zlock DW, Gardner DG. Ras-dependent regulation of hANP gene transcription. 48th Ann Fall Conf and Sci Sess., Council for High Blood Press Res 1994;Abs 9.
81. Grepin C, Dagnino L, Robitaille L, Haberstroh L, Antakly T, Nemer M. A hormone-encoding gene identifies a pathway for cardiac but not skeletal muscle gene transcription. Mol Cell Biol 1994;14:3115–129.
82. Thuerauf DJ, Hanford DS, Glembotski CC. Regulation of rat brain natriuretic peptide transcription. A potential role for GATA-related transcription factors in myocardial cell gene expression. J Biol Chem 1994;269:17,772–17,775.
83. Field LJ. Atrial natriuretic factor-SV40 T antigen transgenes produce tumors and cardiac arrhythmias in mice. Science 1988;239:1029–1033.

84. Steinhelper ME, Lanson NA, Dresdner KP, Delcarpio JB, Wit AL, Claycomb WC, Field LJ. Proliferation in vivo and in culture of differentiated adult atrial cardiomyocytes from transgenic mice. Am J Physiol. 1990;259:H1826–H1834.
85. Lanson NA Jr, Glembotski CC, Steinhelper ME, Field LJ, Claycomb WC. Gene expression and atrial natriuretic factor processing and secretion in cultured AT-1 cardiac myocytes. Circulation 1992;85:1835–1841.
86. Kline RP, Sorota S, Dresdner KP, Steinhelper ME, Lanson NA Jr, Wit AL, Claycomb WC, Field LJ. Spontaneous activity in transgenic mouse heart: comparison of primary atrial tumor with cultured AT-1 atrial myocytes. J Cardiovasc Electrophysiol 1993;4:642–660.
87. Seidman CE, Schmidt EV, Seidman JG. cis-dominance of rat atrial natriuretic factor gene regulatory sequences in transgenic mice. Can J Physiol Pharmacol 1991;69:1486–1492.
88. Knowlton KU, Rockman HA, Itani M, Vovan A, Seidman CE, Chien KR. Divergent pathways mediate the induction of ANF transgenes in neonatal and hypertrophic ventricular myocardium. J Clin Invest 1995;96:1311–1318.
89. Gardner DG, Hane S, Trachewsky D, Schenk D, Baxter JD. Atrial natriuretic peptide mRNA is regulated by glucocorticoids in vivo. Biochem Biophys Res Commun 1986;139:1047–1054.
90. Dananberg J, Grekin RJ. Corticoid regulation of atrial natriuretic factor secretion and gene expression. Am J Physiol 1992;263:H1377–1381.
91. Gardner DG, Gertz BJ, Deschepper CF, Kim DY. Gene for the rat atrial natriuretic peptide is regulated by glucocorticoids in vitro. J Clin Invest 1988;82:1275–1281.
92. Matsubara H, Hirata Y, Yoshimi H, Takata S, Takagi Y, Yamane Y, Umeda Y, Nishikawa M, Inada M. Ventricular myocytes from neonatal rats are more responsive to dexamethasone than atrial myocytes in synthesis of atrial natriuretic peptide. Biochem Biophys Res Commun 1987;148:1030–1038.
93. Shields PP, Dixon JE, Glembotski CC. The secretion of atrial natriuretic factor-(99-126) by cultured cardiac myocytes is regulated by glucocorticoids. J Biol Chem 1988;263:12,619–12,628.
94. Argentin S, Sun YL, Lihrmann I, Schmidt TJ, Drouin J, Nemer M. Distal cis-acting promoter sequences mediate glucocorticoid stimulation of cardiac atrial natriuretic factor gene transcription. J Biol Chem 1991;266:23,315–23,322.
95. Mangelsdorf DJ, Thummel C, Beato M, Herrlich P, Schutz G, Umesono K, Blumberg B, Kastner P, Mark M, Chambon P, et al. The nuclear receptor superfamily: the second decade. Cell 1995;83:835–839.
96. Wang G, Lim AT. Glucocorticoid stimulation of pro-ANF mRNA expression of adult rat hypothalamic neurons in culture: a cAMP-dependent effect. Endocrinology 1996;137:379–382.
97. Deschepper CF, Nguyen KP, Lapointe MC, Zahs KR, Gardner DG. Production and differential endocrine regulation of atrial natriuretic peptide in neuron-enriched primary cultures. Endocrinology 1991;128:5–12.
98. Gardner DG, Gertz BJ, Hane S. Thyroid hormone increases rat atrial natriuretic peptide messenger ribonucleic acid accumulation in vivo and in vitro. Mol Endocrinol 1987;1:260–265.
99. Ladenson PW, Bloch KD, Seidman JG. Modulation of atrial natriuretic factor by thyroid hormone: messenger ribonucleic acid and peptide levels in hypothyroid, euthyroid, and hyperthyroid rat atria and ventricles. Endocrinology 1988;123:652–657.
100. Rundle SE, Fullerton MJ, Funder JW. Induction of ventricular morphogenesis and atrial natriuretic factor synthesis by thyroid hormone. Mol Cell Endocrinol 1990;68:163–168.
101. Ballermann BJ, Bloch KD, Seidman JG, Brenner BM. Atrial natriuretic peptide transcription, secretion, and glomerular receptor activity during mineralocorticoid escape in the rat. J Clin Invest 1986;78:840–843.
102. Metzler CH, Gardner DG, Keil LC, Baxter JD, Ramsay DJ. Increased synthesis and release of atrial peptide during DOCA escape in conscious dogs. Am J Physiol 1987;252:R188–192.
103. Yokota N, Bruneau BG, Kuroski de Bold ML, de Bold AJ. Atrial natriuretic factor significantly contributes to the mineralocorticoid escape phenomenon. Evidence for a guanylate cyclase-mediated pathway. J Clin Invest 1994;94:1938–1946.
104. Wu J, Garami M, Cao L, Li Q, Gardner DG. 1,25(OH)$_2$D$_3$ suppresses expression and secretion of atrial natriuretic peptide from cardiac myocytes. Am J Physiol 1995;268:E1108–1113.
105. Li Q, Gardner DG. Negative regulation of the human atrial natriuretic peptide gene by 1,25-dihydroxyvitamin D$_3$. J Biol Chem 1994;269:4934–4939.
106. Zhou MD, Sucov HM, Evans RM, Chien KR. Retinoid-dependent pathways suppress myocardial cell hypertrophy. Proc Natl Acad Sci USA 1995;92:7391–7395.

107. Wu J, Garami M, Cheng T, Gardner DG. 1,25$(OH)_2$ vitamin $D_3$ and retinoic acid antagonize endothelin-stimulated hypertrophy of neonatal rat cardiac myocytes. J Clin Invest 1996;97:1577–1588.
108. Sei CA, Irons CE, Sprenkle AB, McDonough PM, Brown JH, Glembotski CC. The alpha-adrenergic stimulation of atrial natriuretic factor expression in cardiac myocytes requires calcium influx, protein kinase C, and calmodulin-regulated pathways. J Biol Chem 1991;266:15,910–15,916.
109. Ardati A, Nemer M. A nuclear pathway for alpha 1-adrenergic receptor signaling in cardiac cells. EMBO J 1993;12:5131–5139.
110. Gardner DG, Newman ED, Nakamura KK, Nguyen KP. Endothelin increases the synthesis and secretion of atrial natriuretic peptide in neonatal rat cardiocytes. Am J Physiol 1991;261:E177–182.
111. Shubeita HE, McDonough PM, Harris AN, Knowlton KU, Glembotski CC, Brown JH, Chien KR. Endothelin induction of inositol phospholipid hydrolysis, sarcomere assembly, and cardiac gene expression in ventricular myocytes. A paracrine mechanism for myocardial cell hypertrophy. J Biol Chem 1990;265:20,555–20,562.
112. Thibault G, Doubell AF, Garcia R, Lariviere R, Schiffrin EL. Endothelin-stimulated secretion of natriuretic peptides by rat atrial myocytes is mediated by endothelin A receptors. Circ Res 1994;74:460–470.
113. Irons CE, Murray SF, Glembotski CC. Identification of the receptor subtype responsible for endothelin-mediated protein kinase C activation and atrial natriuretic factor secretion from atrial myocytes. J Biol Chem 1993;268:23,417–23,421.
114. Sadoshima J, Izumo S. Molecular characterization of angiotensin II—induced hypertrophy of cardiac myocytes and hyperplasia of cardiac fibroblasts. Critical role of the AT1 receptor subtype. Circ Res 1993;73:413–423.
115. Sadoshima J, Xu Y, Slayter HS, Izumo S. Autocrine release of angiotensin II mediates stretch-induced hypertrophy of cardiac myocytes in vitro. Cell 1993;75:977–984.
116. Miyata S, Haneda T. Hypertrophic growth of cultured neonatal rat heart cells mediated by type 1 angiotensin II receptor. Am J Physiol 1994;266:H2443–2451.
117. Sadoshima J, Izumo S. Rapamycin selectively inhibits angiotensin II-induced increase in protein synthesis in cardiac myocytes in vitro. Potential role of 70-kD S6 kinase in angiotensin II-induced cardiac hypertrophy. Circ Res 1995;77:1040–1052.
118. Glembotski CC, Irons CE, Krown KA, Murray SF, Sprenkle AB, Sei CA. Myocardial alpha-thrombin receptor activation induces hypertrophy and increases atrial natriuretic factor gene expression. J Biol Chem 1993;268:20,646–20,652.
119. Parker TG, Chow KL, Schwartz RJ, Schneider MD. Differential regulation of skeletal alpha-actin transcription in cardiac muscle by two fibroblast growth factors. Proc Natl Acad Sci USA 1990;87:7066–7070.
120. Tokola H, Salo K, Vuolteenaho O, Ruskoaho H. Basal and acidic fibroblast growth factor-induced atrial natriuretic peptide gene expression and secretion is inhibited by staurosporine. Eur J Pharmacol 1994;267:195–206.
121. LaPointe MC, Deschepper CF, Wu JP, Gardner DG. Extracellular calcium regulates expression of the gene for atrial natriuretic factor. Hypertension 1990;15:20–28.
122. Irons CE, Sei CA, Hidaka H, Glembotski CC. Protein kinase C and calmodulin kinase are required for endothelin-stimulated atrial natriuretic factor secretion from primary atrial myocytes. J Biol Chem 1992;267:5211–5216.
123. Thorburn J, Frost JA, Thorburn A. Mitogen-activated protein kinases mediate changes in gene expression, but not cytoskeletal organization associated with cardiac muscle cell hypertrophy. J Cell Biol 1994;126:1565–1572.
124. Sadoshima J, Izumo S. Mechanical stretch rapidly activates multiple signal transduction pathways in cardiac myocytes: potential involvement of an autocrine/paracrine mechanism. EMBO J 1993;12:1681–1692.
125. Gillespie-Brown J, Fuller SJ, Bogoyevitch MA, Cowley S, Sugden PH. The mitogen-activated protein kinase kinase MEK1 stimulates a pattern of gene expression typical of the hypertrophic phenotype in rat ventricular cardiomyocytes. J Biol Chem 1995;270:28,092–28,096.
126. Puceat M, Hilal-Dandan R, Strulovici B, Brunton LL, Brown JH. Differential regulation of protein kinase C isoforms in isolated neonatal and adult rat cardiomyocytes. J Biol Chem 1994;269:16,938–16,944.
127. Shields PP, Glembotski CC. Regulation of atrial natriuretic factor-(99–126) secretion from neonatal rat primary atrial cultures by activators of protein kinases A and C. J Biol Chem 1989;264:9322–9328.

128. LaPointe MC, Sitkins JR. Phorbol ester stimulates the synthesis and secretion of brain natriuretic peptide from neonatal rat ventricular cardiocytes: a comparison with the regulation of atrial natriuretic factor. Mol Endocrinol 1993;7:1284–1296.
129. Kovacic-Milivojevic B, Gardner DG. Regulation of the human atrial natriuretic peptide gene in atrial cardiocytes by the transcription factor AP-1. Am J Hypertens 1993;6:258–263.
130. Shubeita HE, Martinson EA, Van Bilsen M, Chien KR, Brown JH. Transcriptional activation of the cardiac myosin light chain 2 and atrial natriuretic factor genes by protein kinase C in neonatal rat ventricular myocytes. Proc Natl Acad Sci USA 1992;89:1305–1309.
131. LaMorte VJ, Thorburn J, Absher D, Spiegel A, Brown JH, Chien KR, Feramisco JR, Knowlton KU. Gq- and ras-dependent pathways mediate hypertrophy of neonatal rat ventricular myocytes following alpha 1-adrenergic stimulation. J Biol Chem 1994;269:13,490–13,496.
132. Thorburn A, Thorburn J, Chen SY, Powers S, Shubeita HE, Feramisco JR, Chien KR. HRas-dependent pathways can activate morphological and genetic markers of cardiac muscle cell hypertrophy [published erratum appears in J Biol Chem 1993 Jul 25;268(21):16,082]. J Biol Chem 1993;268:2244–2249.
133. Hunter JJ, Tanaka N, Rockman HA, Ross J, Chien KR. Ventricular expression of a MLC-2v-ras fusion gene induces cardiac hypertrophy and selective diastolic dysfunction in transgenic mice. J Biol Chem 1995;270:23,173–23,178.
134. Bogoyevitch MA, Glennon PE, Andersson MB, Clerk A, Lazou A, Marshall CJ, Parker PJ, Sugden PH. Endothelin-1 and fibroblast growth factors stimulate the mitogen-activated protein kinase signaling cascade in cardiac myocytes. The potential role of the cascade in the integration of two signaling pathways leading to myocyte hypertrophy. J Biol Chem 1994;269:1110–1119.
135. Bogoyevitch MA, Clerk A, Sugden PH. Activation of the mitogen-activated protein kinase cascade by pertussis toxin-sensitive and -insensitive pathways in cultured ventricular cardiomyocytes. Biochem J 1995;309:437–443.
136. Post GR, Goldstein D, Thuerauf DJ, Glembotski CC, Brown JH. Dissociation of p44 and p42 mitogen-activated protein kinase activation from receptor-induced hypertrophy in neonatal rat ventricular myocytes. J Biol Chem 1996;271:8452–8457.
137. Schiebinger RJ, Parr HG, Cragoe EJ Jr. Calcium: its role in alpha 1-adrenergic stimulation of atrial natriuretic peptide secretion. Endocrinology 1992;130:1017–1023.
138. Schiebinger RJ. Calcium, its role in isoproterenol-stimulated atrial natriuretic peptide secretion by superfused rat atria. Circ Res 1989;65:600–606.
139. Okazaki T, Ando K, Igarashi T, Ogata E, Fujita T. Conserved mechanism of negative gene regulation by extracellular calcium. Parathyroid hormone gene versus atrial natriuretic polypeptide gene. J Clin Invest 1992;89:1268–1273.
140. Greenwald JE, Apkon M, Hruska KA, Needleman P. Stretch-induced atriopeptin secretion in the isolated rat myocyte and its negative modulation by calcium. J Clin Invest 1989;83:1061–1065.
141. De Bold ML, De Bold AJ. Effect of manipulation of $Ca^{2+}$ environment on atrial natriuretic factor release. Am. J. Physiol. 1989;256:H1588–H1594.
142. Gardner DG, Schultz HD. Prostaglandins regulate the synthesis and secretion of the atrial natriuretic peptide. J Clin Invest 1990;86:52–59.
143. Rayner TE, Chen BN, McLoughlin JW, Menadue MF, Norman RJ, Oliver JR. Prostaglandin F2 alpha mediates platelet-activating factor-stimulated atrial natriuretic factor release from the isolated rat heart. Endocrinology 1993;133:1108–1015.
144. Kovacic-Milivojevic B, Schultz HD, Gardner DG. Arachidonic acid metabolites regulate the secretion of atrial natriuretic peptide in cultured rat atrial cardiocytes. Can J Physiol Pharmacol 1991;69:1493–1499.
145. Pennica D, King KL, Shaw KJ, Luis E, Rullamas J, Luoh SM, Darbonne WC, Knutzon DS, Yen R, Chien KR, et al. Expression cloning of cardiotrophin 1, a cytokine that induces cardiac myocyte hypertrophy. Proc Natl Acad Sci USA 1995;92:1142–1146.
146. Wollert KC, Taga T, Saito M, Narazaki M, Kishimoto T, Glembotski CC, Vernallis AN, Heath JK, Pennica D, Wood.W.I., Chien KR. Cardiotrophin-1 activates a distinct form of cardiac muscle cell hypertrophy. J Biol Chem. 1996;271:9535–9545.
147. Mukherjee DP, McTiernan CF, Sen S. Myotrophin induces early response genes and enhances cardiac gene expression. Hypertension 1993;21:142–148.
148. Yamamoto A, Kimura S, Hasui K, Fujisawa Y, Tamaki T, Fukui K, Iwao H, Abe Y. Calcitonin gene-related peptide (CGRP) stimulates the release of atrial natriuretic peptide (ANP) from isolated rat atria. Biochem Biophys Res Commun 1988;155:1452–1458.

149. Schiebinger RJ, Santora AC. Stimulation by calcitonin gene-related peptide of atrial natriuretic peptide secretion in vitro and its mechanism of action. Endocrinology 1989;124:2473–2479.
150. Schiebinger RJ, Cragoe EJ Jr. Ouabain. A stimulator of atrial natriuretic peptide secretion and its mechanism of action. Circ Res 1993;72:1035–1043.
151. Bloch KD, Zamir N, Lichtstein D, Seidman CE, Seidman JG. Ouabain induces secretion of proatrial natriuretic factor by rat atrial cardiocytes. Am J Physiol 1988;255:E383–387.
152. Sanchez-Ferrer CF, Burnett JC Jr, Lorenz RR, Vanhoutte PM. Possible modulation of release of atrial natriuretic factor by endothelium-derived relaxing factor. Am J Physiol 1990;259:H982–986.
153. Sadoshima J, Jahn L, Takahashi T, Kulik TJ, Izumo S. Molecular characterization of the stretch-induced adaptation of cultured cardiac cells. An in vitro model of load-induced cardiac hypertrophy. J Biol Chem 1992;267:10,551–10,560.
154. Gardner DG, Wirtz H, Dobbs LG. Stretch-dependent regulation of atrial peptide synthesis and secretion in cultured atrial cardiocytes. Am J Physiol 1992;263:E239–244.
155. Schiebinger RJ, Linden J. The influence of resting tension on immunoreactive atrial natriuretic peptide secretion by rat atria superfused in vitro. Circ Res 1986;59:105–109.
156. Dietz JR. Release of natriuretic factor from rat heart-lung preparation by atrial distension. Am J Physiol 1984;247:R1093–1096.
157. Rockman HA, Ross RS, Harris AN, Knowlton KU, Steinhelper ME, Field LJ, Ross J Jr, Chien KR. Segregation of atrial-specific and inducible expression of an atrial natriuretic factor transgene in an in vivo murine model of cardiac hypertrophy [published erratum appears in Proc Natl Acad Sci USA 1991 Nov 1;88(21):9907]. Proc Natl Acad Sci USA 1991;88:8277–8281.
158. Ruskoaho H, Tholken H, Lang RE. Increase in atrial pressure releases atrial natriuretic peptide from isolated perfused rat hearts. Pflugers Arch 1986;407:170–174.
159. Mangat H, de Bold AJ. Stretch-induced atrial natriuretic factor release utilizes a rapidly depleting pool of newly synthesized hormone. Endocrinology 1993;133:1398–1403.
160. Schiebinger RJ, Greening KM. Interaction between stretch and hormonally stimulated atrial natriuretic peptide secretion. Am J Physiol 1992;262:H78–83.
161. Jiao JH, Baertschi AJ. Synergistic regulation of ANF in isolated rat hearts. Am J Physiol 1995;268:H1405–1411.
162. Mantymaa P, Vuolteenaho O, Marttila M, Ruskoaho H. Atrial stretch induces rapid increase in brain natriuretic peptide but not in atrial natriuretic peptide gene expression in vitro. Endocrinology 1993;133:1470–1473.
163. Brown LA, Nunez DJ, Brookes CI, Wilkins MR. Selective increase in endothelin-1 and endothelin A receptor subtype in the hypertrophied myocardium of the aorto-venacaval fistula rat. Cardiovasc Res 1995;29:768–774.
164. Laine M, Arjamaa O, Vuolteenaho O, Ruskoaho H, Weckstrom M. Block of stretch-activated atrial natriuretic peptide secretion by gadolinium in isolated rat atrium. J Physiol (Lond) 1994;480:553–561.
165. Sadoshima J, Takahashi T, Jahn L, Izumo S. Roles of mechano-sensitive ion channels, cytoskeleton, and contractile activity in stretch-induced immediate-early gene expression and hypertrophy of cardiac myocytes. Proc Natl Acad Sci USA 1992;89:9905–9909.
166. Bilder GE, Siegl PK, Schofield TL, Friedman PA. Chronotropic stimulation: a primary effector for release of atrial natriuretic factor. Circ Res 1989;64:799–805.
167. Schiebinger RJ, Linden J. Effect of atrial contraction frequency on atrial natriuretic peptide secretion. Am J Physiol 1986;251:H1095–1099.
168. Schiebinger RJ, Li Y, Cragoe EJ Jr. Calcium dependency of frequency-stimulated atrial natriuretic peptide secretion. Hypertension 1994;23:710–716.
169. McDonough PM, Glembotski CC. Induction of atrial natriuretic factor and myosin light chain-2 gene expression in cultured ventricular myocytes by electrical stimulation of contraction. J Biol Chem 1992;267:11,665–11,668.
170. McDonough PM, Stella SL, Glembotski CC. Involvement of cytoplasmic calcium and protein kinases in the regulation of atrial natriuretic factor secretion by contraction rate and endothelin. J Biol Chem 1994;269:9466–9472.
171. Lew RA, Baertschi AJ. Mechanisms of hypoxia-induced atrial natriuretic factor release from rat hearts. Am J Physiol 1989;257:H147–156.
172. Toth M, Vuorinen KH, Vuolteenaho O, Hassinen IE, Uusimaa PA, Leppaluoto J, Ruskoaho H. Hypoxia stimulates release of ANP and BNP from perfused rat ventricular myocardium. Am J Physiol 1994;266:H1572–1580.

173. Bloch KD, Scott JA, Zisfein JB, Fallon JT, Margolies MN, Seidman CE, Matsueda GR, Homcy CJ, Graham RM, Seidman JG. Biosynthesis and secretion of proatrial natriuretic factor by cultured rat cardiocytes. Science 1985;230:1168–1171.
174. Wildey GM, Fischman AJ, Fallon JT, Matsueda GR, Zisfein JB, Preibisch G, Seipke G, Homcy CJ, Graham RM. Cellular processing of pro-atrial natriuretic factor (pro-ANF): studies using an antiserum that selectively binds ANF-(99–126) after its cleavage from pro-ANF. Endocrinology 1988;123:2054–2061.
175. Vuolteenaho O, Arjamaa O, Ling N. Atrial natriuretic polypeptides (ANP): rat atria store high molecular weight precursor but secrete processed peptides of 25-35 amino acids. Biochem Biophys Res Commun 1985;129:82–88.
176. Bloch KD, Seidman JG, Naftilan JD, Fallon JT, Seidman CE. Neonatal atria and ventricles secrete atrial natriuretic factor via tissue-specific secretory pathways. Cell 1986;47:695–702.
177. Irons CE, Sei CA, Glembotski CC. Regulated secretion of atrial natriuretic factor from cultured ventricular myocytes. Am J Physiol 1993;264:H282–285.
178. Somlyo AV, Broderick R, Shuman H, Buhle EL Jr, Somlyo AP. Atrial-specific granules in situ have high calcium content, are acidic, and maintain anion gradients. Proc Natl Acad Sci USA 1988;85:6222–6226.
179. Doubell AF, Bester AJ, Thibault G. Annexins V and VI: major calcium-dependent atrial secretory granule-binding proteins. Hypertension 1991;18:648–656.
180. Thibault G, Charbonneau C, Bilodeau J, Schiffrin EL, Garcia R. Rat brain natriuretic peptide is localized in atrial granules and released into the circulation. Am J Physiol 1992;263:R301–309.
181. Suzuki E, Hirata Y, Kohmoto O, Sugimoto T, Hayakawa H, Matsuoka H, Sugimoto T, Kojima M, Kangawa K, Minamino N, et al. Cellular mechanisms for synthesis and secretion of atrial natriuretic peptide and brain natriuretic peptide in cultured rat atrial cells. Circ Res 1992;71:1039–1048.
182. Ando K, Hirata Y, Emori T, Shichiri M, Kurosawa T, Sato K, Marumo F. Circulating forms of human atrial natriuretic peptide in patients with congestive heart failure. J Clin Endocrin Metab 1990;70:1603–1607.
183. Ito T, Toki Y, Siegel N, Gierse JK, Needleman P. Manipulation of stretch-induced atriopeptin prohormone release and processing in the perfused rat heart. Proc Natl Acad Sci USA 1988;85:8365–8369.
184. Greenwald JE, Needleman P, Siegel N, Tetens E, Biel B, Ritter D. Processing of atriopeptin prohormone by nonmyocytic atrial cells. Biochem Biophys Res Commun 1992;188:644–654.
185. Dube GR, Kuroski-de Bold ML, de Bold AJ. Post-translational processing of atrial natriuretic factor by adult rat atrial cardiocytes in culture. Can J Physiol Pharmacol 1993;71:497–505.
186. Corthorn J, Cantin M, Thibault G. Rat atrial secretory granules and pro-ANF processing enzyme. Mol Cell Biochem 1991;103:31–39.
187. Sei CA, Hand GL, Murray SF, Glembotski CC. The cosecretional maturation of atrial natriuretic factor by primary atrial myocytes. Mol Endocrinol 1992;6:309–319.
188. Shiono S, Nakao K, Morii N, Yamada T, Itoh H, Sakamoto M, Sugawara A, Saito Y, Katsuura G, Imura H. Nature of atrial natriuretic polypeptide in rat brain. Biochem Biophys Res Commun 1986;135:728–731.
189. Shields PP, Sprenkle AB, Taylor EW, Glembotski CC. Rat pro-atrial natriuretic factor expression and post-translational processing in mouse pituitary tumor cells. J Biol Chem 1990;265:10,905–10,911.
190. Gibson TR, Shields PP, Glembotski CC. The conversion of atrial natriuretic peptide (ANP)-(1–126) to ANP-(99–126) by rat serum: contribution to ANP cleavage in isolated perfused rat hearts. Endocrinology 1987;120:764–772.
191. Bloch KD, Zisfein JB, Margolies MN, Homcy CJ, Seidman JG, Graham RM. A serum protease cleaves proANF into a 14-kilodalton peptide and ANF. Am J Physiol 1987;252:E147–151.
192. Imada T, Takayanagi R, Inagami T. Atrioactivase, a specific peptidase in bovine atria for the processing of pro-atrial natriuretic factor. Purification and characterization. J Biol Chem 1988;263:9515–9519.
193. Wypij DM, Harris RB. Atrial granules contain an amino-terminal processing enzyme of atrial natriuretic factor. J Biol Chem 1988;263:7079–7086.
194. Rittenhouse J, Moberly L, Ahmed H, Marcus F. Phosphorylation in situ of atrial natriuretic peptide prohormone at the cyclic AMP-dependent site. J Biol Chem 1988;263:3778–3783.
195. Rittenhouse J, Moberly L, O'Donnell ME, Owen NE, Marcus F. Phosphorylation of atrial natriuretic peptides by cyclic AMP-dependent protein kinase. J Biol Chem 1986;261:7607–7610.
196. Bloch KD, Jones SW, Preibisch G, Seipke G, Seidman CE, Seidman JG. Proatrial natriuretic factor is phosphorylated by rat cardiocytes in culture. J Biol Chem 1987;262:9956–9961.
197. Wildey GM, Fischman AJ, Margolies MN, Graham RM, Homcy CJ. Phosphorylation state of pro-atrial natriuretic factor in rat atrial secretory granules. Endocrinology 1990;127:2839–2848.

198. Gagelmann M, Hock D, Forssmann WG. Relaxation of smooth muscle by cardiodilatin/atrial natriuretic peptide is inhibited by cAMP-dependent phosphorylation. FEBS Lett 1987;225:251–254.
199. Olins GM, Mehta PP, Blehm DJ, Patton DR, Zupec ME, Whipple DE, Tjoeng FS, Adams SP, Olins PO, Gierse JK. Phosphorylation of high- and low-molecular-mass atrial natriuretic peptide analogs by cyclic AMP-dependent protein kinase. FEBS Lett 1987;224:325–330.
200. Kubler D, Reinhardt D, Reed J, Pyerin W, Kinzel V. Atrial natriuretic peptide is phosphorylated by intact cells through cAMP-dependent ecto-protein kinase. Eur J Biochem 1992;206:179–186.
201. Chien KR, Knowlton KU, Zhu H, Chien S. Regulation of cardiac gene expression during myocardial growth and hypertrophy: molecular studies of an adaptive physiologic response. FASEB J 1991;5:3037–3046.
202. Marban E, Koretsune Y. Cell calcium, oncogenes, and hypertrophy. Hypertension 1990;15:652–658.
203. Schunkert H, Dzau VJ, Tang SS, Hirsch AT, Apstein CS, Lorell BH. Increased rat cardiac angiotensin converting enzyme activity and mRNA expression in pressure overload left ventricular hypertrophy. J Clin Inves 1990;86:1913–1920.
204. Ito H, Hiroe M, Hirata Y, Fujisaki H, Adachi S, Akimoto H, Ohta Y, Marumo F. Endothelin $ET_A$ receptor antagonist blocks cardiac hypertrophy provoked by hemodynamic overload. Circulation 1994;89:2198–2203.
205. Arai M, Yoguchi A, Iso T, Takahashi T, Imai S, Murata K, Suzuki T. Endothelin-1 and its binding sites are upregulated in pressure overload cardiac hypertrophy. Am J Physiol 1995;268:H2084–2091.
206. Kovacic-Milivojevic B, Gardner DG. Divergent regulation of the human atrial natriuretic peptide gene by c-jun and c-fos. Mol Cell Biol 1992;12:292–301.
207. Bishopric NH, Jayasena V, Webster KA. Positive regulation of the skeletal alpha-actin gene by Fos and Jun in cardiac myocytes. J Biol Chem 1992;267:25,535–25,540.
208. Kovacic-Milivojevic B, Gardner DG. Fra-1, a Fos gene family member that activates atrial natriuretic peptide gene transcription. Hypertension 1995;25:679–682.
209. Kovacic-Milivojevic B, Wong VS, Gardner DG. Selective regulation of the atrial natriuretic peptide gene by individual components of the activator protein-1 complex. Endocrinology 1996;137:1108–1117.
210. Chiu R, Angel P, Karin M. Jun-B differs in its biological properties from, and is a negative regulator of, c-Jun. Cell 1989;59:979–986.
211. Schutte J, Viallet J, Nau M, Segal S, Fedorko J, Minna J. jun-B inhibits and c-fos stimulates the transforming and trans-activating activities of c-jun. Cell 1989;59:987–997.
212. Deng T, Karin M. JunB differs from c-Jun in its DNA-binding and dimerization domains, and represses c-Jun by formation of inactive heterodimers. Genes Dev 1993;7:479–490.
213. Steinhelper ME, Cochrane KL, Field LJ. Hypotension in transgenic mice expressing atrial natriuretic factor fusion genes [see comments]. Hypertension 1990;16:301–307.
214. John SW, Krege JH, Oliver PM, Hagaman JR, Hodgin JB, Pang SC, Flynn TG, Smithies O. Genetic decreases in atrial natriuretic peptide and salt-sensitive hypertension. Science 1995;267:679–681.
215. Lopez MJ, Wong SK, Kishimoto I, Dubois S, Mach V, Friesen J, Garbers DL, Beuve A. Salt-resistant hypertension in mice lacking the guanylyl cyclase-A receptor for atrial natriuretic peptide. Nature 1995;378:65–68.

# 6 Cellular and Molecular Biology of B-Type Natriuretic Peptide

*Christopher C. Glembotski*

*CONTENTS*



## TISSUE-RESTRICTED EXPRESSION OF NATRIURETIC PEPTIDES

The three members of the natriuretic peptide (NP) family, A-, B-, and CNP, are structurally related hormones that serve to decrease plasma volume and blood pressure. Both ANP and BNP appear to be potent natriuretic and diuretic hormones that are also capable of decreasing the activity of the renin/angiotensin/aldosterone system and also possess vasodilatory activities. In contrast, CNP appears to possess more potent vasodilatory activity and less potent diuretic and natriuretic effects than ANP and BNP *(1–6)*.

The NPs are encoded by three genes that are expressed in a tissue-restricted manner. The heart displays the greatest levels of ANP and BNP expression, whereas CNP is virtually undetectable in the heart *(3,7–9)*. Recent run-off transcription analyses in our laboratory have shown that primary neonatal rat ventricular myocytes display high basal- and serum-inducible levels of ANP and BNP transcription; however, CNP transcription is essentially undetectable (Fig. 1). Thus, it is apparent that in cardiac myocytes differential transcription probably serves a major role in determining the differential expression of the three NP genes. The ANP and BNP are also expressed in the central nervous system (CNS), particularly in the hypothalamus/thalamus, septum, midbrain, pons/medulla, and olfactory bulbs *(10)*. Expression levels of ANP and BNP in the CNS are much lower than in the heart. In contrast, CNP is expressed in higher levels in the CNS than in the periphery, leading some to postulate a role for CNP primarily as a neurotransmitter or neuromodulator.

From: *Contemporary Endocrinology: Natriuretic Peptides in Health and Disease*
Edited by: W. K. Samson and E. R. Levin Humana Press Inc., Totowa, NJ

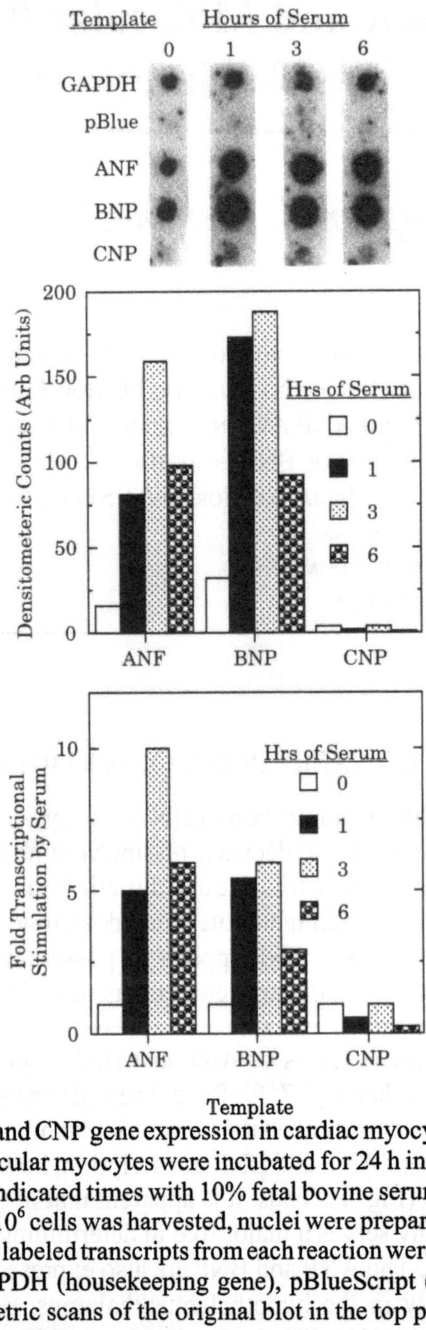


**Fig. 1.** Analyses of A-, B-, and CNP gene expression in cardiac myocytes by run-off transcription. Primary rat neonatal ventricular myocytes were incubated for 24 h in serum-free medium with no additions and then for the indicated times with 10% fetal bovine serum to induce transcription. At each time a culture of $25 \times 10^6$ cells was harvested, nuclei were prepared and used in a nuclear run-off transcription assay. The labeled transcripts from each reaction were used to probe immobilized, denatured cDNAs for GAPDH (housekeeping gene), pBlueScript (plasmid control), rat ANF, BNP, and CNP. Densitometric scans of the original blot in the top panel were used to obtain the graphical data shown in the middle panel. When the densitometric results in the middle panel were all normalized to the respective A-, B-, or CNP values at each time of serum treatment, the fold induction of transcription of each gene by serum was obtained and is shown in the bottom panel.

The NPs have been found outside the heart and the brain, although the levels of expression in these extra cardiac and extra-CNS regions are much lower. For example, CNP expression has been characterized in the adrenals, kidneys, lymph nodes, spleen,

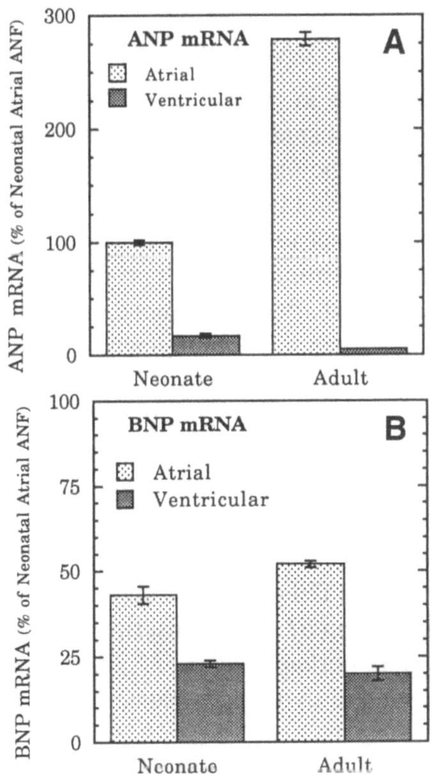


**Fig. 2.** Northern analyses of ANP and BNP mRNA levels in neonatal and adult rat atria and ventricles. Extracts of 1-d-old (neonate) or 3-mo-old (adult) atria and ventricles were prepared and triplicate samples of the resulting RNA was fractionated on a formaldehyde gel. After transfer, the blots were probed for BNP and ANP and after exposure of the blots to phosphorimager screens for 18 h, the relative band intensities were analyzed using the Molecular Dynamics Image Quant analysis program, as described *(33)*. The mean values of triplicate samples are plotted with SEMs. All densitometric data are normalized to neonatal atrial ANP as the 100% value.

and in certain bone cell progenitors *(10–13)*. Also, ANP and BNP have been found in the thyroid gland, lung, and ovaries *(10)*.

In addition to displaying tissue-restricted expression, NP expression in the heart is also regulated in a chamber-specific manner. For example, embryonically, ANP is expressed in both the atria and ventricles. However, in the adult rat, ventricular ANP expression decreases to approx 10% the levels in the neonate, whereas atrial ANP expression increases by 300% over this same time period *(14)* (Fig. 2). Although developmental time courses of expression have not been published for BNP and CNP, it is apparent that BNP is expressed very early in cardiac development, perhaps as soon as embryonic d 8 in the mouse embryo *(15)*. Interestingly, the level of BNP expression in the neonatal rat heart is similar to the expression levels in the adult rat heart. Moreover, whereas ANP expression is always dramatically higher in the atria than the ventricles, atrial BNP expression is only about twofold higher than ventricular BNP expression (Fig. 2), leading to the hypothesis of differential developmental regulation of ANP and BNP expression.

## REGULATION OF NATRIURETIC PEPTIDE TRANSCRIPTION

### *ANP Induction*

It is believed that ANP levels in cardiac myocytes are regulated primarily at the transcriptional level. A variety of studies have addressed the myriad of hormonal and physical treatments, as well as pathological conditions that result in the induction of ANP transcription in primary ventricular myocytes *(4,6,16)*. This ANP induction appears to take place coordinately with the activation of other cardiac genes that are stimulated as part of the myocardial cell hypertrophic growth program *(16)*. Several hormones and growth factors (e.g., glucocorticoids, thyroid hormone, endothelin, norepinephrine, basic-FGF, TGF-β, and several prostaglandins) can augment myocardial cell growth as well as ANP and ANP mRNA levels *(17–23)*. Analyses of the signaling mechanisms responsible for ANP induction in response to such treatments indicate that many of these compounds activate mitogen-activated protein kinase (MAPK), protein kinase C (PKC), and/or calcium calmodulin kinase (CaMK) signaling pathways, and that the effects of the activated kinases converge on several transcriptional enhancing factors, such as serum-response factor (SRF) *(24)*, activator protein-1 (AP-1; fos/jun) *(25)*, SP1-like proteins *(26)*, transcription-enhancement factor-1 (TEF-1) *(27)*, and/or other as of yet unidentified proteins. In the case of the induction of ANP transcription, it has been shown that SRF *(24)*, SP-1-like proteins, and, perhaps, AP-1 *(25,26)* bind to specific regions of the ANP 5'-flanking sequence and in so doing, enhance ANP transcription.

### *BNP Induction*

Since BNP was discovered more recently than ANP, less is known about how its expression is regulated. However, initial studies using phorbol ester and DAG indicated that like ANP, BNP mRNA levels increase in response to PKC activation in primary cardiac myocytes *(28,29)*. Recently, 2.5 kb of the rat BNP 5'-flanking sequence (FS) was cloned *(30)*, and using chimeric BNP/luciferase reporter constructs, it was shown that BNP promoter activity is inducible by phorbol esters (PDBu), Serum (Ser), or $\alpha_1$-adrenergic agonists, such as phenylephrine (PE), which is in part consistent with run-off results (e.g., *see* Fig. 1), and consistent with roles for signaling pathways involving the activation of PKC, MAPK, and/or CaMK. Subsequent studies of the rat BNP promoter have shown that although there are multiple *cis*-elements within 2500 bp of the BNP 5'-FS that can mediate inducible BNP transcription, one region that binds the cardiac transcription enhancement factor-1, TEF-1, is absolutely required for BNP transcription *(31)*. Although TEF-1 does not appear to be involved in regulating ANP transcription, it has been shown to be important for PKC-mediated induction of β-MHC and α-skeletal actin transcription in primary cardiac myocytes *(27)*, providing a potential link by which several cardiac genes can be coordinately activated.

## BNP AND ANP ARE INDUCED WITH DIFFERENT KINETICS

### *BNP Is Induced as a Primary Response Gene*

In more recent studies of BNP induction it has been found that unlike the ANP mRNA, which is quite stable, the BNP mRNA possesses a half-life of approx 60 min *(33)*. Many other short-lived transcripts are encoded by primary-response genes (PRGs), also called immediate-early genes (IEGs) *(32)*. The PRGs are defined as genes encoding transcripts that can be induced very quickly by growth stimuli in a protein synthesis-independent manner. Secondary-response genes (SRGs), also called late-response genes, encode transcripts that are induced later in a protein synthesis-requiring manner. Further studies showed that in cultured myocardial cells, $\alpha_1$-adrenergic agonists increase BNP mRNA levels by at least threefold within as little as 1 h of application, and that this early induction is independent of protein synthesis, both of which are characteristics typical of PRGs. This contrasted with the ANP mRNA, which was increased after 8 h of $\alpha_1$-adrenergic agonist exposure, an event that required protein synthesis *(33)*. Thus, whereas BNP and ANP are both induced by the same agonists, the kinetics of induction differs such that BNP can be induced as a PRG and ANP as a SRG.

### *Possible Roles for BNP as a Primary Response Gene*

This difference of induction kinetics implies that in part, BNP and ANP may have different biological roles, with BNP serving a function separate from its known hemodynamic roles. Many other PRGs code for ubiquitous transcription factors (e.g., *fos*, *jun*, *egr-1*) commonly involved in the initial nuclear responses to growth promoters. However, recently it has been found that other genes, some of which encode secreted cytokines (e.g., plasminogen activator *[34]*, tissue factor *[35]*, parathyroid-hormone-related peptide *[36]*, brain-derived neurotrophic factor *[37]*) are PRGs. Although the purpose underlying the induction of secreted proteins as PRGs is presently unclear, it is possible that like some PRG-encoded transcription factors, PRG-encoded secreted proteins may serve to propagate growth activation signals by ultimately controlling secondary response genes in neighboring cells. The rapid expression and release of BNP from cardiac myocytes in response to hormones (e.g., catecholamines, endothelin-1) that can function as vascular smooth- or endothelial cell mitogens might serve a modulatory role. Indeed, several recent studies have suggested that BNP may actually modulate the growth-promoting effects of several hormones. For example, BNP inhibits PDGF-stimulated proliferation of mesangial cells *(38)*, and it inhibits the biosynthesis and release of endothelin-1 from endothelial cells *(39)*, which would presumably inhibit the known growth-promoting effects of the peptide on endothelial, vascular smooth muscle, and myocardial cells *(40)*.

## MECHANISM OF BNP INDUCTION

### *BNP Transcription*

To date, we have found that all stimuli that induce ANP also induce BNP, the difference being that BNP is induced as a PRG/SRG and ANF as an SRG. This differential of induction kinetics is likely the result of distinct responses of the two genes to signaling pathways activated by the inducers. How could the ANP and BNP genes differentially respond to identical signal transduction pathways? The induction of both ANP and BNP must begin with the activation of specific signaling cascades; for example, previous studies have indicated that many inducers of ANP in cardiac myocytes activate several kinase families, including MAPK, PKC, and CaMK *(42–44)*. Accordingly, it can be proposed that these pathways can influence BNP and ANP gene induction in ways that result in their activation as a PRG and SRG, respectively. For example, these signaling pathways might enhance transcription very early for BNP, in comparison to many other PRGs, whereas ANP transcriptional induction would take place at a later time, consistent

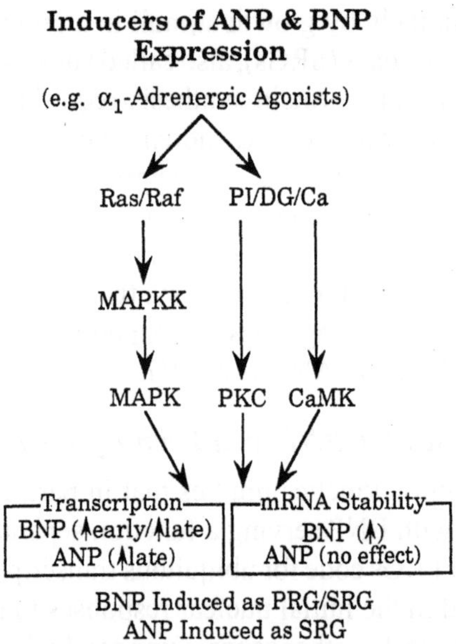


**Fig. 3.** Simplified diagram of a hypothetical mechanism for the hormonal induction of the cardiac natriuretic peptides. Some ANP and BNP inducers (e.g., $\alpha_1$-adrenergic agonists) are known to activate the MAPK, PKC, and CaMK pathways in cardiac myocytes. It is hypothesized that these kinases regulate ANP and BNP transcription and, perhaps, BNP mRNA stability. PI = phosphatidyl inositide hydrolysis; DG = diacylglycerol; MAPKK = mitogen activated protein kinase kinase; MAPK = mitogen activated protein kinase; PKC = protein kinase C; CaMK = Calcium/calmodulin kinase II.

with its SRG character (Fig. 3). Whereas extensive studies have been carried out to characterize the transcriptional induction of ANP, little is known about the timing of ANP or BNP transcriptional induction. The mechanism of the rapid BNP transcriptional induction likely involves known factors (e.g., fos, jun, and egr-1) interacting with consensus AP-1, AP-2, Egr-1, and SRE sites we have identified in the 5'-flanking region of the gene *(44)*.

### *BNP mRNA Stabilization*

In addition to rapid transcriptional induction, a mechanism for rapidly increasing the levels of a short-lived transcript, such as BNP, is to decrease its degradation rate. Accordingly, it is possible that inducers of BNP expression may enhance BNP mRNA half-life. We tested this hypothesis by treating myocardial cell cultures with a transcription inhibitor and then maintaining them for various times in the presence or absence of an inducer, the $\alpha_1$-adrenergic agonist, phenylephrine (PE). In the presence of PE, the rate of degradation of the BNP mRNA decreased by at least fivefold, from a half-life of approx 1 h without PE, to a half-life of approx 5 h with PE *(33,45)*. Thus, it appears as if the activation of $\alpha_1$-adrenergic receptors on cardiac myocytes can lead to the stabilization of the BNP mRNA, an event that likely contributes to its unusual induction kinetics.

Although the mechanism for regulating mRNA degradation in cardiac myocytes has not been studied, studies in other cells have revealed the control of mRNA stability as an important gene-regulatory mechanism *(46–48)*. The transcripts that usually display such regulation are those with short half-lives. Levels of unstable transcripts can be augmented by increases in transcription alone, coordinate changes in transcription and stability, or

**ANP 3'-UTR**

```
TAACAGCCAAATCTGCTCGAGCAGATCGCAAAAGATCCCAAGCCTG
CGGTGTGTCACACAGCTTGGTCGCATTGCCACTGAGAGGTGGTGAA
TACCCTCCTGGAGCTGCAGCTTCCTGTCTTCATCTATCACGATCGA
TGTTAAGTGTAGATGAGTGGTTTAGTGAGGCCTTACCTCTCCCACT
CTGCATATTAAGGTAGATCCTCACCCCTTTCAGAAAGCAGTTGGAA
AAAAATAAATCCGAATAAACTTCAGCACCACGGACAGACGCTGAGG
CCTGGC
```

**BNP 3'-UTR**

```
TAGGAAGACCTCCTGGCTGCAGACTCCGGCTTCTGACTCTGCCTGC
GGCTCTTCTTTCCCCAGCTCTGGGACCACCTCTCAAGTGATCCTGT
TTATTTATTTGTTTATTTATTTATTTTTATGTTGCTGATTTTCTAC
AAGACTGTTTCTTATCTTCCAGCACAAACTTGCCACAGTGTAATAA
ACATAGCCTATTTCTTGCTTTTGG
```

**Fig. 4.** Sequences of the rat ANP and BNP 3'-untranslated regions. The cDNAs of rat ANP and BNP between the termination codon (underlined) and the poly-adenylylation signal (bold underlined) are shown. Boxed = AU-rich elements (AREs; consensus = TTATTTATT) in the BNP 3'-UTR only.

by changes in stability alone *(46,49,50)*. Although it remains unclear in many cases how transcript stability is hormonally modulated, there have been significant studies of the regions of mRNAs that contribute to regulating transcript degradation.

The 5' and 3'-untranslated regions (UTRs), the CAP site, the coding region, and the poly-A tail are each thought to contribute to regulating mRNA stability *(48)*. Among the most thoroughly studied of these structures are the 3'-UTRs. Most notable is the presence of AU-rich elements (AREs = UUAUUUAUU) in the 3'-UTRs of many unstable transcripts, e.g., *fos*, *jun*, *myc*; ARE's have been shown to be critical for controlling the stability of these mRNAs. Unlike ANP, the 3'-UTR of the BNP transcript possesses several consensus destabilization sequences (Fig. 4), consistent with the long- and short half-lives of the ANP and BNP mRNAs, respectively. Thus, it seems probable that the AREs in the BNP 3'-UTR confer instability to that transcript; however, it is unclear whether these regions could be involved in agonist-mediated stabilization of the BNP mRNA.

## *Themes for Hormonal Regulation of Transcript Stability*

A brief analysis of several studies of hormonal regulation of transcript stability and the involvement of the 3'-UTR of mRNAs in noncardiac myocytes reveals three major themes of agonist-regulated mRNA turnover that might apply to BNP mRNA regulation.

### THEME 1

Theme 1 involves ARE-binding proteins (ARE-BPs); in some cases the binding of such ARE-BPs, which are presently uncharacterized, is directly correlated with mRNA stability, whereas in others it is inversely correlated. The first case is typified by a study showing that ARE-BPs that bind to the 3'-UTR of the β-amyloid protein precursor (APP) mRNA in several tumor-cell types are absent under noninducing conditions, and that their lack of binding leads to a short APP mRNA half-life *(51)*. However, when the cells are mitogen-induced, ARE-BP binding activities increase and the transcript becomes more stable. The second case is typified by studies of DDT1 cells where β-adrenergic receptor (AdR) agonists induce the binding of an ARE-BP to the 3'-UTR of the β-AdR mRNA, which then causes decreased transcript stability and receptor downregulation

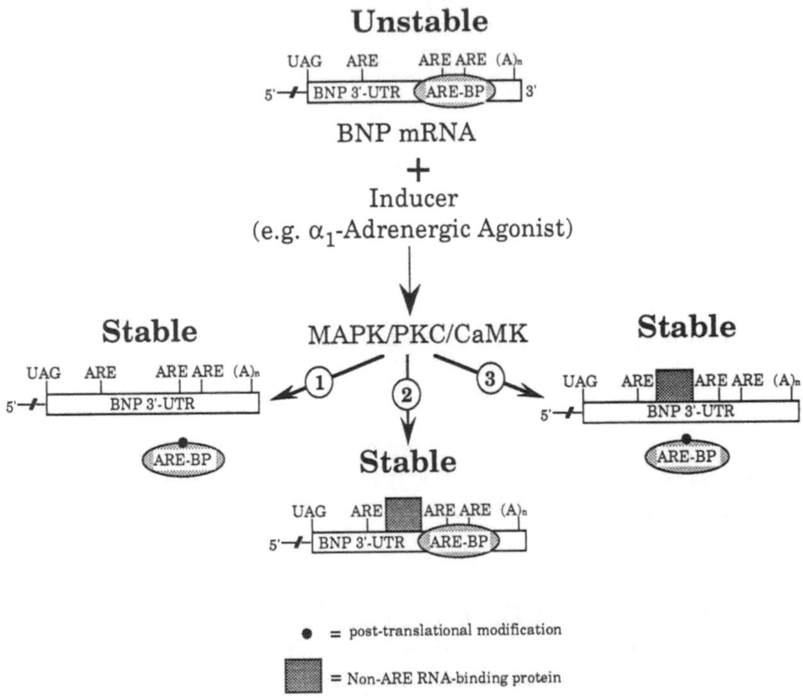


**Fig. 5.** Possible mechanisms of how the 3'-UTR could confer agonist-mediated BNP mRNA stabilization. The BNP 3'-UTR is shown as are the approximate locations of the termination codon (UAG), the AU-rich elements (ARE) and the polyadenylylation signal ([A]n). Also shown are the three themes discussed in the text as possible ways that inducers might increase BNP mRNA stability. Note that in Themes 1 and 3, inducers result in a posttranslational modification of ARE-BPs, which causes a decreased affinity for the ARE in the BNP 3'-UTR. It is also possible that physical interaction of ARE-binding proteins (BP) with the 3'-UTR may be inversely related to transcript stability such that inducers enhance ARE-BP association (not shown). Note that in Theme 2, agonist treatment results in the increased binding of a non-ARE-BP, which contributes to stabilizing the transcript.

*(46)*. Thus, treatments that stabilize the BNP mRNA may first augment certain signaling pathways that would then alter (increase or decrease) the binding of proteins to AREs in the 3'-UTR, an event that could increase transcript half-life (Fig. 5, pathway 1).

### Theme 2

Theme 2 involves the binding of proteins to non-ARE regions of the 3'-UTRs of unstable transcripts (i.e., non-ARE-BPs). Some of the non-ARE-BPs are cell- or transcript-specific, whereas others are ubiquitous. For example, in 3T3 fibroblasts, stabilization of the ribonucleotide reductase R2 mRNA by TGF-β is caused by binding of a novel, 75-kDa TGF-β-inducible protein to a unique 83 nt region of the 3'-UTR and not to the 3'-UTRs of any other unstable transcripts *(52)*. More recently a series of studies have shown that a novel factor from pituitary cells can bind specifically to a 12-nucleotide region of the TSHβ mRNA 3'-UTR in a T3-regulated manner to destabilize that transcript *(54)*. Thus, treatments that stabilize the BNP mRNA might alter the binding of proteins to non-ARE regions in the 3'-UTR, which could increase transcript half-life (Fig. 5, pathway 2).

**THEME 3**

Theme 3 involves non-ARE-BPs that have been shown to collaborate with ARE-BPs in mediating transcript stabilization in response to hormonal treatment *(55)* (Fig. 5, pathway 3).

## SIGNALS MEDIATING HORMONAL STABILIZATION OF THE BNP MRNA

Since transcript stability studies have not been carried out in cardiac myocytes, it is unclear which signal transduction pathways are involved in the modulation of BNP mRNA degradation by the activation of cell-surface receptors. However, studies in other cell types have implicated the involvement of pathways known to be activated during BNP induction in cardiac myocytes. For example, the PKC inhibitors, staurosporine and calphostin, were used to show that TNF induction of IL-1 mRNA in fibroblasts is in part caused by PKC-mediated transcript stabilization *(56)*. In another study in MCF-7 cells, the estrogen receptor mRNA was destabilized by phorbol esters, also implying the involvement of PKC *(57)*. In DDT1 cells, isoproterenol, a β-adrenergic agonist, destabilizes the β-AdR mRNA; this effect can be replicated with cAMP, implying the involvement of PKA *(46,58,59)*. Very recently, elegant studies in pituitary cells have shown that thyroid hormone regulates the stability of the TSHβ mRNA *(53)*. So, although there is some precedent for the involvement of the PKC, MAPK, and CaMK families in hormone-regulated transcript stability, much remains to be learned about this fascinating level of gene regulation in cardiac myocytes as well as many other cell types.

## CONCLUSIONS

Although a great deal is known about physiological activities of the natriuretic peptides, as well as their sites of synthesis and action, with the exception of ANP, surprisingly little is known about the cellular and molecular details of their biosynthesis and regulated, tissue-restricted expression in cardiac and extracardiac neuroendocrine tissues. Moreover, although it might have been believed by some that BNP is just another natriuretic peptide expressed in the heart via the same mechanisms of ANP, recent findings indicate that the regulated expression of these two peptide genes, although shared in some respects, also possess unique characteristics. Indeed, the finding that hormone-induced BNP expression is enhanced as much as 8 h sooner than ANP expression leads to the provocative theory that in part, BNP, like other IEG or PRG products, might serve a function quite different from ANP. Recent studies of the roles of the natriuretic peptides as regulators of cardiac growth and gene expression add to this theory, suggesting that in addition to facilitating a rapid endocrine response to hemodynamic change, the early production of BNP might serve a paracrine or autocrine function in the heart itself. Future studies of the locations of natriuretic peptide receptors in the heart, as well as the fate of all portions of pro-BNP after its synthesis, will provide information required to validate such hypothetical intracardiac roles of the natriuretic peptides as regulators of myocardial cell growth and function.

## ACKNOWLEDGMENTS

The author wishes to acknowledge Donna Thuerauf for performing the run-off transcription assay shown in Fig. 1 and Deanna Hanford for the Northern analyses shown in

Fig. 2. Work in the author's laboratory is supported in part by National Institutes of Health Grants NS-25073 and HL-56861.

## REFERENCES


1. DeBold AJ, Borrenstein HB, Veress AT, Sonnenberg H. A rapid and potent natriuretic response to intravenous injection of atrial myocardial extract in rats. Life Sci 1981;28:89–94.
2. Sudoh T, Kangawa K, Minamino N, Matuso H. A new natriuretic peptide in porcine brain. Nature 1988;332:78–81.
3. Sudoh T, Minomino N, Kangawa K, Matsuo H. C-type natriuretic peptide (CNP): a new member of the natriuretic peptide family identified in porcine brain. Biochem Biophys Res Commun 1990; 168:863–870.
4. Rosenzweig A, Seidman CE. Atrial natriuretic factor and related peptide hormones. Ann Rev Biochem 1991;60:220–255.
5. Brenner BM, Ballerman BJ, Gunning ME, Zeidel ML. Diverse biological actions of atrial natriuretic peptide. Physiol Rev 1990;70:665–699.
6. Ruskoaho H. Atrial natriuretic peptide: synthesis, release and metabolism. Pharm Rev 1992;44: 479–612.
7. Seidman CE, Bloch KD, Klein KA, Smith JA, Seidman JG. Nucleotide sequence of the human and mouse atrial natriuretic factor genes. Science 1984;226:1206–1209.
8. Zivin RA, Condra JH, Dixon RAF, Seidah NG, Chretien M, Nemer M, Chamberland M, Drouin J. Molecular cloning and characterization of DNA sequences encoding rat and human atrial natriuretic factors. Proc Natl Acad Sci USA 1984;81:6325–6329.
9. Minamino N, Makino Y, Tateyama H, Kangawa K, Matsuo H. Characterization of immunoreactive human C-type natriuretic peptide in brain and heart. Biochem Biophys Res Commun 1991;179:535–542.
10. Minamino N, Aburaya M, Kojima M, Miyamoto K, Kangawa K, Matsuo H. Distribution of C-type natriuretic peptide and its messenger RNA in rat central nervous system and peripheral tissue. Biochem Biophys Res Commun 1993;197:326–335.
11. Vollmar AM, Gerbes AL, Nemer M, Schulz R. Detection of C-type natriuretic peptide (CNP) transcript in the rat heart and immuno organs. Endocrinology 1993;132:1872–1874.
12. Totsune K, Takahashi K, Murakami O, Satoh F, Sone M, Saito T, Sasano H, Mouri T, Abe K. Natriuretic peptides in the human kidney. Hypertension 1994;24:758–762.
13. Totsune K, Takahashi K, Murakami O, Itoi K, Sone M, Ohneda M, Satoh F, Miura Y, Mouri T. Immunoreactive C-type natriuretic peptide in human adrenal glands and adrenal tumors. Peptides 1994;15:287–290.
14. Zeller R, Bloch KD, Williams BS, Arceci RJ, Seidman CE. Localized expression of the atrial natriuretic factor gene during cardiac embryogenesis. Genes Dev 1987;1:693–698.
15. Grepin C, Dagnino L, Robitaille L, Haberstroh L, Antakly T, Nemer M. A hormone-encoding gene identifies a pathway for cardiac but not skeletal muscle gene transcription. Mol Cell Biol 1994;14:3115–3129.
16. Chien KR, Zhu H, Knowlton KU, Miller-Hance W, van-Bilsen M, O'Brien TX and Evans SM. Transcriptional regulation during cardiac growth and development. Annu Rev Physiol 1993;55:77–95.
17. Argentin S, Sun YL, Lihrmann I, Schmidt TJ, Drouin J, Nemer M. Distal cis-acting promoter sequences mediate glucocorticoid stimulation of cardiac ANF gene transcription. J Biol Chem 1991;266:23,315–23,322.
18. Shields PP, Dixon JE, Glembotski CC. The secretion of ANF-(99–126) by cultured cardiac myocytes is regulated by glucocorticoids. J Biol Chem 1988;263:12,619–12,628.
19. Gardner DG, Gertz BJ, Hane S. Thyroid hormone increases rat ANP mRNA accumulation in vivo and in vitro. Mol Endo 1987;1:260–265.
20. Shubeita HE, McDonough PM, Harris AN, Knowlton KU, Glembotski CC, Brown JH, Chien KR. ET induction of PI hydrolysis, sarcomere assembly, and cardiac gene expression in ventricular myocytes. J Biol Chem 1990;265:20,555–20,562.
21. Sei CA, Irons CE, Sprenkle AB, McDonough PM, Brown JH, Glembotski CC. Alpha-adrenergic regulation of ANF message and peptide levels in primary ventricular myocardial cells: potential roles for calcium, CaM, and protein kinases. J Biol Chem 1991;266:15,910–15,916.
22. Knowlton KU, Ross RS, Baracchini E, Harris AN, Henderson SA, Evans SM, Glembotski CC, and Chien KR. Co-regulation of the rat cardiac MLC-2 and ANF genes during α-adrenergic stimulation of neonatal myocardial cells. J Biol Chem 1991;266:7759–7768.

23. Parker TG, Chow K-L, Schwartz RJ Schneider MD. Differential regulation of skeletal α-actin transcription in cardiac muscle by two FGFs. Proc Natl Acad Sci USA 1990;87:507–514.
24. Sprenkle AB, Murray SF, Glembotski CC. Involvement of multiple cis-elements in basal- and α-adrenergic agonist-inducible ANF transcription: roles for serum response elements and an SP-1-like element. Circ Res 1995;77:1060–1069.
25. Kovacic-Milivojevic B, Gardner DG. Divergent regulation of the human ANP gene by c-Jun and c-Fos. Mol Cell Biol 1992;12:292–301.
26. Ardati A, Nemer M. A nuclear pathway for $\alpha_1$-adrenergic receptor signaling in cardiac cells. EMBO J 1993;12:5131–139.
27. Kariya K, Farrance IKG, Simpson PC. TEF-1 in cardiac myocytes interacts with an $\alpha_1$-adrenergic and β-PKC-inducible element in the rat α-MHC promoter. J Biol Chem 1993;268:26,658–26,662.
28. LaPointe MC, Sitkins JR. Phorbol ester stimulates the synthesis and secretion of brain natriuretic peptide from neonatal rat ventricular cardiocytes: comparison with regulation of atrial natriuretic factor. Mol Endo 1993;7:1284–1296.
29. Suzuki E, Hirata Y, Kohmoto O, Sugimoto T, Hayakawa H, Matsuoka H, Suigomoto T, Kojima M, Kangawa K, Minamino N, and Matsuo H. Cellular mechanisms for synthesis and secretion of a trial natriuretic peptide and brain natriuretic peptide in cultured rat atrial cells. Circ Res 1992;71:1039–1048.
30. Thuerauf DJ, Hanford DS, Glembotski CC. Regulation of rat BNP transcription: A potential role for the transcription factor GATA-4 in cardiac myocyte gene expression. J Biol Chem 1994;269: 17,772–17,775.
31. Thuerauf DJ, Glembotski CC. Differential effects of protein kinase C, ras and raf-1 kinase on the induction of the cardiac B-type natriuretic peptide gene through a critical promoter-proximal M-CAT element. J Biol Chem 1997;272:7464–7472.
32. Herschman HR. Primary response genes induced by growth factors and tumor promoters. Annu Rev Biochem 1991;60:281–319.
33. Hanford DS, Thuerauf DJ, Murray SF, Glembotski CC. Brain natriuretic peptide is induced by $\alpha_1$-adrenergic agonists as a primary response gene in cultured rat cardiac myocytes. J Biol Chem 1994;269:26,227–26,233.
34. Qian Z, Gilbert ME, Colicos MA, Kandel ER, Kuhl D. Tissue-plasminogen activator is induced as an immediate-early gene during seizure, kindling and long-term potentiation. Nature 1993;361:453–457.
35. Cui MZ, Parry GC, Edgington TS, Mackman N. Regulation of tissue factor gene expression in epithelial cells. Induction by serum and phorbol 12-myristate 13-acetate. Arterioscler Thromb 1994;14:807–814.
36. Holt EH, Lu C, Dreyer BE, Dannies PS, Broadus AE. Regulation of parathyroid hormone-related peptide gene expression by estrogen in GH4C1 rat pituitary cells has the pattern of a primary response gene. J Neurochem 1994;62:1239–1246.
37. Hughes P, Beilharz E, Gluckman P, Drugnow M. Brain-derived neurotrophic factor is induced as an immediate early gene following N-methyl-D-aspartate receptor activation. Neuroscience 1993; 57:319–328.
38. Kohno M, Ikeda M, Johchi M, Horio T, Ysunari K, Kurihara N, Takeda T. Interaction of PDGF and natriuretic peptides on mesangial cell proliferation and endothelin secretion. Am J Physiol 1993;265:E673–E679.
39. Emori T, Hirata Y, Imai T, Eguchi S, Kanno K, Marumo F. Cellular mechanism of natriuretic peptides-induced inhibition of endothelin-1 biosynthesis in rat endothelial cells. Endocrinology 1993;133:2474–2480.
40. Itoh H, Pratt RE, Ohno M, Dzau VJ. Atrial natriuretic polypeptide as a novel antigrowth factor of endothelial cells. Hypertension 1992;19:758–761.
41. Boygoyevitch MA, Clerk A, Sugden P. Activation of the MAPK cascade by pertussis toxin-sensitive and -insensitive pathways in cultured ventricular cardiocytes. Biochem J 1995;309:437–443.
42. Irons CE, Sei CA, Hidaka H, Glembotski CC. PKC and CaMK are required for ET-stimulated ANF secretion from primary atrial myocytes. J Biol Chem 1992;267:5211–5216.
43. Thorburn J, Frost JA, Thorburn A. MAPKs mediate changes in gene expression, but not cytoskeletal organization associated with cardiac muscle cell hypertrophy. J Cell Biol 1994;126:1565–1572.
44. Thuerauf DJ, Hanford DS, Glembotski CC. Regulation of rat BNP transcription: a potential role for the transcription factor GATA-4 in cardiac myocyte gene expression. J Biol Chem 269: 1994;17,772–17,775.

45. Hanford DS, Glembotski CC. Stabilization of the B-type natriuretic peptide mRNA in cardiac myocytes by α adrenergic receptor activation: potential roles for protein kinase C and mitogen-activated protein kinase. Mol Endocrinol 1996; 10:1719–1727.
46. Huang L-Y, Tholanikunnel BG, Vakalopoulou E, Malbon CC. The Mr 35,000 β-adrenergic receptor mRNA-binding protein induced by agonists requires both an AUUUA pentamer and U-rich domains for RNA recognition. J Biol Chem 1993;268:25,769–25,775.
47. Schiavi SC, Belasco JG, Greenberg ME. Regulation of proto-oncogene mRNA stability. Biochem Biophys Acta 1992;1114:95–106.
48. Sachs AB. Messenger RNA degradation in eukaryotes. Cell 1993;4:413–421.
49. Nielsen DA, Shapiro DJ. Insights into hormonal control of mRNA stability. Mol Endo 1990; 7:953–956.
50. Chen M, Schnermann J, Smart AM, Brosius FC, Killen PD, Briggs JP. cAMP selectively increases renin mRNA stability in cultured JG Cells. J Biol Chem 1993;268:24,138–24,144.
51. Zaidi SHE, Denman R, Malter JS. Multiple proteins interact at a unique cis-element in the 3'-untranslated region of amyloid precursor protein mRNA. J Biol Chem 1994;269:24,000–24,006.
52. Amara FM, Chen FY, Wright JA. A novel TGF-β responsive cytoplasmic trans-acting factor binds selectively to the 3'-UTR of mammalian ribonucleotide reductase R2 mRNA: role in message stability. Nucleic Acids Res 1993;21:4803–4809.
53. Leedman PJ, Stein AR, Chin WR. Regulated specific protein binding to a conserved region of the 3'-UTR of thyrotropin β-subunit mRNA. Mol Endo 1995;9:375–387.
54. Zubiaga AM, Belasco JG, Greenberg ME. The nonamer UUAUUUAUU is the key AU-rich sequence motif that mediates mRNA degradation. Mol Cell Biol 1995;15:2219–2230.
55. Williams DL, Sensel M, McTigue M, Binder R. Hormonal and developmental regulation of mRNA turnover. In: Belasco JG, Brawerman G, eds. Control of mRNA Stability. Academic, San Diego, 1993; pp. 161–197.
56. Gorospe M, Kumar S, Balioni C. TNF increases stability of IL-1 mRNA by activating PKC. J Biol Chem 1993;268:6214–6220.
57. Aceda M, Knabbe C, Dickson RB, Lippman ME, Bronzert, D, Lindsey, RK, Gottardis, MM, Martin, MB. Post-transcriptional destabilization of estrogen receptor mRNA in MCF-7 cells by 12-O-tetradecanoylphorbol-13-acetate. J Biol Chem 1991;266:17,808–17,814.
58. Port JD, Huang L-Y, Malbon CC. β-Adrenergic agonists that down-regulate receptor mRNA up-regulate a Mr 35,000 protein(s) that selectively binds to β-adrenegic receptor mRNAs. J Biol Chem 1992;267:24,103–24,108.
59. Tholanikunnel BG, Granneman JG, Malbon CC. The Mr 35,000 β-adrenergic receptor mRNA-binding protein binds transcripts of G-protein-linked receptors which undergo agonist-induced destabilization. J Biol Chem 1995;270:12,787–12,793.

# 7 Cellular and Molecular Aspects of C-Type Natriuretic Peptide (CNP)

*Hiroshi Itoh, Shinichi Suga,*
*Yoshihiro Ogawa, Yasato Komatsu,*
*Naohisa Tamura, Toshio Igaki,*
*Jun Yamashita, Kentaro Doi,*
*Tae-Hwa Chun, Mayumi Inoue,*
*Ken Masatsugu, Tadashi Ikeda,*
*Gotaro Shirakami, Katsuhiko Matsuda,*
*Goro Katsuura, Makiko Ueda,*
*Toshihiko Ban, and Kazuwa Nakao*

**CONTENTS**



## EXPRESSION AND DISTRIBUTION OF CNP

### *Structure of CNP*

C-type natriuretic peptide (CNP) was originally isolated from porcine brain extracts *(1)*. CNP has a primary structure similar to α-ANP and BNP consisting of 22 amino acids,

From: *Contemporary Endocrinology: Natriuretic Peptides in Health and Disease*
Edited by: W. K. Samson and E. R. Levin Humana Press Inc., Totowa, NJ

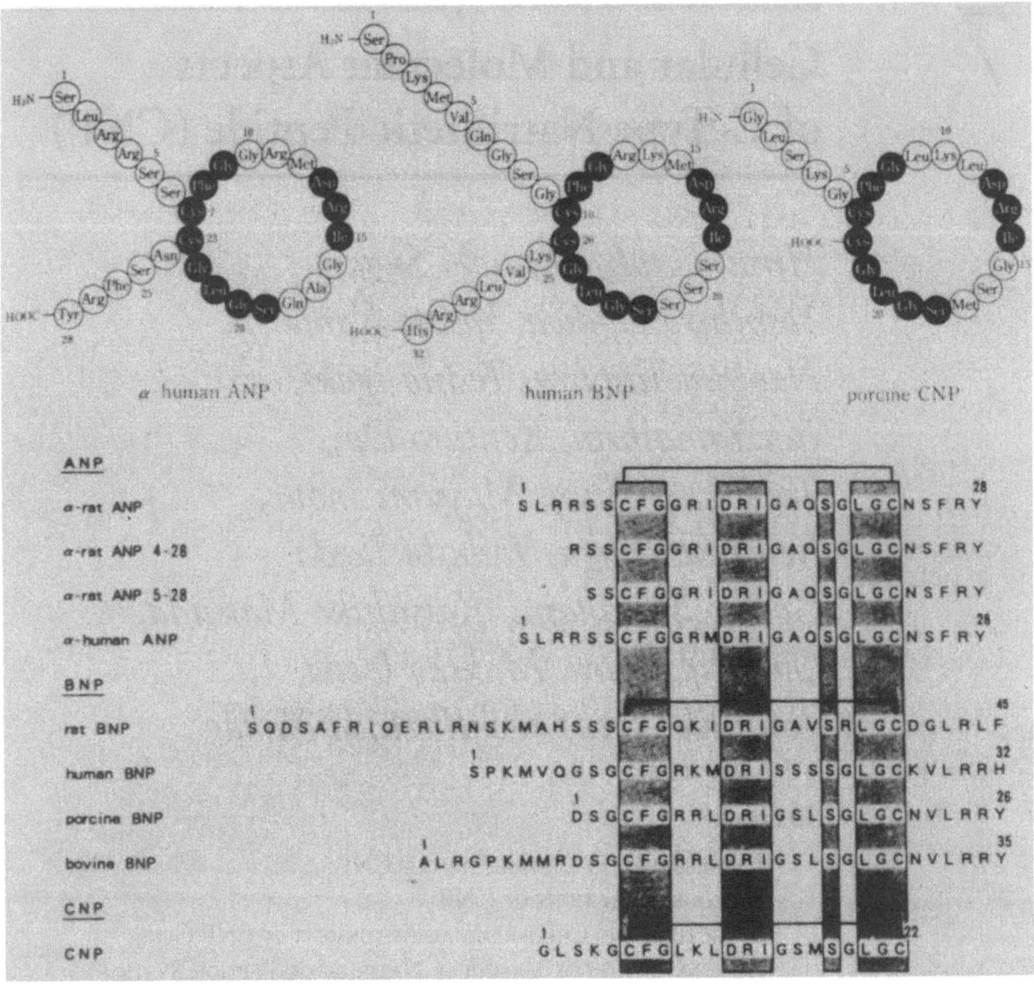

**Fig. 1.** Amino acid sequences and primary structures of the natriuretic peptide family. Shadowed area shows homology among mammalian ANP, BNP, and CNP.

and the ring portion consisting of 17 amino acids is highly homologous to α-ANP and BNP (Fig. 1). Different from ANP and BNP, CNP lacks the C-terminal tail, and has a Cys residue at the C-terminus. Another species of CNP is CNP-53, which has the N-terminal extension of 31 amino acids compared with CNP-22 *(2)*. Molecular cloning of the CNP precursor in the pig, rat, human, and mouse has revealed that the primary structure of CNP-22 is identical in these species *(1,3–5)*. However, two amino acid substitutions are noted in CNP-53 between human and porcine/rat/mouse precursors. ProCNP consists of 103 amino acids and a single arginine residue precedes CNP-53 and CNP-22. Using reverse-phase high-performance liquid chromatography (HPLC), we demonstrated that the major molecular forms in the human brain are CNP-22 and CNP-53 *(4)*.

## *Distribution of CNP*

Using the specific radioimmunoassay (RIA) for CNP, it was elucidated that immunoreactive CNP was detected in the porcine, rat, and human brains, but not in the periph-

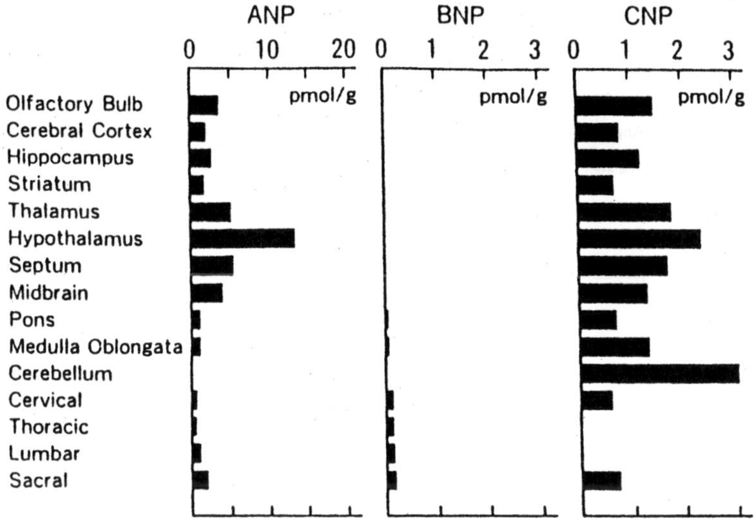


**Fig. 2.** Regional distribution of ANP-, BNP-, and CNP-like immunoreactivities in rat brain areas.

Table 1
ANP-, BNP-, and CNP-Like Immunoreactivities (LI) in Discrete Regions of the Human Brain[a]

| | *Natriuretic peptide* | | |
|---|---|---|---|
| *Brain regions* | *ANP-LI* | *BNP-LI* | *CNP-LI* |
| Cerebral cortex | <0.07 | <0.06 | 0.61 |
| Thalamus | <0.07 | <0.06 | 2.90 |
| Hypothalamus | 0.13 | <0.06 | 4.02 |
| Midbrain | <0.07 | <0.06 | 3.28 |
| Pons | 0.15 | <0.06 | <0.4 |
| Medulla oblongata | 0.08 | <0.06 | 2.78 |
| Cerebellar cortex | <0.07 | <0.06 | 0.85 |

[a]All values are pmol/g of wet tissue.

eral organs, including the heart *(4–6)*. As shown in Fig. 2, the CNP concentration in the rat brain is highest in the cerebellum, followed by the hypothalamus, thalamus, septum, and olfactory bulb. In the human brain, the highest concentration was found in the hypothalamus, followed by the midbrain and thalamus (Table 1). As shown in Table 1, the concentration of CNP in the human brain is higher than that of ANP or BNP by at least one order of magnitude. Reflecting the high level of CNP in the brain, the CNP concentration in the human cerebrospinal fluid is about 10 times higher than the ANP or BNP concentration *(7)*. CNP is, therefore, considered to be the major natriuretic neuropeptide. By using Northern blotting, CNP mRNA could be detected in the porcine and rat brains, but not in the heart *(1,3)*. In the rat brain, CNP mRNA is abundantly expressed in the anteroventral preoptic area and arcuate nucleus *(8)*. CNP is also highly expressed in the human, rat, and porcine pituitary glands *(1,3,4,6)*.

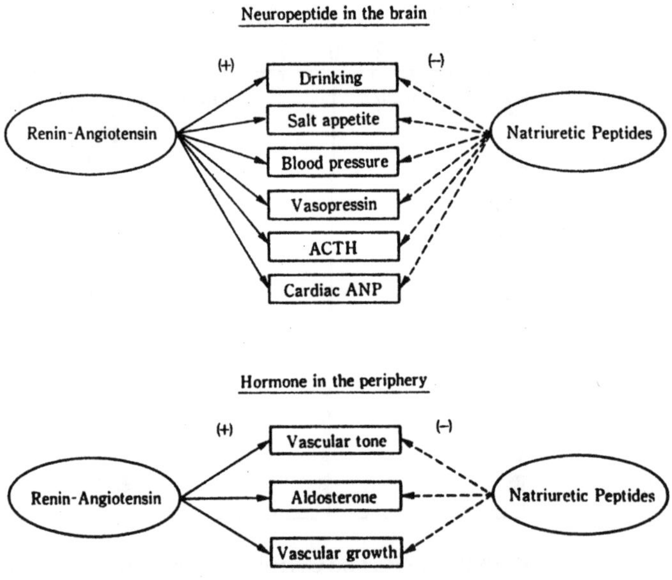


**Fig. 3.** Antagonistic relationship between the natriuretic peptide system and the renin–angiotensin system in the central nervous system and the periphery.

# BIOLOGICAL ACTIONS OF CNP

## *Central Actions of CNP*

By the intracerebroventricular (icv) injection of ANP and BNP into unrestrained conscious rats, we and others have demonstrated that natriuretic peptides can exert inhibitory actions on water drinking, induction of salt appetite, centrally mediated blood pressure increase, and vasopressin and ACTH secretions *(9–12)*. These actions of natriuretic peptides are antagonistic to those of angiotensin II (AII). In the periphery, AII causes vasoconstriction and enhances aldosterone secretion, thus acting to increase body fluid and raise blood pressure. ANP induces vasodilation, diuresis, and natriuresis and inhibits aldosterone secretion, hence acting to decrease body fluid and lower blood pressure. Thus, the renin–angiotensin system and the natriuretic peptide system have similar actions in both the central nervous system (CNS) and the periphery and regulate body fluid and cardiovascular homeostasis in an antagonistic way (Fig. 3) *(13,14)*.

Among the natriuretic peptide receptors, it was elucidated that guanylyl cyclase-B (GC-B) is selectively activated by CNP and considered to be the specific receptor for CNP, whereas guanylyl cyclase-A (GC-A) shows high affinity to ANP and BNP *(15,16)*. Figure 4 shows the mRNA expressions of three subtypes of the natriuretic peptide receptor in the rat brain by the *in situ* hybridization technique. Reflecting the abundant expression of GC-B in the CNS, the icv injection of CNP is demonstrated to exert similar inhibitory action on AII-induced vasopressin secretion *(17)*. However, the potency of centrally administered CNP was about one-tenth that of ANP. Figure 5 shows the effect of the icv injection of the antisense oligonucleotide directed against rat GC-B mRNA in conscious rats on AII-induced vasopressin secretion, to block the endogenous expression of GC-B. As clearly shown in Fig. 5, the increase of plasma vasopressin level induced by AII was significantly potentiated by the administration of the GC-B antisense oligonucleotide compared to the control oligonucleotide. This result indicates the physiological role

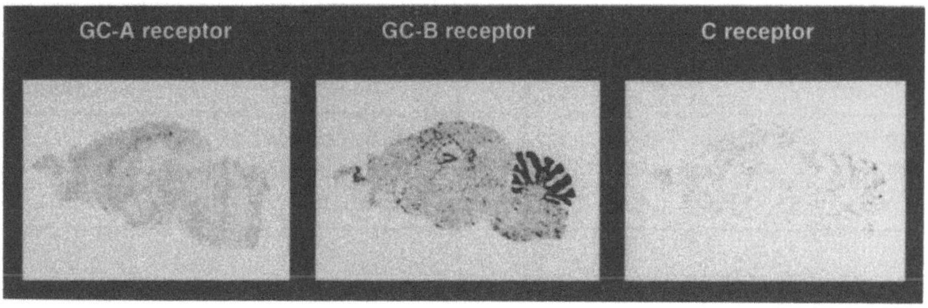


**Fig. 4.** Regional expression of three subtypes of the natriuretic peptide receptor (GC-A, GC-B, and C-receptor) mRNAs in the rat brain by *in situ* hybridization.

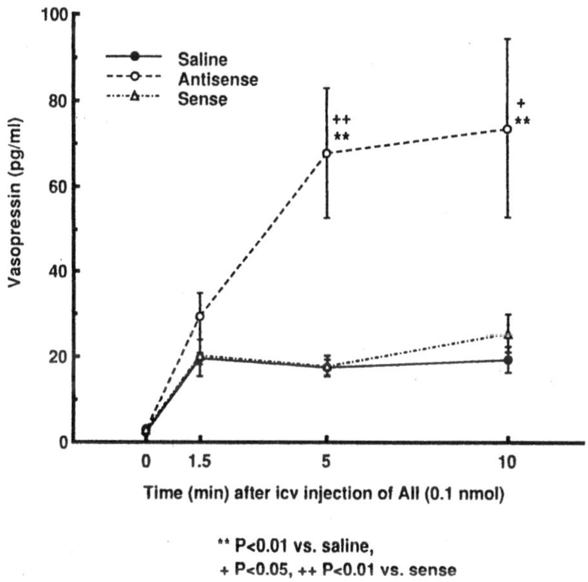


**Fig. 5.** Effect of the icv injection of the antisense oligonucleotide against GC-B mRNA on angiotensin II-induced vasopressin secretion in conscious, unrestrained rats.

of CNP/GC-B in modulating vasopressin secretion in the brain. As shown in Fig. 4, the distribution of GC-B is not the same of GC-A. Furthermore, differential expression of GC-B and GC-A in neuronal and glial cells is also demonstrated *(18)*. Therefore, the different roles of CNP in the CNS is well expected.

## *Peripheral Actions of CNP*

To examine whether CNP can elicit biological actions in the periphery, we investigated the effect of synthetic CNP administered intravenously in humans and rats and compared it with that of ANP *(19,20)*. The iv injection of CNP (2 nmol/rat) into the conscious rat increased the plasma level of cyclic GMP (cGMP) from 8.8 ± 1.6 pmol/mL to 39 ± 2.2 pmol/mL 5 min after CNP injection. In contrast, the iv administration of ANP at the same dose increased the plasma cGMP level up to 258 ± 15 pmol/mL 5 min after the injection. In humans, the administration of CNP (0.43 nmol/kg) elicited significant

Table 2
Chromosomal Localization of Natriuretic Peptide System

| *Ligand* | *Mouse/human* | *Receptor* | *Human* |
|---|---|---|---|
| ANP | 4/1 | ANP-A receptor | 1 |
| BNP | 4/1 | ANP-B receptor | 9 |
| CNP | 1/2 | Clearance receptor | 5 |

increase of plasma cGMP level (from 4.77 ± 1.25 to 8.33 ± 1.59 pmol/mL 15 min after administration) and of urinary cGMP excretion (from 30.7 ± 4.3 to 74.9 ± 13.4 nmol/30 min). Intravenously administered CNP exerted significant diuretic, natriuretic, kalliuretic, and chloriuretic actions with the increase of endogenous creatinine clearance. The CNP also elicited significant hypotensive actions (ΔBPs/ΔBPd: –4.3 ± 1.3/–4.1 ± 1.0 mmHg) with concomitant increase of heart rate (+7.6 ± 2.6 bpm). Plasma aldosterone concentration significantly decreased from 45.4 ± 2.3 to 35.4 ± 4.9 pg/mL 30 min after administration. The biological activities of CNP administered intravenously in humans and rats were, therefore, weaker than that of ANP *(21–24)*, suggesting the reduced abundancy of GC-B, to which intravenously injected CNP can access.

The CNP was also demonstrated to exert potent antigrowth effect on cultured vascular smooth muscle cells (VSMC) *(14,25–27)* or migrating and proliferating VSMC in the neointima after balloon-injury in vivo that possess synthetic phenotype and highly express GC-B as described below *(28)*.

## STRUCTURE AND CHROMOSOMAL ASSIGNMENT OF CNP GENE (TABLE 2)

We cloned mouse and human CNP genes *(4,5)*. The mouse CNP gene is composed of at least two exons and one intron. The 5'-flanking region contains an array of *cis*-acting regulatory elements, including Y box, GC box, and CRE-like sequence as well as a dinucleotide CA repeat (microsatellite). On the basis of the polymerase chain reaction (PCR)-analyzed microsatellite-length polymorphisms among recombinant inbred strains of mice, the CNP gene *(Nppc)* was assigned to mouse chromosome 1 (cen-*Acrg*-0.79 ± 0.82c *M-Nppcl(Sag)*-3.33 ± 2.67c *M-ln*). In contrast, using somatic hybrid cell methodology, the human CNP gene *(Nppc)* was assigned to chromosome 2 (2q24-qter). Thus, chromosomal assignment of the CNP locus in mice and humans has added another locus to the conserved syntenic group in mice and humans. It has been reported that both the ANP and the BNP genes, which are expressed predominantly in the heart, are localized on human chromosome 1 *(29,30)*. Furthermore, it is also demonstrated that these genes are tightly linked on mouse chromosome 4 *(29–31)*. In contrast, CNP does not act as a cardiac hormone but as a local regulator or a neuropeptide. The CNP gene is shown to be localized on mouse chromosome 1/human chromosome 2, which is physically separated from the ANP and BNP genes. Therefore, CNP is functionally and evolutionally distinct from ANP and BNP. Given the remarkable structural conservation of CNP among species, and even among nonmammals, it is tempting to speculate that CNP is a prototype of the natriuretic peptide family, from which ANP and BNP originated through gene duplication.

## DEMONSTRATION OF VASCULAR NATRIURETIC PEPTIDE SYSTEM

### *Secretion of CNP from Vascular Endothelial Cells*

Using specific RIA for CNP, we detected CNP-like immunoreactivity (-LI) in the culture medium conditioned with endothelial cells derived from the bovine carotid artery *(32)*. The CNP-LI in the conditioned medium was accumulated in a time-dependent manner and reached $17.8 \pm 1.6$ fmol/$10^6$ cells/24 h. The endothelial cellular content of CNP-LI was $0.15 \pm 0.05$ fmol/$10^6$ cells. High-performance gel-permeation chromatographic analysis revealed that CNP-LI in the condition medium from the endothelial cells consisted of two components corresponding to synthetic CNP and CNP-53. Northern blot analysis detected CNPmRNA with a size of 1.2 kb in the cultured endothelial cells. These results demonstrate the synthesis and secretion of CNP by the cultured endothelial cells.

### *Occurrence of CNP in the In Vivo Blood Vessels*

We examined whether CNP gene expression occurs in the in vivo blood vessel wall by reverse transcription and polymerase chain reaction (RT-PCR) *(19,33)*. Our RT-PCR revealed that CNP gene transcripts were detected in blood vessels in vivo of a wide range of mammals, including not only the bovine but also human, rabbit, rat, and hamster aortae. Furthermore, the gene transcript for GC-B was simultaneously detected with that for CNP in human arteries at various sites *(33,34)*. More recently, by the immunohistochemical method using our monoclonal antibody (MAb) against CNP (KY-CNP-I) *(35)*, we demonstrated the presence of CNP in the endothelial cells in human coronary arteries *(36)*. Detection of the expressions of CNP and GC-B in in vivo vascular walls strongly indicates the presence of the "vascular natriuretic peptide system," in which CNP serves as a local regulator of vascular tone and growth (Fig. 6), like "the vascular renin–angiotensin system" *(37)*. The CNP can, therefore, be considered to be novel endothelium-derived relaxing factor (EDRF) *(19,38)*.

### *Regulation of Endothelial CNP Secretion in the Interaction of Endothelial Cells/Vascular Smooth Muscle Cells/Macrophages (Fig. 7)*

The endothelial secretion of CNP was regulated by various vasoactive substances, including growth factors and cytokines. Among them, transforming growth factor-β (TGF-β), one of the crucial factors involved in vascular remodeling, induced more than two orders of magnitude increase of endothelial CNP secretion *(32)*. In addition, the endothelial secretion of CNP is potently stimulated by tumor necrosis factor-α (TNF-α), one of the central cytokines for inflammation and also by lipopolysaccharide (LPS), the endotoxin itself *(39)*. We also demonstrated that shear stress at the physiological range augmented CNP expression in human, bovine, and mouse endothelial cells *(40)*. Actually, within the mouse CNP gene we cloned *(5)* are NF-κ β (which is crucial for TNF-α induced gene expression), TGF-β, and shear-stress responsive elements. In contrast, endothelial CNP secretion was potently inhibited by oxidized LDL *(41)*, insulin *(42)*, and vascular-endothelial growth factor (VEGF), a novel endothelium-specific angiogenic factor *(43)*.

We investigated the interaction of endothelial cells (EC), VSMC, and macrophages (M ϕ) for the regulation of endothelial CNP production, its molecular mechanisms, and CNP action on vascular growth, using the coculture system *(44)*. The concentration of CNP in the medium of EC/VSMC coculture was increased 60-fold, compared with that

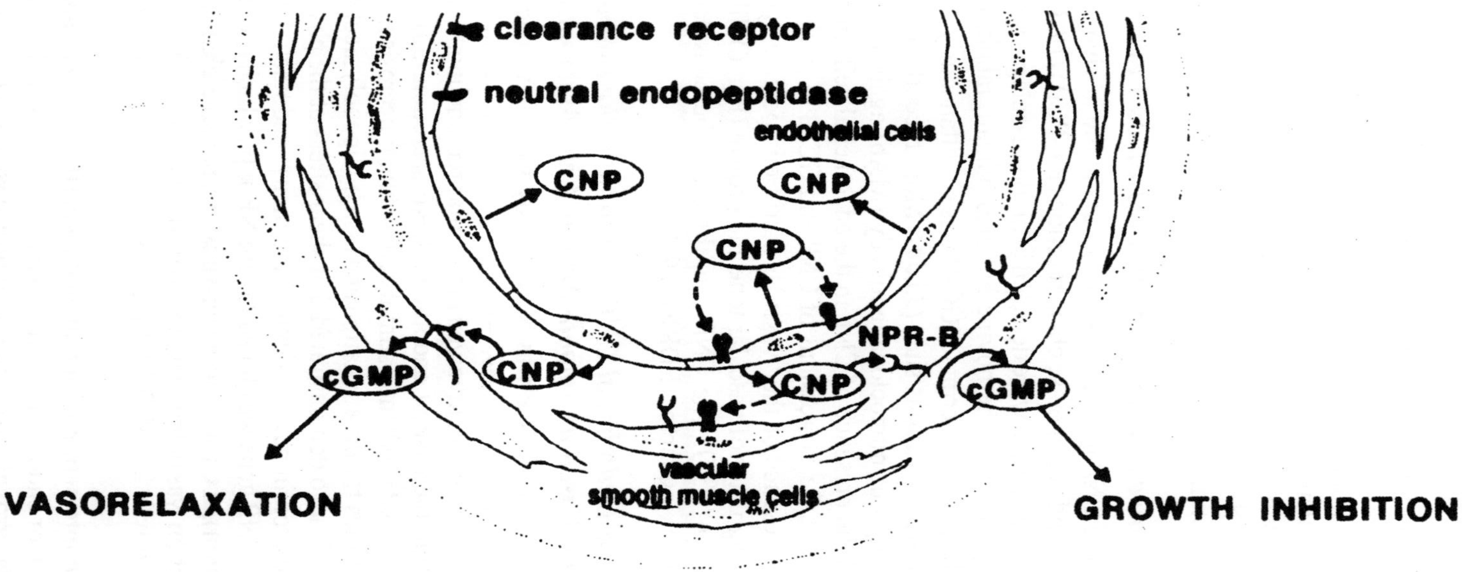


**Fig. 6.** Existence of the vascular natriuretic peptide system.

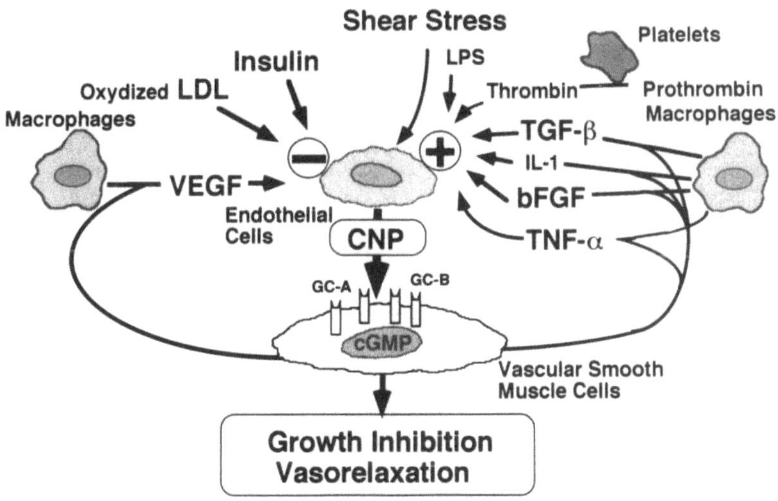


**Fig. 7.** Regulation of endothelial secretion of CNP.

of EC alone. The endothelial production of CNP was also augmented 10-fold in EC/M ϕ (THP-1, a human monocytic leukemia cell line, treated with phorbol ester) coculture (Fig. 8A). As shown in Fig. 8B, Northern blot analysis revealed augmented expression of CNPmRNA in EC/M ϕ cocultures. The accumulation of intracellular cGMP in the medium was increased in EC/VSMC coculture. This response was blocked by the anti-CNP MAb, and HS-142-1, the nonpeptide natriuretic-peptide receptor antagonist *(45)*. The concentration of biologically active TGF-β in the medium of EC/VSMC coculture was elevated to the level to stimulate the endothelial production of CNP. Actually, the neutralizing antibody against TGF-β abrogated the cGMP accumulation in the coculture. In contrast, the augmentation of the endothelial production of CNP in EC/M ϕ coculture was suppressed by the neutralizing antibody against TNF-α. These results show that the endothelial production of CNP in EC/VSMC and EC/M ϕ cocultures is at least in part regulated by TGF-β and TNF-α, respectively. Furthermore, the conditioned medium from EC was demonstrated to have the growth inhibitory effect on VSMC, which was abolished by the anti-CNP MAb and HS-142-1. The treatment of the anti-CNP MAb and HS-142-1 also significantly increased the cell number of the EC/VSMC coculture. These results reveal the pathophysiological significance of endothelial CNP as a novel paracrine/autocrine vascular local regulator for vascular growth in the interaction of EC, VSMC, and M ϕ.

## DETECTION OF PLASMA CNP AND ITS REGULATION

We examined the presence of CNP in circulation in humans and rats *(19,46)*. The dilution curves of the plasma extracts from both humans and rats were parallel to the standard curve of CNP in the RIA. The plasma CNP-LI level of CNP in healthy persons was 1.4 ± 0.6 fmol/mL and that of male Wistar rats (200–300 g) was 1.2 ± 0.2 fmol/mL. Whereas there was no alteration in the plasma CNP-LI level in patients with congestive heart failure or hypertension, the plasma level of CNP was increased in septic patients and further elevated in septic shock (Fig. 9). TNF-α is recognized to be the central mediator

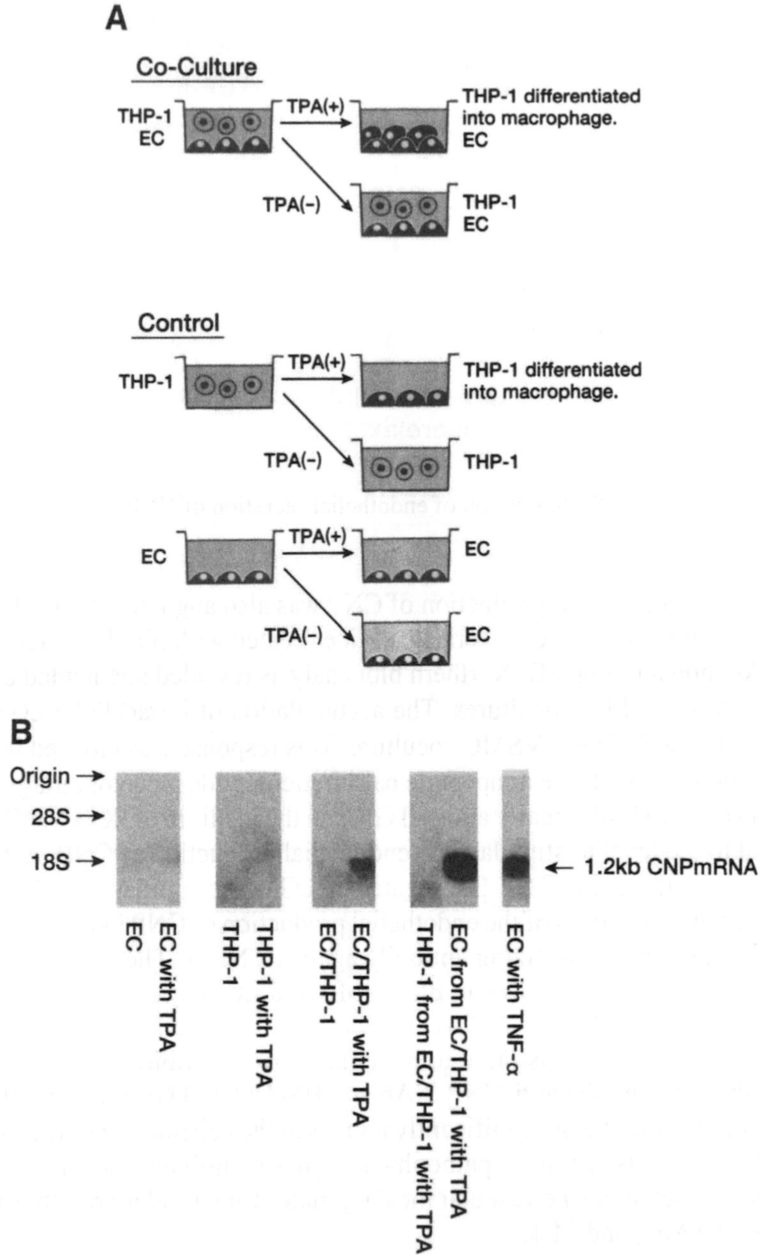


**Fig. 8.** Endothelial production of CNP in the interaction of endothelial cells and macrophages. **(A)** Experimental protocol of the coculture of endothelial cells and THP-1 treated with TPA (which differentiated into macrophages). **(B)** Expression of CNPmRNA in endothelial cells/THP-1 with and without the treatment of TPA.

of the septic shock and its production is enhanced in the syndrome. The increased plasma level of CNP in sepsis is, therefore, at least in part the consequence of augmented CNP secretion from vascular endothelial cells, since TNF-α is one of the potent stimulators of

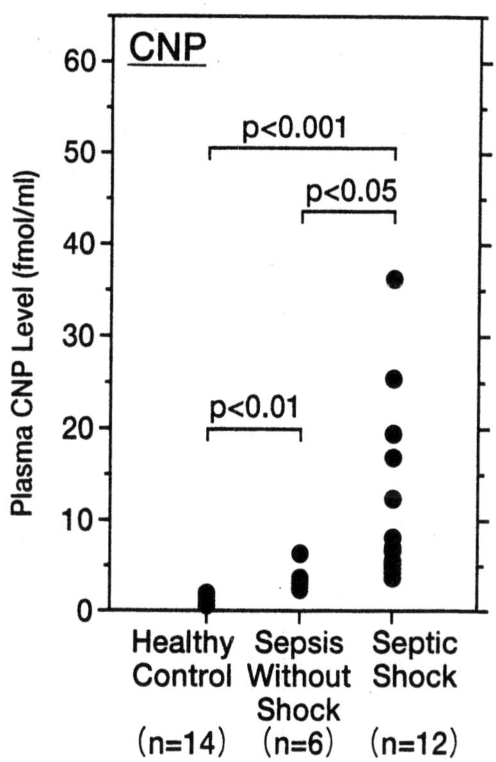


**Fig. 9.** Plasma CNP level in septic patients.

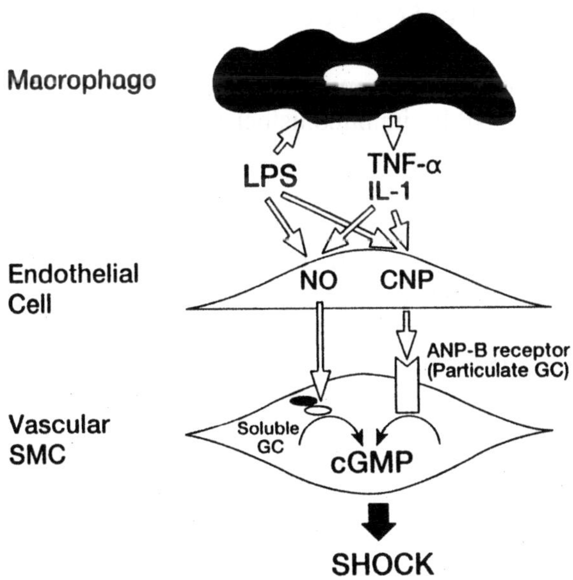


**Fig. 10.** Possible pathophysiological implication of endothelial CNP in endotoxin shock.

endothelial CNP secretion *(39)*. So far, nitric oxide (NO), EDRF itself, is recognized to be the central mediator of septic shock. Our results suggest that CNP can also be involved in the pathophysiology of shock in concert with NO (Fig. 10).

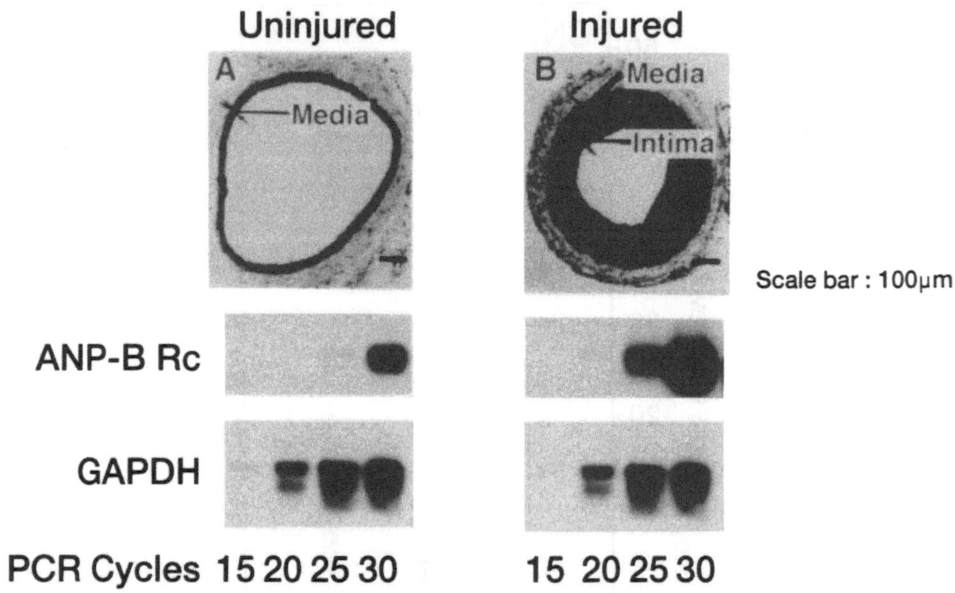


**Fig. 11.** GC-B (ANP-BRc) gene expression in rat balloon-injured carotid arteries evaluated by RT-PCR/Southern blotting (2 wk after the balloon injury).

## SIGNIFICANCE OF VASCULAR NATRIURETIC SYSTEM IN VASCULAR REMODELING

To elucidate the role of CNP in human proliferative vascular lesions, we examined vascular expression of CNP by immunohistochemical analysis *(36)*, using the anti-CNP MAb (KY-CNP-I), which we developed *(35)*. In normal coronary arteries, intense CNP staining was observed in EC. In contrast, atherosclerotic lesions showed a marked decrease in positivity for staining of EC. However, M $\phi$ and VSMC, which were recruited in the lesions, showed positive staining for CNP, indicating CNP production in these cells. Actually, besides EC, these are some reports that show the production of CNP in cultured human and murine macrophages *(47,48)*. In restenotic coronary lesions after PTCA, regenerating EC also exhibited attenuated CNP expression, whereas infiltrating M $\phi$ and VSMC with synthetic phenotype showed CNP positivity.

The gene expression of natriuretic peptide receptor subtypes was also investigated. In VSMC obtained from the media of the intact rat aorta that possess the contractile phenotype, GC-A is highly expressed. In contrast, in cultured VSMC that show the synthetic phenotype, GC-A expression is downregulated and, in turn, GC-B and C-receptor expressions are much augmented *(49)*. The subtype switching of the natriuretic-peptide receptor in the phenotype change of VSMC was assessed in intimal-thickening lesions after carotid injury in rats. A Fogarty catheter was introduced into the left carotid artery, inflated, and withdrawn to produce endothelial denudation on male SD rats. The mRNA expression of GC-B was significantly augmented in the injured carotid arteries (Fig. 11) *(14,28)*. There is also a report that shows the upregulation of C-receptor expression in balloon-injured lesions *(50)*.

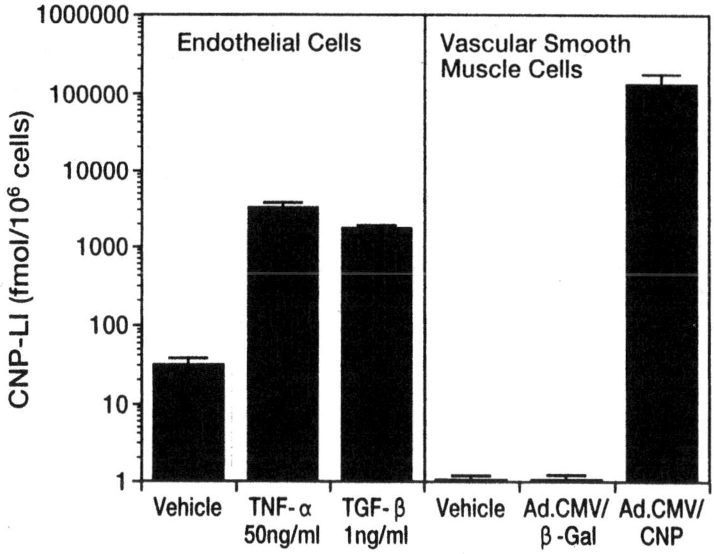


**Fig. 12.** CNP secretion from vascular smooth muscle cells transacted with adenovirus overexpressing CNP gene in comparison with CNP secretion from endothelial cells. Ad.CMV/CNP: adenovirus expressing CNP gene. Ad.CMV/β-Gl: control adenovirus expressing β-galactosidase gene.

## IN VIVO GENE TRANSFER OF CNP INTO BLOOD VESSELS— APPLICATION OF VASCULAR NATRIURETIC PEPTIDE SYSTEM TO GENE THERAPY FOR PROLIFERATIVE VASCULAR LESIONS

As described above, we revealed that in human atheromatous arteries, VSMC with synthetic phenotype overexpress GC-B. In contrast, CNP expression in EC was reduced in these lesions. Therefore, to selectively suppress the growth of VSMC with synthetic phenotype in vascular lesions, we aimed at overexpression of CNP gene in VSMC by adenovirus-mediated gene transfer, expecting the autocrine/paracrine action of CNP. Rat aortic VSMC ($5 \times 10^3$ cells/well) were infected with a replication-deficient recombinant adenovirus containing cytomegalovirus promoter-driving rat CNP cDNA, (advCMV/CNP) (40 MOI). The advCMV/CNP-transacted VSMC produced $122 \pm 44$ pmol/$10^4$ cells of CNP, which was 4000 times higher than EC ($32 \pm 5$ fmol/$10^4$ cells) (Fig. 12).

The augmentation of cGMP production and almost complete growth arrest was observed (84% reduction) in AdCMV/CNP-treated VSMC. Furthermore, CNP overexpression at the site of balloon injury of the rabbit femoral artery actually significantly suppressed the formation of the neointima in vivo. These results indicate the profound autocrine/ paracrine growth-inhibitory property of CNP chronically secreted from adenoviral-transfected VSMC and suggest the usefulness of CNP as an autocrine/paracrine growth inhibitor for gene therapy of proliferative vascular lesions.

## REFERENCES


1. Sudoh T, Minamino N, Kangawa K, Matsuo H. C-type natriuretic peptide (CNP): a new member of natriuretic peptide family identified in porcine brain. Biochem Biophys Res Commun 1990;168:863–870.
2. Minamino N, Kangawa K, Matsuo H. N-terminally extended form of C-type natriuretic peptide (CNP-53) identified in porcine brain. Biochem Biophys Res Commun 1990;170:973–979.

3. Kojima M, Minamino N, Kangawa K, Matsuo H. Cloning and sequence analysis of a cDNA encoding a precursor for rat C-type natriuretic peptide (CNP). FEBS Lett 1990;276:209–213.
4. Ogawa Y, Nakao K, Nakagawa O, Komatsu Y, Hosoda K, Suga S, Arai H, Nagata N, Yoshida N, Imura H. Human C-type natriuretic peptide: characterization of the gene and peptide. Hypertension 1992;19:809–813.
5. Ogawa Y, Itoh H, Yoshitake Y, Inoue M, Yoshimasa T, Serikawa T, Nakao K. Molecular cloning and chromosomal assignment of the mouse C-type natriuretic peptide (CNP) gene (Nppc): comparison with the human CNP gene (NPPC). Genomics 1994;24:383–387.
6. Komatsu Y, Nakao K, Suga S, Ogawa Y, Mukoyama M, Arai H, Shirakami G, Hosoda K, Nakagawa O, Hama N, Kishimoto I, Imura H. C-type natriuretic peptide (CNP) in rats and humans. Endocrinology 1991;129:1104–1106.
7. Kaneko T, Shirakami G, Nakao K, Nagata I, Nakagawa O, Hama N, Suga S, Miyamoto Y, Kubo H, Hirai O, Kikuchi H, Imura H. C-type natriuretic peptide (CNP) is the major natriuretic peptide in human cerebrospinal fluid. Brain Res 1993;612:104–109.
8. Gundlach A, Knobe KE. Distribution of preproatrial natriuretic peptide mRNA in rat brain detected by in situ hybridization of DNA oligonucleotides: enrichment in hypothalamic and limbic regions. J Neurochem 1992;59:758–761.
9. Imura H, Nakao K, Itoh H. The natriuretic peptide system in the brain: implications in the central control of cardiovascular and neuroendocrine functions. Frontiers Neuroendocrinol 1992;13:217–249.
10. Itoh H, Nakao K, Imura H. Natriuretic peptide system in the brain and its implication in central cardiovascular regulation. In: Kunos G, Ciriello J, eds. Central Neural Mechanisms in Cardiovascular Regulation volume 2. Birkhauser Publishers, Boston, 1993, pp. 266–279.
11. Samson WK. Atrial natriuretic factor and the central nervous system. Clin Endocrinol Metab 1987;16:145–161.
12. Itoh H, Nakao K, Katsuura G, Morii N, Yamada T, Shiono S, Sakamoto M, Sugawara A, Saito Y, Matsushita A, Imura H. Centrally infused atrial natriuretic polypeptide attenuates exaggerated salt appetite in spontaneously hypertensive rats. Circ Res 1986;59:342–347.
13. Nakao K, Morii N, Itoh H, Yamada T, Shiono S, Sugawara A, Saito Y, Mukoyama M, Arai H, Sakamoto M, Imura H. Atrial natriuretic polypeptide in brain—implication of central cardiovascular control. J Hypertension 1986;4(Suppl 6):S492–S496.
14. Itoh H, Nakao K. Antagonism between vascular renin-angiotensin and natriuretic peptide systems in vascular remodeling. Blood Pressure 1994;3(Suppl. 5):49–53.
15. Koller KJ, Lowe DG, Bennett GL, Minamino N, Kangawa K, Matsuo H, Goeddel DV. Selective activation of the B natriuretic peptide receptor by C-type natriuretic peptide (CNP). Science 1991;252:120–123.
16. Suga S, Nakao K, Hosoda K, Mukoyama M, Ogawa Y, Shirakami G, Arai H, Saito Y, Kambayashi Y, Inoue K, Imura H. Receptor selectivity of natriuretic peptide family, atrial natriuretic peptide, brain natriuretic peptide, and C-type natriuretic peptide. Endocrinology 1992;130:229–239.
17. Shirakami G, Itoh H, Suga S, Komatsu Y, Hama N, Mori K, Nakao K. Central action of C-type natriuretic peptide on vasopressin secretion in conscious rats. Neurosci Lett 1993;159:25–28.
18. Sumners C, Tang W. Atrial natriuretic peptide receptor subtypes in rat neuronal and astrocyte glial cultures. Am J Physiol 1992;262:C1134–1143.
19. Nakao K, Itoh H, Yoshimasa T, Imura H. C-type natriuretic peptide (CNP) as a novel endothelium-derived relaxing peptide: occurrence of vascular natriuretic peptide system. In: Masaki T, ed. Endothelium-Derived Factors and Vascular Functions. Elsevier Science Publishers, B.V. Amsterdam, 1994, pp. 23–129.
20. Igaki T, Itoh H, Suga S, Hama N, Ogawa Y, Komatsu Y, Mukoyama M, Sugawara A, Yoshimasa T, Tanaka I, Nakao K. C-type natriuretic peptide in chronic renal failure and its action in humans. Kidney Int 1996;49 Suppl 55:S144–S147.
21. Sugawara A, Nakao K, Morii N, Sakamoto M, Horii K, Shimokura M, Kiso Y, Nishimura K, Ban T, Kihara M, Yamori Y, Kangawa K, Matsuo H, Imura H. Significance of α-human atrial natriuretic polypeptide as a hormone in humans. Hypertension 1986;8[Suppl I]:I-151–155.
22. Itoh H, Nakao K, Mukoyama M, Shiono S, Morii N, Sugawara A, Yamada T, Saito Y, Arai H, Hosoda K, Kambayashi Y, Inoue K, Imura H. Effects of intravenously administered β-human atrial natriuretic polypeptide in humans. Hypertension 1988;11:697–702.
23. Hunt PJ, Richards A, Espiner E, Nicholls M, Yandle T. Bioactivity and metabolism of C-type natriuretic peptide in normal man. J Clin Endocrinol Metab 1994,78:1428–1435.

24. Cargill R, Struthers A, Lipworth B. Human C-type natriuretic peptide: effects on the haemodynamic and endocrine responses to angiotensin II. Cardiovasc Res 1995;29:108–111.
25. Itoh H, Pratt RE, Dzau VJ. Atrial natriuretic polypeptide inhibits hypertrophy of vascular smooth muscle cells. J Clin Invest 1990;86:1690–1697.
26. Furuya M, Yoshida M, Hayashi Y, Ohnuma N, Minamino N, Kangawa K, Matsuo H. C-type natriuretic peptide is a growth inhibitor of rat vascular smooth muscle cells. Biochem Biophys Res Commun 1991;177:927–931.
27. Porter JG, Catalanor R, McEnroe G, Lewicki JA, Protter AA. C-type natriuretic peptide inhibits growth factor-dependent DNA synthesis in smooth muscle cells. Am J Physiol 1992;263:C1001–C1006.
28. Furuya M, Aisaka K, Miyazaki T, Honbou N, Kawashima N, Ohno T, Tanaka S, Minamino N, Kangawa K, Matsuo H. C-type natriuretic peptide inhibits intimal thickening after vascular injury. Biochem Biophys Res Commun 1993;193:248–253.
29. Yang-Feng TL, Floyd-Smith G, Nemer M, Drouin J, Francke U. The pronatriodilatin gene is located on the distal short arm of human chromosome 1 and on mouse chromosome 4. Am J Hum Genet 1985,37:1117–1128.
30. Tamura N, Ogawa Y, Yasoda A, Itoh H, Saito Y, Nakao K. Two cardiac natriuretic peptide genes (atrial natriuretic peptide and brain natriuretic peptide) are organized in tandem in the mouse and human genomes. J Mol Cell Cardiol 1996;28:1811–1815.
31. Steinhelper ME. Structure, expression, and genomic mapping of the mouse natriuretic peptide type-B gene. Circ Res 1992;72:984–992.
32. Suga S, Nakao K, Itoh H, Komatsu Y, Ogawa Y, Hama N, Imura H. Endothelial production of C-type natriuretic peptide and its marked augmentation by transforming growth factor- β—Possible existence of "vascular natriuretic peptide system." J Clin Invest 1992;90:1145–1149.
33. Komatsu Y, Nakao K, Itoh H, Suga S, Ogawa Y, Imura H. Vascular natriuretic peptide. Lancet 1992;340:622.
34. Ikeda T, Itoh H, Komatsu Y, Hanyu M, Yoshimasa T, Matsuda K, Nakao K, Ban T. Natriuretic peptide receptors in human arterial and venous coronary bypass vessels and rabbit vein grafts. Hypertension 1996;27:833–837.
35. Hama N, Itoh H, Suga S, Komatsu Y, Yoshimasa T, Nakao K. A monoclonal antibody to C-type natriuretic peptide—preparation and application to radioimmunoassay and neutralization experiment. J Endocrinol 1994;141:473–479.
36. Naruko T, Ueda M, van der Wal AC, van der Loos CM, Itoh H, Nakao K, Becker AK. C-type natriuretic peptide expression in human coronary atherosclerotic lesions. Circulation 1996;94:3103–3108.
37. Dzau VJ. Circulating versus local renin-angiotensin system in cardiovascular homeostasis. Circulation 1988;77(Suppl I):I4–I13.
38. Komatsu Y, Itoh H, Suga S, Igaki T, Ogawa Y, Kishimoto I, Nakagawa O, Yoshimasa T, Nakao K. Regulation of secretion and clearance of C-type natriuretic peptide in the interaction of vascular endothelial cells and smooth muscle cells. J Hypertens 1996;14:585–592.
39. Suga S, Itoh H, Komatsu Y, Ogawa Y, Hama N, Yoshimasa T, Nakao K. Cytokine-induced C-type natriuretic peptide (CNP) secretion from vascular endothelial cells. Evidence for CNP as a novel autocrine/paracrine regulator from endothelial cells. Endocrinology 1993;133:3038–3041.
40. Chun T-H, Itoh H, Suga S, Ogawa Y, Tamura N, Takaya K, Igaki T, Yamashita J, Doi K, Inoue M, Masuda I, Saito Y, Korenaga R, Ando J, Nakao K. Augmented expression of C-type natriuretic peptide mRNA in endothelial cells under physiological shear stress. J Hypertens 1996;14(Suppl 1):S12.
41. Sugiyama S, Kugiyama K, Matsumura T, Suga S, Itoh H, Nakao K, Yasue H. Lipoproteins regulate C-type natriuretic peptide secretion from cultured vascular endothelial cells. Arterioscler Thromb Vasc Biol 1995;15:1968–1974.
42. Igaki T, Itoh H, Suga S, Komatsu Y, Ogawa Y, Doi K, Yoshimasa T, Nakao K. Insulin suppresses endothelial secretion of C-type natriuretic peptide, a novel endothelium-derived relaxing peptide. Diabetes 1996;45(Suppl 3):S62–S64.
43. Doi K, Itoh H, Komatsu Y, Igaki T, Chun T-H, Takaya K, Yamashita J, Inoue M, Yoshimasa T, Nakao K. Vascular endothelial growth factor potently suppresses CNP secretion from endothelial cells. Hypertension 1996;27:811–815.
44. Komatsu Y, Itoh H, Suga S, Ogawa Y, Hama N, Kishimoto I, Nakagawa O, Igaki T, Doi K, Yoshimasa T, Nakao K. Regulation of endothelial production of C-type natriuretic peptide in coculture with vascular smooth muscle cells—the role of vascular natriuretic peptide system in vascular growth inhibition. Circ Res 1996;78:606–614.

45. Toki S, Morishita Y, Sano T, Matsuda Y. HS-142-I, a novel non-peptide ANP antagonist, blocks the cyclic GMP production elicited by natriuretic peptides in PC12 and NG108-15 cells. Neurosci Lett 1992;175:117–120.
46. Hama N, Itoh H, Shirakami G, Suga S, Komatsu Y, Yoshimasa T, Tanaka I, Mori K, Nakao K. Detection of C-type natriuretic peptide (CNP) in human circulation and marked increase of plasma CNP level in septic shock patients. Biochem Biophys Res Commun 1994;198:1177–1182.
47. Ishizaka Y, Kangawa K, Minamino K, Ishii S, Takano T, Eto T, Matsuo H. Isolation and identification of C-type natriuretic peptide in human monocytic cell line, THP-1. Biochem Biophys Res Commun 1992;189:697–704.
48. Vollmar AM, Schulz R. Expression and differential regulation of natriuretic peptide in mouse macrophages. J Clin Invest 1995;95:2442–2450.
49. Suga S, Nakao K, Kishimoto I, Hosoda K, Mukoyama M, Arai H, Shirakami G, Ogawa Y, Komatsu Y, Nakagawa O, Imura H. Phenotype-related alteration in expression of natriuretic peptide receptors in aortic smooth muscle cells. Circ Res 1992;71:34–39.
50. Brown J, Chen Q. Regional expression of natriuretic peptide receptors during the formation of arterial neointima in the rabbit. Circ Res 1995;77:906–918.

# 8 Physiology of Natriuretic Peptides

*Cardiovascular Function*

*Eric A. Espiner*

**CONTENTS**



## NATRIURETIC PEPTIDES AND CARDIOVASCULAR HOMEOSTASIS

De Bold's discovery 16 yr ago of an atrial tissue factor *(1)* with strong natriuretic activity understandably captured the imagination of renal physiologists seeking the elusive "third factor" *(2)* and new insights into sodium regulation. However, as well as natriuresis, de Bold's original experiments showed that the atrial extract also caused vasorelaxation and lowered blood pressure—suggesting potentially important actions intrinsic to cardiovascular homeostasis, independent of natriuresis. Subsequent work in the decade after ANP's* discovery has substantiated this view to the extent that some observers regard the physiological actions of natriuretic peptides as primarily cardiovascular, with little if any role for ANP in the day-to-day regulation of sodium excretion in normal health *(3,4)*. Whereas the debate continues to this day on the relative importance of hemodynamic vs renal actions of natriuretic peptides, it is now abundantly clear that a family of hormones, including ANP, brain natriuretic peptide (BNP), and C-type natriuretic peptide (CNP), participate as both paracrine and classical endocrine (ANP and BNP) hormones in the regulation of blood pressure and fluid homeostasis. This view is based on multiple layers of evidence—and in particular on the close coupling of intracardiac pressure with hormone secretion (ANP largely from the atrial myocyte, BNP from the ventricular myocyte) *(5)*, the demonstration of hemodynamic effects of these

*ANP (atrial natriuretic peptide) is synonymous with ANF (atrial natriuretic factor) in this review.

From: *Contemporary Endocrinology: Natriuretic Peptides in Health and Disease*
Edited by: W. K. Samson and E. R. Levin Humana Press Inc., Totowa, NJ

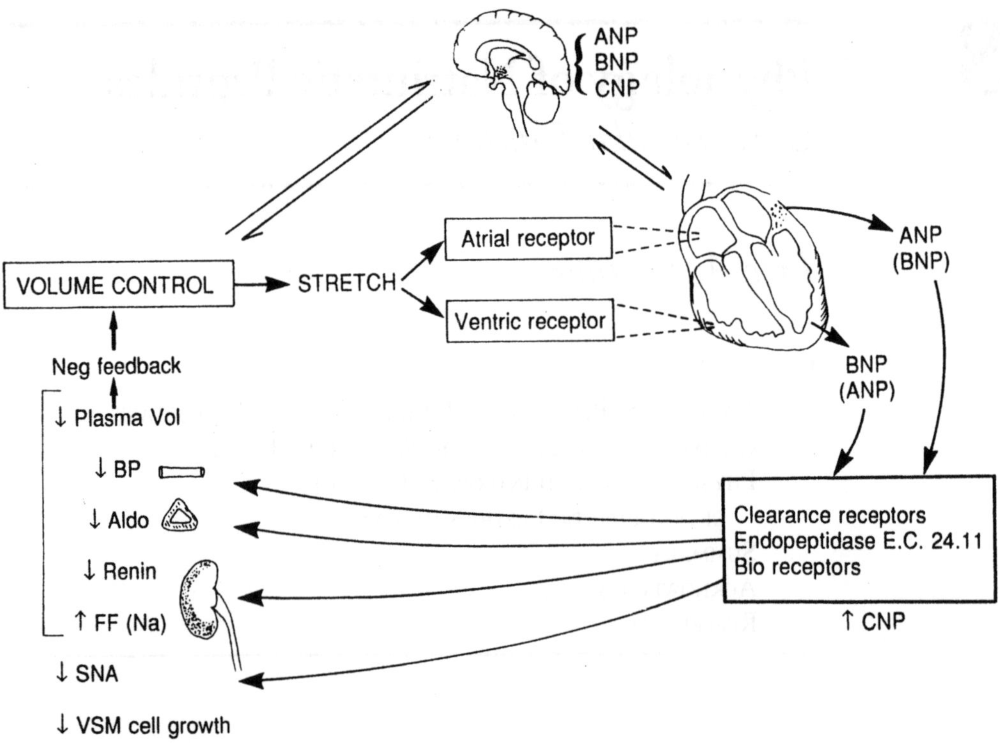


**Fig. 1.** The regulation and actions of natriuretic peptides. Note: atrial and ventricular "receptors" have not as yet been clarified. Brain production of atrial natriuretic peptide (ANP), brain natriuretic peptide (BNP), and C-type natriuretic peptide (CNP) have paracrine roles in the regulation of systemic blood pressure by the CNS. Other paracrine actions of natriuretic peptide (also indicated by ↑ CNP) on vessel wall, kidney, and so on are omitted for reasons of clarity. BP, blood pressure: aldo, aldosterone: FF, filtration fraction: SNA, sympathetic nervous activity; VSM, vascular smooth muscle. (Modified from Espiner EA. Physiology of natriuretic peptides. J Intern Med 1994;235:527–541, with permission of Blackwell.)

circulating hormones (ANP and BNP) at physiological and pathophysiological concentrations (and which can be inhibited by administration of appropriate hormone antagonists or inhibitors), and more recently the findings from transgenic models amplifying or inhibiting the expression or actions of these hormones.

Before considering the details of cardiovascular action, a broad concept of how the endocrine and paracrine effects of natriuretic peptides mediate cardiovascular homeostasis is needed (Fig. 1). In health, in response to an acute increase in central blood volume/pressure, the secretion of the circulating hormones, particularly ANP (and to a lesser extent BNP), is increased, exerting widespread effects (mediated largely by the guanylyl cyclase receptor NPR-A) *(6)*—the combined actions of which reduce cardiac filling pressures (preload) and restore hormone secretion toward normal. Although many different tissues contribute, increased natriuresis, increased flux of plasma into the interstitial space, inhibition of renin, and suppression of aldosterone secretion are all clearly demonstrable effects of small (physiological) increments in plasma ANP or BNP concentration. All of these actions, if not themselves primarily "cardiovascular," importantly affect the integrity of the circulation and must be considered together when interpreting the cardiovascular actions of natriuretic peptides. For example, a sustained natriuretic

Table 1
Mechanisms Contributing to Cardiovascular Actions of Chronic Increases in Natriuretic Peptides (ANP/BNP/CNP)

| |
|---|
| Plasma volume shrinkage (ANP/BNP)[a] |
| Leftward shift in pressure-natriuresis curve (ANP/BNP) |
| Reduced peripheral vascular resistance (ANP/BNP) |
| Increased cardiac (ANP/BNP) and venous (CNP, ?ANP) compliance |
| Inhibition of SNA (ANP/?BNP/?CNP) |
| Inhibition of renin-AII-aldosterone (ANP/BNP) |
| Inhibition of central AII (ANP, ?BNP) |
| Vascular remodelling (CNP, ANP, ?BNP) |

[a]ANP, atrial natriuretic peptide; BNP, brain natriuretic peptide; CNP, C-type natriuretic peptide; SNA, sympathetic nervous activity; AII, angiotensin II. Principal hormone(s) determining activity are shown in parentheses.

effect of ANP (or BNP) is in part dependent on hemodynamic factors—especially the maintenance of renal perfusion pressure and other intrarenal factors affected by renal blood flow *(7)*. Recent work in both human and experimental animals shows that there is a complex interaction between both circulating hormones (ANP and BNP) and CNP (currently thought to be local or paracrine in action). Thus, an increase in plasma ANP concentration can increase the steady-state plasma concentration of BNP and vice versa *(8)*. Such complex interactions among all three natriuretic peptides could be as important as changes in cardiac secretion in determining the biological (and cardiovascular) effects of these hormones. Also affecting the tissue response to natriuretic peptides will be the concentration (and hormone occupancy) of biological receptors (NPR-A in the case of ANP and possibly BNP, NPR-B for CNP), which are themselves subject to regulation *(7,9)*. Many other factors will also affect the hormone response, including the activity of tissue enzymes and clearance receptors involved in the hormone's degradation (Fig. 1) and the activity of other vasoactive peptides, particularly those of the renin angiotensin system, which (as discussed below) oppose many, if not all, of the actions of natriuretic peptides.

Although such a model is a basis for understanding the acute cardiovascular effects of natriuretic peptides, knowledge of the long-term actions and how they are mediated is harder to obtain. Potentially important cardiovascular sites of actions have been documented throughout the circulation, ranging from the capillaries to the heart itself (Table 1). Just as the brain/renin/angiotensin system is known to be important in blood pressure regulation *(10,11)*, there is now good evidence for a central (brain) natriuretic peptide hormone system (Fig. 1) *(12)*. A number of brain nuclei, both within and outside the blood–brain barrier and subserving cardiovascular regulation in experimental animals, are responsive to natriuretic peptides. Since this chapter is primarily concerned with the systemic actions of natriuretic peptides, readers are referred to previous reviews *(12,13)* for further details of the specific effects of natriuretic peptides within the central nervous system (CNS) (*see also* Chapters 13–15).

## CARDIOVASCULAR ACTIONS OF NATRIURETIC PEPTIDES

### *ANP*

The hemodynamic effects of ANP are multiple and complex, in keeping with the abundance and diverse distribution of NPR-A and NPR-C *(6,14,15)* within the vascular and nervous systems. Initially, the acute hypotensive effect of ANP, observed after iv injection of large doses of the hormone, was attributed to arterial vasorelaxation. This finding was consistent with in vitro observations of significant vasorelaxant effects of ANP on isolated precontracted blood vessels—particularly angiotensin II (AII)-induced vasoconstriction *(16)*. Subsequent in vivo studies, using large doses of ANP and producing variable vasodepression but consistent increase in heart rate (at least in humans *[17,18]*), showed that vascular resistance actually increased in most vascular beds *(19)*—presumably because of counter-neurohormonal responses to the initial and abrupt fall in blood pressure *(20)*. Despite this, a number of studies support the view that natriuretic peptides may have important "buffering" effects on potent arterial vasoconstrictors, including AII *(13,21)*—particularly within the renal preglomerular vessels where specific ANP receptors have been identified in vascular-membrane preparations *(22)*. Inhibition by ANP (and/or BNP) of endothelin I (ET-1) synthesis *(23)* and/or angiotensin-converting enzyme activity *(24)* is a further potential mechanism mediating the long-term hypotensive actions of natriuretic peptides. Although direct vasodilatation occurs during large-dose infusions of ANP in humans *(25,26)*, the physiological importance of such effects remain to be determined.

It is now agreed that the main factor causing fall in arterial pressure after acute (systemic) administration of ANP in normal health is reduction in cardiac output—mediated by a fall in cardiac filling pressure. Several factors contribute to the fall in filling pressure (preload), including shrinkage of plasma volume, atrial and venodilation, and (later) further fall in plasma volume from natriuresis and diuresis—but the shift of plasma fluid to the extra vascular space appears to be the most important. In humans, increase in venous blood hematocrit and albumin concentration are prompt and reproducible responses to small increments of ANP (or BNP) and presumably follow from an action of the hormone to increase capillary permeability to fluids (both water and solute), without affecting selectivity of the capillary barrier itself *(27,28)*. The site at which fluid shifts across the capillary is unclear, but there is evidence in humans receiving large doses of ANP that capillaries of the upper body segments and/or splanchnic sites may be the most affected *(29)*. The precise mechanism responsible for the increased capillary permeability is unknown but may involve an increase in efferent vascular tone within the capillary bed—analogous to the pressure changes induced by ANP across the glomerulus *(30)*. The possible contribution of ANP to venodilatation of capacitance vessels *(31)* and/or increased resistance to venous return *(32)* has been difficult to confirm but should not be dismissed—particularly in view of the known effects of AII on venous return *(33)*. Although cardiac muscle contractility *(34)* and coronary blood flow *(35)* are largely unaffected by ANP, a direct effect of the hormone on cardiac myocytes, reducing cell volume *(36)* (as well as tension and excitability *[37]*) has been proposed and may be a means whereby ANP modulates (inhibits) its own secretion. In this context ANP and/or BNP *(38)* (acting via NPR-A) may mediate changes in cardiac compliance and serve to minimize increases in end diastolic pressure in states of cardiac overload (Table 1).

Other direct effects on the heart (and other vascular smooth muscle tissues) relate to the antimitogenic action of ANP. Many factors, including AII and ET-1, appear to promote cardiac fibroblastic proliferation in the setting of cardiac hypertrophy and remodeling—actions that can be inhibited (possibly via NPR-C *[39]*) by all three natriuretic peptides *(40)*. Thus, natriuretic peptides could act as important modulators of vascular smooth-muscle-cell growth *(41)* not only within the heart *(42)*, but also within the arterial wall *(43)*. Although the physiological significance of these findings is difficult to assess, these actions are likely to be important in hypertension, heart disease, and heart failure and represent further examples of the remarkable antagonism between circulating (and paracrine) pressor (AII, ET-1) and dilator (natriuretic peptides and adrenomedullin, *44*) hormones.

One of the most elusive but potentially important vasodepressor actions of ANP involves the autonomic nervous system *(45)*. Inhibition of the arterial baroreceptor response *(46)* and/or inhibition of sympathetic nerve outflow *(47)* are reproducible effects of ANP in several species, including humans *(45)*. ANP appears to sensitize vagal afferents *(48)* (leading to reflex bradycardia), an effect that in humans can be prevented by angiotensin-converting enzyme inhibitors *(49)*. The mechanism and site(s) of action are complex, involving arterial and nonarterial *(48)* baroreceptor afferents as well as central and peripheral sites within the autonomic nerve *(45)*. Unexpectedly, it appears that the peripheral site of action may be potentiated by ANP in some circumstances *(50)*. Additional neural actions (vasodepressor) of ANP have been shown using microinjections into specific regions of the solitary tract nucleus *(51,52)*. These findings stand in contrast to AII, central injections of which raise systemic pressure *(53)*. These and many other central actions of ANP on the regulation of salt and water balance and arginine vasopressin secretion *(12)* illustrate the central interactions between AII and natriuretic peptides *(12,13,54)*.

Finally, but by no means unimportantly, it is obvious that the renal (natriuretic/diuretic) and endocrine (inhibition of renin/AII, inhibition of AII-induced aldosterone) actions of ANP will have major effects on cardiovascular homeostasis by reducing the volume of the vascular compartment, in addition to inhibiting the concentration of plasma AII itself. In this context a further important long-term effect of ANP is the leftward shift in the pressure natriuresis curve as discussed below, allowing sodium balance to be maintained at lower levels of arterial pressure *(55)*.

## *BNP*

Although clearly a circulating hormone, the secretion of which is subject to subacute change in central blood volume *(7)*, the cardiovascular actions of BNP have been less well-studied than ANP—possibly because (in contrast to ANP) the amino-acid sequence of BNP has strong species specificity. Thus, the biological actions of administered BNP in cross-species studies are much more variable than those of ANP and are difficult to predict *(56)*. For these reasons, it is essential that the actions of BNP are studied in the species of origin. Even if a homologous system is used, caution is necessary in extrapolating the findings in experimental animals to humans, where in several respects the actions (and metabolism) of the two hormones differ (Table 2). Early reports of porcine BNP showed similar hypotensive and muscle-relaxant effects to those reported for ANP *(57)*. Similarly, hypotensive and vasodilator activity of BNP was confirmed in both normotensive and spontaneously hypertensive rats *(58)*. The potency of porcine BNP in inhibiting AII-induced ET-1 release from cultured porcine endothelial cells was also similar to that of ANP *(59)*. These results, together with findings of similar affinity of ANP/BNP for NPR-A, similar cyclic guanosine monophosphate (cGMP) production (except humans; *see* Physiological and Integrative Effects of Natriuretic Peptides In Vivo), and similar inhibitory effects on both renin and aldosterone secretion as well as central vasodepressor activity *(60)*, encouraged the view that ANP and BNP were equally

Table 2
Comparative Physiology of ANP and BNP in Normal Humans

| | *ANP*[a] | *BNP* |
|---|---|---|
| Source of secretion | Atrium | Ventricle |
| Venous concentration | 4–27 pmol/L | 2–9 pmol/L |
| Circulating form | Low molecular weight | Mostly high molecular weight |
| Metabolic clearance rate | 3–4 L/min | 3–4 L/min |
| Half life | 3 min | 22 min |
| Femoral arteriovenous ratio | 1.5 | 1.06 |
| G-C activity[b] | 0.3 nmol cGMP/pmol ANP | 0.1 nmol cGMP/pmol BNP |
| Cleavage by endopeptidase | ++ | +– |
| Biological effects | | |
| Natriuresis | ++ | ++ |
| Renin inhibition | ++ | ++ |
| Fall in plasma volume | ++ | ++ |
| Inhibition of AII/aldo. | ++ | +– |
| Vasodepression[c] | + | ++ |

[a]ANP, atrial natriuretic peptide; BNP, brain natriuretic peptide; AII, angiotensin II; aldo, aldosterone; cGMP, cyclic guanosine monophosphate.
[b]G-C (guanylate cyclase) activity as determined from rise in plasma cGMP during exogenous infusions.
[c]From 5 h infusions in hypertensive humans *(110)*.

involved in cardiovascular homeostasis. Less well studied, however, are the direct neural (autonomic) and cardiac actions of BNP. In normal *(61)* and hypertensive humans *(62)*, incremental infusions of BNP reduce isovolemic relaxation time, suggesting an action to improve diastolic performance *(38)*, but whether such actions are specific to BNP is not known. Both ANP and BNP in high dosage induce increases in coronary blood flow *(63)*, but there is no evidence that these hormones increase coronary flow at hormone levels approximating physiological concentrations. The same can be said for the pulmonary vasorelaxant activity of ANP and BNP in humans *(64)*, although in the setting of hypoxic pulmonary vasoconstriction, reduction in pulmonary vascular resistance appeared to be greater during BNP than ANP infusions *(65)*. As yet no truly unique action (specific to BNP) has been identified in any species. Antimitogenic effects of BNP (e.g., paracrine actions within the heart) appear to be similar to those of ANP *(40,42)*, but much more research is required to fully evaluate such actions and their physiological significance.

## *CNP*

Unlike ANP and BNP, present evidence favors the view that CNP's actions are autocrine or paracrine rather than truly endocrine, since plasma levels are low (if detectable at all) and do not appear to be subject to regulation *(7)*. Nor is there good evidence for CNP synthesis within the heart. Nonetheless, findings that specific CNP receptors (NPR-B) and CNP gene transcripts exist within the vascular wall, the kidney *(66)*, and within the central and peripheral nervous system strongly point to a regulatory role for CNP in cardiovascular homeostasis. Systemic CNP-22 is vasorelaxant in dogs *(67)* (acting presumably via its specific receptor NPR-B *[68]*) and when tested in isolated canine vessels *(69)* was relatively more potent as a venodilator than ANP. Further, in canine arterial vessel preparations, there were differential responses to ANP and CNP; CNP was more active in saphenous arterial rings and less active in renal artery rings *(69)*. In human vascular preparations *(70)*, CNP has both veno and arterial dilator actions—in contrast to ANP, which is more potent than CNP as an arterial dilator but has no venodilator activity in vitro. Intrabrachial arterial infusions of CNP in humans also induce much less vasodilatation than does ANP *(71)*. Other cardiovascular actions include coronary vasorelaxation *(72)*, inhibition of AII-stimulated ET-1 release in porcine endothelial cells *(59)* (where CNP was more potent than either ANP or BNP), and inhibition of vascular angiotensin-converting enzyme activity *(73)*. CNP-22 inhibits vascular smooth-muscle cell growth in tissue culture *(74)* and inhibits intimal thickening after vascular injury *(75)*, raising the possibility that CNP may have an important antimitogenic role in the prevention of atheroma. Taken together, these findings support the existence of a vascular natriuretic peptide system *(76)* in which CNP participates as an endothelium-derived autocrine/paracrine regulator of vascular tone and remodeling (Fig. 1). Finally, neural effects of CNP may have significance for cardiovascular regulation. CNP suppressed peripheral sympathetic neurotransmission (independent of guanylyl cyclase activation) in rabbit isolated vas deferens preparations *(77)*. In rats, CNP is more potent than ANP in stimulating guanylyl cyclase in brain microvascular endothelial cells *(78)*, whereas in sheep, intracerebroventricular (icv) infusion of CNP-22 reduces systemic arterial pressure, in contrast to icv/ANP, which has no effect *(79)*. Again the physiological significance of these findings is quite unknown and awaits specific studies of mRNA CNP gene regulation in states of disordered cardiovascular homeostasis, and/or study of specific inhibitors of the hormone's action.

### *Other Natriuretic Peptides*

Large molecular weight forms (proANP, proBNP) presumably have similar biological actions to those of the respective mature forms, but have not been well studied for obvious reasons of availability. A number of other natriuretic peptides and/or metabolites are found in plasma and tissue extracts *(7)*. Urodilatin (proANP [95–126]), first identified from extracts of human urine, is an amino terminally extended form of ANP and is thought to be synthesized within renal tubular cells *(80)*. Urodilatin has not been detected in the circulation. Nevertheless, its increased production by the kidney in hypervolemic states may act to increase natriuresis *(80)* and thereby contribute to extracellular fluid and cardiovascular homeostasis *(4)*. Although the amino terminal metabolites of the prohormones (for example, proANP [1–98], proBNP [1–76] *[81]*) are present at relatively high concentration (e.g., amino-terminal proANP [1–98], 1 nmol/L in normal human plasma), specific actions and any physiological role of these "metabolites" are still debated. When infused in large doses, some of the smaller fragments (e.g., proANP [1–30], proANP [31–67], proANP [79–98]) are reported to show biological activity, including vasodepression and/or natriuresis and kaliuresis *(82–84)*. Contrary views notwithstanding *(85)*, recent evidence fails to confirm any significant levels of these smaller fragments in human plasma *(86)* and therefore their functions and roles in health and disease remain uncertain.

## PHYSIOLOGICAL AND INTEGRATIVE EFFECTS OF NATRIURETIC PEPTIDES IN VIVO

Following on from the above, it is clear that natriuretic peptides have a wide range of possible actions within the cardiovascular system. Whether these are realized in vivo is more difficult to prove, particularly since effects of hormone changes (occurring naturally or from exogenous infusions) will depend on dose, duration of stimulus, resting vascular tone, vascular compartment filling pressures, and counter regulatory neurohumoral (vasoconstrictor) activity—among many other factors. As emphasized in previous reviews *(7,87)*, deducing the physiological role of natriuretic peptides from studies using large infusions of hormone is difficult because of the interdependent nature of many of the organ responses and the counterregulatory response that is often evoked. Nonetheless, defining the physiological effects of natriuretic peptides is possible using several approaches. First, the effects of changes in blood hormone levels, commensurate with those found in health (or attained in states of disordered function, as in pathophysiological states) can be studied by infusing the exogenous hormone or administering drugs that affect the hormone's rate of degradation. Alternatively, specific inhibitors or blockers of natriuretic peptide action may be used to illuminate chronic effects. Both approaches have limitations in that physiological status is changed by the intervention—and in the direction opposite to that occurring in normal homeostasis. For example, small infusions of ANP in euvolemic states may raise plasma levels to those seen in sodium-loaded (mild) hypervolemic states but the effects of the intervention (increase in plasma ANP) will induce sodium depletion (and possibly vasodepression)—changes that are opposite to those seen in volume-loaded states that normally initiate the hormone's secretion. To this extent, many of the actions of infused hormone—particularly those on the kidney—are underestimated by interventions of this type. Similarly, the effects of blocking agents (reducing the activity of natriuretic peptide, therefore increasing extracellular fluid volume by antinatriuresis and possibly raising systemic arterial pressure) will be opposite to the set of circumstances that normally initiate a reduction in natriuretic peptide secretion. Nonetheless, these types of studies provide some insights into the actions and role of natriuretic peptide and complement numerous other contrived and ingenious interventions *(88–90)* aimed to address these issues.

In normal humans, short-term (3 h) infusions of ANP—sufficient to raise venous plasma ANP levels by 8 pmol/L (approximating the increase observed when dietary sodium is increased from normal to high intake)—induces significant natriuresis, fall in systolic arterial pressure, and inhibition of renin/AII as well as plasma aldosterone *(91)* (Fig. 2). Plasma catecholamines are unaffected. Similar effects occurred in subjects with hypertension (Fig. 2) except that blood pressure did not change *(87)*. Recent evidence in normal humans suggests that sympathetic nerve traffic to muscle and heart is also inhibited by changes in plasma ANP concentrations within the physiological range *(92)*. Fall in plasma volume *(93)*, inhibition of renin–aldosterone, and natriuresis are all consistent effects of small increases in ANP (and BNP, *see below*) in normal humans, whereas fall in blood pressure (as measured by sphygmomanometry) is much less consistent in short-term studies. In studies in which low-dose ANP infusions were given for 12 h to normal and hypertensive subjects at doses insufficient to increase venous plasma ANP, hematocrit was increased in both groups, but systolic blood pressure fell (approx 5%) only in the hypertensives *(94)*. Longer term (5-d) low-dose infusions of ANP in hypertensive humans, sufficient to double plasma levels of ANP, lower blood pressure (10%) after 12–24 h—pressures returning slowly toward preinfusion levels over 3 d after cessation of infusions *(95)*. The infusions also induced a significant early diuresis and natriuresis that waned as arterial pressure fell. A new steady state of sodium balance was then sustained at a lower level of arterial pressure *(95)*. In these studies, hematocrit was also

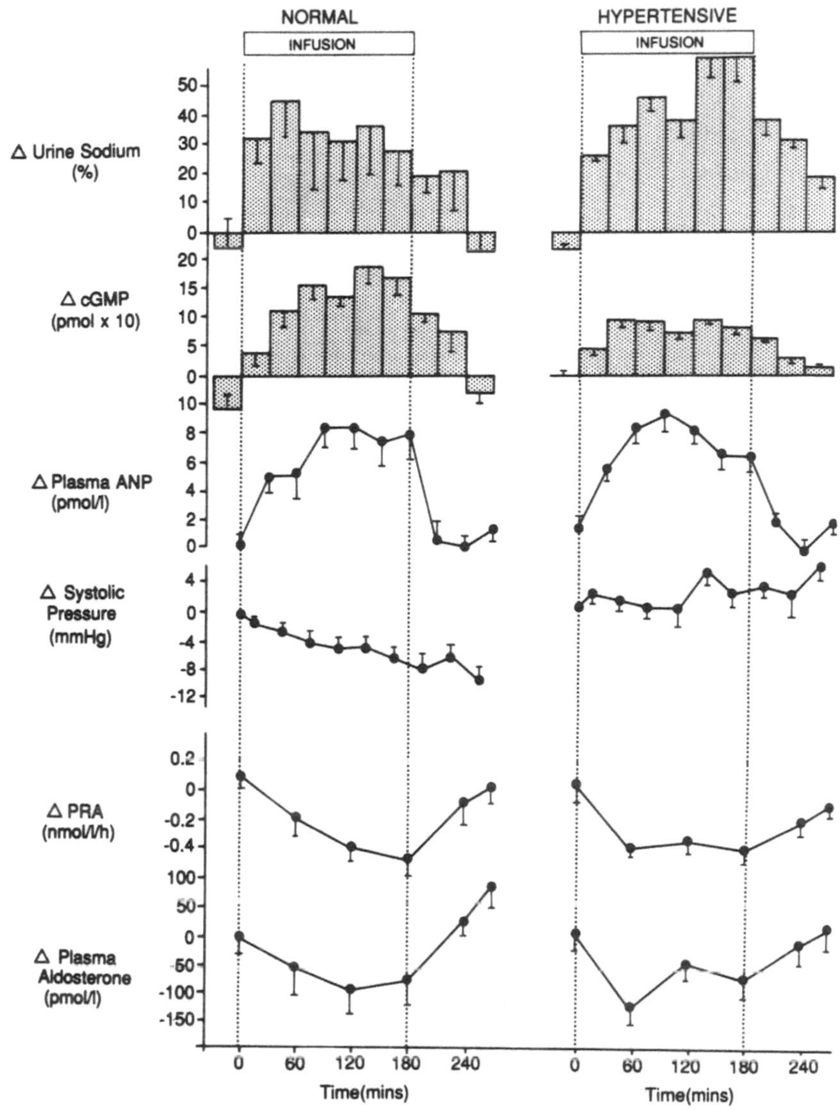


**Fig. 2.** Response to iv ANP (0.75 pmol/kg/min for 180 min) in six normotensive **(left)** and six hypertensive subjects **(right)** equilibrated on a daily sodium intake of 150 mmol. Values are plotted as change (Δ) from time-matched levels on control day. Change in urine sodium excretion is expressed as percent change from baseline (mean of first three collection periods before infusions = 100%). Change in systolic intra-arterial pressure is shown relative to values integrated for 30 min immediately before infusions in these subjects in whom intra-arterial pressure was continuously monitored. cGMP, cyclic guanosine monophosphate; ANP, atrial natriuretic peptide; PRA, plasma renin activity. (From Espiner EA, Richards AM. Atrial natriuretic peptide. Lancet 1989;i:707–709, with permission.)

increased. Similar hypotensive effects of chronic ANP infusions have been observed in normotensive primates *(96)*. In sheep, constant ANP infusion for 24 h, increasing plasma ANP from 10 to 30 pmol/L, significantly reduces systolic and mean arterial pressure, right atrial pressure, and plasma volume, yet induces only a transient natriuresis *(97)*.

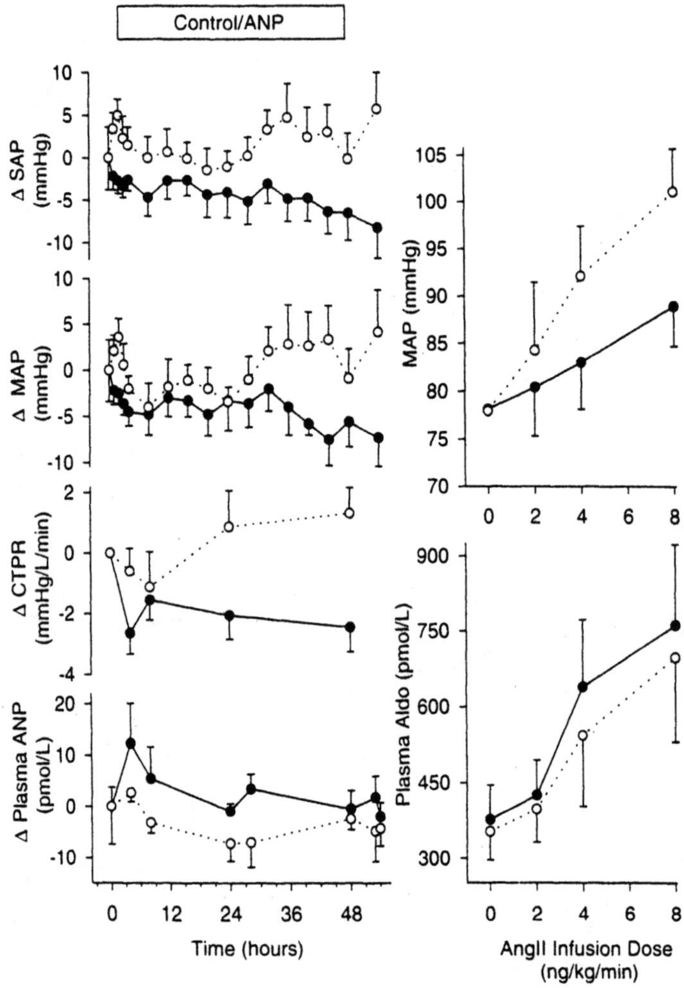


**Fig. 3.** Cardiovascular and endocrine effects of prolonged low dose (0.8 pmol/kg/min for 53 h) infusions of ANP (atrial natriuretic peptide, filled symbols) and placebo control (open circles) in eight normal sheep. **(Left)** Hemodynamic response. Values are shown as mean changes (Δ) from baseline ± SE. SAP, systolic arterial pressure; MAP, mean arterial pressure; CTRP, calculated total peripheral resistance. **(Right)** Pressor and plasma aldosterone response to stepped 30 min infusions of angiotensin II (AII), 0, 2, 4, 6, and 8 ng/kg/min after 48–50 h infusions of natriuretic peptide (filled circles) or placebo control (open circles). Values are mean ± SE in eight sheep. MAP, mean arterial pressure; plasma aldo, plasma aldosterone. (From Charles CJ, Espiner EA, Richards AM. Cardiovascular action of ANF: contributions of renal, neurohumoral and hemodynamic factors in sheep. Am J Physiol 1993;264:R533–R538, with permission.)

Lower doses given for 53 h, insufficient to significantly change plasma venous ANP levels above control observations (Fig. 3), similarly reduce arterial pressure as well as calculated peripheral vascular resistance *(21)*. No significant inhibition of basal renin/angiotensin or aldosterone was observed, but there was a significant inhibition of the pressor response to exogenous AII (Fig. 3), without change in the autonomic (baroreceptor) responsiveness to exogenous stimulation *(21)*. In normal sheep receiving small doses of ANP over a 5-d period *(98)*, similar changes were observed, with an early and transient fall in cardiac output followed later by sustained vasodepression as peripheral vascular resistance declined and cardiac output returned toward normal. Taken together, these studies support a chronic vasodepressor (mainly systolic) action of ANP, characterized by plasma volume contraction without neurohumoral reflex activation or increase in renin–angiotensin. Also contributing to the sustained lowering of arterial pressure is the leftward shift in the pressure natriuresis curve, allowing the kidney to maintain sodium balance at a lower level of arterial pressure *(55,99)*.

It should be emphasized that these biological effects of natriuretic peptides apply to normal physiology in which the integrity of the circulation is largely intact. Quite different responses may occur in pathological states, such as chronic heart failure. Although there are few reports on the effect of small incremental changes in plasma natriuretic peptide levels in heart failed states, large dose infusions *(100)*—raising venous plasma ANP levels from approx 100 to 600 pmol/L—lowered preload, mean arterial pressure, and peripheral vascular resistance, and *increased* cardiac output without changing heart rate or sympathetic nervous activity *(100)*. Similar observations have been made by others *(101)*. These "beneficial" effects on the overloaded heart presumably follow from peripheral vasodilatation (reducing afterload) as well as from reduction in preload (? veno dilatation and/or fall in plasma volume). The lack of neurohormonal activation (particularly renin-AII), despite substantial fall in mean arterial pressure, suggests neurohormonal inhibition by ANP—actions that may further contribute to the beneficial effect in heart failure. Again, defining precise mechanisms of action whereby natriuretic peptide elevations exert their effects requires further work. However, it is interesting to note that in contrast to the characteristic renal unresponsiveness of sodium-retaining states to exogenous infusions of ANP *(100)* (and BNP *[102]*), significant vasodepression still occurs—despite raised levels of endogenous plasma levels of natriuretic peptide in such situations as congestive heart failure *(103)*.

Whereas BNP is also a circulating hormone whose secretion is increased by raised intracardiac (ventricular) pressure and/or stretch *(7)*, the effects of small ("physiological") increments in plasma BNP have been less well-studied. Possibly because of the availability of the human form (BNP-32), and the availability of suitable antisera for human BNP assays, more studies have been done in humans than in experimental animals. In normotensive humans, short-term (2–3 h) infusions of BNP-32, sufficient to achieve plasma levels observed after acute myocardial infarction (20–30 pmol/L *[104]*) or found in mild heart failure (30–40 pmol/L *[105]*), induce renin–aldosterone inhibition, natriuresis, and fall in plasma volume without significant vasodepression *(106)*. Although these effects are similar to those of equimolar infusions of ANP *(107)*, the increase in plasma (and urine *[108]*) cGMP is much less (approx 25% of that observed with ANP)—suggesting that the mechanism of action of the two hormones differs in humans (Table 2). Also different is the effect on AII-induced aldosterone secretion *(107)*. Compared to placebo infusions, ANP infusions achieving plasma levels of 30–40 pmol/L significantly inhibited the plasma aldosterone response to incremental infusions of AII, whereas BNP infusions achieving the same plasma concentration did not *(107)*. Similar observations have been made by others at much higher plasma natriuretic peptide hormone concentrations *(108)*. These differential effects may relate to the greater affinity of ANP (compared to BNP) for NPR-A as well as NPR-C in humans *(109)*, but do not explain the comparable potency of the two hormones on sodium excretion, plasma volume, and basal renin–aldosterone concentrations in the face of much less stimulation of plasma cGMP by BNP. More prolonged BNP infusions (5 h) in subjects with hypertension were more vasodepressor than equimolar infusions of ANP and induced significantly greater natriuresis despite much less stimulation of plasma and urine cGMP *(110)*. Short-term, low-dose infusion of BNP, sufficient to increase basal plasma levels 1.5–3-fold in normal subjects, also induced natriuresis and inhibited renin–aldosterone, but did not affect hematocrit or blood pressure *(111)*. Biological activity resulting from small changes in plasma BNP has been confirmed by recent studies in which BNP increments as little as 4 pmol/L (twofold above resting level and quite within the normal range for healthy subjects) significantly reduced basal levels of plasma renin activity (unpublished observations). Taken together, it is clear that short-term, small increases in plasma BNP have significant biological actions that are likely to have important hemodynamic consequences, both in normal and hypertensive states. These findings mean that both ANP and BNP must be taken into account in assessing the circulation's response to chronic volume overload, and that earlier reports measuring only changes in ANP underestimated the heart's endocrine response.

When BNP is coinfused with ANP in normal humans, additive rather than synergistic effects are observed *(108,112)*. In these and similar studies, systolic arterial pressure was lowered by the addition of BNP—an effect that continued well beyond the cessation of infusion *(110,112)*. Although synergistic effects of the two hormones have not been shown over short periods of time *(8,110,112)*, a functional synergism is a possible outcome of more sustained periods of hormone hypersecretion, particularly since plasma (and presumably tissue) levels of other endogenous natriuretic peptides are increased by changes in any one individual hormone *(113)*. As yet few, if any, studies have examined the cardiovascular or other biological effects of chronic elevations in BNP. Thus, the functional contributions made by BNP over and above those of ANP in states of chronic cardiac overload are currently unknown.

Although there is little if any evidence that CNP is a circulating hormone, increase in endogenous plasma CNP concentration is observed during infusion of BNP *(113)* and is presumably analogous to the rise in plasma ANP observed during infusion of CNP *(114)*. Cardiovascular actions of exogenous CNP infusions have been studied in dogs, sheep, and humans. Initial reports in anesthetized dogs suggested that CNP was more vasodepressor than ANP *(115)*, possibly because CNP induced vasorelaxation of both arteries and veins *(69,70)*. However, comparative studies of infusions (Fig. 4) achieving levels of natriuretic peptide in the range of 20–30 pmol/L in normal conscious sheep indicate that CNP is less vasoactive in this species than equimolar infusions of either ANP or BNP *(113)*. All three hormones have similar pharmacokinetics in sheep. Compared to ANP and BNP—which were equipotent in stimulating an increase in plasma cGMP and in lowering mean arterial pressure and cardiac output, CNP infusions were less effective—yet all three hormone infusions were equipotent in suppressing plasma aldosterone concentration in sheep (Fig. 4) *(113)*. Since infusions of CNP in sheep also increase the endogenous level of BNP (and ANP), some or all of the bioactivity from exogenous CNP infusions may be attributed to the action of other natriuretic peptides *(113,116)*. Similarly in humans, CNP infusions sufficient to raise the plasma level to 60 pmol/L *(114)* were neither vasodepressor nor natriuretic but did act to inhibit plasma aldosterone. Since plasma ANP was significantly increased by these CNP infusions, fall in aldosterone during CNP infusion may have been mediated by other natriuretic peptides. However, since CNP has some inhibitory actions on aldosterone production in vitro, a direct effect on the adrenal glomerulosa cannot be excluded *(117)*. Plasma volume was unaffected by

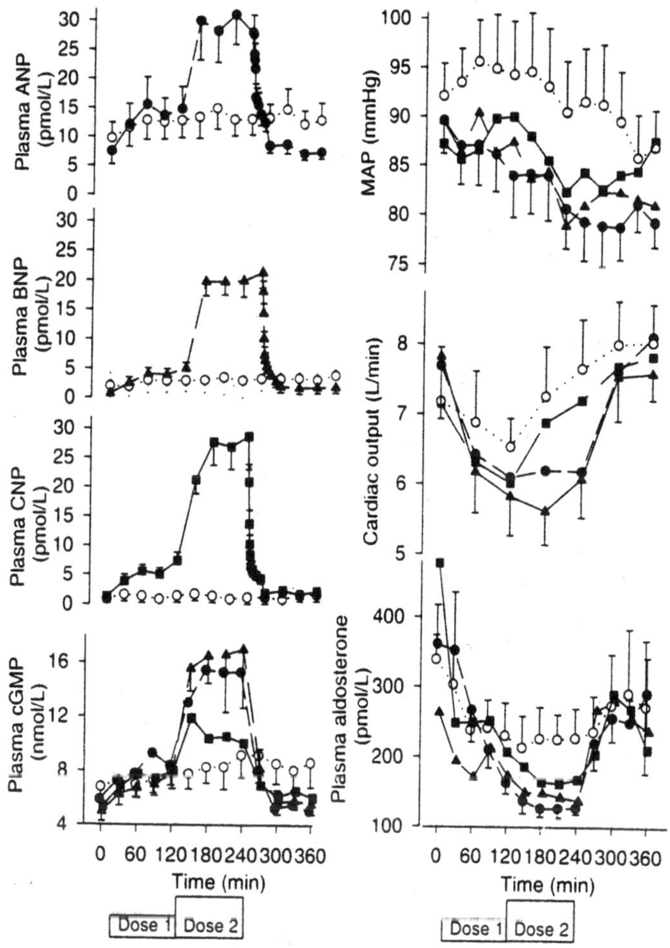


**Fig. 4.** Comparative effects of natriuretic peptide infusions (ANP, filled circles; BNP, filled triangles; CNP, filled squares) and placebo control (open circles) in eight normal sheep. Sheep received two separate doses (dose 1, 0.5 pmol/kg/min for 2 h; dose 2, 2.5 pmol/kg/min for 2 h) in succession. **(Left)** Plasma natriuretic peptide values and plasma cGMP response to infusions of ANP (atrial natriuretic peptide); BNP (brain natriuretic peptide); and CNP (C-type natriuretic peptide). **(Right)** Hemodynamic and plasma aldosterone response to infusions. MAP, mean arterial pressure. (From Charles CJ, Espiner EA, Richards AM, Nicholls MG, Yandle TG. Comparative bioactivity of atrial, brain and C-type natriuretic peptides in conscious sheep. Am J Physiol 1996;270:R1324–1331, with permission).

CNP infusions in these studies in humans and sheep. Whereas these studies do not illuminate the role of endogenous CNP, they do indicate the complexity of interactions among the family of natriuretic hormones and the potential for multiple sites of action in cardiovascular homeostasis. Although prolonged CNP infusions in rats have been shown to inhibit intimal thickening after vascular injury *(118)*, the physiological significance of these findings is not yet known. As yet there have been no studies of the hemodynamic effects (if any) of chronic elevations in plasma CNP.

Cardiovascular actions of natriuretic peptides can also be deduced from the effects of drugs inhibiting the hormone's clearance or metabolism. Numerous studies in humans and experimental animals *(5,119)* show that inhibitors of the enzyme endopeptidase

EC.24.11—an enzyme that inactivates ANP by cleaving the Cys 105 Phe 106 bond—induce increases in plasma ANP and commensurate physiological effects including natriuresis, inhibition of renin-aldosterone, fall in systolic arterial pressure *(120)*, and inhibition of sympathetic nervous activity *(92)*. Since the enzyme also has the capacity to hydrolyse CNP and BNP (variably, depending on the species form of BNP *[121]*), biological responses to enzyme inhibition may represent the actions of more than one hormone. Further, a variety of other peptides (including AII, bradykinin, and substance P), are substrates for the enzyme, so caution is necessary before accepting that effects from inhibition of the enzyme are specific to the promotion of the actions of tissue natriuretic peptides. Nonetheless, as reviewed in detail elsewhere *(5,119,122)*, the sum of the evidence from acute administration of enzyme inhibitors strongly supports findings obtained using exogenous infusions of natriuretic peptide, and are consistent with the view that small increases in ANP (and possibly other natriuretic peptides) have potentially important actions on the circulation. Deducing actions of natriuretic peptides from longer term use of enzyme inhibitors is more difficult since increases in plasma ANP concentrations are less well sustained *(119)*, possibly because of reducing cardiac secretion. Even so, subtle but significant and potentially important cardiovascular effects have been observed *(119)*, including a delayed but sustained fall in blood pressure over a 4-d period in mildly hypertensive humans. Reports *(123)* that chronic administration of endopeptidase inhibitors in rats exposed to chronic hypoxia reduced pulmonary vascular remodeling (as well as reducing right ventricular hypertrophy) are consistent with an important antimitogenic action of natriuretic peptides. These findings require confirmation using other models of cardiovascular overload, as well as evidence that the effects are mediated by increases in tissue levels of natriuretic peptides.

A major determinant of natriuretic peptide disappearance from plasma is specific uptake (clearance) via the NPR-C, which (at least in the rat) is the most numerous of the natriuretic receptors *(15)*. At physiological levels, receptor occupancy is reported to be <5% *(15)*. Administration of the truncated ANP analog C-ANP (4–23)—which binds to the NPR-C but not to NPR-A or NPR-B—raises endogenous ANP levels in blood with consequential biological effects *(15)*. In general, studies in experimental animals show that such increments in plasma ANP have cardiovascular effects similar to those expected from exogenous infusions of the hormone *(124)*. However, in view of the possible mediation of NPR-C in some biological actions of natriuretic peptides *(125)*, caution is again necessary before attributing all the actions of the NPR-C ligand in vivo to effects of natriuretic peptides mediated by NPR-A or -B. Similarly, since all three natriuretic peptides bind to NPR-C with high (and similar affinity) in many species, biological effects following from administration of C-ANP (4–23) in vivo may be mediated by one or more of these hormones. Short-term infusions of C-ANP (4–23) in normal conscious sheep lower arterial pressure, reduce cardiac output and stroke volume, and slightly increase peripheral vascular resistance *(126)*, as is also observed during short-term exogenous infusions of ANP. Dose-dependent biological effects of the NPR-C ligand and EC.24.11 inhibitor, both alone and in combination, have recently been studied in conscious sheep *(127)*. When given alone, both "inhibitors" induced similar increments in plasma ANP and BNP, showing that (at least in sheep) both of these hormones are similarly metabolized and that the contribution of the two degradative pathways (NPR-C and EC.24.11) are similar. When given in combination (Fig. 5), increments of plasma ANP and BNP were associated with a progressive fall in mean arterial pressure, increase

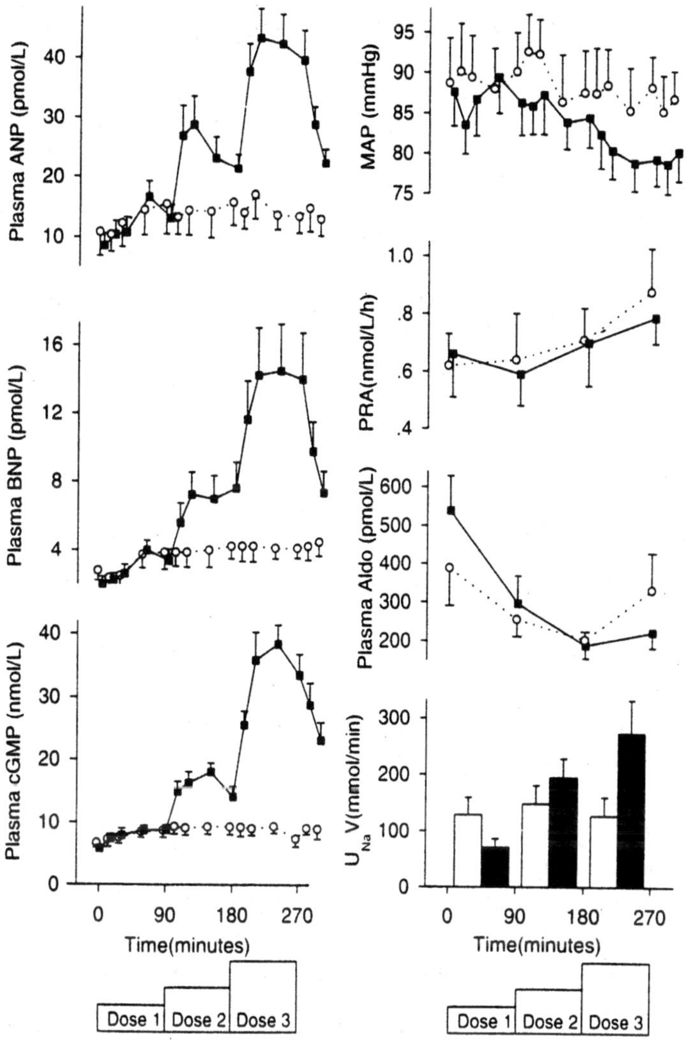


**Fig. 5.** Hormone and hemodynamic responses to inhibition of natriuretic peptide degradation by natriuretic peptide receptor subtype C (NPR-C) ligand (C-ANP4–23) and endopeptidase inhibitor (SCH 32615). Three incremental doses, each a combination of C-ANP (4–23) and SCH 32615, were given at intervals of 90 min to eight normal sheep. **(Left)** Plasma ANP, BNP, and cGMP responses to incremental combination of inhibitors (filled squares) or placebo infusions (open symbols). Values are mean ± SE ($n$ = 8). **(Right)** Hemodynamic, hormonal, and renal responses to incremental combination of inhibitors (filled symbols) or placebo infusion (open symbols). MAP, mean arterial pressure; PRA, plasma renin activity; plasma aldo, plasma aldosterone; $U_{Na}V$, urinary sodium excretion. Values are mean ± SE ($n$ = 8). Dose 1, SCH 32615 1 mg/kg bolus plus C-ANP4–23 20 pmol/kg/min. Dose 2, SCH 32615 5 mg/kg bolus plus C-ANP4–23 100 pmol/kg/min. Dose 3, SCH 32615 25 mg/kg bolus plus C-ANP4–23 500 pmol/kg/min. (From Charles CJ, Espiner EA, Nicholls MG, Richards AM, Yandle TG, Protter A, Kosoglou T. Clearance receptors and endopeptidase 24.11: equal role in natriuretic peptide metabolism in conscious sheep. Am J Physiol 1996;271:R373–R380 with permission.)

in heart rate, increasing natriuresis, and hemoconcentration. Despite these significant hemodynamic effects in normal sheep, renin was not increased and plasma aldosterone actually fell. These responses are concordant with responses achieved during exogenous infusions of natriuretic peptide *(97,113)*. In contrast to endogenous levels of ANP and BNP (both of which increased approximately fourfold) (Fig. 5), no increase in endogenous plasma CNP was detected during these studies in which hormone degradation pathways were blocked. It needs to be noted that species variations may well determine the response to either NPR-C ligand or EC.24.11 inhibition. As already discussed, in contrast to many of the findings in sheep, the actions (and metabolism) of ANP and BNP differ in humans *(107)* (Table 2), as does the affinity of the two hormones for NPR-A and NPR-C *(109)* and the enzyme EC.24.11 *(121)*. Therefore, specific studies are needed in humans to determine the cardiovascular actions consequent to NPR-C ligand administration, enzyme inhibition, and combination of these inhibitors. However, overall the findings from studies of interventions blocking metabolism/degradation of natriuretic peptides are remarkably similar in both experimental animals and humans, and provide further strong support for the cardiovascular actions demonstrated using exogenous natriuretic peptide infusions.

Deducing the likely long-term physiological effects of natriuretic peptides has been possible using specific blockers of hormone action. For example, administration of monoclonal antibody raised against ANP has induced significant cardiovascular responses in rats, including acceleration of hypertension *(128)* and increase in cardiac preload in the setting of experimental heart failure *(129)*. The advent of a specific guanylyl cyclase inhibitor—HS-142-1 *(130)* has allowed further dissection of the chronic effects of natriuretic peptides in vivo. Although only a relatively weak competitive antagonist at the receptor level (both NPR-A and NPR-B), the drug decreased absolute and fractional sodium excretion, and increased plasma renin activity in normal (anesthetized) dogs when given directly by bolus injection into the renal artery *(131)*. Similar findings of antinatriuresis have been reported in normal (euvolemic) rats *(132)* after systemic bolus injection of the antagonist. Blood pressure was unaffected. Similar renal effects are reported in rat models of heart failure *(133)* and in "remnant" kidney failure *(134)*. Using larger (fourfold) doses of HS-142-1 in anesthetized rats with congestive heart failure, it was possible to show important hemodynamic effects *(135)*, including increase in mean arterial pressure and renal vascular resistance. Glomerular filtration rate (GFR), renal plasma flow, and fractional sodium excretion were all reduced by the antagonist. These findings support the view that natriuretic peptides play an important role in maintaining GFR and sodium excretion in heart failure and in addition have vasodilator actions. Further, in conscious dogs with pacing-induced heart failure *(136)*, repeated bolus injections of HS-142-1 induced approximately twofold increases in plasma renin activity, aldosterone, and norepinephrine when compared to control studies. Dose-dependent reduction (30–40% fall) in both plasma and urine cGMP was also observed, but in these studies, blood pressure was unaffected. Significantly, in a model of mild left ventricular dysfunction developed in the conscious dog *(137)*, the guanylyl cyclase antagonist decreased plasma and renal generation of cGMP—changes that were associated with abrupt increases in PRA, mean arterial pressure, and fractional reabsorption of sodium at both proximal and distal renal tubular levels. Again, these findings are consistent with the known effects of natriuretic peptides and provide strong evidence that these hormones have important cardiovascular and homeostatic roles in chronic disorders, such as mild left ventricular dysfunction.

Insight to the chronic effects of natriuretic peptides in experimental animals can be gained by gene manipulations. The first transgenic model in mice *(138)*, in which there is constitutive expression of the murine ANP gene targeted to hepatocytes through the transthyretin promoter, demonstrates fivefold increases in plasma ANP (compared to nontransgenic litter mates) and associated vasodepression (mean arterial pressure 24 mmHg lower than litter mates) and fall (21%) in peripheral vascular resistance *(139)*. A similar experimental murine model *(140)* in which hepatocytes express a human serum amyloid P component/mouse BNP fusion gene shows that greatly increased plasma BNP levels (10–100-fold) are associated with a significantly lower level of blood pressure (mean arterial systolic blood pressure 100 compared to 125 mmHg in nontransgenic litter mates). Plasma cGMP was increased two- to sixfold *(140)*. Two other models involving genetic manipulations *reducing* the synthesis or action of natriuretic peptides have shown exactly opposite effects. Homozygous mutants *(141)* carrying a disruption of the proANP gene had no detectable ANP in blood (nor granules in the cardiac atrium) and showed blood pressure increases of 8–23 mmHg—increases that were, in part, salt-intake dependent. Heart size was increased and hematocrit reduced in homozygotes. Recently a salt-resistant form of hypertension has been described in transgenic mice in which the NPR-A gene was disrupted *(142)*. Both homo and heterozygous NPR-A mutants showed an elevated systolic blood pressure (27 and 10 mmHg, respectively) compared to control mice. These blood pressure changes were unaffected by high salt intake, and renal and aldosterone responses to changing dietary sodium were similar to those observed in controls. Further, plasma ANP levels, kidney function, and body weight and hematocrit were not different from control animals. The difference in the degree of blood pressure response (and sensitivity to salt) in these two models may relate to their differential effects on ANP and BNP—both of which may be rendered ineffective when disruption of the NPR-A gene occurs, whereas the presence of BNP activity may partially compensate when the proANP gene is disrupted. In any event, all of these four transgenic models—two enhancing and two reducing the contributions of natriuretic peptides—show important hemodynamic consequences with appropriate (and intact) mechanisms of renal compensation. Such findings are consistent with the known physiological actions of these hormones. Importantly, these experiments highlight the role of natriuretic peptides in long-term cardiovascular homeostasis.

## SUMMARY

Findings from both short-term and longer term exogenous infusions of natriuretic peptides in humans or experimental animals, as well as from studies in which hormone degradation is inhibited, confirm many in vitro actions of natriuretic peptides at multiple sites within tissues regulating cardiovascular homeostasis. Most, if not all, these actions, including those within the brain, oppose those of the renin–angiotensin system. Major initial effects of the circulating hormones ANP and BNP include contraction of plasma volume, natriuresis, inhibition of renin–aldosterone, and suppression of sympathetic nervous activity. Whereas ANP and BNP appear to be equipotent in some species, recent findings point to differential effects and mechanism of action of ANP and BNP in humans. Evidence for sustained actions of natriuretic peptides is now supported by findings from transgenic studies and experiments using specific inhibitors of the "biological" natriuretic peptide receptor. The sustained effects of natriuretic peptides are likely to include vasorelaxation and leftward shift in the pressure-natriuresis curve, and possibly inhibition of vascular smooth muscle cell growth—all of which can contribute to the maintenance of sodium homeostasis at lower levels of arterial pressure. Despite these advances in knowledge, major questions remain—not least being the specific action of the paracrine hormone, CNP, the possible contributions of the amino terminal prohormone "metabolites" of ANP and BNP, and the roles of BNP over and above those of ANP. Some of these questions, made more difficult by knowledge of the complex interaction among all three natriuretic peptides at the cellular level, may be answered by genetic manipulations, such as ablations of specific genes and/or their putative receptors.

## ACKNOWLEDGMENTS

Studies undertaken from the Departments of Endocrinology (Espiner EA, Yandle TG, Charles C, Cameron V) and Cardiology (Richards AM, Nicholls MG, Crozier I) Christchurch Hospital, and quoted in this review were generously supported by the Health Research Council of New Zealand and the National Heart Foundation. Susan Isaacs's assistance with manuscript preparation is gratefully acknowledged.

## REFERENCES

1. de Bold AJ, Borenstein HB, Veress AT, Sonnenberg H. A rapid and potent natriuretic response to intravenous injection of atrial myocardial extract in rats. Life Sci 1981;28:89–94.
2. de Wardener HE. Kidney, salt intake, and Na+, K+-ATPase inhibitors in hypertension. Hypertension 1991;17:830–836.
3. Goetz KL. Evidence that atriopeptin is not a physiological regulator of sodium excretion. Hypertension 1990;15:9–19.
4. Drummer C, Franck W, Heer M, Forssmann W-G, Gerzer R, Goetz K. Postprandial natriuresis in humans: further evidence that urodilatin, not ANP, modulates sodium excretion. Am J Physiol 1996;270:F301–F310.
5. Ruskoaho H. Atrial natriuretic peptide: synthesis, release, and metabolism. Pharmacol Rev 1992;44:479–602.
6. Brenner BM, Ballermann BJ, Gunning ME, Zeidel ML. Diverse biological actions of atrial natriuretic peptide. Physiol Rev 1990;70:665–699.
7. Espiner EA, Richards AM, Yandle TG, Nicholls MG. Natriuretic hormones. Endocrinol Metab Clin North Am 1995;24:481–509.
8. Hunt PJ, Espiner EA, Richards AM, Yandle TG, Frampton C, Nicholls MG. Interactions of atrial and brain natriuretic peptides at pathophysiological levels in normal men. Am J Physiol 1995;269:1397–1403.
9. Cao L, Wu J, Gardner DG. Atrial natriuretic peptide suppresses the transcription of its guanylyl cyclase-linked receptor. J Biol Chem 1995;270:24,891–24,897.
10. Bunneman B, Fuxe K, Ganten D. The renin–angiotensin system in the brain: an update 1993. Regulatory Peptides 1993;46:497–509.
11. Rademaker MT, Fitzpatrick MA, Charles CJ, Frampton CM, Richards AM, Nicholls MG, Espiner EA. Central angiotensin II AT1-receptor antagonism in normal and heart failed sheep. Am J Physiol 1995;269:H425–H432.
12. Imura H, Nakao K, Itoh H. The natriuretic peptide system in the brain: implications in the central control of cardiovascular and neuroendocrine functions. Frontiers Neuroendocrinol 1992;13:217–249.
13. Schiffrin EL, Garcia R, Konrad EM. Interactions of angiotensin II and atrial natriuretic peptide. In: Raizada MK, Phillips MI, Sumners C, eds. Cellular and Molecular Biology of the Renin–Angiotensin System. CRC, Boca Raton, FL 1993;7, pp. 433–468.
14. Suga S, Nakao K, Hosoda K, Mukoyama M, Ogawa Y, Shirakami G, Arai H, Saito Y, Kambayashi Y, Inouye K. Receptor selectivity of natriuretic peptide family, atrial natriuretic peptide, brain natriuretic peptide, and C-type natriuretic peptide. Endocrinology 1992;130:229–239.
15. Maack T, Okolicany J, Koh GY, Price DA. Functional properties of atrial natriuretic factor receptors. Semin Nephrol 1993;13:50–60.
16. Garcia R, Thibault G, Cantin M, Genest J. Effects of a purified atrial natriuretic factor on rat and rabbit vascular strips and vascular beds. Am J Physiol 1984;247:R34.
17. Richards AM, Nicholls MG, Ikram H, Webster MW, Yandle TG, Espiner EA. Renal, haemodynamic, and hormonal effects of human alpha natriuretic peptide in healthy volunteers. Lancet 1985;1:545–549.

18. de Zeeuw D, Janssen WMT, de Jong PE. Atrial natriuretic factor: its (patho)physiological significance in humans. Kidney Intern 1992;41:1115–1133.
19. Lappe RW, Smits JF, Todt JA, Debets JJ, Wendt RL. Failure of atriopeptin II to cause arterial vasodilation in the conscious rat. Circ Res 1985;56:606–612.
20. Shapiro JT, Deleonardis VM, Needleman P, Hintze TH. Integrated cardiac and peripheral vascular response to atriopeptin 24 in conscious dogs. Am J Physiol 1986;251:H1292–H1297.
21. Charles CJ, Espiner EA, Richards AM. Cardiovascular action of ANF: contributions of renal, neurohumoral and hemodynamic factors in sheep. Am J Physiol 1993;264:R533–R538.
22. de Leon H, Bonhomme M-C, Thibault G, Garcia R. Localization of atrial natriuretic factor receptors in the mesenteric arterial bed. Circ Res 1995;77:64–72.
23. Kohno M, Yokokawa K, Horio T, Yasunari K, Murakawa K-I, Takeda T. Atrial and brain natriuretic peptides inhibit the endothelin-1 secretory response to angiotensin II in porcine aorta. Circ Res 1992;70:241.
24. Kawaguchi H, Sawa H, Yasuda H. Effect of atrial natriuretic factor on angiotensin converting enzyme. J Hyper 1990;8:749–753.
25. Bolli P, Muller FB, Linder L, Raine AEG. Resink TJ. The vasodilator potency of atrial natriuretic peptide in man. Circulation 1987;75:221–228.
26. Branten AJW, Smits P, Jansen TLThA, Wollersheim H, Thien Th. Effect of atrial natriuretic factor on skin microcirculation versus skeletal muscle blood flow. J Cardiovasc Pharmacol 1996;27:303–306.
27. McKay MK, Huxley VH. ANP increases capillary permeability to protein independent of perfusate protein composition. Am J Physiol 1995;268:H1139–H1148.
28. Wijeyaratne CN, Moult PJA. The effect of α human atrial natriuretic peptide on plasma volume and vascular permeability in normotensive subjects. J Clin Endocrinol Metab 1993;76:343–346.
29. Watenpaugh DE, Vissing SF, Lane LD, Buckey JC, Firth BG, Erdman W, Hargens AR, Blomqvist CG. Pharmacologic atrial natriuretic peptide reduces human leg capillary filtration. J Cardiovasc Pharmacol 1995;26:414–419.
30. Dunn BR, Ichikawa I, Pfeffer JM, Troy JL, Brenner BM. Renal and systemic hemodynamic effects of synthetic atrial natriuretic peptide in the anesthetized rat. Circ Res 1986;59:237–246.
31. Villanueva MM, Nunes JP, Soares-da-Silva P. Relaxant effects of α human atrial natriuretic peptide on venous smooth muscle. J Auton Pharmacol 1991;11:139–145.
32. Chien Y-W, Frohlich ED, Trippodo NC. Atrial natriuretic peptide increases resistance to venous return in rats. Am J Physiol 1987;252:H894–H899.
33. Warner TD. Simultaneous perfusion of rat isolated superior mesenteric arterial and venous beds: comparison of their vasoconstrictor and vasodilator responses to agonists. Br J Pharmacol 1990;99:427–433.
34. Semigran MJ, Aroney CN, Herrmann HC, Dec GW, Boucher CA, Fifer MA. Effects of atrial natriuretic peptide on left ventricular function in hypertension. Hypertension 1994;24:271–279.
35. Bach RJ, Dai X-Z, Schwartz JS, Chen DG. Effects of atrial natriuretic peptide in the canine coronary circulation. Circ Res 1988;62:178–183.
36. Clemo HF, Baumgarten CM. cGMP and atrial natriuretic factor regulate cell volume of rabbit atrial myocytes. Circ Res 1995;77:741–749.
37. Meulemans AL, Sipido ER, Sys SV, Brutsaert DL. Atriopeptin III induces early relaxation of isolated mammalian papillary muscle. Circ Res 1988;62:1171–1174.
38. Clarkson PBM, Wheeldon NG, MacLeod C, Coutie W, MacDonald TM. Brain natriuretic peptide: effect of left ventricular filling patterns in healthy subjects. Clin Sci 1995;88:159–164.
39. Cahill PA, Hassid A. Clearance receptor-binding atrial natriuretic peptides inhibit mitogenesis and proliferation of rat aortic smooth muscle cells. Biochem Biophys Res Comm 1991;179:1606–1613.
40. Cao L, Gardner DG. Natriuretic peptides inhibit DNA synthesis in cardiac fibroblasts. Hypertension 1995;25:227–234.
41. Appel RG. Growth-regulatory properties of atrial natriuretic factor. Am J Physiol 1992;262:F911–F918.
42. Fujisaki H, Ito H, Hirata Y, Tanaka M, Hata M, Lin M, Adachi S, Akimoto H, Marumo F, Hiroe M. Natriuretic peptides inhibit angiotensin II-induced proliferation of rat cardiac fibroblasts by blocking endothelin-1 gene expression. J Clin Invest 1995;96:1059–1065.
43. Brandt RR, Heublein DM, Mattingly, MT, Pittelkow MR, Burnett JC. Presence and secretion of atrial natriuretic peptide from cultured human aortic endothelial cells. Am J Physiol 1995;268:H921–H925.
44. Schell DA, Vari RC, Samson WK. Adrenomedullin: a newly discovered hormone controlling fluid and electrolyte homeostasis. Trends Endocrinol Metab 1996;7:7–13.

45. Clemo HF, Baumgarten CM, Ellenbogen KA, Stambler BS. Atrial natriuretic peptide and cardiac electrophysiology: autonomic and direct effects. J Cardiovasc Electrophysiol 1996;7:149–162.
46. Allen DE, Gellai M. Cardioinhibitory effect of atrial peptide in conscious rats. Am J Physiol 1987;252:R610–R616.
47. Floras JS. Sympathoinhibitory effects of atrial natriuretic factor in normal humans. Circulation 1990;81:1860–1873.
48. Woods RL, Courneya C-A, Head GA. Nonuniform enhancement of baroreflex sensitivity by atrial natriuretic peptide in conscious rats and dogs. Am J Physiol 1994;267:R678–R686.
49. Volpe M, Lembo G, Condorelli G, De Luca N, Lamenza F, Indolfi C, Trimarco B. Converting enzyme inhibition prevents the effects of atrial natriuretic factor on baroreflex responses in humans. Circulation 1990;82:1214–1221.
50. Jansen TLThA, Smits P, Lenders JWM, Jacobs M-CGS, Willemsen J, Thien Th. Atrial natriuretic factor potentiates the human forearm vasoconstrictor response to sympathetic stimulation. Clin Sci 1994;86:275–283.
51. Ermirio R, Ruggeri P, Cogo CE, Molinari C, Calaresu FR. Neuronal and cardiovascular responses to ANF microinjected into the solitary nucleus. Am J Physiol 1986;256:R577.
52. McKitrick DJ, Calaresu FR. Cardiovascular responses to microinjection of ANF into dorsal medulla of rats. Am J Physiol 1988;255:R182.
53. Casto R, Phillips MI. Mechanisms of pressor effects by angiotensin in the nucleus tractus solitarius of rats. Am J Physiol 1984;247:R575.
54. Oparil S, Wyss JM. Atrial natriuretic factor in central cardiovascular control. NIPS 1993;8:223–228.
55. Lohmeier TE, Mizelle HL, Reinhart GA. Role of atrial natriuretic peptide in long-term volume homeostasis. Clin Exper Pharmacol Physiol 1995;22:55–61.
56. Kamabayshi Y, Nakao K, Kimura H, Kawabata T, Nakamura M, Inouye K, Yoshida N, Imura H. Biological characterization of human brain natriuretic peptide (BNP) and rat BNP: species-specific actions of BNP. Biochem Biophys Res Commun 1990;173:599–605.
57. Sudoh T, Kangawa K, Minamino N, Matsuo H. A new natriuretic peptide in porcine brain. Nature 1988;332:78–80.
58. Kita T, Kida O, Kato J, Nakamura S, Eto T, Minamino N, Kangawa K, Matsuo H, Tanaka K. Natriuretic and hypotensive effects of brain natriuretic peptide (BNP) in spontaneously hypertensive rats. Life Sci 1989;44:1541–1545.
59. Kohno M, Horio T, Yokokawa K, Kurihara N, Takeda T. C-type natriuretic peptide inhibits thrombin- and angiotensin II-stimulated endothelin release via cyclic guanosine 3',5'-monophosphate. Hypertension 1992;19:320–325.
60. Lang CC, Choy A-M, Struthers AD. Atrial and brain natriuretic peptides: a dual natriuretic peptide system potentially involved in circulatory homeostasis. Clin Sci 1992;83:519–527.
61. Lazzeri C, La Villa G, Bisi G, Boddi V, Messeri G, Strazzulla G, Franchi F. Cardiovascular function during brain natriuretic peptide infusion in man. Clin Pharmacol 1995;86:396–401.
62. La Villa G, Bisi G, Lazzeri C, Fronzaroli C, Stefani L, Barletta G, Del Bene R, Messeri G, Strazzulla G, Franchi F. Cardiovascular effects of brain natriuretic peptide in essential hypertension. Hypertension 1995;25:1053–1057.
63. Okumura K, Yasue H, Fujii H, Kugiyama K, Matsuyama K, Yoshimura M, Jougasaki M, Kikuta K, Kato H, Tanaka H, Sumida H, Nakao K. Effects of brain (B-type) natriuretic peptide on coronary artery diameter and coronary hemodynamic variables in humans: comparison with effects of systemic hemodynamic variables. J Am Coll Cardiol 1995;25:342–348.
64. Cargill RI, Lipworth BJ. The role of the renin-angiotensin and natriuretic peptide systems in the pulmonary vasculature. Br J Clin Pharmacol 1995;40:11–18.
65. Cargill RI, Lipworth BJ. Acute effects of ANP and BNP on hypoxic pulmonary vasoconstriction in humans. Br J Clin Pharmacol 1995;40:585–590.
66. Lohe A, Yeh I, Hyver T, Pratt R, Jamison R. Natriuretic peptide B receptor and C-type natriuretic peptide in the rat kidney. J Am Soc Nephrol 1995;6:1552–1558.
67. Stingo AJ, Clavell AL, Aarhus LL, Burnett JC Jr. Cardiovascular and renal actions of C-type natriuretic peptide. Am J Physiol 1992;262:H308–H312.
68. Drewett JG, Fendly BM, Garbers DL, Lowe DG. Natriuretic peptide receptor-B (guanylyl cyclase-B) mediates C-type natriuretic peptide relaxation of precontracted rat aorta. J Biol Chem 1995;270:4668–4674.
69. Wei C-M, Aarhus LL, Miller VM, Burnett JC Jr. Action of C-type natriuretic peptide in isolated canine arteries and veins. Am J Physiol 1993;264:H71–H73.

70. Wei C-M, Miller VM, Schaff HV, Burnett JC Jr. Specificity of C-type and atrial natriuretic peptides in human blood vessels. Endothelium 1996;4:23–28.
71. Nakamura M, Arakawa N, Yoshida H, Makita S, Hiramori K. Vasodilatory effects of C-type natriuretic peptide on forearm resistance vessels are distinct from those of atrial natriuretic peptide in chronic heart failure. Circulation 1994;90:1210–1214.
72. Wei C-M, Shiling HU, Miller VM, Burnett JC Jr. Vascular actions of C-type natriuretic peptide in isolated porcine coronary arteries and coronary vascular smooth muscle cells. Biochem Biophys Res Commun 1994;205:765–771.
73. Davidson NC, Barr CS, Struthers AD. C-type natriuretic peptide: an endogenous inhibitor of vascular angiotensin-converting enzyme activity. Circulation 1996;93:1155.
74. Furuya M, Yoshida M, Hayashi Y, Ohnuma N, Minamino N, Kangawa K, Matsuo H. C-type natriuretic peptide is a growth inhibitor of rat vascular smooth muscle cells. Biochem Biophys Res Commun 1991;177:927–931.
75. Furuya M, Aisaka K, Miyazaki T, Honbou N, Kawashima K, Ohno T, Tanaka S, Minamino N, Kangawa K, Matsuo H. C-type natriuretic peptide inhibits intimal thickening after vascular injury. Biochem Biophys Res Commun 1993;193:248–253.
76. Suga S, Nakao K, Itoh H, Komatsu Y, Ogawa Y, Hama N, Imura H. Endothelial production of C-type natriuretic peptide and its marked augmentation by transforming growth factor β: possible existence of "vascular natriuretic peptide system". J Clin Invest 1992;90:1145–1149.
77. Tachte GJ, Drewett JG. C-type natriuretic peptide neuromodulates independently of guanylyl cyclase activation. Hypertension 1994;23:38–43.
78. Vigne P, Frelin C. C-type natriuretic peptide is a potent activator of guanylate cyclase in endothelial cells from brain microvessels. Biochem Biophys Res Commun 1992;183:640–644.
79. Charles CJ, Richards AM, Espiner EA. Central C-type natriuretic peptide but not atrial natriuretic factors lowers blood pressure and adrenocortical secretion in normal conscious sheep. Endocrinology 1992;131:1721–1726.
80. Forssmann W-G. Urodilatin (Ularitide, INN): a renal natriuretic peptide. Nephron 1995;69:211–222.
81. Hunt PJ, Yandle TG, Nicholls MG, Richards AM, Espiner EA. The amino-terminal portion of pro-brain natriuretic peptide (proBNP) circulates in human plasma. Biochem Biophys Res Commun 1995;214:1175–1183.
82. Vesely DL, Douglass MA, Dietz JR, Gower WR, McCormick MT, Rodriguez-Pas G, Shocken DD. Three peptides from the atrial natriuretic factor prohormone amino terminus lower blood pressure and produce diuresis, natriuresis, and/or kaliuresis in humans. Circulation 1994;90:1129–1140.
83. Vesely DL. Atrial natriuretic hormones originating from the N-terminus of the atrial natriuretic factor prohormone. Clin Exper Pharmacol Physiol 1995;22:108–114.
84. Benjamin BA. Effects of ANF prohormone peptides in conscious primates. Clin Exper Pharmacol Physiol 1995;22:125–129.
85. Gower WR, Chiou S, Skolnick KA, Vesely DL. Molecular forms of circulating atrial natriuretic peptides in human plasma and their metabolites. Peptides 1994;15:861–867.
86. Buckley MG, Markandu ND, Sagnella GA, MacGregor GA. N-terminal atrial natriuretic peptide and atrial natriuretic peptide in human plasma: investigation of plasma levels and molecular circulating form(s) using radioimmunoassays for pro-atrial natriuretic peptide (31–67), pro–atrial natriuretic peptide (1–30) and atrial natriuretic peptide (99–126). Clin Sci 1994;87:311–317.
87. Espiner EA, Richards AM. Atrial natriuretic peptide. Lancet 1989;i:707–710.
88. Lee ME, Miller WL, Edwards BS. Role of endogenous atrial natriuretic factor in acute congestive heart failure. J Clin Invest 1989;84:1962–1966.
89. Lohmeier TE, Shin Y, Reinhart GA. Angiotensin and ANP secretion during chronically controlled increments in atrial pressure. Am J Physiol 1994;266:R989–R996.
90. Westenfelder C, Birth FM, Baranowski RL. Volume homeostasis in calves with artificial atria and ventricles. Am J Physiol 1990;258:F1005–F1017.
91. Richards AM, McDonald D, Fitzpatrick MA, Nicholls MG, Espiner EA, Ikram H, Jans S, Grant S, Yandle T. Atrial natriuretic hormone has biological effects in man at physiological plasma concentrations. J Clin Endocrinol Metab 1988;67:1134–1139.
92. Ando S, Rahman MA, Butler GC, Sen BL, Floras JS. Comparison of candoxatril and atrial natriuretic factor in healthy men. Hypertension 1995;26:1160–1166
93. Tonolo G, Richards AM, Manunta P, Pazzola A, Pala F, Gloriso N. Low dose infusion of atrial natriuretic factor in mild essential hypertension. Circulation 1989;80:893–902.

94. Cusson JR, Thibault G, Cantin M, Larochelle P. Cardiovascular, renal and endocrine responses to low doses of atrial natriuretic factor in mild essential hypertension. J Hum Hypertens 1989;3:89–96.
95. Janssen WMT, de Zeeuw D, van der Hem GK, de Jong PE. Antihypertensive effect of a 5-day infusion of atrial natriuretic factor in humans. Hypertension 1989;13:640–646.
96. Hamet P, Testaert E, Palmour R, Larochelle P, Cantin M, Gutkowska J, Langlois Y, Ervin F, Tremblay J. Effect of prolonged infusion of ANF in normotensive and hypertensive monkeys. Am J Hypertens 1989;2:690–695.
97. Charles CJ, Espiner EA, Cameron VA, Richards AM. Hemodynamic, renal, and endocrine actions of ANF in sheep: effect of 24-h low-dose infusions. Am J Physiol 1990;258:R1279–R1285.
98. Parkes DG, Coghlan JP, McDougall JG, Scoggins BA. Long-term hemodynamic actions of atrial natriuretic factor (99–126) in conscious sheep. Am J Physiol 1988;254:H811–H815.
99. Garcia R, Thibault G, Gutkowska J, Horky K, Hamet P, Cantin M, Genest J. Chronic infusion of low-doses of atrial natriuretic factor reduces blood pressure in conscious SHR without apparent changes in sodium excretion. Proc Soc Exp Biol Med 1985;179:396–401.
100. Crozier IG, Nicholls MG, Ikram H, Espiner EA, Gomez HJ, Warner NJ. Haemodynamic effects of atrial peptide infusion in heart failure. Lancet 1986;2:1242–1245.
101. Cody RJ, Atlas SA, Laragh JH, Kubo SH, Covit AB, Ryman KS, Shaknovich A, Pondolfino K, Clark M, Camargo MJ. Atrial natriuretic factor in normal subjects and heart failure patients—plasma levels and renal, hormonal, and hemodynamic responses to peptide infusion. J Clin Invest 1986;78: 1362–1374.
102. Clavell AL, Stingo AJ, Aarhus LL, Burnett JC Jr. Biological actions of brain natriuretic peptide in thoracic inferior vena caval constriction. Am J Physiol 1993;265:R1416–R1422.
103. Nicholls MG. The natriuretic peptides in heart failure. J Intern Med 1994;235:515–526.
104. Foy SG, Crozier IG, Richards AM, Nicholls MG, Turner JG, Frampton CM. Neurohormonal change after acute myocardial infarction relationships with haemodynamic induces and effect of ACE inhibitor. Eur Heart J 1995;16:770–778.
105. Yandle TG, Richards AM, Gilbert A, Fisher S, Holmes S, Espiner EA. Assay of brain natriuretic peptide (BNP) in human plasma: evidence for high molecular weight BNP as a major plasma component in heart failure. J Clin Endocrinol Metab 1993;76:832–838.
106. Holmes SJ, Espiner EA, Richards AM, Yandle TG, Frampton. Renal, endocrine, and hemodynamic effects of human brain natriuretic peptide in normal man. J Clin Endocrinol Metab 1993;76:91–96.
107. Hunt PJ, Espiner EA, Nicholls MG, Richards AM, Yandle TG. 1996 Differing biological effects of equimolar atrial natriuretic and brain natriuretic infusions in normal man. J Clin Endocrinol Metab 1996;81:3871–3876.
108. Cargill RI, Struthers AD, Lipworth BJ. Comparative effects of atrial natriuretic peptide and brain natriuretic peptide on the aldosterone and pressor responses to angiotensin II in man. Clin Sci 1995;88:81–86.
109. Koller KJ, Goeddel DV. Molecular biology of the natriuretic peptides and their receptors. Circulation 1992;86:1081–1088.
110. Pidgeon GB, Richards AM, Nicholls MG, Espiner EA. Yandle TG, Frampton C. Differing metabolism and bioactivity of atrial and brain natriuretic peptides in essential hypertension. Hypertension 1996;27:906–913.
111. La Villa G, Stefani L, Lazzeri C, Zurli C, Guerra CT, Barletta G, Bandinelli R, Strazzulla G, Franchi F. Acute effects of physiological increments of brain natriuretic peptide in humans. Hypertension 1995;26:623–633.
112. Florkowski CM, Richards AM, Espiner EA, Yandle TG, Frampton C. Renal, endocrine, and hemodynamic interactions of atrial and brain natriuretic peptides in normal men. Am J Physiol 1994;266: R1244–R1250.
113. Charles CJ, Espiner EA, Richards AM, Nicholls MG, Yandle TG. Comparative bioactivity of atrial, brain and C-type natriuretic peptides in conscious sheep. Am J Physiol 1996;270;R1324–1331.
114. Hunt PJ, Richards AM, Espiner EA, Nicholls MG, Yandle TG. Bioactivity and metabolism of C-type natriuretic peptide in normal man. J Clin Endocrinol Metab 1994;78:1428–1435.
115. Clavell AL, Stingo AJ, Wei C-M, Heublein DM, Burnett JC Jr. C-type natriuretic peptide: a selective cardiovascular peptide. Am J Physiol 1993;264:R290–R295.
116. Charles CJ, Espiner EA, Richards AM, Nicholls MG, Yandle TG. Biological actions and pharmacokinetics of C-type natriuretic peptide in conscious sheep. Am J Physiol 1995;268:R201–R207.

117. Kawai M, Naruse M, Yoshimoto T, Naruse K, Shionoya K, Tanaka M, Morishita Y, Matsuda Y, Demura R, Demura H. C-type natriuretic peptide as a possible local modulator of aldosterone secretion in bovine adrenal zona glomerulosa. Endocrinology 1996;137:42–46.
118. Shinomiya M, Tashiro J, Saito Y, Yoshida S, Furuya M, Oka N, Tanaka S, Kangawa K, Matsuo H. C-type natriuretic peptide inhibits intimal thickening of rabbit carotid artery after balloon catheter injury. Biochem Biophys Res Commun 1994;205:1051–1056.
119. Richards AM. The natriuretic peptides and hypertension. J Intern Med 1994;235:543–560.
120. Charles CJ, Espiner EA, Cameron VA, Richards AM, Yandle TG, Sybertz EJ. Hemodynamic and hormonal effects of neutral endopeptidase inhibitor SCH 39370 in sheep. Hypertension 1991;17: 643–651.
121. Kenny AJ, Bourne A, Ingram J. Hydrolysis of human and pig brain natriuretic peptides, urodilatin, C-type natriuretic peptide and some C-receptor ligands by endopeptidase-24.11. Biochem J 1993;291:83–88.
122. Seymour AA, Abboa-Offei BE, Smith PL, Mathers PD, Asaad MM, Rogers WL. Potentiation of natriuretic peptides by neutral endopeptidase inhibitors. Clin Exper Pharmacol Physiol 1995;22:63–69.
123. Thompson JS, Sheedy W, Morice AH. Effects of the neutral endopeptidase inhibitor, SCH 42495, on the cardiovascular remodelling secondary to chronic hypoxia in rats. Clin Sci 1994;87:109–114.
124. Maack T, Suzuki M, Almeida FA, Nussenzveig, Scarborough RM, McEnroe GA, Lewicki JA. Physiological role of silent receptors of atrial natriuretic factor. 1987;238:675–678.
125. Levin ER. Natriuretic peptide C-receptor: more than a clearance receptor. Am J Physiol 1993;264: E483–E489.
126. Parkes DG, Coghlan JP, Lewicki JA, Scarborough RM, Scoggins BA. Hemodynamic effects of atrial natriuretic factor clearance receptor occupancy in conscious sheep. Am J Hypertens 1990;3:829–832.
127. Charles CJ, Espiner EA, Nicholls MG, Richards AM, Yandle TG, Protter A, Kosoglou T. Clearance receptors and endopeptidase 24.11: equal role in natriuretic peptide metabolism in conscious sheep. Am J Physiol 1996;271:R373–R380.
128. Itoh H, Nakao K, Mukoyama M, Yamada T, Hosada K, Shirakami G. Chronic blockade of endogenous atrial natriuretic polypeptide (ANP) by monoclonal antibody against ANP accelerates the development of hypertension in spontaneously hypertensive and deoxycorticosterone acetate-salt-hypertensive rats. J Clin Invest 1989;84:145–154.
129. Drexler H, Hirth C, Stasch HP, Neuser D, Just H. Vasodilatory action of endogenous atrial natriuretic factor in a rat model of chronic heart failure as determined by monoclonal ANF antibody. Circ Res 1990;66:1371–1380.
130. Morishita Y, Sano T, Ando K, Saitoh Y, Kase H, Yamada K, Matsuda Y. Microbial polysaccharide, HS-142-1, competitively and selectively inhibits ANP binding to its guanylyl cyclase-containing receptor. Biochem Biophys Res Commun 1991;176:949–957.
131. Stevens TL, Wei C-M, Aahrus LL, Heublein DM, Kinoshita M, Matsuda Y, Burnett JC Jr. Modulation of exogenous and endogenous atrial natriuretic peptide by a receptor inhibitor. Hypertension 1994;23:613–618.
132. Zhang PL, Mackenzie HS, Guo J, Troy JL, Brenner BM. HS-142-1, a potent antagonist of natriuretic peptides in vitro and in vivo. J Am Soc Nephrol 1994;5:1099–1105.
133. Nishikimi T, Miura K, Minamino N, Takeuchi K, Takeda T. Role of endogenous atrial natriuretic peptide on systemic and renal hemodynamics in heart failure rats. Am J Physiol 1994;267:H182–H186.
134. Zhang PL, Mackenzie HS, Troy JL, Brenner BM. Effects of natriuretic peptide receptor inhibition on remnant kidney function in rats. Kidney Intern 1994;46:414–420.
135. Zhang PL, Mackenzie HS, Tosune K, Troy JL, Brenner BM. Renal effects of high-dose natriuretic peptide receptor blockade in rats with congestive heart failure. Circ Res 1995;77:1240–1245.
136. Wada A, Tsutamoto T, Matsuda Y, Kinoshita M. Cardiorenal and neurohumoral effects of endogenous atrial natriuretic peptide in dogs with severe congestive heart failure using a specific antagonist for guanylate cyclase-coupled receptors. Circulation 1994;89:2232–2240.
137. Stevens TL, Burnett JC Jr, Kinoshita M, Matsuda Y, Redfield MM. A functional role for endogenous atrial natriuretic peptide in a canine model of early left ventricular dysfunction. J Clin Invest 1995;95:1101–1108.
138. Steinhelper ME, Cochrane KL, Field LJ. Hypotension in transgenic mice expressing atrial natriuretic factor fusion genes. Hypertension 1990;16:301–307.
139. Barbee RW, Perry BD, Re RN, Murgo JP, Field LJ. Hemodynamics in transgenic mice with overexpression of atrial natriuretic factor. Circ Res 1994;74:747–751.

140. Ogawa Y, Itoh H, Tamura N, Suga S, Yoshimasa T, Uehira M, Matsuda S, Shiono S, Nishimoto H, Nakao K. Molecular cloning of the complementary DNA and gene that encode mouse brain natriuretic peptide and generation of transgenic mice that overexpress the brain natriuretic peptide gene. J Clin Invest 1994;93:1911–1921.
141. John SWM, Krege JH, Oliver PM, Hagaman JR, Hodgin JB, Pang SC, Flynn TG, Smithies O. Genetic decreases in atrial natriuretic peptide and salt–sensitive hypertension. Science 1995;267:679–681.
142. Lopez MJ, Wong SKF, Kisimoto I, Dubois S, Mach V, Friesen J, Garbers DL, Beuve A. Salt-resistant hypertension in mice lacking the guanylyl cyclase-A receptor for atrial natriuretic peptide. Nature 1995;378:65–68.

# 9 Renal Actions of Atrial Natriuretic Peptide

*Markus Meyer and Wolf-Georg Forssmann*

*CONTENTS*



## PHYSIOLOGICAL SIGNIFICANCE OF ANP IN THE KIDNEY

As of 1956, Henry and Gauer *(1)* stated that the heart is involved in renal functions. This was supported in 1969 by Lockett and coworkers, who mentioned for the first time a hormonal influence of the heart on the kidney *(2)*. In 1976, Marie and coworkers *(3)* observed that the granular index of myoendocrine cells is affected during experiments, altering the body fluid and electrolyte homeostasis. By applying atrial extracts, deBold and Sonnenberg *(4)* reported for the first time an up to then unknown strong diuresis and natriuresis. Then, a further biological effect of atrial extracts, relaxation of vascular smooth muscle, was discovered *(5–8)*. Based on these two main biological effects, the first isolation of atrial natriuretic peptides was performed. Depending on the bioassay used for isolation, the peptides were designated *cardiodilatin* (from the pig) *(8)*, *cardionatrin* (from the rat) *(9)*. Kangawa and coworkers *(10)* were the first to isolate the hormone from the human heart. They designated the peptide human α atrial natriuretic peptide (h α ANP), which will also be used in the following text. Since then, remarkable advances have been made regarding the characterization of its gene and protein nature, synthesis, intracellular processing, release, metabolism, second messenger systems, and receptors *(11–14)*. The main function of ANP is perceived to be an action on the kidney to induce natriuresis and diuresis *(15–17)*. Furthermore, close correlations between plasma ANP and the load of extracellular fluid were reported *(18–21)* and evidence exists

From: *Contemporary Endocrinology: Natriuretic Peptides in Health and Disease*
Edited by: W. K. Samson and E. R. Levin Humana Press Inc., Totowa, NJ

that ANP plays an important role in body-fluid homeostasis when concentrations of ANP not higher than those secreted and circulating in the bloodstream under physiological conditions are administered intravenously *(15–17)*. However, it should be noted that the biological effects and the functional role of ANP, when present in physiological concentrations, are still far from being clarified *(22,23)*. A vast number of studies have used supraphysiological concentrations of ANP trying to characterize the physiological significance of ANP *(24–28)*. These doses were demonstrated to induce, besides a brisk diuresis and natriuresis, increases in $PO_4^{2-}$, $Ca^{2+}$, $Mg^{2+}$, $Cl^-$, and cGMP excretion (25–30).

Approaches have been made to overcome this problem: i.e., cardiac appendectomy was performed to prevent ANP release *(31)*. However, the problems in this setting are apparent since the missing normal cardiac function results in a lack of physiological reflexes like cardiovascular reflexes, which are normally induced by atria. Another approach has been raised using antibodies against the circulating ANP *(32)*. However, even if results suggest the monoclonal antibody directed against ANP is a useful tool for the physiological characterization of ANP, problems are still raised because of nonspecific effects of the antibody and/or antigen-antibody complex. These difficulties could be at least partly overcome since a specific nonpeptide antagonist for the guanylyl cyclase-coupled receptors, called HS-142-1, was established *(33–37)*. Morishita and coworkers *(33)* demonstrated that when ANP was injected intravenously into rats, HS-142-1 abolished ANP-induced diuresis and natriuresis. Furthermore, to determine if defects in the ANP system can cause hypertension, mice were generated with a disruption of the proANP gene *(38)*. Whereas homozygous mutants had no circulating ANP and elevated blood pressures receiving standard and intermediate salt diets, heterozygotes had normal amounts of circulating ANP and normal blood pressures on standard salt diets. However, on high salt diets they were hypertensive, indicating that genetically reduced production of ANP can lead to salt-sensitive hypertension. Another approach for the characterization of the ANP system was the knockout model of the biologically active guanylyl cyclase receptor gene *(39)*. The ANP is thought to mediate its natriuretic and vasorelaxant effects through the biologically active guanylyl cyclase receptor *(40)*. The authors reported that disruption of the receptor gene results in chronic elevations of blood pressure in mice on a normal salt diet. However, the blood pressure remained elevated and unchanged in response to either minimal or high salt diets, whereas ANP concentrations were not affected by the genotype. These results might be contradictory to those observed by John and coworkers *(38)* who found no increase in blood pressure in the model of the knockout ANP gene when mice were fed normal diet. This obvious discrepancy could be caused by a binding of ligands to the receptor other than ANP like brain natriuretic peptide (BNP) *(41)*. Taken together, these results still indicate that ANP plays a role in the fluid and sodium homeostasis and that human genetic variants of the ANP system may affect regulation of blood pressure.

The excretory effects of sodium and water were often associated with increases in glomerular filtration rate (GFR), which was initially suggested to be solely responsible for the renal actions of ANP *(26,42,43)*. However, ANP also has direct and indirect effects on the tubules (proximal tubule, cortical, and inner medullary collecting duct) and even causes natriuresis without alterations in GFR at a lower dose (Table 1) *(30,44,45)*. Therefore, a combination of both hemodynamic effects and tubular transport processes appears to be responsible for the ANP-induced natriuresis and diuresis. In this review, we

Table 1
Renal Actions of ANP[a]

| | |
|---|---|
| Direct actions of ANP on different sites of the nephron | |
| Renal artery | RBF variable |
| Glomerulus | Increased GFR and FF |
| Proximal tubule | Decreased angiotensin-induced sodium and water reabsorption |
| Cortical collecting duct | Decreased vasopressin-induced hydrosmosis |
| Inner medullary collecting duct | Decreased amiloride-sensitive $Na^+$ channel |
| | Decreased $Na^+$-$K^+$-$2Cl^-$ cotransporter |
| Vasa recta | Increased flow and hydraulic pressure |
| Medulla | Decreased solute content |
| Juxtaglomerular cells | Decreased renin secretion |
| Indirect actions of ANP on different sites of the nephron | |
| Proximal tubule | Decreased reabsorption by increased filtrate delivery |
| Inner medullary collecting duct | Decreased reabsorption by decreased solute content and increased vasa recta pressure |

[a]Direct and indirect intrarenal actions of ANP at several nephron sites.
(RBF = renal blood flow; GFR = glomerular filtration rate; FF = filtration fraction).

therefore wish to focus at first on the hemodynamic changes induced by ANP and then report on the action of ANP on the different segments of the tubular system.

## EFFECTS OF ANP ON RENAL HEMODYNAMICS

### *Glomerular Filtration Rate and Renal Blood Flow*

Variable effects in renal blood flow (RBF) induced by ANP infusions are reported *(25,26,46–49)*. If RBF is augmented, the increase is only transient and short-lived *(25,47)* following onset of continuous ANP administration. However, brisk natriuresis is sustained despite the return of RBF to a control level *(25,46,50)*. These results are also supported by other working groups, indicating that increased RBF does not necessarily contribute to the enhanced renal solute excretion *(26,51–53)*. Conflicting results exist regarding the relative cortical/medullary distribution of blood flow with studies reporting a decline *(54,55)*, no change *(46,56)*, or even increases *(57)*. These findings could be due to different actions of ANP on distinct segments of the renal vascular tree.

Furthermore, there are a number of studies reporting increases in GFR and filtration fraction after ANP administration *(26,42,43,52,58)*. To clarify the mode of action of ANP leading to an increased GFR, Marin Grez and coworkers *(59)* directly studied the effects of ANP on blood vessels in the kidney of anesthetized rats. Using quantitative videomicroscopy, the authors investigated the diameter of renal microvasculature before and during administration of ANP and observed dose-dependent dilation of arcuate, interlobular, and proximal afferent vessels and constriction of efferent arterioles leading to an increased hydraulic pressure within glomerular capillaries. This finding was also supported by other groups *(60,61)*. Veldkamp and coworkers *(61)* investigated the effects of ANP on renal vasculature. Using the in vitro blood-perfused juxtamedullary nephron technique, the authors were able to visualize the changes of lumen diameter of arcuate arteries and afferent and efferent arterioles. They reported a vasodilation of the preglomerular vessels induced by ANP while—in contrast to Marin Grez *(59)*—the efferent arteriolar diameter remained unaltered.

Mesangial cells appear to be important for the stabilization of the glomerular tuft considering that glomerular capillaries are perfused at a high hydraulic pressure *(62)*. ANP may act to relax glomerular mesangial cells, which possess specific, high-affinity ANP receptors and respond to ANP with striking elevations in cGMP accumulation *(63)*. How ANP may induce relaxation was further investigated by Cermak and coworkers *(64)*. They examined the influence of ANP on membrane voltages by patch-clamp technique and found that this hormone hyperpolarized rat mesangial cells, probably because of an increase of a $K^+$ conductance suggesting a first step in the cascade of relaxation. Whether cytosolic $Ca^{2+}$ is at least modulated by ANP treatment is debated *(65–68)*.

These observations are of interest with regard to the effect of ANP on glomerular function *(69)*. Relaxation of mesangial cells leads to an increase in ultrafiltration coefficient ($K_f$), a further characteristic of GFR and filtration fraction. These effects were demonstrated by several groups. Using the isolated perfused glomerulus technique, Fried and coworkers *(70)* reported an increase in the glomerular capillary hydrostatic pressure and also a significant increase in $K_f$, leading to an enhanced GFR. Furthermore, glomeruli freshly isolated from rats and precontracted with angiotensin II were shown to dilate when exposed to ANP, suggesting an ANP-induced relaxation of mesangial cells *(71)*. This group also reported that $^{125}$I-ANP uptake is mainly localized on mesangial cells, followed by epithelial visceral cells and the endothelium of capillaries, suggesting that ANP may regulate $K_f$ essentially by two mechanisms: modulation of glomerular permeability, and surface area.

### *Is the Increment of GFR Solely Responsible for the ANP-Induced Natriuresis and Diuresis?*

Since the discovery of ANP, there has been substantial controversy regarding the mechanisms by which enhanced urinary solute excretion develops. Several investigators have suggested that the principal, if not the sole, means by which ANP effects a solute diuresis is hemodynamic, by increasing GFR *(43,50,70,72,73)*. Others, however, have proposed that the ANP-induced natriuresis is caused by direct inhibition of tubular transport processes, independent of alterations in GFR *(30,44,45)*. There are studies in which infusion of ANP at low doses resulted in natriuresis and diuresis without detectable changes in GFR, and only at higher doses was an increase in GFR obtained *(25,52,74–76)*. However, these findings do not exclude the hemodynamic influence. Indeed, even if an explanation for the ANP-induced natriuresis and diuresis is difficult to find, the majority of studies favor a combination of both effects, hemodynamic and tubular *(30,44,45,75,77–83)*. Therefore, in the next paragraphs we focus our interest on the direct effects of ANP on segmental tubular transport processes beyond the glomerulus.

## EFFECTS OF ANP ON TUBULAR TRANSPORT PROCESSES

### *Effects of ANP on Proximal Tubule*

Variable data exist regarding an influence of ANP on the ion transport in the proximal tubule. Even if clearance studies have shown increased $Li^+$ and $PO_4^{2-}$ excretion in response to ANP, indicating that the hormone may inhibit proximal tubule solute transport *(84,85)*, many studies have demonstrated that ANP is devoid of transport effects in the proximal tubule. In vivo and in vitro microperfusion studies of proximal convoluted and straight tubules in the rat or rabbit have failed to find any effects of ANP on sodium, bicarbonate, and chloride transport *(73,86–88)*. Cogan and coworkers *(42,43)* demonstrated that the solute delivery out of the proximal tubule increased by approx 30–35%. When single-nephron and whole-kidney GFR were reduced back to a control level by aortic constriction, proximal transport reverted to normal values despite persistent ANP administration, suggesting that ANP has no direct effect on reabsorption in the superficial proximal convoluted tubule independent of changes in filtration rate, whereas Van de Stolpe and coworkers *(89)* demonstrated that ANP reduces proximal $Na^+$ reabsorption independent of changes in GFR. Harris and coworkers *(90)* reported that the stimulatory effect of angiotensin II on fluid reabsorption was inhibited by peritubular ANP at physiological concentrations and abolished by higher concentrations of ANP. In contrast to the results of Cogan and coworkers *(42,43)*, the authors concluded, that at physiological concentrations, ANP acts within the kidney to decrease proximal reabsorption by inhibition of angiotensin II-stimulated sodium and water transport, which was also supported by Garvin and coworkers *(91)*.

The conflicting results concerning the biological effects of ANP at the proximal tubule are also reflected by reports demonstrating either no or only minimal stimulation of cGMP from proximal tubules following ANP incubation *(84–94)*. In contrast, our group demonstrated that ANP could induce intracellular cGMP generation using a new cell line derived from human proximal tubule *(95)*. However, these contradictory data might reflect the lack of biologically active natriuretic peptide receptors in this portion of the nephron as reported by some *(94)*, but also the presence of this receptor by others *(96)*.

Aperia and coworkers *(97)* recently investigated the molecular mechanisms by which ANP regulates renal sodium metabolism considering the role of $Na^+$, $K^+$-ATPase. They found that the angiotensin II-induced stimulation of $Na^+$, $K^+$-ATPase is blocked by ANP and cGMP. Since the activity of this enzyme can be regulated by phosphorylation and dephosphorylation, these results suggest that ANP may modulate an intracellular protein phosphorylation cascade. Thus, a regulatory function of ANP on sodium transport in the proximal tubules also seems to be likely *(98)*. The hypothesis that ANP influences ion transport might even gain more credibility since Darvish and coworkers *(99)* demonstrated a novel cGMP-activated $Cl^-$ channel in proximal tubules which are activated by ANP and cGMP.

## *Effects of ANP on Henle's Loop*

The functional responses of ANP in Henle's loop are unknown. Administration of pharmacological amounts of ANP resulted in an increase in urinary magnesium excretion *(17)*. Magnesium is principally reabsorbed in the cortical thick ascending limb and urinary magnesium excretion has been used as a marker for loop function *(100)*. Accordingly, it is inferred that ANP may alter salt transport in this segment. However, other studies have shown little effect of ANP in the loop. Peterson and coworkers *(101)* investigated the effects of these hormones on thick ascending limb $Na^+Cl^-$ reabsorption in vivo. Using the microstop-flow conductivity technique, despite the presence of a dramatic natriuretic response, ANP did not significantly increase the minimum $Na^+Cl^-$ concentration reached by the thick ascending limb. In contrast, vasopressin increased the rate at which luminal $Na^+$ concentration declined by 23%. Thus, a significant direct peritubular effect of ANP on thick ascending limb $Na^+Cl^-$ permeability appears to be unlikely *(102)*. However, the blunted or even missing effect of ANP is possibly related to the lack of biologically active natriuretic peptide receptors *(103)*, which might also explain the low response of the cells of Henle's loop in cGMP formation following ANP incubation *(81,93,104,105)*.

Comparable to other segments of the nephron, where ANP could modulate intracellular $Ca^{2+}$ levels *(65,66,68)*, a putative physiological importance of ANP could also be demonstrated in this nephron portion *(100)*. Using primary cell cultures, this group reported an ANP-induced receptor-mediated $Ca^{2+}$ transient in cortical thick ascending limb cells resulting from internal release and influx across the plasma membrane. The authors discussed whether this receptor is most likely to be the clearance receptor, that initiates $Ca^{2+}$ signals. They concluded that these ANP-induced $Ca^{2+}$ transients mediated by the putative clearance receptor may have significant physiological actions within Henle's loop.

### *Effects of ANP on the Collecting Duct System*

There is strong evidence that ANP effects natriuresis and diuresis in part by inhibiting net $Na^+$ and water reabsorption in the collecting duct. In the following, the actions of ANP investigated in in vivo experiments are presented first.

In 1979 Jamison et al. *(106)* reviewed the day-to-day regulation of fluid and electrolyte balance, in which the collecting tubule system "appears to occupy a paramount position among segments of the renal tubule." But at the same time, the authors referred to the discrepancies in reported results due to different methods and they concluded that "controversy has arisen concerning the quantitative contribution by the collecting tubule system to the regulation of individual solute excretion, which in part may be due to differences among the investigative techniques (microcatheterization technique and the micropuncture method) employed."

In a fundamental work, Sonnenberg and coworkers *(44)* described the effects of ANP on the tubular system. Micropuncture and microcatheterization were performed in anesthetized rats to collect tubular fluid from the proximal and distal tubules and from the outer medullary collecting duct. Samples were taken from the same sites before and after iv administration of atrial tissue extract. Sodium excretion increased 17-fold, after atrial extract was injected, whereas clearances of inulin and single nephron filtration rates did not change significantly. The infusions of ANP did not alter the delivery of sodium and water to Henle' loop, but led to increases in sodium and water delivery out of the outer medullary duct. Since between the outer medullary collecting duct and the final urine there was a large increment of fractional sodium excretion, it appeared likely that the atrial extracts interfered with the capacity of solute load reabsorption in the inner medullary collecting duct. The same working group *(45)* reported that infusion of ANP was associated with increased sodium delivery and reduced fractional reabsorption in the duct, whereas $K^+Cl^-$ infusion had no effect on fractional reabsorption, suggesting that ANP has a highly specific inhibitory effect on net sodium transport in this part of the nephron.

### *Further Mechanisms Contributing to ANP-Induced Natriuresis and Diuresis*

Even if the in vivo studies strongly indicated the actions of ANP by inhibition of net $Na^+$ and water reabsorption in the collecting duct, one cannot exclude that these actions are also caused by alterations of transepithelial driving forces. The ANP has been shown to preferentially increase blood flow to juxtamedullary nephrons and to augment vasa recta blood flow. Mendez and coworkers *(107)* demonstrated that ANP-induced alterations in transepithelial driving forces in the renal medulla cause unfavorable hydraulic gradients for net $Na^+$ reabsorption in the inner medullary collecting duct (IMCD). These findings suggest that ANP can increase the hydraulic gradient from vasa recta to papillary collecting duct, favoring reduced net volume and $Na^+$ reabsorption in IMCD.

### *Effects of ANP on the Cortical Collecting Duct*

Water reabsorption takes place in the cortical collecting duct (CCD) cells, induced by the peptide hormone vasopressin. Increase in 3'5'-cyclic adenosine monophosphate (cAMP) levels in these cells leads to an increasing permeability of the CCD apical membrane to water *(108)*. The ANP-induced diuresis is probably because of direct inhibition of renal tubular epithelial water transport that was first investigated by Dillingham and coworkers *(108)*. The results that ANP leads to an inhibition of the hydraulic conductivity response to the hormone vasopressin but not to either cAMP or forskolin suggested that ANP acts proximally to cAMP formation. In contrast, Umemura and coworkers *(109)* studied the effect of ANP on cAMP levels in single nephron segments and in glomeruli from the rat and reported that ANP does not inhibit cAMP generation in any of these nephron segments. Consistent results were obtained in primary cultures of cortical collecting tubule cells immunodissected with a monoclonal antibody, where the effect of ANP on cyclic nucleotide production was investigated *(46)*. ANP caused a dose-dependent stimulation in cGMP but had no effect on cAMP. Thus, the conflicting results mentioned above are difficult to interpret and it still remains unclear whether ANP acts via a direct mechanism on the reduction of cAMP accumulation or a step distal or proximal to cAMP formation leading to an increase in osmotic water permeability.

What is the effect of ANP on ion transport processes at this nephron site? Until now, no conclusive data demonstrating an ANP-induced inhibition of $Na^+$ reabsorption in the cortical collecting tubule exist. However there are studies demonstrating an increase in natriuresis towards the IMCD *(44,45)* and there are also studies reporting increased cGMP formation following ANP stimulation *(81,103)* even though only low levels of receptor expression could be detected in this nephron segment *(103,110)*. In contrast, other groups could not show any effects of ANP on $Na^+$ transport in the CCD *(111,112)* neither a regulation of the electrogenic electrolyte transport in principal cells of rat CCD *(113)*.

### *Effects of ANP on the Inner Medullary Collecting Duct*

There is strong evidence that most of the natriuretic effects of ANP contribute to the inhibition of $Na^+$ reabsorption in the IMCD. The working group of Zeidel has made great efforts to clarify the mechanisms by which ANP acts and induces the natriuretic response in the IMCD. They demonstrated that ANP inhibits transport-dependent oxygen consumption ($QO_2$) in the IMCD yet has no effect on the outer medullary collecting duct *(82)*. Further studies investigating the interaction of ANP with amiloride, ouabain, and amphotericin using $QO_2$ measurements suggest that ANP inhibits $Na^+$ entry into these cells *(82)*. ANP reduced isotopic $Na^+$ uptake in suspensions of IMCD cells by two-thirds at concentrations identical to those found effective in oxygen consumption studies *(114)*. To characterize precisely the ANP-induced effect on $Na^+$ reabsorption in IMCD, the patch-clamp technique was used *(115)*. This study revealed that ANP and cGMP inhibited an apically located cation channel in cell-attached patches. Thus, it seems likely that the natriuretic action of ANP is related in part to cGMP-mediated inhibition of electrogenic $Na^+$ absorption by the IMCD. This finding was also supported by Nonoguchi and coworkers *(81)*, demonstrating that ANP inhibits $Na^+Cl^-$ and fluid absorption in cortical collecting duct of rat kidney, actions which closely mimicked cGMP formation.

Another effect of ANP in IMCD is the action on water permeability. Increased cAMP levels in these cells led to increased permeability of the apical membrane to water comparable to the mechanisms already discussed for cortical collecting duct cells *(80,110)*. Studies revealed strong evidence that ANP acts at a site distal of cAMP formation since ANP did not alter cAMP accumulation in this segment in response to vasopressin *(80,110)*. Regarding the cell site from which ANP acts, Nonoguchi and coworkers *(80)* found that ANP added only to the basolateral and not to the apical site-inhibited vasopressin-stimulated water flow in isolated-perfused rat IMCD *(80)*.

### *Is cGMP a Mediator for the Actions of ANP in the Inner Medullary Dollecting Duct?*

There is strong evidence that cGMP is the second messenger by which the actions of ANP are mediated in the IMCD. This is supported by the fact that in the IMCD, a receptor is expressed that is likely to be the ANP receptor *(110,116–122)*. Many studies indicated that ANP increases intracellular cGMP in a dose-response relationship *(105,110,123)* in IMCD cells, and even addition of the cGMP analog 8-Br-cGMP mimics the inhibitory effect of ANP on transport-dependent oxygen consumption and on uptake of $Na^+$ via the apically located conductive channel *(114,123)*. Zeidel and coworkers *(124)* discussed that this cation channel is regulated either by direct binding of cGMP to the cation channel or by cGMP-induced activation of a protein kinase resulting in the inhibition of the opening time for $Na^+$ (Fig. 1).

### *Indirect Effects of ANP on Renal Function by Alterations in Neuroendocrine Function*

ANP inhibits biosynthesis of renin of juxtaglomerular cells of the kidney and aldosterone in the adrenal gland *(25,50)*. Several studies demonstrated that ANP infusion markedly lowers renin secretory rate and plasma renin concentration *(125,126)*. It has been shown that a cGMP-specific phosphodiesterase inhibitor enhances the suppressive effect of ANP on renin release, and methylene blue, a guanylyl cyclase inhibitor, inhibits the suppressive effect of ANP, suggesting that this is a cGMP-dependent process *(127)*. In contrast, there are studies that do not support these findings. ANP failed to show effects on stimulated renin release by diltiazem, or on suppressed renin release by angiotensin II *(128)*. Furthermore, Rodriguez Puyol and coworkers *(129)* could not demonstrate a regulatory effect of ANP on renin release in isolated glomeruli either. In addition, ANP also suppresses the activity of the sympathetic nervous system and reduces vasopressin release and action *(80,108)* and these neuroendocrine effects mediated by ANP can also affect the ion-transport along the nephron. Thus, changes in renal function resulting from alterations in hormone levels may at least contribute to the effects evoked by direct ANP-induced alterations in GFR and/or tubular transport processes.

## PHYSIOLOGICAL EVIDENCE FOR A FURTHER NATRIURETIC PEPTIDE SYSTEM

Much experimental evidence is consistent with the concept that ANP is an important physiological regulator of renal sodium excretion, particularly regarding the results of the ANP- and guanylyl cyclase receptor-knockout models *(38,39)*. At the same time, at least

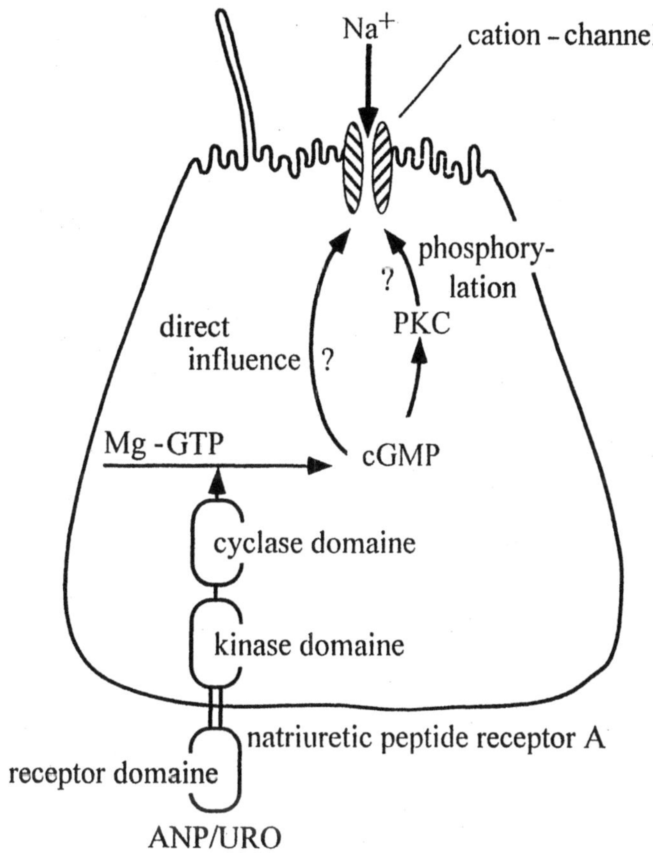


**Fig. 1.** Schematical drawing of an IMCD cell and the proposed signal transduction pathway leading to an inhibition of $Na^+$ reabsorption from the apical site.

in some points, the contradictory results of several groups and some open questions may reflect the uncertainties and difficulties to precisely characterize and define the role of ANP in sodium homeostasis. Some authors would even question the functional role of ANP on the kidney and favor a regulatory role primarily on the cardiovascular system *(22,130–132)*. However, this view is difficult to understand particularly with regard to the data of the knockout models. Even if ANP plays a fundamental role in sodium homeostasis, there is evidence of a further member of the natriuretic peptide family which may act either in a synergistic fashion with or independently from ANP. This peptide also exerts effects on renal sodium excretion during normal day-to-day living conditions and thus participates in the concert of the other fluid and sodium homeostasis-regulating hormones *(130)*. This hypothesis is based on a number of common physiological, experimental, and pathophysiological studies in which there is no direct correlation between sodium excretion and ANP plasma levels (for review, *see* ref. *130*). Therefore, in the following, we discuss the potential physiological role of ANP in the light of a new natriuretic peptide hormone that is presumably synthesized in the kidney itself and exerts its renal effects in a paracrine manner.

### *Urodilatin, a Renal Natriuretic Peptide System?*

Immunohistochemical and/or tissue extraction studies, along with the demonstration of mRNA for ANP, indicate the presence and synthesis of ANP in organs in addition to the heart. These organs include, among others the brain *(133)*, adrenals *(134)*, gastrointestinal tract *(135)*, the immunomodulatory system *(37)*, and especially the kidneys *(136–138)*.

Recently, Ritter and coworkers *(57)* reported that cultured neonatal kidney cells secrete ANP. Furthermore, Greenwald and coworkers *(55,97)* demonstrated an ANP-like peptide in the mature rat kidney and increased immunostaining for ANP prohormone in the distal tubule of nephrotic rats that could be later supported by other groups *(13,136–139)*. These data gave the first hints for synthesis or storage of a natriuretic peptide in the kidney. Indeed, in 1988, a new member of the natriuretic peptide family, was discovered by Schulz-Knappe and coworkers *(140,141)* from human urine. A purification procedure from 1000 L of human urine was performed and resulted in a peptide that was fully homologous to the C-terminal 32 amino acids found in the prohormone ANP and exhibited an N-terminal extension by four amino acids in comparison to the circulating ANP. This peptide was called Urodilatin (URO) (Fig. 2). The N-terminal extension suggests that the urinary peptide URO discovered does not originate from the circulation, but is probably synthesized in the kidney and excreted into the urine. This view is also supported by the fact that using a specific URO radioimmunoassay, the peptide could not be detected in the circulation but only in urine *(142)*. However, demonstration of a specific mRNA is still required. Thus, the working hypothesis is that URO is derived from ANP gene product with a different processing pattern compared to that in the heart. Then, after secretion into the lumen, it may interact with luminally located natriuretic peptide receptors in the IMCD, exerting its natriuretic and diuretic effects (Fig. 3) *(143–145)*.

## PHYSIOLOGICAL SIGNIFICANCE OF URODILATIN

To characterize the relative physiological importance of URO and ANP, these two peptides were measured in different experimental settings. Several observations suggest that URO plays a key role in the physiological regulation of renal function, especially in the control of renal sodium and water excretion *(131,146)*. Our group *(147)* measured the effect of long-term sodium load in healthy volunteers. We found a close correlation between natriuresis and URO excretion. A stepwise increase in sodium intake induced a concomitant increase in URO excretion parallel to sodium excretion. These results are supported by earlier observations of Heer and coworkers *(148)*, who performed a similar study and also found a close correlation between URO and sodium excretion. Furthermore, an increased URO and sodium excretion was observed after acute volume load by saline infusion *(149)*. Drummer and coworkers *(149)* investigated the role of URO in six healthy volunteers during a period of nine days and nights. In this period, an acute isotonic saline infusion was performed. The closest relation with $Na^+$ excretion was observed not for ANP plasma levels but for URO excretion into the urine, which was considerably increased during the long-term period of up to 22 h postinfusion. These results suggest that URO might participate in renal response to an acute saline infusion. Furthermore, Goetz and coworkers *(131)* investigated the endogenous production of URO and ANP in conscious dogs following balloon dilatation of the left atrium after cardiac denervation compared to normal dogs. Only URO excretion in urine paralleled the natriuresis induced by balloon dilatation, whereas the plasma levels of ANP showed no positive correlation. Emmeluth and coworkers *(146)* showed an increase in URO excretion after a selective increase in cephalic sodium concentration, suggesting that cephalic sodium concentration receptors regulate the rate of sodium excretion via URO. Norsk and coworkers *(150)* studied the effects of 12 h head-out water immersion in seated normal human subjects and

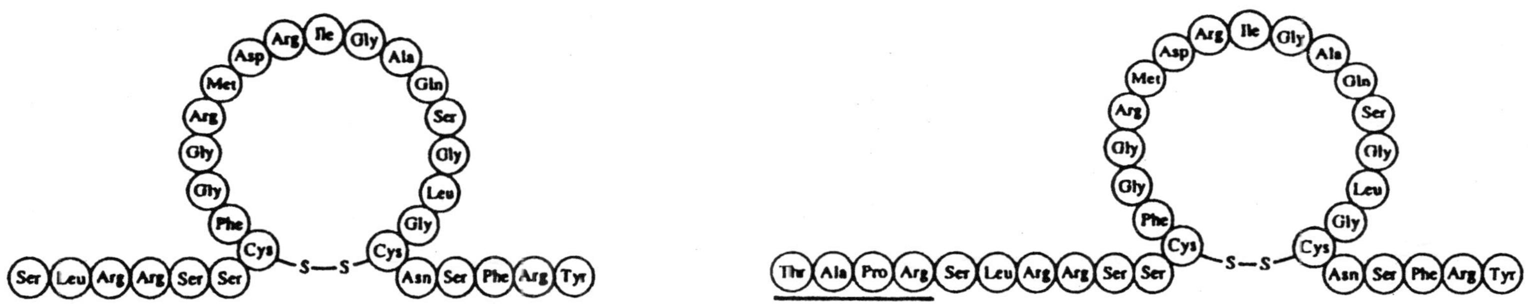


human A type of natriuretic peptide

circulating form (CDD/ANP - 99 - 126)

Urodilatin

(CDD/ANP - 95 - 126)

**Fig. 2.** Amino acid sequence of circulating ANP-99-126 and URO (ANP 95–126). URO differs from ANP by an N-terminal extension of four amino acids (indicated by the bar).

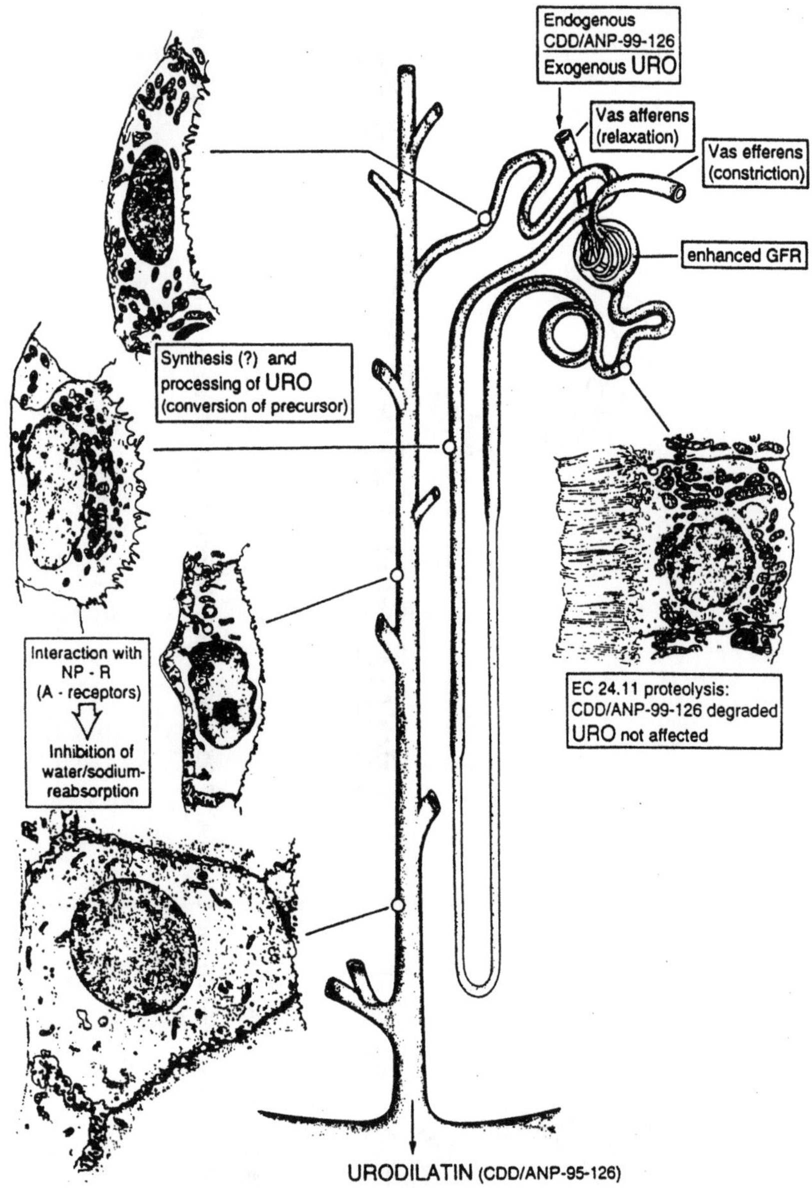


**Fig. 3.** Proposed mechanism of physiological and pharmacological effects of the natriuretic peptides ANP and URO. In the proximal tubule, enzymatical degradation of filtered CDD/ANP-99–126 takes place preventing an interaction of ANP with luminal receptors located in the collecting duct. In contrast, URO is luminally secreted from distal tubule cells and can exert its renal effects interacting with these receptors in the collecting duct.

found that the excretion of URO followed a temporal pattern similar to the increases in urine flow and sodium excretion produced by immersion. Drummer and coworkers *(151)* investigated the role of URO in the regulation of sodium homeostasis and they demonstrated that the circadian rhythm of urinary sodium excretion paralleled URO excretion. Acute saline infusion *(149)* induced natriuresis which correlated closer to URO excretion than to ANP plasma levels. Similar results were obtained by Hansell and coworkers

Table 2
Natriuretic Peptide Receptors Interacting with ANP and URO as Well as the Induced cGMP Formation at Different Nephron Sites[a]

| | *Natriuretic peptide receptors* | *cGMP formation following ANP stimulation* |
|---|---|---|
| Glomerulus | + | + |
| Proximal tubule | – | – |
| Henle's loop | | – |
| CCD | – | – |
| IMCD | + | + |

[a](CCD = cortical collecting duct, IMCD = inner medullary collecting duct.

*(152)*, who showed that natriuresis occurring after intracerebroventricular application of hypertonic saline is not associated with an increase in ANP plasma levels. Emmeluth and coworkers *(153)* demonstrated that sodium load induced a natriuretic response, but did not change ANP plasma levels. Taken together, these observations indicate that URO may be the natriuretic factor primarily responsible for sodium and water regulation.

URO closely mimics the receptor-binding and agonist properties of ANP and binds to natriuretic peptide receptors at several nephron sites, including especially the glomerulus and the IMCD (Table 2) *(154,155)*. Valentin *(155)* showed similar cGMP generation characteristics of URO and ANP in glomeruli and IMCD cells, whereas Saxenhofer *(154)* reported a lower potency of URO for cGMP stimulation compared to ANP in papillary collecting-duct homogenates. However, these results suggest that URO interacts basically with the same receptors and signaling pathways that mediate the actions of ANP.

The stimuli for URO synthesis and secretion in the kidney are, however, still far from being clarified. Emmeluth and coworkers *(146)* investigated the role of cephalic $Na^+$ concentration and URO in control of renal $Na^+$ excretion. Using the split-infusion technique, the authors infused hypertonic saline into both carotid arteries in order to increase the concentration of sodium supplying the head. This procedure caused an increase in sodium excretion that was paralleled with an increase in URO excretion. The authors concluded that cephalic sodium concentration receptors regulate the rate of excretion of sodium via URO. Whether the URO secretion is a humoral or neural-mediated response is unclear. To further clarify this issue, the same working group performed similar experiments, but this time in dogs with surgically denervated kidneys. Again, sodium excretion and URO excretion increased. However, this URO excretion was not significantly different to that induced by isotonic saline infusion in control animals, which tended to be higher. Thus, the natriuretic response modulated by URO, if there is one, appears to be independent of renal nerves favoring a humoral factor. However, the authors could demonstrate a strong correlation between natriuresis and endothelin excretion *(156)*, suggesting that the role of URO is still unclear.

Comparable to the efforts that had been undertaken to characterize the renal function of ANP, the exact mechanisms by which URO acts in the kidney were investigated by administering URO intravenously. Initiation of profound diuresis and natriuresis as well as slight reduction in blood pressure are the most prominent effects *(157–159)*. The strength and duration of these effects differed considerably from those of ANP. In bolus injection in healthy volunteers, the resulting natriuresis and diuresis was double that of

ANP *(157–160)*, which was confirmed in a study by Bestle and Bie *(160)*. Here, the authors demonstrated in volume-expanded dogs that URO induced a stronger diuresis and natriuresis than that seen by ANP.

The precise way in which URO interacts in the kidney and leads to diuresis and natriuresis was investigated in pharmacological studies. Although the available literature about URO is not extensive compared to that of ANP, the mode of action of URO in the kidney appears to be well established. Based on the similar affinity of URO and ANP to intrarenal natriuretic peptide receptors *(154,155)*, most of the renal actions of URO can be explained in analogy to those induced by ANP. Using hydronephrotic rat kidneys, Endlich and coworkers *(161)* demonstrated that URO leads to a dilation of preglomerular vessels and to a constriction of efferent arterioles, leading to an increased glomerular filtration rate. These results could also be confirmed by other groups in men *(157)* and animals *(158)*, whereas Hildebrandt and coworkers *(159)* could not find any changes in GFR after injection into the renal artery of dogs. The same group found no changes in fractional lithium excretion as a marker for proximal tubular reabsorption. Therefore, the induced diuresis and natriuresis is most probably due, similarly to ANP, to an enhancement of glomerular filtration rate and an inhibition of $Na^+$ reabsorption in the IMCD.

### *On Which Cell Site of the Inner Medullary Collecting Duct Cell Do ANP and Urodilatin Exhibit their Renal Actions?*

There is still controversy about the cell site where the natriuretic peptides interact with the functional bioactive receptors resulting in intracellular cGMP formation. In first studies using rabbit IMCD cultures grown on filters, evidence arose that the natriuretic peptides bind to basolaterally located receptors in the kidney and exert their natriuretic effects by inhibiting amiloride-sensitive apical $Na^+$ uptake *(82)*. These results are consistent with those found in isolated perfused tubule preparations of cortical collecting duct and IMCD. Here, ANP administered to the basolateral and not to the apical site inhibited vasopressin-stimulated water flow and $Na^+$ reabsorption *(80,81,108)*. In contrast, Sonnenberg *(162)* has demonstrated that urinary sodium excretion is increased by the intraluminal administration of ANP in the rat kidney. Furthermore, Hashimoto and coworkers *(163)* demonstrated that the addition of ANP to the apical side of LLC-PK1 cells—extensively employed as a model for the analysis of epithelial functions in the proximal and distal tubules *(164)*—produced a concentration-dependent increase in cGMP formation, the extent being similar to that produced by the addition to the basolateral side. The authors concluded that the biologically active natriuretic receptor is expressed in both the apical and the basolateral membranes. Moreover, the administration of inhibitors of a metabolizing enzyme of ANP in rats has been shown to induce the increase in immunoreactive ANP excreted into urine and to produce a much higher level of diuresis-natriuresis than that expected from the circulating levels of ANP *(163)*.

Considering these new insights, the mode of action of URO might be explained. Presumably, URO is synthesized in the distal tubule and secreted into the lumen, then the hormone passes further downstream in the urine in the tubule system and interacts with luminally located natriuretic peptide receptors in the IMCD. Here, the peptide exhibits its natriuretic and diuretic effects. The intracellular mechanisms by which URO and ANP may lead to an inhibition of $Na^+$ reabsorption, i.e., by direct interaction of the receptor with the apically located amiloride-sensitive $Na^+$ channel and even the stimuli leading to an increase of URO from distal tubules, still remains unclear.

### *Metabolism of ANP and Urodilatin in the Kidney and its Physiological and Pharmacological Significance*

Removal of natriuretic peptides from circulation occurs via different mechanisms: besides binding of circulating natriuretic peptides to clearance receptors, enzymatic degradation in lung, liver, and kidney takes place *(165–169)*. The main enzyme responsible for degradation is the metalloendoprotease E.C.3.4.24.11. One of the main location sites is the brush border of the proximal tubule. This enzyme cleaves the loop structure of ANP between position Cys-105 and Phe-106, reducing receptor affinity and thus biological activity *(170–173)*. Thus, under physiological conditions at least, the filtered ANP reaching the proximal tubule may be degraded and is unlikely to be detected as an intact molecule further downstream in the urine. Evidence that URO is synthesized in the distal tubule and probably secreted into the lumen could therefore explain the isolation of the natriuretic peptide URO in urine *(13,22,136–139,155)*.

From the pharmacological point of view, the following finding was very important: in contrast to ANP, studies demonstrate a high resistance of URO to enzymatic degradation when both peptides were incubated with kidney cortex membranes from the dog, which are known to be rich of endoproteases *(170)*. This relative resistance of URO to the endoproteases may be due to its N-terminal extension of four additional amino acids *(170,171,174)*. This structural difference may induce configurational changes preventing the enzyme to attack cleavage site. Thus, administered intravenously, URO could reach nephron sites that are located distal from the proximal tubule whereas ANP is inactivated there. This could also explain the stronger natriuretic and diuretic effects of URO compared to ANP described by several groups *(157–159,175)*. Bub and coworkers *(143)* infused radioactive labeled URO and ANP *in situ* into the renal artery prior to freeze sectioning and autoradiography. They found binding sites for both peptides predominantly located in the cortex related to glomeruli and arterial blood vessels (Fig. 4A). But furthermore, URO was also labeled in the medullary structures indicating a binding at the IMCD receptors (Fig. 4B). This finding supports the data of the relative resistance of URO to the endoproteases and additionally appears to be of great clinical importance when considering the potential renal effects of intravenously administered ANP and URO.

## SUMMARY

There is substantial evidence that the primary sites of action of ANP are the beginning and the end of the nephron; the glomerulus and the IMCD. By dilating afferent and constricting efferent arterioles of the renal glomerular vessel supply, ANP increases GFR and filtration fraction, leading to increased hydraulic pressure in the glomerular capillaries; this increases the driving force for ultrafiltration. Mesangial changes may contribute to this mechanism. Furthermore, ANP disrupts glomerulus proximal tubule balance such that increased delivery of filtrate to the proximal tubule is not reabsorbed proportionately. An increased load of $Na^+$ and water is passed through ANP-insensitive descending and ascending Henle limbs and the distal convoluted tubule to the cortical and medullary collecting duct, where ANP reduces $Na^+$ and vasopressin-dependent water reabsorption. Furthermore, hormonal changes may also contribute or at least modulate these effects by inhibition of renin and aldosterone synthesis and by antagonizing the action of angiotensin II. Even if substantial controversy persists regarding the relative importance of the two major sites of action of ANP (glomerulus and IMCD), the understanding so far is that

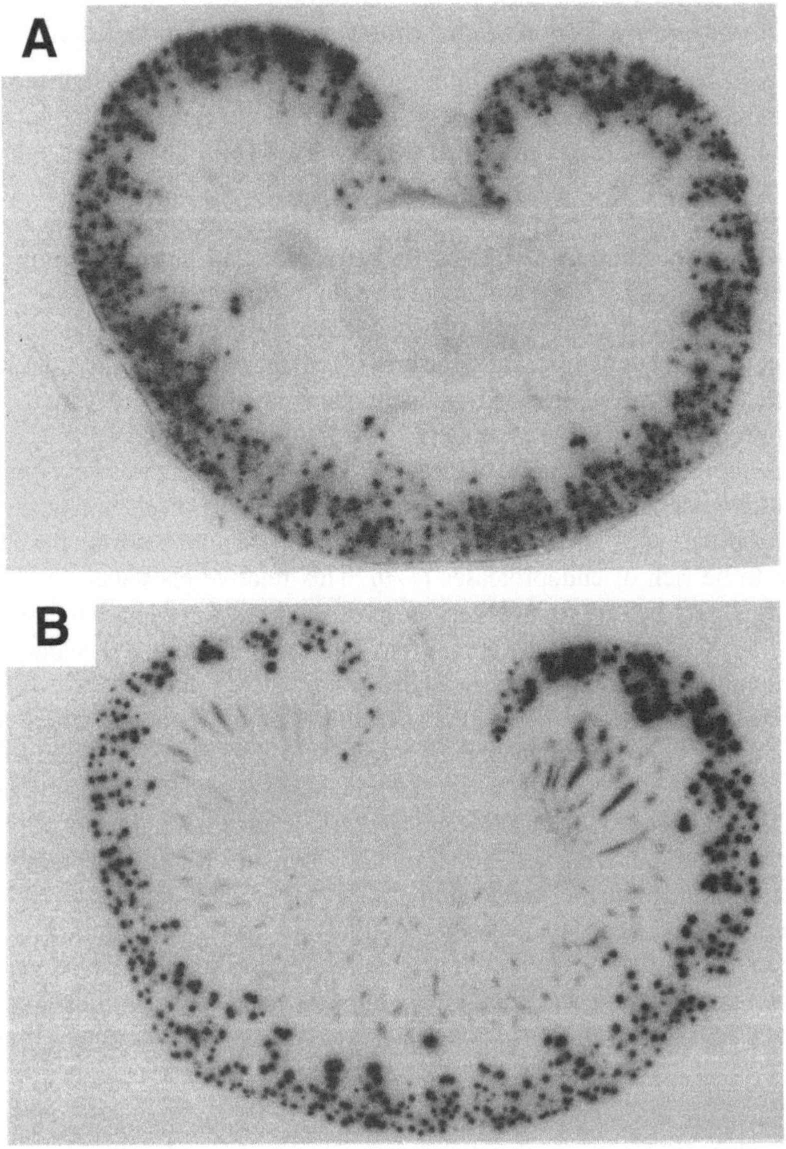

**Fig. 4. (A)** Autoradiograph of $^{125}$I-ANP binding to renal kidney. Binding sites are predominantly located in the cortex related to glomeruli and arterial blood vessels. **(B)** Autoradiograph of $^{125}$I-URO binding to renal kidney. Besides binding sites in the cortex also the medullary structures are labelled, indicating a binding at the inner medullary collecting duct receptors (with permission of Bub *[143]*).

the natriuretic and diuretic effects of ANP are provoked by a combination of hemodynamic and tubular effects.

Recent results using the ANP and guanylyl cyclase receptor knockout models indicate that ANP plays a role in regulation in fluid and sodium homeostasis. Further evidence came mainly from results demonstrating that infusion of ANP in animals and humans provokes brisk diuresis and natriuresis. This evidence gained more importance when atrial distension was associated with elevated ANP plasma levels and sodium excretion.

However, since the discovery of or even before the renal natriuretic peptide URO was known, this fundamental role of ANP in the concert of other fluid and sodium homeostasis-regulating hormones was, at least partly, questionable. The isolation of URO from human urine and its presumable synthesis site in the kidney opened up a new field of research objectivizing the role of ANP in fluid and sodium homeostasis. The working hypothesis, which is based on many studies, is that URO is synthesized in the distal tubule, then secreted into the lumen and binds to luminally located natriuretic peptide receptors that are located in the IMCD. Here, the peptide exhibits its diuretic and natriuretic effects. In the meantime, many physiological studies have been performed to further elucidate the role of ANP and URO and it is not too provocative to postulate that ANP serves primarily as a regulator of the cardiovascular system with relatively little effect on the kidney, whereas URO may be the more important candidate in the natriuretic peptide family, serving as a physiological ligand for the receptors in the IMCD relating to the intrarenal regulation of sodium excretion.

## REFERENCES


1. Henry JP, Gauer OH, Reeves JL. Evidence of the atrial location of receptors influencing urine flow. Circulation Res 1956;4:85–90.
2. Lockett MF. A hormonal influence of the heart on the kidney. Isr J Med Sci 1969;5:525–529.
3. Marie JP, Guillemot H, Hatt P-Y. Le degre de granulation des cardiocytes auriculaires. Path Biol 1976;24:549–554.
4. deBold AJ, Borenstein HB, Veress AT, Sonnenberg H. A rapid and potent natriuretic response to intravenous injection of atrial myocardial extract in rats. Life Sciences 1981;28:89–94.
5. Deth RC, Wong K, Fukoyawa S, Rocco R, Smart JL, Lynch CJ, Wawad R. Inhibition of rat aorta contractile response by natriuresis inducing extract of rat atrium. Fed Proc 1982;41:983 (Abstract).
6. Grammer RT, Fukumi H, Inagami T, Misono KS. Rat atrial natriuretic factor: purification and vasorelaxant activity. Biochem Biophys Res Comm 1983;116:696–703.
7. Currie MG, Geller DM, Cole BR, Boylan JG, YuSheng W, Holmberg SW, Needleman P. Bioactive cardiac substances: potent vasorelaxant activity in mammalian atria. Science 1983;221:71–73.
8. Forssmann WG, Hock D, Lottspeich F, Henschen A, Kreye V, Christmann M, Reinecke M, Metz J, Carlquist M, Mutt V. The right auricle of the heart is an endocrine organ. Anat Embryol 1983;168:307–313.
9. Flynn TG, deBold ML, deBold AJ. The amino acid sequence of an atrial peptide with potent diuretic and natriuretic properties. Biochem Biophys Res Commun 1983;117:859–865.
10. Kangawa K, Matsuo H. Purification and complete amino acid sequence of alpha-human atrial natriuretic polypeptide (alpha-hANP). Biochem Biophys Res Commun 1984;118:131–139.
11. Appel RG, Wang J, Simonson MS, Dunn MJ. A mechanism by which atrial natriuretic factor mediates its glomerular actions. Am J Physiol 1986;251:F1036–1042.
12. Forssmann WG. Cardiac hormones. I. Review on the morphology, biochemistry, and molecular biology of the endocrine heart. Eur J Clin Invest 1986;16:439–451.
13. Forssmann WG, Feller SM, Meyer M, Rippegather R, Schulz-Knappe P. Morphology of the myoendocrine cardiac cell and extra-auricular systems producing cardiac hormones. In: Kaufmann W, Wambach G, eds. Endocrinology of the heart. Berlin-Heidelberg-New York: Springer-Verlag, 1989, pp. 3–26.
14. Rosenzweig A, Seidman CE. Atrial natriuretic factor and related peptide hormones. Annu Rev Biochem 1991;60:229–255.
15. Espiner EA. Physiology of natriuretic peptides. J Int Med 1994;235:527–541.
16. Goetz KL. Physiology and pathophysiology of the atrial peptides. Am J Physiol 1988;254:E1–E15.
17. Gunning ME, Brenner BM. Natriuretic peptides and the kidney: current concepts. Kidney Int 1992;42:127–133.
18. Sagnella GA, Markandu ND, Shore AC, MacGregor GA. Raised circulating levels of atrial natriuretic peptides in essential hypertension. Lancet 1986;1:179–181.
19. Kohno M, Murakawa K, Yasunari K, Nishizawa Y, Morii H, Takeda T. Circulating atrial natriuretic peptides in hyperthyroidism and hypothyroidism. Am J Med 1987;83:648–652.

20. Iimura O, Shimamoto K, Ando T, Ura N, Ishida H, Nakagawa M, Yokoyama T, Fukuyama S, Yamaguchi Y, Yamaji I. Plasma levels of human atrial natriuretic peptide in patients with hypertensive diseases. Can J Physiol Pharmacol 1987;65:1701–1705.
21. Yamaji T, Ishibashi, M., Sekihara, H., Takaku, F., Nakaoka, H., Fujii, J. Plasma levels of atrial natriuretic peptide in primary aldosteronism and essential hypertension. J Clin Endocrinol Metab 1986;63:815–818.
22. Goetz KL. Renal natriuretic peptide (urodilatin?) and atriopeptin: evolving concepts. Am J Physiol 1991;261:F921–932.
23. Blaine EH. Atrial natriuretic factor plays a significant role in body fluid homeostasis. Hypertension 1990;15:2–8.
24. Briggs JP, Soejima H, Schnermann J. Effect of atrial natriuretic peptide on chloride absorption along papillary collecting duct. Proc Tenth Int Cong Nephrol 1988;10:207.
25. Burnett JC Jr, Granger JP, Opgenorth TJ. Effects of synthetic atrial natriuretic factor on renal function and renin release. Am J Physiol 1984;247:F863–866.
26. Camargo MJ, Kleinert HD, Atlas SA, Sealey JE, Laragh JH, Maack T. Ca-dependent hemodynamic and natriuretic effects of atrial extract in isolated rat kidney. Am J Physiol 1984;246:F447–456.
27. Keeler R, Azzarolo AM. Effects of atrial natriuretic factor on renal handling of water and electrolytes in rats. Can J Physiol Pharmacol 1983;61:996–1002.
28. Wong KR, Xie MH, Shi LB, Liu FY, Huang CL, Gardner DG, Cogan MG. Urinary cGMP as biological marker of the renal activity of atrial natriuretic factor. Am J Physiol 1988;255:F1220–1224.
29. Anderson JV, Donckier J, Payne NN, Beacham J, Slater JD, Bloom SR. Atrial natriuretic peptide: evidence of action as a natriuretic hormone at physiological plasma concentrations in man. Clin Sci 1987;72:305–312.
30. Briggs JP, Steipe B, Schubert G, Schnermann J. Micropuncture studies of the renal effects of atrial natriuretic substance. Pflugers Arch 1982;395:271–276.
31. Schwab TR, Edwards BS, Heublein DM, Burnett JC Jr. Role of atrial natriuretic peptide in volume-expansion natriuresis. Am J Physiol 1986;251:R310–313.
32. Stasch JP, Hirth Dietrich C, Kazda S, Neuser D. Role of endogenous ANP on endocrine function investigated with a monoclonal antibody. Peptides 1990;11:577–582.
33. Morishita Y, Sano T, Kase H, Yamada K, Inagami T, Matsuda Y. HS-142-1, a novel nonpeptide atrial natriuretic peptide (ANP) antagonist, blocks ANP-induced renal responses through a specific interaction with guanylyl cyclase-linked receptors. Eur J Pharmacol 1992;225:203–207.
34. Oda S, Sano T, Morishita Y, Matsuda Y. Pharmacological profile of HS-142-1, a novel nonpeptide atrial natriuretic peptide (ANP) antagonist of microbial origin. II. Restoration by HS-142-1 of ANP-induced inhibition of aldosterone production in adrenal glomerulosa cells. J Pharmacol Exp Ther 1992;263:241–245.
35. Ohyama Y, Miyamoto K, Morishita Y, Matsuda Y, Saito Y, Minamino N, Kangawa K, Matsuo H. Stable expression of natriuretic peptide receptors: effects of HS-142-1, a non-peptide ANP antagonist. Biochem Biophys Res Commun 1992;189:336–342.
36. Sano T, Imura R, Morishita Y, Matsuda Y, Yamada K. HS-142-1, a novel polysaccharide of microbial origin, specifically recognizes guanylyl cyclase-linked ANP receptor in rat glomeruli. Life Sci 1992;51:1445–1451.
37. Sano T, Morishita Y, Matsuda Y, Yamada K. Pharmacological profile of HS-142-1, a novel nonpeptide atrial natriuretic peptide antagonist of microbial origin. I. Selective inhibition of the actions of natriuretic peptides in anesthetized rats. J Pharmacol Exp Ther 1992;260:825–831.
38. John SW, Krege JH, Oliver PM, Hagaman JR, Hodgin JB, Pang SC, Flynn TG, Smithies O. Genetic decreases in atrial natriuretic peptide and salt-sensitive hypertension [published erratum appears in Science 1995 Mar 24;267(5205):1753]. Science 1995;267:679–681.
39. Lopez MJ, Wong SK, Kishimoto I, Dubois S, Mach V, Friesen J, Garbers DL, Beuve A. Salt-resistant hypertension in mice lacking the guanylyl cyclase-A receptor for atrial natriuretic peptide. Nature 1995;378:65–68.
40. Maack T. Receptors of atrial natriuretic factor. Annu Rev Physiol 1992;54:11–27.
41. Sudoh T, Minamino N, Kangawa K, Matsuo H. Brain natriuretic peptide-32: N-terminal six amino acid extended form of brain natriuretic peptide identified in porcine brain. Biochem Biophys Res Commun 1988;155:726–732.
42. Cogan MG. Atrial natriuretic factor can increase renal solute excretion primarily by raising glomerular filtration. Am J Physiol 1986;250:F710–714.

43. Huang CL, Lewicki J, Johnson LK, Cogan MG. Renal mechanism of action of rat atrial natriuretic factor. J Clin Invest 1985;75:769–773.
44. Sonnenberg H, Cupples WA, de Bold AJ, Veress AT. Intrarenal localization of the natriuretic effect of cardiac atrial extract. Can J Physiol Pharmacol 1982;60:1149–1152.
45. Sonnenberg H, Honrath U, Chong CK, Wilson DR. Atrial natriuretic factor inhibits sodium transport in medullary collecting duct. Am J Physiol 1986;250:F963–966.
46. Takezawa K, Cowley AW Jr, Skelton M, Roman RJ. Atriopeptin III alters renal medullary hemodynamics and the pressure-diuresis response in rats. Am J Physiol 1987;252:F992–1002.
47. Hintze TH, Currie MG, Needleman P. Atriopeptins: renal-specific vasodilators in conscious dogs. Am J Physiol 1985;248:H587–591.
48. Koike H, Sada T, Miyamoto M, Oizumi K, Sugiyama M, Inagami T. Atrial natriuretic factor selectively increases renal blood flow in conscious spontaneously hypertensive rats. Eur J Pharmacol 1984;104:391–392.
49. Borenstein HB, Cupples WA, Sonnenberg H, Veress AT. The effect of a natriuretic atrial extract on renal haemodynamics and urinary excretion in anaesthetized rats. J Physiol (Lond) 1983;334:133–140.
50. Maack T, Marion DN, Camargo MJ, Kleinert HD, Laragh JH, Vaughan ED Jr, Atlas SA. Effects of auriculin (atrial natriuretic factor) on blood pressure, renal function, and the renin-aldosterone system in dogs. Am J Med 1984;77:1069–1075.
51. Pollock DM, Arendshorst WJ. Effect of atrial natriuretic factor on renal hemodynamics in the rat. Am J Physiol 1986;251:F795–801.
52. Seymour AA, Blaine EH, Mazack EK, Smith SG, Stabilito, II, Haley AB, Napier MA, Whinnery MA, Nutt RF. Renal and systemic effects of synthetic atrial natriuretic factor. Life Sci 1985;36:33–44.
53. Smits JF, van Essen H, Struyker Boudier HA, Lappe RW. Lack of renal vasodilation during intrarenal infusion of synthetic atriopeptin II in conscious intact SHR. Life Sci 1986;38:81–87.
54. Fujioka S, Tamaki T, Fukui K, Okahara T, Abe Y. Effects of a synthetic human atrial natriuretic polypeptide on regional blood flow in rats. Eur J Pharmacol 1985;109:301–304.
55. Hirata Y, Ganguli M, Tobian L, Iwai J. Dahl S rats have increased natriuretic factor in atria but are markedly hyporesponsive to it. Hypertension 1984;6:I148–155.
56. Kiberd BA, Larson TS, Robertson CR, Jamison RL. Effect of atrial natriuretic peptide on vasa recta blood flow in the rat. Am J Physiol 1987;252:F1112–1117.
57. Jaschke W, Sievers RS, Lipton MJ, Cogan MG. Cine-computed tomographic assessment of regional renal blood flow. Acta Radiol 1990;31:77–81.
58. Yukimura T, Ito K, Takenaga T, Yamamoto K, Kangawa K, Matsuo H. Renal effects of a synthetic alpha-human atrial natriuretic polypeptide (alpha-hANP) in anesthetized dogs. Eur J Pharmacol 1984;103:363–366.
59. Marin Grez M, Fleming JT, Steinhausen M. Atrial natriuretic peptide causes pre-glomerular vasodilatation and post-glomerular vasoconstriction in rat kidney. Nature 1986;324:473–476.
60. Aalkjaer C, Mulvany MJ, Nyborg NC. Atrial natriuretic factor causes specific relaxation of rat renal arcuate arteries. Br J Pharmacol 1985;86:447–453.
61. Veldkamp PJ, Carmines PK, Inscho EW, Navar LG. Direct evaluation of the microvascular actions of ANP in juxtamedullary nephrons. Am J Physiol 1988;254:F440–444.
62. Kriz W, Elger M, Lemley KV, Sakai T. Mesangial cell-glomerular basement membrane connections counteract glomerular capillary and mesangium expansion. Am J Nephrol 10 Suppl 1990;1:4–13.
63. Ballermann BJ, Hoover RL, Karnovsky MJ, Brenner BM. Physiologic regulation of atrial natriuretic peptide receptors in rat renal glomeruli. J Clin Invest 1985;76:2049–2056.
64. Cermak R, Kleta R, Forssmann WG, Schlatter E. Natriuretic peptides increase a K+ conductance in rat mesangial cells. Pflugers Arch 1996;431:571–577.
65. Hassid A. Atriopeptins decrease resting and hormone-elevated cytosolic Ca in cultured mesangial cells. Am J Physiol 1987;253:F1077–1082.
66. Appel RG, Dubyak GR, Dunn MJ. Effect of atrial natriuretic factor on cytosolic free calcium in rat glomerular mesangial cells. FEBS Lett 1987;224:396–400.
67. Kremer S, Troyer D, Kreisberg J, Skorecki K. Interaction of atrial natriuretic peptide-stimulated guanylate cyclase and vasopressin-stimulated calcium signaling pathways in the glomerular mesangial cell. Arch Biochem Biophys 1988;260:763–770.
68. Meyer Lehnert H, Tsai P, Caramelo C, Schrier RW. ANF inhibits vasopressin-induced Ca2+ mobilization and contraction in glomerular mesangial cells. Am J Physiol 1988;255:F771–780.

69. Singhal PC, DeCandido S, Satriano JA, Schlondorff D, Hays RM. Atrial natriuretic peptide and nitroprusside cause relaxation of cultured rat mesangial cells. Am J Physiol 1989;257:C86–93.
70. Fried TA, McCoy RN, Osgood RW, Stein JH. Effect of atriopeptin II on determinants of glomerular filtration rate in the in vitro perfused dog glomerulus. Am J Physiol 1986;250:F1119–1122.
71. Bianchi C, Gutkowska J, Thibault G, Garcia R, Genest J, Cantin M. Distinct localization of atrial natriuretic factor and angiotensin II binding sites in the glomerulus. Am J Physiol 1986;251:F594–602.
72. Cogan MG, Huang CL, Liu FY, Madden D, Wong KR. Effect of atrial natriuretic factor on acid-base homeostasis. J Hypertens Suppl 1986;4:S31–34.
73. Liu FY, Cogan MG. Atrial natriuretic factor does not inhibit basal or angiotensin II-stimulated proximal transport. Am J Physiol 1988;255:F434-437.
74. Zimmerman RS, Schirger JA, Edwards BS, Schwab TR, Heublein DM, Burnett JC Jr. Cardio-renal-endocrine dynamics during stepwise infusion of physiologic and pharmacologic concentrations of atrial natriuretic factor in the dog. Circ Res 1987;61:63–69.
75. Murray RD, Itoh S, Inagami T, Misono K, Seto S, Scicli AG, Carretero OA. Effects of synthetic atrial natriuretic factor in the isolated perfused rat kidney. Am J Physiol 1985;249:F603–609.
76. Pamnani MB, Clough DL, Chen JS, Link WT, Haddy FJ. Effects of rat atrial extract on sodium transport and blood pressure in the rat. Proc Soc Exp Biol Med 1984;176:123–131.
77. Ortola FV, Ballermann BJ, Anderson S, Mendez RE, Brenner BM. Elevated plasma atrial natriuretic peptide levels in diabetic rats. Potential mediator of hyperfiltration. J Clin Invest 1987;80:670–674.
78. Ortola FV, Ballermann BJ, Brenner BM. Endogenous ANP augments fractional excretion of Pi, Ca, and Na in rats with reduced renal mass. Am J Physiol 1988;255:F1091–1097.
79. Itoh H, Nakao K, Mukoyama M, Yamada T, Hosoda K, Shirakami G, Morii N, Sugawara A, Saito Y, Shiono S, et al. Chronic blockade of endogenous atrial natriuretic polypeptide (ANP) by monoclonal antibody against ANP accelerates the development of hypertension in spontaneously hypertensive and deoxycorticosterone acetate-salt-hypertensive rats. J Clin Invest 1989;84:145–154.
80. Nonoguchi H, Sands JM, Knepper MA. Atrial natriuretic factor inhibits vasopressin-stimulated osmotic water permeability in rat inner medullary collecting duct. J Clin Invest 1988;82:1383–1390.
81. Nonoguchi H, Sands JM, Knepper MA. ANF inhibits NaCl and fluid absorption in cortical collecting duct of rat kidney. Am J Physiol 1989;256:F179–186.
82. Zeidel ML, Seifter JL, Lear S, Brenner BM, Silva P. Atrial peptides inhibit oxygen consumption in kidney medullary collecting duct cells. Am J Physiol 1986;251:F379–383.
83. Lee J, Malvin RL. Natriuretic response to homologous heart extract in aglomerular toadfish. Am J Physiol 1987;252:R1055–1058.
84. Hammond TG, Yusufi AN, Knox FG, Dousa TP. Administration of atrial natriuretic factor inhibits sodium-coupled transport in proximal tubules. J Clin Invest 1985;75:1983–1989.
85. Hammond TG, Haramati A, Knox FG. Synthetic atrial natriuretic factor decreases renal tubular phosphate reabsorption in rats. Am J Physiol 1985;249:F315–318.
86. Ando R, Sasaki S, Shiigai T, Takeuchi J. Lack of effect of alpha-human atrial natriuretic polypeptide on volume reabsorption and p-aminohippuric acid secretion in the rabbit proximal straight tubule. Jpn J Physiol 1987;37:81–91.
87. Baum M, Toto RD. Lack of a direct effect of atrial natriuretic factor in the rabbit proximal tubule. Am J Physiol 1986;250:F66–69.
88. McGowan JA, Pitts TO, Rose ME, Puschett JB. The effects of atrial natriuretic peptide on whole-kidney and proximal straight tubular function in the rabbit. Proc Soc Exp Biol Med 1987;185:62–68.
89. Van de Stolpe A, Blouch K, Jamison RL. Effect of atrial natriuretic peptide on the superficial proximal tubule and loop of Henle. Kidney Int 1988;33 (Abstract).
90. Harris PJ, Thomas D, Morgan TO. Atrial natriuretic peptide inhibits angiotensin-stimulated proximal tubular sodium and water reabsorption. Nature 1987;326:697,698.
91. Garvin JL. Inhibition of Jv by ANF in rat proximal straight tubules requires angiotensin. Am J Physiol 1989;257:F907–911.
92. Stokes TJ Jr, McConkey CL Jr, Martin KJ. Atriopeptin III increases cGMP in glomeruli but not in proximal tubules of dog kidney. Am J Physiol 1986;250:F27–31.
93. Koike J, Nonoguchi H, Terada Y, Tomita K, Marumo F. Effect of urodilatin on cGMP accumulation in the kidney. J Am Soc Nephrol 1993;3:1705–1709.
94. Ritter D, Dean AD, Gluck SL, Greenwald JE. Natriuretic peptide receptors A and B have different cellular distributions in rat kidney. Kidney Int 1995;48:5758–5766.

95. Hirsch JR, Kruhøffer M, Meyer M, Forssmann WG, Mollerup S, Cermak R, Ankorina-Stark I, Schlatter E. Natriuretic peptides inhibit a K+ conductance in immortalized human kidney epithelial cells (IHKE-1). 11th International Symposium on Regulatory Peptides 1996; 64(1–3):70 (Abstract).
96. Kruhøffer M, Meyer M, Forssmann WG, Mollerup S, Herter P, Weber G, Cermak R, Ankorina I, Schlatter E, Hirsch JR. Natriuretic peptides inhibit a K+ conductance in immortalized human kidney epithelial cells (IHKE-1) (submitted).
97. Aperia A, Holtback U, Syren ML, Svensson LB, Fryckstedt J, Greengard P. Activation/deactivation of renal Na+,K(+)-ATPase: a final common pathway for regulation of natriuresis. FASEB J 1994;8:436–439.
98. Reddy S, Gyory AZ, Dyne M, Salipan Moore N, Pollock C, Field MJ, Cockayne DJ. Effect of atrial natriuretic peptide on cellular element concentrations in rat proximal tubules: evidence for inhibition of the sodium pump. Clin Exp Pharmacol Physiol 1994;21:775–780.
99. Darvish N, Winaver J, Dagan D. A novel cGMP-activated Cl- channel in renal proximal tubules. Am J Physiol 1995;268:F323–329.
100. Dai LJ, Quamme GA. Atrial natriuretic peptide initiates Ca2+ transients in isolated renal cortical thick ascending limb cells. Am J Physiol 1993;265:F592–597.
101. Peterson LN, De Rouffignac C, Sonnenberg H, Levine DZ. Thick ascending limb response to dDAVP and atrial natriuretic factor in vivo. Am J Physiol 1987;252:F374–381.
102. Kondo Y, Imai M, Kangawa K, Matsuo H. Lack of direct action of alpha-human atrial natriuretic polypeptide on the in vitro perfused segments of Henle's loop isolated from rabbit kidney. Pflugers Arch 1986;406:273–278.
103. Butlen D, Mistaoui M, Morel F. Atrial natriuretic peptide receptors along rat and rabbit nophrons: (125I) Alpha-rat atrial natriuretic peptide binding in microdissected glomeruli and tubules. Pflugers Arch 1987;408:356.
104. Chabardes D, Montegut M, Mistaoui M, Butlen D, Morel F. Atrial natriuretic peptide effects on cGMP and cAMP contents in microdissected glomeruli and segments of the rat and rabbit nephrons. Pflugers Arch 1987;408:366–372.
105. Tremblay J, Gerzer R, Vinay P, Pang SC, Beliveau R, Hamet P. The increase of cGMP by atrial natriuretic factor correlates with the distribution of particulate guanylate cyclase. FEBS Lett 1985;181:17–22.
106. Jamison RL, Sonnenberg H, Stein JH. Questions and replies: role of the collecting tubule in fluid, sodium, and potassium balance. Am J Physiol 1979;237:F247–261.
107. Mendez RE, Dunn BR, Troy JL, Brenner BM. Modulation of the natriuretic response to atrial natriuretic peptide by alterations in peritubular Starling forces in the rat. Circ Res 1986;59:605–611.
108. Dillingham MA, Anderson RJ. Inhibition of vasopressin action by atrial natriuretic factor. Science 1986;231:1572–1573.
109. Umemura S, Smyth DD, Pettinger WA. Lack of inhibition by atrial natriuretic factor on cyclic AMP levels in single nephron segments and the glomerulus. Biochem Biophys Res Commun 1985;127: 943–949.
110. Nonoguchi H, Knepper MA, Manganiello VC. Effects of atrial natriuretic factor on cyclic guanosine monophosphate and cyclic adenosine monophosphate accumulation in microdissected nephron segments from rats. J Clin Invest 1987;79:500–507.
111. Rouch AJ, Chen L, Troutman SL, Schafer JA. Na+ transport in isolated rat CCD: effects of bradykinin, ANP, clonidine, and hydrochlorothiazide. Am J Physiol 1991;260:F86–95.
112. Hawk CT, Schafer JA. Clonidine, but not bradykinin or ANP, inhibits Na+ and water transport in Dahl SS rat CCD. Kidney Int 1993;44:30–35.
113. Schlatter E, Cermak R, Forssmann WG, Hirsch JR, Kleta R, Kuhn M, Sun D, Schafer JA. cGMP activating peptides do not regulate electrogenic electrolyte transport in principal cells of rat CCD. Am J Physiol 1996;271:F1158–1165.
114. Zeidel ML, Kikeri D, Silva P, Burrowes M, Brenner BM. Atrial natriuretic peptides inhibit conductive sodium uptake by rabbit inner medullary collecting duct cells. J Clin Invest 1988;82:1067–1074.
115. Light DB, Schwiebert EM, Karlson KH, Stanton BA. Atrial natriuretic peptide inhibits a cation channel in renal inner medullary collecting duct cells. Science 1989;243:383–385.
116. Gunning ME, Ballermann BJ, Silva P, Brenner BM, Zeidel ML. Characterization of ANP receptors in rabbit inner medullary collecting duct cells. Am J Physiol 1988;255:F324–330.
117. Bianchi C, Gutkowska J, Thibault G, Garcia R, Genest J, Cantin M. Radioautographic localization of 125I-atrial natriuretic factor (ANF) in rat tissues. Histochemistry 1985;82:441–452.

118. Chai SY, Sexton PM, Allen AM, Figdor R, Mendelsohn FA. In vitro autoradiographic localization of ANP receptors in rat kidney and adrenal gland. Am J Physiol 1986;250:F753–757.
119. Healy DP, Fanestil DD. Localization of atrial natriuretic peptide binding sites within the rat kidney. Am J Physiol 1986;250:F573–578.
120. Koseki C, Hayashi Y, Torikai S, Furuya M, Ohnuma N, Imai M. Localization of binding sites for alpha-rat atrial natriuretic polypeptide in rat kidney. Am J Physiol 1986;250:F210–216.
121. Mantyh CR, Kruger L, Brecha NC, Mantyh PW. Localization of specific binding sites for atrial natriuretic factor in peripheral tissues of the guinea pig, rat, and human. Hypertension 1986;8:712–721.
122. Murphy KM, McLaughlin LL, Michener ML, Needleman P. Autoradiographic localization of atriopeptin III receptors in rat kidney. Eur J Pharmacol 1985;111:291,292.
123. Zeidel ML, Silva P, Brenner BM, Seifter JL. cGMP mediates effects of atrial peptides on medullary collecting duct cells. Am J Physiol 1987;252:F551–559.
124. Zeidel ML. Regulation of collecting duct Na+ reabsorption by ANP 31–67. Clin Exp Pharmacol Physiol 1995;22:121–124.
125. Opgenorth TJ, Burnett JC Jr, Granger JP, Scriven TA. Effects of atrial natriuretic peptide on renin secretion in nonfiltering kidney. Am J Physiol 1986;250:F798–801.
126. Henrich WL, McAlister EA, Smith PB, Lipton J, Campbell WB. Direct inhibitory effect of atriopeptin III on renin release in primate kidney. Life Sci 1987;41:259–264.
127. Kurtz A, Della Bruna R, Pfeilschifter J, Taugner R, Bauer C. Atrial natriuretic peptide inhibits renin release from juxtaglomerular cells by a cGMP-mediated process. Proc Natl Acad Sci USA 1986;83:4769–4773.
128. Henrich WL, Needleman P, Campbell WB. Effect of atriopeptin III on renin release in vitro. Life Sci 1986;39:993–1001.
129. Rodriguez Puyol D, Arriba G, Blanchart A, Santos JC, Caramelo C, Fernandez Cruz A, Hernando L, Lopez Novoa JM. Lack of a direct regulatory effect of atrial natriuretic factor on prostaglandins and renin release by isolated rat glomeruli. Biochem Biophys Res Commun 1986;138:496–501.
130. Goetz KL. Is urodilatin (rather than atriopeptin) the primary natriuretic peptide of the ANP family? J Cardiovasc Pharmacol 22 Suppl 1993;2:S84–85.
131. Goetz KL. Evidence that atriopeptin is not a physiological regulator of sodium excretion. Hypertension 1990;15:9–19.
132. Goetz K, Drummer C, Zhu JL, Leadley R, Fiedler F, Gerzer R. Evidence that urodilatin, rather than ANP, regulates renal sodium excretion. J Am Soc Nephrol 1990;1:867–874.
133. Gardner DG, Vlasuk GP, Baxter JD, Fiddes JC, Lewicki JA. Identification of atrial natriuretic factor gene transcripts in the central nervous system of the rat. Proc Natl Acad Sci USA 1987;84:2175–2179.
134. Inagaki S, Kubota Y, Kito S, Kangawa K, Matsuo H. Immunoreactive atrial natriuretic polypeptides in the adrenal medulla and sympathetic ganglia. Regul Pept 1986;15:249–260.
135. Vuolteenaho O, Arjamaa O, Vakkuri O, Maksniemi T, Nikkila L, Kangas J, Puurunen J, Ruskoaho H, Leppaluoto J. Atrial natriuretic peptide (ANP) in rat gastrointestinal tract. FEBS Lett 1988;233:79–82.
136. Flügge G, Inagami T, Fuchs E. Atrial natriuretic peptide detected by immunocytochemistry in peripheral organs of Tupaia belangeri. Histochemistry 1987;86:479–483.
137. Figueroa CD, Lewis HM, MacIver AG, Mackenzie JC, Bhoola KD. Cellular localisation of atrial natriuretic factor in the human kidney. Nephrol Dial Transplant 1990;5:25–31.
138. McKenzie JC, Tanaka I, Misono KS, Inagami T. Immunocytochemical localization of atrial natriuretic peptide in the kidney, adrenal medulla, pituitary and atrium of the rat. J Histochem Cytochem 1985;33:828–832.
139. Forssmann WG, Nokihara K, Gagelmann M, Hock D, Feller SM, Schulz-Knappe P, Herbst F. The heart is the center of a new endocrine, paracrine and neuroendocrine system. Arch Histol Cytol 1989;52:293–315.
140. Forssmann WG, Meyer M, Schulz-Knappe P. Urodilatin: from cardiac hormones to clinical trials. Exp Nephrol 1994;2:318–323.
141. Schulz-Knappe P, Forssmann K, Herbst F, Hock D, Pipkorn R, Forssmann WG. Isolation and structural analysis of "urodilatin", a new peptide of the cardiodilatin-(ANP)-family, extracted from human urine. Klin Wochenschr 1988;66:752–759.
142. Drummer C, Fiedler F, Bub A, Kleefeld D, Dimitriades E, Gerzer R, Forssmann WG Development and application of a urodilatin (CDD/ANP-95–126)-specific radioimmunoassay. Pflugers Arch 1993;423:372–377.
143. Bub A, Marxen P, Forssmann WG. Urodilatin (Uro) binding sites in rat kidney. Anat Rec 1993;[Suppl 1]:41 (Abstract).

144. Koller KJ, Lowe DG, Bennett GL, Minamino N, Kangawa K, Matsuo H, Goeddel DV. Selective activation of the B-natriuretic peptide receptor by C-type natriuretic peptide (CNP). Science 1991;252:120–123.
145. Koller KJ, Lowe DG, Minamino n, Matsuo H, Kangawa KD, Goeddel DV. Differential activation of the human natriuretic peptide receptor guanylyl cyclases. In: Imura H, Matsuo H, Masaki T eds. Peptide Regulation of Cardiovascular Function. Osaka, Takeda Science Foundation, 1991; pp 91–99.
146. Emmeluth C, Drummer C, Gerzer R, Bie P. Roles of cephalic $Na^+$ concentration and urodilatin in control of renal $Na^+$ excretion. Am J Physiol 1992;262:F513–F516.
147. Meyer M, Richter R, Brunkhorst R, Wrenger E, Schulz-Knappe P, Kist A, Mentz P, Brabant G, Koch KM, Rechkemmer G, Forssmann WG. Urodilatin is involved in sodium homeostasis and exerts sodium-state dependent natriuretic and diuretic effects. Am J Physiol 1996;271:489–497.
148. Heer M, Drummer C, Baisch F, Gerzer R. Long-term elevations of dietary sodium produce parallel increases in the renal excretion of urodilatin and sodium. Pflugers Arch 1993;425:390–394.
149. Drummer C, Gerzer R, Heer M, Molz B, Bie P, Schlossberger M, Stadaeger C, Röcker L, Strollo F, Heyduck B, Bauer K, Warberg J, Baisch F, Christensen NJ, König A, Norsk P. Effects of an acute saline infusion on fluid and electrolyte metabolism in humans. Am J Physiol 1993;262:F744–F754.
150. Norsk P, Drummer C, Johansen LB, Gerzer R. Effect of water immersion on renal natriuretic peptide (urodilatin) excretion in humans. J Appl Physiol 1993;74:2881–2885.
151. Drummer C, Fiedler F, König A, Gerzer R. Urodilatin, a kidney-derived natriuretic factor, is excreted with a circadian rhythm and stimulated by saline infusion in man. J Am Soc Nephrol 1991;2:1109–1113.
152. Hansell P, Goransson A, Leppaluoto J, Arjamaa O, Vakkuri O, Ulfendahl HR. CNS-induced natriuresis is not mediated by the atrial natriuretic factor. Acta Physiol Scand 1987;129:221–227.
153. Emmeluth C, Schütten HJ, Knigge U, Warberg J, Bie P. Increase in plasma sodium enhances natriuresis in response to a sodium load unable to change plasma atrial peptide concentration. Acta Physiol Scand 1989;10:93–95.
154. Saxenhofer H, Fitzgibbon WR, Paul RV. Urodilatin: binding properties and stimulation of cGMP generation in rat kidney cells. Am J Physiol 1993;264:F267–273.
155. Valentin JP, Humphreys MH. Urodilatin: a paracrine renal natriuretic peptide. Hypertension 1993;21:432–438.
156. Emmeluth C, Goetz, KL, Drummer C, Gerzer R, Forssmann WG, Bie P. Natriuretic caused by increased carotid (Na) after renal denervation. Am J Physiol 1996;270:F510–F517.
157. Saxenhofer H, Raselli A, Weidmann P, Forssmann WG, Bub A, Ferrari P, Shaw SG. Urodilatin, a natriuretic factor from kidneys, can modify renal and cardiovascular function in men. Am J Physiol 1990;259:F832–F838.
158. Abassi ZA, Powell JR, Golomb E, Keiser HR. Renal and systemic effects of urodilatin in rats with high-output heart failure. Am J Physiol 1992;262:F615–621.
159. Hildebrandt DA, Mizelle HL, Brands MW, Hall JE. Comparison of renal actions of urodilatin and atrial natriuretic peptide. Am J Physiol 1992;262:R395–R399.
160. Bestle MH, Bie P. Renal effects of urodilatin and atrial natriuretic peptide in volume expanded conscious dogs. Acta Physiol Scand 1993;149:77–83.
161. Endlich K, Forssmann WG, Steinhausen M. Effects of urodilatin in the rat kidney: comparison with ANF and interaction with vasoactive substances. Kidney Int 1995;47:1558–1568.
162. Sonnenberg H, Honrath U, Wilson DR. In vivo microperfusion of inner medullary collecting duct in rats: effect of amiloride and ANF. Am J Physiol 1990;259:F222–226.
163. Hashimoto Y, Ozaki J, Yasuhara M, Hori R, Suga S, Itoh H, Nakao K, Inui K. Functional evidence for an apical ANP receptor in LLC-PK1 kidney epithelial cells. Eur J Pharmacol 1994;268:443–445.
164. Inui K, Saito, H, Matsukawa Y, et al. Specific binding activities and cyclic GMP responses by atrial natriuretic polypeptide in kidney epithelial cell line (LLC-PK1). Biochem Biophys Res Commun 1985;132:253–260.
165. Crozier IG, Nicholls MG, Ikram H, Espiner EA, Yandle TG, Jans S. Atrial natriuretic peptides in humans, production and clearance by various tissues. Hypertension 1986;8 [Suppl. II]:II11–II15.
166. Erdös EG, Skidgel RA. Neutral endopetidase 24.11(enkephalinase) and related regulators of peptide hormones. FASEB 1989;J3:145–151.
167. Espiner EA, Nicholls MG, Yandle TG, Crozier IG, Cuneo RC, McCormick D, Ikram H. Studies on the secretion, metabolism, and action of atrial natriuretic peptide in man. J Hypertension 1986;[Suppl II]:85–91.

168. Hollister AS, Rodeheffer RJ, White FJ, Potts JR, Imada T, Inagami T. Clearance of atrial natriuretic factor by lung, liver, and kidney in human subjects and the dog. J Clin Invest 1989;83:623–628.
169. Perrella MA, Margulies KB, Chi-Ming W, Aarhus LL, Heublein DM, Burnett JC. Pulmonary and urinary clearance of atrial natriuretic factor in acute congestive heart failure in dogs. J Clin Invest 1991;87:1649–1655.
170. Gagelmann M, Hock D, Forssmann WG. Urodilatin (CDD/ANP-95-126) is not biologically inactivated by a peptidase from dog kidney cortex membranes in contrast to atrial natriuretic peptide/cardiodilatin (alpha-hANP/CDD-99–126). FEBS Lett 1988;233:249–254.
171. Kenny AJ, Bourne A, Ingram J. Hydrolysis of human and pig brain natriuretic peptides, urodilatin, C-type natriuretic peptide and some C-receptor ligands by endopeptidase-24.11. Biochem J 1993;291:83–88.
172. Olins GM, Spear KL, Siegel NR, ZurcherNeely HA. Inactivation of atrial natriuretic factor by the renal brush border. Biochem Biophys Acta 1987;901:97–100.
173. Stephenson SL, Kenny AJ. The hydrolysis of alpha-human atrial natriuretic peptide by pig kidney microvillar membranes is initiated by endopeptidase-24.11. Biochem J 1987;243:183–187.
174. Feller SM, Bub A, Gagelmann M, Forssmann WG. Natriuretic peptides from the heart, brain, and kidney: localization, processing, vasoactivity, and proteolytic degradation. In: Lewis BS, Kimchi A, eds. Heart failure mechanisms and management. Berlin-Heidelberg: Springer-Verlag, 1991; pp. 398–407.
175. Villarreal D, Freeman RH, Johnson RA. Renal effects of ANF (95-126), a new atrial peptide analogue, in dogs with experimental heart failure. Am J Hypertens 1991;4:508–515

# 10 Physiology of the Natriuretic Peptides

*Gonadal Function*

*Kailash N. Pandey*

***Contents***



## INTRODUCTION

Atrial natriuretic peptide (ANP) is the first described member in the natriuretic peptide hormone family *(1)*. Two other members, brain natriuretic peptide (BNP) and C-type natriuretic peptide (CNP), closely related to ANP, have been identified and isolated with molecular and biochemical characteristics similar to ANP, but each derived from separate genes *(2)*. Although these three types of natriuretic peptides (ANP, BNP, CNP) have highly homologous structure, they have distinct sites of synthesis, bind to specific cell-surface receptors, and probably elicit discrete biological functions *(3,4)*. Among all three natriuretic peptides, ANP is the most extensively studied member of this group and is considered an important hormone that regulates the maintenance of salt and water homeostasis *(3–7)*. The ANP gene is expressed in several extracardiac tissues, where it has been reported to function in autocrine and/or paracrine regulation, and its existence is by no means confined only to the heart *(9,10)*. Among various cells and tissues examined, natriuretic peptides have been found to be present in a number of extracardiac tissues, including adrenal glands *(11)*, brain *(12)*, pituitaries *(13)*, placenta *(14)*, spleen *(15)*, thymus *(16,17)*, testis *(18,19)*, ovary *(20,21)*, and aortic endothelial cells *(23–25)*. Thus, in addition to the effects on renal excretion of sodium and water and relaxation of the vascular smooth-muscle cells, ANP has been shown to inhibit the release of aldosterone from adrenal gland *(26–28)*, renin from kidney *(29,30)*, and vasopressin from posterior pituitary *(31,32)*, and to stimulate the release of androgen from Leydig cells *(33,34)*,

From: *Contemporary Endocrinology: Natriuretic Peptides in Health and Disease*
Edited by: W. K. Samson and E. R. Levin Humana Press Inc., Totowa, NJ

progesterone from granulosa cells *(35)*, and luteinizing hormone from anterior pituitary *(36)*. In this chapter, the current understanding of the biology of natriuretic peptides, their receptors, and physiological roles in both male and female gonads will be discussed. In addition, the potential areas of interest of ongoing research on the natriuretic peptides and intracellular signaling mechanisms by which they act in reproductive tissues will be summarized.

## NATRIURETIC PEPTIDES AND ROLES IN THE MALE REPRODUCTIVE SYSTEM

Since the discovery of ANP in the mammalian heart, most research has focused on its potent natriuretic, diuretic, and vasorelaxant activities. However, there is increasing evidence that natriuretic peptides also serve a number of additional physiological functions *(3)*. One receptor-mediated effect of ANP and BNP is the stimulation of testosterone production in Leydig cells. Among various tissues examined, testis showed the highest stimulation of the particulate guanylyl cyclase in response to ANP *(37)*. Soon after, it was found that the testis contains immunoreactive ANP-like materials and a high density of ANP receptors. ANP and BNP dramatically stimulate testicular steroidogenesis with concomitant increases in the intracellular accumulation of cGMP and a decrease in cAMP. These findings suggested that natriuretic peptide hormones play paracrine and/or autocrine role(s) in testicular cells.

### *Isolation and Localization of Natriuretic Peptides in Testis*

The characterization of both pro-ANP and processed ANP molecules in mouse and rat testes homogenates and immunohistochemical localization of ANP in the spermatids and sperm cells suggested that synthesis and processing of prohormone to biologically active peptide hormone occurs in the male germ line *(18,19,38)*. Quantitative analyses of ANP by radioimmunoassay showed that testes of adult rats and mice contain an average of 8–12 ng of ANP/g wet tissue, respectively *(19)*. This concentration range is approx 20–30-fold higher than in the pituitary but almost 50- to 100-fold lower than in the cardiac atrium *(7,39)*. The characterization of ANP from the extracts of mouse and rat testes, using high-performance gel filtration on high-pressure liquid chromatography (HPLC), revealed that partially purified ANP corresponded to molecular weights ($M_r$) of 18,000 and 3300 (Fig. 1). Further examination of the material, corresponding to the $M_r$ of 3300 by the reverse-phase HPLC established that the elusion position of this peptide corresponded to a 31-residue ANP molecule *(19)*. It is clearly a larger molecule than the circulating 28-amino-acid residue ANP. These observations suggested that ANP found in the testis is most likely the product of local synthesis and processing intrinsic to the testis *(18,19)*. The presence of immunoreactive ANP in the testis was supported by demonstration that ANP mRNA was detected in rat testis that hybridized to a 580-bp ANP cRNA probe *(18)*.

In both mouse and rat testes, ANP-specific immunohistochemical staining was observed in round and elongated spermatids and in spermatozoa (Fig. 2). In rat testis, staining was first detected in the acrosomal caps of step 3 spermatids *(40)*, persisted in the acrosome up to step 16 spermatids, after which staining decreased *(19)*. No immunoreactive staining could be detected in the spermatid cytoplasm or any other cell types. The immunolocalization of ANP-like material in both the acrosome and the flagellum of the spermatids was unexpected. The presence of other antigenic sites in the acrosome and the flagellum has also been reported *(41,42)*. Although the function of ANP in the mam-

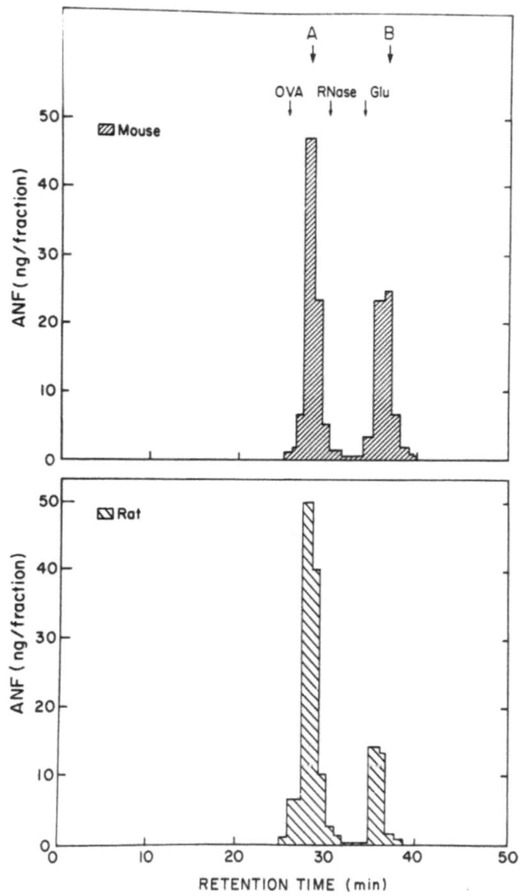


**Fig. 1.** Gel filtration chromatography of mouse and rat testes extracts. Samples of testis extracts were applied on high-performance gel filtration chromatography. The amount of immunoreactive ANP in each fraction was determined by radioimmunoassay. Elution positions of the standard proteins indicated by arrows: ovalbumine (Ova; $M_r$ = 45,000), ribonuclease (RNase; $M_r$ = 12,000), and glucagon (Glue; $M_r$, 3500). Arrows A and B indicate the elusion positions of pro-ANP and synthetic ANP, respectively. The upper and lower panels represent the chromatographic elusion patterns of pro-ANP and ANP (96–26), from mouse and rat testes, respectively. Reproduced with permission from ref. *19*.

malian spermatozoa is beginning to emerge, it is expected to participate in capacitation, sperm motility, and/or sperm development *(43,44)*.

The ANP can induce acrosome reaction in both capacitated and noncapacitated spermatozoa *(45)*. The capacitated spermatozoa underwent same maximal response to ANP as noncapacitated spermatozoa. Furthermore, the acrosome reaction was essentially equal in magnitude when induced with ANP or calcium ionophore A23187. However, higher concentrations of ANP were required to induce acrosome in capacitated as compared with noncapacitated spermatozoa. These studies suggested that ANP-induced human acrosome reaction does not require physiological concentrations of extracellular calcium. Acrosome reaction is known to involve various extracellular signals, including prostaglandins *(46)*, cAMP *(47)*, cGMP *(48,49)*, $Ca^{2+}$ *(50)*, inositol phosphates *(51)*,

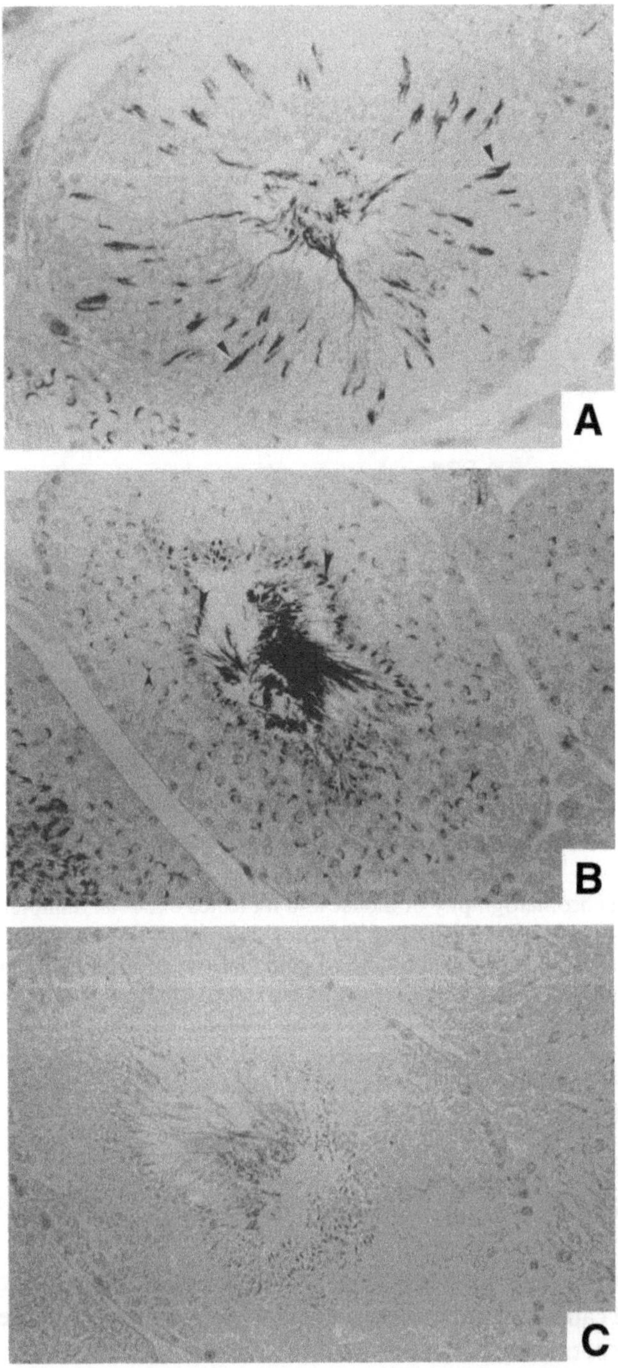


**Fig. 2.** Immunohistochemical localization of ANP in mouse and rat testes. **(A)** Cross-section of a stage 1 seminiferous tubule in rat testis. Note the unstained step 1 round spermatids and the staining of the heads of elongated spermatids in bundles (arrow) and of their flagella in the lumen. No counter stain, ×350. **(B)** Cross-section through a stage VII seminiferous tubule in mouse testis. Note the staining on the acrosomal cap of round spermatids (small arrow), the heads (large arrow),

diacylglycerol *(52)*, and activators of tyrosine protein kinase *(53)* and $Ca^{2+}$/calmodulin-dependent protein kinase *(54)*. It is expected that ANP-dependent acrosome reaction is likely mediated by second messenger cGMP, since analogs of cGMP are known to induce acrosome in mammalian spermatozoa *(48,49)*. The ANP may be chemotactic toward human spermatozoa *(54)*, however, more detailed studies are needed to extend these observations. Sea urchin egg-derived peptide (speract) is known to stimulate homologous spermatozoa, including increased respiration, intracellular alkalization, calcium influx, and actin polymerization with concomitant increase in cGMP accumulation *(55)*. The sea urchin peptides bind to homologous receptors, activate guanylyl cyclase, and increase cGMP production by similar mechanisms as displayed by natriuretic peptides and their receptors *(44)*.

Recent studies have also suggested that seminal plasma contains very high concentrations of CNP *(56)*. The concentrations of CNP in porcine seminal plasma and seminal vesicle fluid was about 2000- and 100,000-fold higher than that reported in porcine brain *(57)*. The CNP in seminal plasma was identified and characterized by reverse-phase HPLC and subsequent amino acid sequence analysis *(56)*. The immunoreactive-CNP contents in testis or epididymis was extremely low. The marked differences in the CNP content of epididymal fluid, seminal vesicle fluid, and seminal plasma suggested that the primary source of CNP is most likely the seminal vesicle rather than testis or epididymis. The CNP in seminal plasma appears to arise by local production; however, the cell type that produces the CNP has not been identified. Similarly, the exact function of CNP in the seminal plasma remains to be clarified.

### *Identification and Characterization of Natriuretic Peptide Receptors in Testis*

The paradigm for the relationship between natriuretic peptide hormones and their cellular, biochemical, and physiological effects, is the interaction of these molecules with specific cell-surface receptors. Natriuretic peptides (ANP, BNP, CNP) and their interaction with distinct receptors, most likely accounts for extensive functional diversity in a wide range of target cells. A great deal of studies have been carried out with natriuretic peptide receptor systems to understand their structure-function relationship and cellular and molecular basis of transmembrane signal transduction. Initial crosslinking and photoaffinity labeling studies showed the existence of ANP receptors of the 66–70- and the 120–180-kDa proteins *(3,7,58)*. The high affinity binding and guanylyl cyclase activity resided in the 135–180-kDa disulfide unlinked receptor molecules *(59–62)*. Simultaneously, the 70-kDa, disulfide-linked receptor was also isolated, which did not contain guanylyl cyclase activity *(63)*. Initially, ANP receptors were characterized based on their differential binding characteristics and abilities to generate cGMP in response to various lengths of ANP peptides, truncated at the amino and/or carboxyl terminal of 28-residues ANP *(64,65)*. The full-length ANP (28-residues) seems to provide the structural frame whose specificity is defined by the unique amino-acid sequences of Asn-Phe-Arg-Tyr at the carboxyl terminus of ANP molecule. The lack of these residues markedly

and the flagella of elongated spermatids. No counter stain, ×350, **(C)** Cross-section of a seminiferous tubule in mouse testis. Normal rabbit serum was used as a control. Note the absence of staining. No counter stain. ×350. Reproduced with permission from ref. *19*.

reduced the binding of the truncated peptide (atriopeptin I, API) to 135–180-kDa ANP receptor molecules. Based on the biological activity of different ANP analogs, initially ANP receptors were classified as biologically active and silent (clearance) receptors *(66)*. Subsequently, it was suggested that prevalence of the different subtypes of ANP receptors was specific to different target cells *(67,68)*. Membrane-bound forms of guanylyl cyclase represent biologically active natriuretic peptide receptors (Nprs) that have been cloned and sequenced. The cloning and expression of cDNAs have identified two different forms of natriuretic peptide guanylyl cyclase receptors, designated as GC-A and GC-B, also referred to as Npra and Nprb, respectively *(4)*. The evidence suggest that both ANP and BNP selectively activate Npra, whereas CNP has been shown to stimulate primarily Nprb *(24,69,70)*.

Mouse Leydig tumor (MA-10) cell line contains predominantly very high density of Npra *(71)*. The binding of $^{125}$I-ANP to MA-10 cells was specific, rapid, and saturable. The scatchard analysis of the binding data indicated a Kd of $1 \times 10^{-9} M$ and binding density ($B_{max}$) of $1 \times 10^6$ receptor sites/cell than any other cell or tissue source reported to date *(67,72)*. The incubation of MA-10 cells with photoaffinity ligand, 4-azidobenzoyl *N*-hydroxyl succinimide-$^{125}$I-ANP (AZB-$^{125}$I-ANP) followed by photoactivation yielded essentially a single monomeric radiolabeled protein band with an apparent $M_r$ of 135,000 as analyzed by SDS-gel electrophoresis and autoradiography (Fig. 3). The labeling of 135 kDa ANP receptor in MA-10 cells was completely abolished by unmodified ANP. In contrast, the truncated ANP (API) was unable to protect the 135-kDa-labeled receptor band. The radioactivity present in the 135-kDa-labeled receptor band was cut out from the dried gel; the amount of $^{125}$I-ANP moiety incorporated per $1 \times 10^6$ cells was estimated to be approx 10 fmol. On the basis of the estimated $B_{max}$, of $1–2 \times 10^6$ receptor sites/cell, this amount represents an affinity-labeling yield of approx 0.5–1%, indicating a very high density of ANP receptors in mouse Leydig tumor (MA 10) cells *(72)*.

The presence of high density, nonreducible 135-kDa ANP receptors in the MA-10 cell line has provided a unique model system to study cellular, biochemical, and molecular properties of endogenous guanylyl cyclase coupled-ANP receptor (Npra) molecules. The cDNA libraries representing the mRNA from MA-10 cells were constructed, which yielded a large number of Npra cDNA clones *(71)*. After the analysis of $2.5 \times 10^5$ recombinants from an unamplified pool of λ ZAPII cDNA library, 31 positively hybridizing clones were obtained, indicating the high abundance of Npra message in MA-10 cells. The nucleotide sequence analysis showed that murine Npra cDNA constituted the entire coding region of 3171 bp with 5'- and 3'-untranslated sequences, predicting a protein of 1057 amino-acid residues with molecular mass *($M_r$)* of 119,223 and an isoelectric point of 6.53. Hydrophobic analysis predicted a 21-residue transmembrane domain that divides the entire protein into a 469 amino-acid extracellular domain and a 567-amino-acid residue of intracellular protein kinase-like and guanylyl cyclase catalytic domains *(71)*. The murine Npra cDNA clone from MA-10 cells showed 95 and 97% sequence identity to human placenta and rat brain cDNA clones, respectively *(73,74)*.

Because of the unique property that Leydig tumor (MA-10) cells predominantly contain Npra in high density, these cells have also provided an ideal model system for the pharmacodynamic studies of this receptor protein. The kinetic analyses of internalization, recycling, and redistribution of ANP-receptor in MA-10 cell line indicated that Npra-ligand complexes are rapidly internalized, processed, and redistributed into the intracellular compartments *(75)*. The measurement of the stoichiometric distribution of

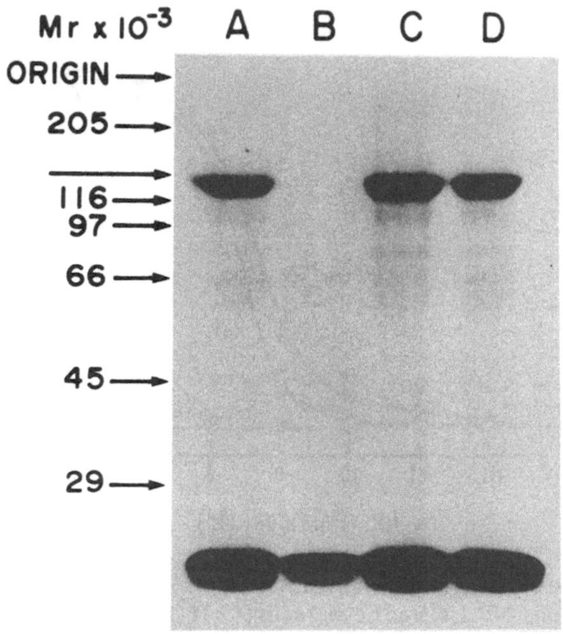


**Fig. 3.** Autoradiogram showing photoaffinity labeling of ANP receptor on cultured mouse Leydig tumor (MA-10) cells. The cells were allowed to bind 4-azidobenzoyl-$^{125}$I-ANP (AZB-$^{125}$I-ANP) in the dark in the absence (lane A), or in the presence of unmodified ANP (lane B), atriopeptin I (lane C), or angiotensin II (lane D), each at 1 μ*M* concentration. After washing with fresh medium, cells were photolyzed. The labeled proteins in the cell extracts were resolved in the presence of 100 m*M* dithiothreitol by sodium dodecylsulfate polyacrylamide gel electrophoresis (SDS-PAGE) and autoradiography. Arrow indicates the 135-kDa radioactive ANP receptor (Npra) specifically labeled in MA-10 cells. Molecular weight markers of the standard proteins include myosine (205 kDa), β-galactosidase (119 kDa), phosphorylase b (97 kDa), BSA (66 kDa), ovalbumine (45 kDa), and carbonic anhydrase (29 kDa). Reproduced with permission from ref. *72*.

$^{125}$I-ANP from cell surface to the cell interior and into the culture media provided the means to directly examine the dynamic processes of ANP-mediated translocation and distribution of biologically active Npra in MA-10 cells. Using tryptic proteolysis of the photoaffinity-labeled receptors, it was found that a population of internalized ligand-receptor complexes recycled back to the plasma membranes *(75)*.

Natriuretic peptides have been shown to specifically bind the cell surface receptors in isolated normal murine Leydig cells. The binding of radiolabeled peptides was displaced with unlabeled ligands in a dose-dependent manner. The Scatchard analysis of the binding data indicated high-affinity and low-capacity binding sites for ANP (Kd = $2 \times 10^{-10}$ *M*, $B_{max}$ 20 fmol $1 \times 10^5$ cells) and BNP (Kd = $3.1 \times 10^{-10}$ *M*, $B_{max}$ 19 fmol $1 \times 10^5$ cells). The binding of $^{125}$I-CNP was significantly lower than ANP and BNP in isolated normal Leydig cells *(76)*. Autoradiographic studies showed that in whole tissue section of Wistar Kyoto (WKY) rat testis, $^{125}$I-ANP binding was approx 22-fold higher than $^{125}$I-CNP binding *(77)*. The results from these studies also showed that $^{125}$1-ANP binding in WKY rat testis was approx eightfold greater than spontaneously hypertensive rat (SHR) testis. Also, $^{125}$I-ANP binding in the spermatids of WKY rat testis was approx 2.5-fold higher than SHR. Furthermore, these studies showed a higher level of Npra mRNA expression

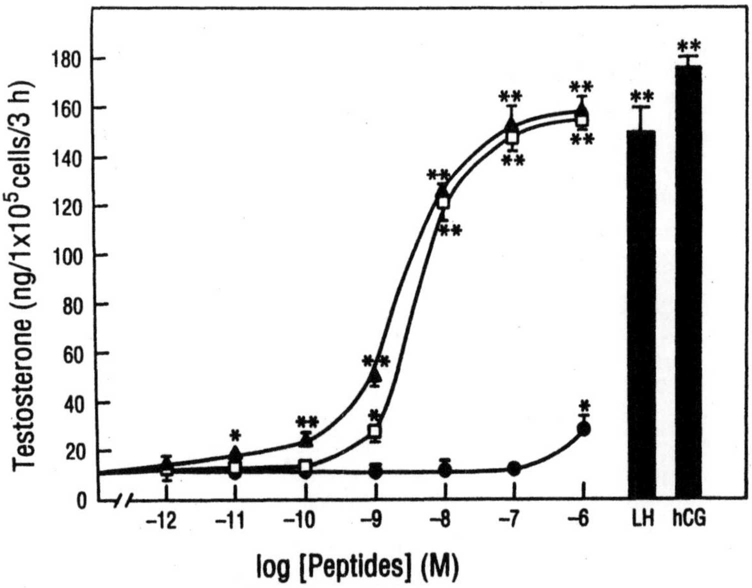


**Fig. 4.** Dose response curve of stimulation of testosterone production in purified mouse Leydig cells by ANP, BNP, and CNP. Isolated Leydig cells ($1 \times 10^5$) were incubated in 1 mL assay medium in the presence of increasing concentrations of ANP (□), BNP (▲) and CNP (●) for 3 h at 37°C in an atmosphere of 5% $CO_2$ and 95% $O_2$. Testosterone produced was estimated by direct radioimmunoassay. Bars indicate the testosterone production by LH (100 ng/mL) and hCG (25 ng/mL). Reproduced with permission from ref. *76*.

in WKY rat testis as compared with SHR. The binding of $^{125}$I-ANP was also reported in the interstitial cells of whole tissue sections of Sprague-Dawley rat testis *(78)*. Similarly, specific binding sites for $^{125}$I-ANP have been detected and localized in viable human spermatozoa through radioreceptor analysis and autoradiography with a Kd of 2.5 n*M* and $B_{max}$ of 1.03 fmol/$10^6$ sperm 2.5 min$^{-1}$, corresponding to about 620 receptor sites/sperm *(79)*. The crosslinking of $^{125}$I-ANP exhibited a single nonreducible 135–140-kDa receptor protein specific to sperm cells. Evidence suggests that in testis and testicular cells, including Leydig and sperm cells, the 135–180-kDa guanylyl cyclase/ANP receptor (Npra) seems to be the predominant ANP-binding protein *(72,76,79–81)*. The specific binding of $^{125}$I-ANP and localization of the 135-kDa radiolabeled proteins in Leydig cells and spermatozoa indicated that natriuretic peptides may play a critical role in male gamete physiology.

### *Natriuretic Peptide-Dependent Signal Transduction and Testicular Steroidogenesis*

Several reports have shown that ANP stimulates the synthesis and release of testosterone in a dose-dependent manner in isolated and purified normal Leydig cells *(34,72,76,82,83)*. ANP stimulated testosterone production 15–30-fold in mouse Leydig cells and the concentration of ANP required for the half maximal response ($ED_{50}$) was $4 \times 10^{-9}$*M* (Fig. 4). Both ANP and luteinizing hormone (LH) in combination exerted synergistic effect on testosterone stimulation in Leydig cells; however, at lower concentrations of ANP ($1 \times 10^{-11}$ to $1 \times 10^{-10}$*M*), a slight inhibitory effect on LH-stimulated testosterone production was observed *(72)*. In contrast, at higher concentrations of ANP, testosterone production was stimulated by >75% over LH-stimulated levels. The ANP elicited the accumulation of cGMP in both time- and dose-dependent manner. The extent of cGMP stimulation was >40-fold over the controls. The cAMP accumulation was inhibited by 15–20% in ANP-treated normal mouse Leydig cells, as compared with controls. These studies suggested that the effects of ANP on normal Leydig cell steroidogenesis do not occur via the classic mechanisms of cAMP-mediated regulation of steroidogenic activity by gonadotropins. The BNP also stimulated testosterone production in Leydig cells, both in a time- and dose-dependent manner. The stimulated levels of testosterone production by ANP, BNP, or gonadotropins were comparable, whereas CNP was a very weak stimulator of testosterone production in Leydig cells *(76)*. Both ANP and BNP elicited the higher accumulation of intracellular cGMP at a comparable levels; however, CNP stimulated the production of cGMP only minimally in isolated Leydig cells. The c-ANF, which is a truncated form of ANP, did not affect either testosterone production or cGMP accumulation in these cells. The treatment of Leydig cells with ANP and BNP stimulated a very high level intracellular cGMP that directly supports a mediatory role of cGMP in the natriuretic peptide-dependent steroidogenic signal transduction in testis.

ANP significantly increased the production of progesterone (P4; twofold), 17-α-hydroxy progesterone (17-α-OHP4; 2.4-fold), androstenedione, (adione; 18.7-fold); pregnenolone ($P_5$; twofold); 17-α-hydroxy pregnenolone (17-α-$OHP_5$; 2.4-fold), and dehydroepiandrosterone sulfate (DHES, 9.5-fold) as compared with controls. Similarly, BNP also significantly stimulated the production of $P_4$ (5.8-fold), 17-α-$OHP_4$ (ninefold), adione (26.8-fold), $P_5$ (twofold), 17-α-$OHP_5$ (2.8-fold), and DHES (9.7-fold). On the other hand, the stimulation of testosterone precursors by CNP was only minimal. After treatment of Leydig cells with ANP or BNP, the conversion of exogenous pregnenolone to testosterone was significantly stimulated. Similarly, the conversion of exogenous progesterone to testosterone was also enhanced by the natriuretic peptide hormones *(76)*. These studies established that treatments of Leydig cells with ANP or BNP resulted in the stimulation of both $\Delta^4$- and $\Delta^5$-pathways of the steroidogenesis. LY83583, an inhibitor of guanylyl cyclase, attenuated natriuretic peptide-stimulated levels of testosterone production at lower dosages. However, it almost completely blocked the steroidogenic responsiveness at 10 $\mu M$ concentration.

Since CNP-dependent stimulation of testosterone as well as the precursors involved in the $\Delta^4$- and $\Delta^5$-pathways of steroidogenesis were extremely low, as compared to ANP or BNP, this suggested that ANP and BNP are the major natriuretic peptides involved in the steroidogenic responsiveness of Leydig cells *(76)*. The results of these studies indicated that natriuretic peptides stimulate steroidogenesis, beginning at the first step involved in the conversion of cholesterol to pregnenolone. Cycloheximide completely blocked the stimulatory effect of natriuretic peptides on testosterone synthesis, suggesting that these peptides probably stimulate testosterone production by affecting the protein synthesis in the steroidogenic pathway. Cycloheximide is also known to inhibit the steroidogenesis by preventing the synthesis of a labile protein involved in the transfer of cholesterol from outer to the inner mitochondrial membrane *(84–86)*.

Intriguing was the finding that ANP ($1 \times 10^{-8} M$) stimulated four- to fivefold increased levels of testosterone production in isolated Leydig cells from normotensive WKY rats as compared with those from genetic strain SHR *(77)*. These findings support a new physiological role of ANP in Leydig cells and in testis, in which a functional relationship seems to exist between testicular ANP receptor expression and testosterone production and the state of hypertension in SHR. Since ANP regulates the steroidogenic responsiveness in SHR testis, these results support the concept that a defect at one or more loci in ANP and/or its receptor system in testis may contribute to the development of hypertension in SHR.

The natriuretic peptides stimulate the testicular steroidogenesis by transmembrane signal transduction mechanisms involving the generation of intracellular cGMP. Both ANP and BNP were shown to stimulate guanylyl cyclase activity in Leydig cells, whereas CNP was less effective *(76)*. These findings suggested that the stimulation of steroidogenesis by natriuretic peptides in Leydig cells involves the activation of particulate guanylyl cyclase and generation of intracellular cGMP. The receptor binding selectivity of these natriuretic peptides confirms the presence of GC-A (Npra) and GC-B (Nprb) on the cell surface of freshly prepared mouse Leydig cells, which seems to interact with three natriuretic peptide ligands *(76)*. The ANP and BNP are known to stimulate Npra and CNP activates Nprb *(24,70,76)*. Since ANP and BNP rapidly stimulated the testosterone production, guanylyl cyclase activation, and intracellular accumulation of cGMP, as compared to CNP, it can be inferred that either ANP and BNP are potent hormones in stimulating the testosterone production by interacting with Npra, or the two guanylyl cyclase receptors (Npra and Nprb) differentially regulate steroidogenic responsiveness in Leydig cells.

The control mechanisms of steroidogenesis in testis may involve not only the polypeptide hormones, but also paracrine and autocrine factors. Gonadotropins are well known to regulate steroidogenesis through cAMP-dependent protein kinase in testis, ovary, and placenta *(86–88)*. Similarly, adrenocorticotropic hormone (ACTH) and angiotensin II regulate the mineralocorticoid synthesis through protein kinase A and $Ca^{2+}$-stimulated protein kinase C in adrenal glomerulosa cells, respectively *(89)*. In adrenal tumor (Y-1) cells, the steroidogenesis is stimulated by 8-bromo cAMP through protein kinase A and is inhibited by phorbol 12-myristate 13-acetate plus the calcium ionophore A23187 through protein kinase C *(90)*. The molecular mechanisms by which natriuretic peptides affect the testicular steroidogenesis are not yet completely understood. ANP and 8- bromo cGMP stimulated the phosphorylation of several proteins in plasma membrane preparations of Leydig tumor (MA-10) cells and adrenal glomerulosa cells *(91)*. Since cGMP is considered a second messenger of ANP, it is implied that intracellular cGMP produced by ANP probably activates cGMP-dependent protein kinase (PKG), which in turn phosphorylates cellular proteins during biological action of ANP. On the other hand, ANP exerted an inhibitory effect on the phosphorylation of protein kinase C in MA-10 cells *(92)* as well in adrenal glomerulosa cells *(93)*. It has also been shown that cGMP generated in response to ANP in isolated normal Leydig cells exerted a stimulatory effect on cAMP-dependent protein kinase *(94)*. However, ANP has been demonstrated to inhibit or to have no effect on the intracellular accumulation of cAMP in Leydig and other cell systems *(33,95–97)*.

It is apparent that ANP can affect testicular steroidogenesis in two ways: First, ANP alone can potently stimulate testosterone production in Leydig cells by more than 30-fold, and second, ANP can exert an apparent biphasic effect on steroidogenesis in Leydig cells, under stimulation of gonadotropins. At lower concentrations, ANP slightly inhibited the LH-stimulated levels of testosterone production in normal Leydig cells; however, with higher concentrations of ANP, testosterone production was dramatically increased *(72)*. The overall effect of ANP on testosterone production was stimulatory with concomitant

Table 1
ANP-Dependent Inhibition of Luteinizing Hormone-Stimulated Progesterone Secretion in Murine Leydig Tumor (MA-10) Cells[a]

| | *Progesterone (ng/mL)* | |
|---|---|---|
| *ANF(M)* | *Control* | *bLM* |
| None | 1.1 ± 0.2 | 385.9 ± 34.0 |
| $1 \times 10^{-12}$ | 1.3 ± 0.1 | 414.4 ± 10.8 |
| $1 \times 10^{-10}$ | 1.2 ± 0.2 | 346.0 ± 43.3 |
| $1 \times 10^{-9}$ | 1.1 ± 0.2 | 230.4 ± 44.0 |
| $10 \times 10^{-8}$ | 1.0 ± 0.3 | 232.2 ± 1912 |
| $10 \times 10^{-7}$ | 1.1 ± 0.1 | 236.3 ± 14.56 |
| $10 \times 10^{-6}$ | 1.1 ± 0.1 | 316.7 ± 38.74 |

[a]Confluent cultured Leydig tumor cells were incubated with varying concentrations of ANP alone or in combination with bovine luteinizing hormone (bLH) in the presence of 0.2 m*M* 3-isobutyl-1-methyl xanthin (IBMX). All cultures were incubated at 37°C for 90 min and media were collected for progesterone assay. Reproduced with permission from ref. *95*.

increase in the intracellular levels of cGMP. Testosterone production in mouse Leydig cells is also stimulated by 8-bromo cGMP in a dose-dependent manner and high concentrations of 8-bromo cGMP were found to stimulate Leydig cell steroidogenesis to a similar extent as that obtained by saturating concentrations of gonadotropins *(34,76)*. In Leydig tumor (MA-10) cells, ANP elicited the accumulation of intracellular cGMP by 1500- to 2000-fold and reduced the level of cAMP by >75–80% as compared with the controls *(95)*. Furthermore, ANP inhibited the gonadotropin-induced progesterone secretion by approx 50% and cAMP production by 80–90%, respectively, in these mouse Leydig tumor (MA-10) cells. The inhibitory effect of ANP on the progesterone production was maximum at the peptide concentration of $1 \times 10^{-9}M$, but at higher dosages of ANP ($1 \times 10^{-7}$ to $1 \times 10^{-6}M$), the inhibitory effects on progesterone production were diminished (Table 1). It is likely that the inhibitory effect of ANP on gonadotropin-stimulated progesterone secretion is a reflection of its inhibitory effect on cAMP production in MA-10 cells. The mechanisms by which ANP exerts the biphasic-positive and -negative effects on testicular steroidogenesis in normal and transformed Leydig cells and the physiological relevance of these responses have not been clearly elucidated.

Originally, cAMP and cGMP were thought to act solely as adversaries *(98)*. However, interaction between the two cyclic nucleotide second messenger systems has been suggested in certain cell types *(99–101)*. The synergistic effect of ANP and LH on testosterone production in Leydig cells has been demonstrated *(72,76,102)*. Probably, the ANP-dependent steroidogenic responsiveness in Leydig cells is triggered by the second messenger cGMP and a cGMP-dependent protein kinase in this process. Evidence also suggests that ANP/cGMP may interact cooperatively with LH/cAMP system in the stimulatory control of testosterone production in Leydig cells and that the site of synergistic interaction may be the activation of cAMP-dependent protein kinase *(94,103)*. The degree of compartmentalization or crosstalk between second messenger cAMP and cGMP has not been adequately addressed in most tissue systems. This is at least in part, because of the complexity of the systems and the difficulty in selectively examining them in the intact cells. Our initial findings demonstrated that relatively high concentrations of ANP reduced the gonadotropin-stimulated cAMP accumulation in normal and transformed mouse Leydig cells *(76,95)*. However, the steroidogenic responses of normal and tumor Leydig cells to relatively high stimulatory levels of LH was inhibited by low concentrations of ANP *(72,95,102)*. Nevertheless, the effect of coexposure of Leydig cells to minimally stimulatory levels of LH and ANP have not been adequately addressed. Recent studies have demonstrated that ANP-dependent stimulation of guanylyl cyclase activity is potentiated by GTPγS in MA-10 cells and the involvement of G-protein has been postulated in this process *(81,104)*. Very recently, it has also been shown that ANP inhibited the agonist-stimulated hydrolysis of phosphoinositide in MA-10 cells *(105)*. However, it is not yet clear to what extent these mechanisms play regulatory role(s) in ANP-dependent steroidogenic responsiveness in normal Leydig cells.

Hitherto, natriuretic peptides can potently stimulate androgen production in mouse and rat Leydig cells, involving both $\Delta^4$- and $\Delta^{-5}$-pathways of steroidogenesis with concomitant production of second messenger cGMP *(76)*. The ANP- and BNP-mediated stimulation of androgen output can rival that of highly stimulatory concentrations of luteinizing/chorionic gonadotropin hormones (LH/CG) in Leydig cells *(33,34,76)*. However, unlike LH/CG, the Leydig cell steroidogenesis in response to ANP is mediated through the generation of cGMP, without any increase in cAMP production, suggesting that it is not only the gonadotropins that regulate the testosterone production by rat and mouse Leydig cells, but there seems to be a complex system that also involves a family of natriuretic peptide-dependent new regulatory system in male gonads.

## NATRIURETIC PEPTIDES AND ROLES IN THE FEMALE REPRODUCTIVE SYSTEM

The presence of ANP and its receptors have been reported in ovarian cells and the possibility that the natriuretic peptides play an important physiological role in the female gonad has been previously suggested *(20,22,35)*. Increasing evidence strongly supports that natriuretic peptides are present and probably locally synthesized in ovarian cells of different mammalian species. The reports of relatively high concentrations of ANP in ovary and follicular fluids raised the possibility that ANP is synthesized and released from the ovarian cells. ANP receptors coupled with guanylyl cyclase activity have been demonstrated in these cells. Taken together, it raises the possibility that the steroid-producing cells of corpora lutea and other populations of intraovarian cells may be the sites for the endocrine and/or paracrine function of the natriuretic peptides and their receptors in female gonads.

### *Isolation and Localization of Natriuretic Peptides in Ovarian Cells*

The presence of immunoreactive ANP was reported in bovine corpus-luteal extracts *(20)*, rabbit ovary *(21)*, human ovarian follicular fluid *(106)*, and pig as well as trout ovarian follicles *(21,107)*. Concurrent detection of large molecular weight species of ANP and its mRNA in ovary suggested the *de novo* synthesis of ANP in female gonad *(22)*. In the ovary, the sites of ANP synthesis were shown to be granulosa cells *(108)* and also oocytes *(109)*. ANP was detected in the ovarian interstitial cells; however, it was absent in theca cells *(22)*. These findings implied that the presence of ANP in the interstitial cells that completely surrounds the ovarian follicle promotes the release of ANP to neighboring theca cells. The localization of ANP in the ovarian interstitial cells is intriguing because of the undefined role of these cells in ovarian steroidogenesis and/or homeostasis. It may, however, indicate that ANP synthesized in the interstitial cells acts in a paracrine manner to affect other hormones produced by the ovary.

ANP content in rat ovaries was estimated to be approx 240 pg/mg protein, which was comparable to that found in the hypothalamus and chromaffin cells of the adrenal medulla, however, approx 500-fold lower than in cardiac atrium *(22)*. The immunoreactive ANP was isolated and characterized from rat ovarian homogenates by reverse-phase HPLC, which revealed the presence of 28-amino-acid circulating peptide as well as 126-amino-acid prohormone, suggesting that ovaries are the site of local synthesis of ANP. To identify the cell types containing ANP immunoreactivity within ovaries, immunocytochemical staining of rat ovarian tissue sections revealed dense staining in interstitial cells, surrounding the mature Graafian follicles in which numerous cells showed strong ANP-like immunoreactivity. The endothelial cells of capillaries located within the interstitium, granulosa cells of mature and immature follicles, as well as cells of both the theca interma and exterma were devoid of ANP immunostaining. However, a weak but specific staining was observed in the ovum *(22)*. To confirm whether the ovaries indeed synthesized ANP, the ANP-specific mRNA transcripts were detected from rat ovaries that were of identical size to that from the heart *(22,110)*. The quantitative comparisons of mRNA signals indicated that the concentrations of ANP transcripts in ovary was 300–400-fold lower than in the heart ventricles. ANP-like materials have also been detected in the follicular fluids of human, pig, and rabbit; however, the ANP contents in different species was highly variable. The concentrations of immunoreactive-circulating ANP *(99–126)*, and pro-ANP *(1–98)* in human follicular fluid was estimated at approx 12 and 300 pmol/L, respectively *(111)*. The amount of extracted immunoreactive ANP-like material in the follicular fluid of pig ovarian follicle was approx 18–20 pg/mL, which was lower than the plasma concentrations of ANP *(21)*.

It is clear that the follicular fluid, ovary, and ovarian perfusates contain both high- and low-mol-wt forms of immunoreactive ANP. Both gel filtration chromatography and reverse-phase HPLC data revealed that a higher content of large $M_r$ form of ANP was found in ovary than the circulating form of ANP *(21)*. However, it has been realized that in the follicular fluid and ovarian perfusates, the proportion of low and high $M_r$ forms of ANP was not always consistent *(109)*. It is possible that these variations might be caused by individual differences, specifically the different ovarian cycle of the animals. These postulates are supported by the findings that the ovary locally synthesizes ANP and the concentrations of ovarian ANP seem to have a synchronized cycle change with the estrous cycle *(108)*.

### *Ligand Binding and Characterization of Natriuretic Peptide Receptors in Ovarian Cells*

ANP-binding sites have been detected in human granulosa-lutein cells *(35)*, bovine carpus luteum *(20)*, and bovine ovaries *(112)*. Using the affinity crosslinking technique, a single 120-kDa ANP receptor band was detected in the rat ovarian plasma membranes and intact cells *(22)*. The addition of cold ANP completely abolished the labeling of the 120-kDa protein band, whereas c-ANF (truncated peptide) specific to the ANP clearance receptor did not compete for binding. The competition binding experiments further indicated that the binding of $^{125}$I-ANP to ovarian membranes was displaced equally with

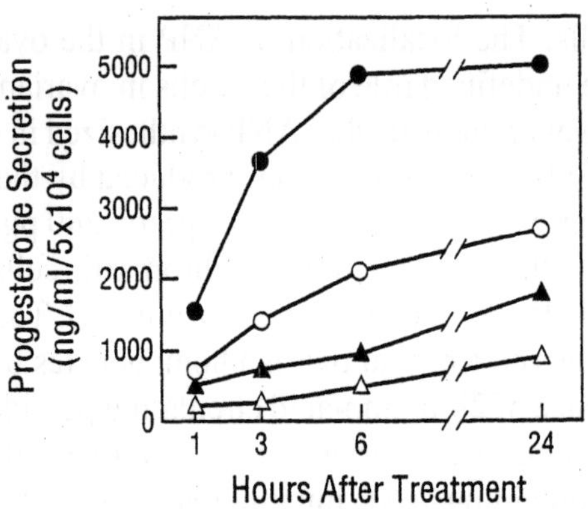


**Fig. 5.** Time-course of the stimulation of progesterone secretion from cultured human granulosa-lutein cells in response to ANP. Isolated granulosa-lutein cells were plated in 24-well plates. After 24 h, cells were incubated in the absence of any addition (△) or in the presence of luteinizing hormone (LH, 200 ng/mL, ▲), ANP (10 n*M*, ○) and LH + ANP (●) for indicated time periods. Progesterone in the medium was determined by radioimmunoassay. Reproduced with permission from ref. *35*.

unlabeled ANP and BNP; however, more than 1000-fold excess of CNP was required for 50% inhibition of binding. It is noteworthy that both ANP and BNP, but not CNP, bound to the ovarian receptor, pointing to the Npra predominating in ovary *(22)*. ANP binding to ovarian plasma membranes was not displaced by c-ANF, indicating the absence of the ANP clearance receptor (Nprc) in the ovary. Interestingly, the receptor crossreactivity was not observed with CNP, which has high sequence homology with ANP. Since both ANP and BNP but not CNP bound to ovarian membranes, this suggests the presence of the Npra subtype in the ovary *(22)*. Thus, the existence of high affinity guanylyl cyclase-coupled transducing receptors clearly indicates a functional role of Npra in the ovaries.

### *Natriuretic Peptide-Dependent Signal Transduction and Steroidogenesis in Ovarian Cells*

Initially, the specific receptor-mediated stimulation of progesterone and cGMP production by ANP was observed in human granulosa cells *(35)*. The treatment of granulosa cells with ANP ($1 \times 10^{-8} M$) resulted in a three- to fourfold increase in progesterone secretion (Fig. 5). Furthermore, ANP in combination with LH increased the synthesis of progesterone more than sixfold as compared with controls. Although ANP has been shown to stimulate the bovine luteal guanylyl cyclase activity and cGMP production *(113)*, there were no acute effects of ANP on the regulation of progesterone production by these cells *(114)*. ANP-dependent progesterone secretion was also measured in human follicular fluid. Higher levels of ANP and progesterone were detected in follicular fluids, but a significant correlation has not been established *(106)*. As previously shown, there is a synchrony in follicular development, and, accordingly, the pattern of steroid synthesis differs in individual follicles *(115)*. The knowledge of how the ovarian ANP activity changes throughout the follicular development up to the time of ovulation may explain some definitive correlation between follicular ANP production and progesterone secretion.

Recently it was also demonstrated that ANP evoked a biphasic effect on progesterone production in both differentiated and undifferentiated rat granulosa cells *(116)*. The results of these studies showed that ANP caused two- and threefold increase in progesterone and 20-α-OH-P production relative to untreated controls, respectively. However, a higher concentration of ANP in combination with porcine follicle-stimulating hormone (FSH) resulted in a biphasic response on progesterone production. These authors demonstrated that a maximal inhibition in progesterone production was achieved at $1 \times 10^{-9} M$ ANP, followed by a stimulation to the levels comparable with FSH-stimulated controls at high dosages of ANP examined. On the other hand, in FSH-stimulated differentiated granulosa cells, ANP evoked a threefold increase in progesterone production and a 60% inhibition in 20-α-OH-P production. The data from these studies suggested that ANP indeed can directly modulate progestin steroidogenesis in both undifferentiated and differentiated rat ovarian granulosa cells. Studies on the effect of ANP on progesterone production have revealed that infusion of human ANP in Yucantan micropigs induced a dose-dependent increase in the progesterone concentration in ovarian venous blood *(117)*. These authors also suggested a duality in the ovarian response to ANP, such as a decrease in progesterone secretion in response to low concentrations of ANP and a stimulated progesterone secretion in response to high concentrations of this peptide hormone. ANP has also been shown to exert a mutual relationship on granulosa cell steroidogenesis alone and in combination with endothelin and angiotensin II *(118)*. Because of its well-established effects on water and electrolyte balance, ANP may play a role in ovarian follicular fluid homeostasis directly or indirectly by interacting with local ovarian hormones *(119,120)*. Indeed, further studies are needed to understand the physiological significance, the regulatory factors involved in the secretion, and the exact origin of ANP production and function from ovary to follicular fluids.

Intriguing was the finding that ANP inhibited the spontaneous maturation (resumption of meiosis) in rat oocyte-cumulus complexes *(121)*. The inhibition was dose-dependent and persistent over the time period examined, which correlated with the effects of exogenous cGMP analogs. ANP also stimulated the intracellular cGMP accumulation, which corresponded with the inhibitory effect of ANP on oocyte maturation. Data from those previous studies demonstrated that ANP influenced the oocyte meiosis via elevation of cGMP. The truncated ANP analog (c-ANF) did not affect either spontaneous oocyte maturation or cGMP accumulation in oocyte–cumulus complexes. The findings of these studies correlated with earlier results that have suggested that cGMP derivatives also inhibited the spontaneous maturation of oocyte-cumulus complexes in the hamster *(122)* and in denuded rat oocytes *(123)*. Both exogenous cAMP analogs as well as cAMP-elevating agents can also prevent the spontaneous maturation of oocytes *(124–126)*. Conversely, ANP inhibits adenylyl cyclase and cAMP accumulation *(97)*, suggesting that both cGMP and cAMP probably affect the oocyte maturation through independent signaling pathways.

## SUMMARY AND CONCLUSION

There is overwhelming evidence that natriuretic peptides are locally synthesized and may play important physiological roles in both male and female gonads. The ovary and testis share homology of structure and function in gamete production and in the regulation of steroidogenesis that may provide new model systems for the study of endocrine, paracrine, and intracrine mechanisms of natriuretic peptide hormone action in gonads. The recent experimental results, however, suggest that ANP and BNP stimulate the production of testosterone in Leydig cells and progesterone in granulosa cells with a concomitant increase in cGMP production via Npra, which seems to be the predominant natriuretic peptide receptor in gonads. The Nprb in both testis and ovary is present at a very low density; consequently the effect of CNP on steroidogenic responses in these tissues is either absent or only minimal. Importantly, natriuretic peptides stimulate the steroidogenesis in Leydig and granulosa cells, through the generation of cGMP, without any increase in cAMP production. This pathway exists in addition to the previously known cAMP-generating systems of LH/CG and FSH in testis and ovary, respectively.

Natriuretic peptides have been localized specifically in spermatids, mature spermatozoa, and seminal fluid and are expected to participate in sperm motility, development, and acrosome reaction. Similarly, ANP inhibits spontaneous maturation of oocytes and thus seems to play a critical role in the ovum development. Research progress to date has revealed that natriuretic peptides may become key regulators and/or modulators of hormonal action in gonads. Further elucidation of the regulation and the mechanisms of action of natriuretic peptide hormone family in testis and ovary will provide insights into the expanding roles of these hormones in gonads. As yet, our understanding of the intracellular signaling mechanisms used by the natriuretic peptides in gonads are still limited. The complexity of the signal transduction of the natriuretic peptides in gonads is likely to be extended by the potential crosstalk by other hormones at numerous steps in the signaling cascade.

## ACKNOWLEDGMENTS

The research of the author's laboratory has been supported by the National Institutes of Health Grant HD25527 and the American Heart Association Established Investigatorship Award (900260) and Grant-in-Aids from the National Center (920-1011) and Georgia affiliate.

## REFERENCES

1. de Bold AJ. Atrial natriuretic factor: a hormone produced by the heart. Science 1985;230:767–770.
2. Rosenzweig A, Seidman CE. Atrial natriuretic factor and related peptide hormones. Annu Rev Biochem 1991;60:229–255.
3. Brenner BM, Ballerman BJ, Gunning ME, Zeidel ML. Diverse biological actions of atrial natriuretic peptide. Physiol Rev 1990;70:665–699.
4. Garbers DL. Guanylyl cyclase receptors and their endocrine, paracrine, and autocrine ligands. Cell 1992;71: 1–4.
5. Genest J, Cantin M. The atrial natriuretic factor. Its physiology and biochemistry 1988;110:1–45.
6. Baxter JD, Lewicki JA, Gardner DG. Structure, expression and function of atrial natriuretic factor in extraatrial tissues. Endocrine Rev 1988;10:519–536.
7. Inagami T. Atrial natriuretic factor. J Biol Chem 1989;264:3043–3046.
8. Gardner DG, Deschepper CF, Ganong WF, Hane S, Fidds J, Baxter JD, Lewicki J. Extra-atrial expression of the gene for atrial natriuretic factor. Proc Natl Acad Sci USA 1986;83:6697–6701.
9. Gutkowska J, Nemer M. Structure, expression and function of atrial natriuretic factor in extratrial tissues. Endocrine Rev 1989;10:519–536.
10. Vollmer AM. Atrial natriuretic peptide in peripheral organs other than the heart. Klin Wachenschr 1990;68:699–708.
11. Morel G, Chabot JG, Garcia Caballero T, Gossard F, Dihl F, Belles IM, Heisler S. Synthesis, internalization and localization of atrial natriuretic peptide in rat adrenal medulla. Endocrinology 1988; 123: 148–158.
12. Gardner DG, Vlasuk GP, Baxter JD, Fidds JC, Lewicki JA. Identification of atrial natriuretic factor gene transcripts in the central nervous system of the rat. Proc Natl Acad Sci USA 1987;84:2175–2179.
13. Gutkowska J, Cantin M. Bioactive atrial natriuretic factor-like peptides in rat anterior pituitary. Can J Physiol Pharmacol 1988;66:270–275.

14. Huang W, Lee D, Yang J, Carley DJ, Trosby M, Copolou DL, Johnston CI, Lim AT. Evidence for ANP production by rat placental cytotrophoblasts. Endocrinology 1992;131:919–924.
15. Throsby M, Lee D, Huang W, Yang Z, Copolou DL, Lim AT. Evidence for ANP (5–28), production by macrophages of the rat spleen. An immunochemical and non-radioactive *in situ* hybridization approach. Endocrinology 1991;129:991–1000.
16. Vollmer AM, Schulz R. Evidence for the presence of ANP precursor material in the rat thymus. Biochem Biophys Res Commun 1988;155:700–708.
17. Vollmer AM, Schulz R. Atrial natriuretic peptide is synthesized in the human thymus. Endocrinology 1990;126:2277–2280.
18. Vollmer AM, Friedrich A, Schulz R. Atrial natriuretic peptide precursor material in rat testis. J Androl 1990;11:471–475.
19. Pandey KN, Orgebin-Crist MC. Atrial natriuretic factor in mammalian testes: immunological detection in spermatozoa. Biochem Biophys Res Commun 1991;180:437–444.
20. Vollmer AM, Mytzka C, Arendt RM, Schulz R. Atrial natriuretic peptide in bovine corpus luteum. Endocrinology 1988;123:762–767.
21. Kim SH, Cho KW, Seul KH, Ryu H, Koh GY. Presence of immunoreactive atrial natriuretic peptide in follicular fluid, ovary and ovarian perfusates. Life Sci 1989;45:1581–1589.
22. Gutkowska J, Tremblay J, Antakly T, Meyer R, Mukaddam-Dahere S, Nemer M. The atrial natriuretic peptide system in rat ovaries. Endocrinology 1993;132:693–700.
23. Stingo AJ, Clavell AJ, Heublein DM, Wei CM, Pittelkom MR, Burnett JC Jr. Presence of C-type natriuretic peptide in cultured human endothelial cells and Plasma. Am J Physiol 1992;263:H1318–H1321.
24. Suga S, Nakao K, Itoh H, Komatsu Y, Ogawa Y, Hama N, Imura H. Endothelial production of C-type natriuretic peptide and its marked augmentation by transforming growth factor β. J Clin Invest 1992;90:1145–1149.
25. Nazario B, Hu R-M, Pedram A, Prins B, Levin E. Atrial and brain natriuretic peptides stimulate the production and secretion of C-type natriuretic peptide from bovine aortic endothelial cells. J Clin Invest 1995;261:R453–R457.
26. Atarashi K, Mulrow PJ, Franco-Saenz R, Snajdar R, Rupp J. Inhibition of aldosterone production by an atrial extract. Science 1984;224:992–994.
27. DeLean A, Racz K, Gutkowska J, Nguyen TT, Cantin M, Genest J. Specific receptor-mediated inhibition by atrial natriuretic factor of hormone-stimulated steroidogenesis in cultured bovine adrenal cells. Endocrinology 1984;115:1636–1638.
28. Goodfriend TL, Elliott ME, Atlas SA. Actions of synthetic atrial natriuretic factor on bovine adrenal glomerulosa. Life Sci 1984;35:1675–1682.
29. Burnett JC Jr, Granger JP, Opgenorth TJ. Effects of synthetic atrial natriuretic factor on renal function and renin release. Am J Physiol 1984;247:F863–F866.
30. Obana K, Naruse M, Naruse K, Sakurai Y, Demura H, Inagami T. Synthetic atrial natriuretic factor inhibits *in vitro* and *in vivo* renin secretion in rats. Endocrinology 1985;117:1282–1284.
31. Samson WK. Atrial natriuretic factor inhibits dehydration and hemorrhage-induced vasopressin release. Neuroendocrinology 1985;40:277–279.
32. Obana K, Naruse M, Inagami T, Brown AB, Naruse K, Kurimoto F, Sakurai H, Demura H, Shizume K. Atrial natriuretic factor inhibits vasopressin secretion from rat posterior pituitary. Biochem Biophys Res Commun 1985;132:1088–1094.
33. Pandey KN, Pavlou SN, Kovacs WJ, Inagami T. Atrial natriuretic factor regulates steroidogenic responsiveness and cyclic nucleotide levels in mouse Leydig cells *in vitro*. Biochem Biophys Res Commun 1986;138:399–404.
34. Mukhopadhyay AK, Bohnet HG, Leidenberger FA. Testosterone production by mouse Leydig cells is stimulated *in vitro* by atrial natriuretic factor. FEBS Lett 1986;202:111–116.
35. Pandey KN, Osteen KG, Inagami T. Specific receptor-mediated stimulation of progesterone secretion and cGMP accumulation by rat atrial natriuretic factor in cultured human granulosa-lutein (G-L) cells. Endocrinology 1987;121:1195–1197.
36. Horvath J, Ertl T, Schally AV. Effect of atrial natriuretic peptide on gonadotropin release in superfused rat pituitary cells. Proc Natl Acad Sci USA 1986;83:3444–3446.
37. Waldman SA, Rappoport RM, Murad F. Atrial natriuretic factor selectively activates particulate guanylate cyclase and elevates cyclic GMP in rat tissues. J Biol Chem 1984;259:14,332–14,334.
38. Vesely DL, Palmer PA, Giordano AT. Atrial natriuretic factor prohormone peptides are present in a variety of tissues. Peptides 1992;13:165–170.

39. Miyata A, Kangawa K, Toshimeri T, Hatoh T, Matsuo H. Molecular forms of atrial natriuretic polypeptides in mammalian tissues and plasma. Biochem Biophys Res Commun 1985;129:248–255.
40. Lebland CP, Clermont Y. Spermatogenesis of rat, mouse hamster and guinea pig as revealed by the peptide acid-fuchsin sulfuric acid technique. Am J Anat 1952;90:167–206.
41. Segretain D, Roussel C. Endocytic origin for periaxonemal vesicles along the flagellum during mouse spermiogenesis. Gamete Res 1988;21:451–463.
42. Kierszenbaum AL, Ueda H, Ping L, Abdullah M, Tres L. Antibodies to rat Sertoli cell secretory proteins recognize antigenic sites in acrosome and tail of developing spermatids and sperm. J Cell Sci 1988;91:145–153.
43. Garbers DL. Guanylyl cyclase, a cell surface receptor. J Biol Chem 1989;264:9103–9106.
44. Garbers DL. Identification of cell surface receptor common to germ and somatic cells. Biol Reprod 1991;44:225–230.
45. Anderson RA, Feathergill KA, Drisdel RC, Rawlins RG, Mack SR, Zaneveld LJD. Atrial natriuretic peptide (ANP) as a stimulus of the human acrosome reaction and a component of ovarian follicular fluid: correlation of follicular ANP content with *in vitro* fertilization outcome. J Androl 1994;15: 61–70.
46. Joyce C, Nuzzo N, Wilson L, Zaneveld L. Evidence for a role of cyclooxygenase (prostaglandins) in the sperm acrosome reaction and fertilization. J Androl 1987; 8:74–82.
47. Anderson R, Feathergill K, De Jonge C, Mack S, Zaneveld L. Facilitative effect of pulsed addition of dibutyryl cAMP on the acrosome reaction of noncapacitated human spermatozoa. J Androl 1992;13:398–408.
48. Santos-Sacchi J, Gordon M. Induction of acrosome reaction in guinea pig spermatozoa by cGMP analogues. J Cell Biol 1980;85:798–803.
49. Komatsu S, Endo Y, Suzaki S. Changes in protein phosphorylation during acrosome reaction of mouse sperm. Yakugaku Zasshi 1990; 110:325–331.
50. Yanagimachi R. Mechanisms of fertilization in mammals. In: Mastroianni L, Biggers JD, eds. Fertilization and Embryonic Development *In Vitro*. Plennum, New York, 1981, pp. 81–181.
51. Thomas P, Meizel S. Phosphatidylinositol 4,5-bisphosphate hydrolysis in human sperm stimulated with follicular fluid or progesterone is dependent upon $Ca^{2+}$ influx. Biochem J 1989;264:539–546.
52. Breitbart H, Lax J, Rotem R, Naor J. Role of protein kinase C in the acrosome reaction of mammalian spermatozoa. Biochem J 1992;281:473–476.
53. Leyton L, Saling P. 95 kDa sperm protein bind ZP3 and serve as tyrosine kinase substrates in response to zone binding. Cell 1989;57:1123–1130.
54. Feathergill K, Anderson J, Zaneveld L. Human acrosome reaction requiring extracellular $Ca^{2+}$ occurs by $Ca^{2+}$/Calmodulin-dependent protein kinase II-dependent and-independent pathways. 48th Annual Meet (Abstract). Amer Fert Soc 1992;S181.
55. Garbers DL. Molecular basis of signaling in spermatozoa. J Androl 1989;10:99–107.
56. Chrisman TD, Schulz S, Potter LR, Garbers DL. Seminal plasma factors that cause large elevations in cellular cyclic GMP are C-type natriuretic peptides. J Biol Chem 1993;268:3698–3703.
57. Ueda S, Minamino N, Aburaya M, Kangawa K, Matsuaka S, Matsuo H. Distribution and characterization of immunoreactive porcine C-type natriuretic peptide. Biochem Biophys Res Commun 1991;175:759–767.
58. Macck T. Receptors for atrial natriuretic factor. Ann Rev Physiol 1992;54:11–27.
59. Kuno T, Andresen JW, Kamisaki Y, Waldman SA, Chang LY, Saheki S, Leitman DC, Nakane M, Murad F. Co-purification of an atrial natriuretic factor receptor and particulate guanylate cyclase from rat lung. J Biol Chem 1986;261:5817–5823.
60. Paul AK, Marala RB, Jaiswal RK, Sharma RK. Coexistence of guanylate cyclase and atrial natriuretic factor receptor in a 180-kD protein. Science 1987;235:1224–1226.
61. Takayanagi R, Snajdar RM, Imada T, Tamura M, Pandey KN, Misono KS, Inagami T. Purification and characterization of two types of atrial natriuretic factor receptors from bovine adrenal cortex: guanylate cyclase-linked and cyclase-free receptors. Biochem Biophys Res Commun 1987;144: 244–250.
62. Meloche S, McNicoll N, Liu B, Ong H. DeLean A. Atrial natriuretic factor R1 receptor from bovine adrenal glomerulosa: purification, characterization and modulation by amiloride. Biochemistry 1988;27:8151–8158.
63. Schenk DB, Phelps MH, Porter JG, Fuller F, Cordell B, Lewicki JA. Purification and subunit composition of atrial natriuretic peptide. Proc Natl Acad Sci USA 1987;84:1521–1525.

64. Leitman DC, Andresen JW, Kuno T, Kamisaki Y, Chang JK, Murad F. Identification of multiple binding sites for atrial natriuretic factor by affinity cross-linking in cultured endothelial cells. J Biol Chem 1986;261:11,650–11,655.
65. Scarborough RM, Schenk DB, McEnroe GA, Arfsten A, Kang L-L, Schwartz K, Lewicki JA. Truncated atrial natriuretic peptide analogs. Comparison between receptor binding and stimulation of cyclic GMP accumulation in cultured vascular smooth muscle cells. J Biol Chem 1986;261:12,960–12,964.
66. Maack T, Suzuki M, Alemida FA, Nussenzveig D, Scarborough RM, McEnroe GA, Lewicki JA. Physiological role of silent receptors of atrial natriuretic factor. Science 1987;238:675–678.
67. Pandey KN, Pavlou SN, Inagami T. Identification and characterization of three distinct atrial natriuretic factor receptors. Evidence for tissue-specific heterogeneity of receptor subtypes in vascular smooth muscle, kidney tubular epithelium, and Leydig tumor cells by ligand binding, photoaffinity labeling, and tryptic proteolysis. J Biol Chem 1988;263:13,406–13,413.
68. Leitman DC, Andresen JW, Catalano RM, Waldman SA, Tuan JJ, Murad F. Atrial natriuretic peptide binding, cross-linking, and stimulation of cyclic GMP accumulation and particulate guanylate cyclase activity in cultured cells. J Biol Chem 1988;263:3720–3728.
69. Koller KJ, Lowe DG, Bennett GL, Minamino N, Kangawa K, Matsuo H, Goeddel DV. Selective activation of the B-natriuretic peptide receptor by C-type natriuretic peptide (CNP). Science 1991;252:120–123.
70. Drewett JG, Garbers DL. The family of guanylyl cyclase receptors and their ligands. Endocr Rev 1994;15:135–162.
71. Pandey KN, Singh S. Molecular cloning and expression of murine guanylate cyclase/atrial natriuretic factor receptor cDNA. J Biol Chem 1990;265:12,342–12,348.
72. Pandey KN, Inagami T, Misono KS. Atrial natriuretic factor receptor on cultured Leydig tumor cells: ligand binding and photoaffinity labeling. Biochemistry 1986;25:8467–8472.
73. Chinkers M, Garbers DL, Chang M-S, Lowe DG, Chin HM, Goeddel DV, Schulz S. A membrane form of guanylate cyclase is an atrial natriuretic peptide receptor. Nature 1989;338:78–83.
74. Lowe DG, Chang M-S, Hellmiss R, Chen E, Singh S, Garbers DL, Goeddel DV. Human atrial natriuretic peptide receptor defines a new paradigm for second messenger signal transduction. EMBO J 1989;8:1377–1384.
75. Pandey KN. Stoichiometric analysis of internalization, recycling and redistribution of photoaffinity-labeled guanylate cyclase/atrial natriuretic factor receptors in cultured murine Leydig tumor cells. J Biol Chem 1993;268:4382–4390.
76. Khurana ML, Pandey KN. Receptor-mediated stimulatory effect of atrial natriuretic factor, brain natriuretic peptide and C-type natriuretic peptide on stimulation of testosterone production in purified mouse Leydig cells: activation of cholesterol side-chain cleavage enzyme. Endocrinology 1993;133:2141–2149.
77. Kapasi AA, Kumar R, Pauly JR, Pandey KN. Differential expression and autoradiographic localization of atrial natriuretic peptide receptor in spontaneously hypertensive and normotensive rat testes. Hypertension 1996;28:847–853.
78. Pelletier G. Radioautographic localization of atrial natriuretic factor receptors in the rat testes. J Androl 1988;9:352–355.
79. Silverstroni L, Palleschi S, Guglielmi R, Croce CT. Identification and localization of atrial natriuretic factor receptor in human spermatozoa. Archiv Androl 1992;28:75–82.
80. Marala RB, Sharma RK. Characterization of atrial natriuretic factor receptor-coupled membrane guanylate cyclase from rat and mouse testes. Biochem J 1988;251:301–304.
81. Khurana ML, Pandey KN. Catalytic activation of guanylate cyclase/atrial natriuretic factor receptor by combined effects of ANF and GTPγS in plasma membranes of Leydig tumor cells. Arch Biochem Biophys 1995;316:392–398.
82. Bex F, Corbin A. Atrial natriuretic factor stimulates testosterone production by mouse interstitial cells. Eur J Pharmacol 1985;115:125,126.
83. Foresta C, Mioni R, Careto A. Specific binding and steroidogenic effects of atrial natriuretic factor in Leydig cells of rats. Arch Androl 1987;19:253–259.
84. Privalle CT, McNamara BC, Dhariwal MS, Jefcoate GR. ACTH control of cholesterol side-chain cleavage at adrenal mitochondrial cytochrome $P450_{SCC}$ regulation of intramitochondrial cholesterol transfer. Mol Cell Endocrinol 1987;53:87–101.
85. Pedersen RC, Brownie AC. Steroidogenesis-activator polypeptide isolated from a rat Leydig cell tumor. Science 1987;36:188–190.

86. Miller WL. Molecular biology of steroid hormone synthesis. Endocr Rev 1988;9:295–318.
87. Miller WL. Regulation of mRNAs for human steroidogenic enzymes. Endocr Res 1989;15:1–16.
88. Waterman MR, Simpson ER. Regulation of steroid hydroxylase gene expression is multifactorial in nature. Recent Prog Horm Res 1989;45:533–566.
89. Barrett P, Bollag WB, Isales CM, McCarthy RT, Rasmussen H. The role of calcium in angiotensin II-mediated aldosterone secretion. Endocrine Rev 1989;10:496–518.
90. Black SM, Szklarz GD, Harikrishna JA, Lin D, Wolf CR, Miler WL. Regulation of proteins in the cholesterol side-chain cleavage system in JEG3 and Y-1 cells. Endocrinology 1993;132:539–545.
91. Pandey KN. Stimulation of protein phosphorylation by atrial natriuretic factor in plasma membranes of adrenal cortical cells. Biochem Biophys Res Commun 1989;162:988–994.
92. Pandey KN. Atrial natriuretic factor inhibits the phosphorylation of protein kinase C in plasma membrane preparations of cultured Leydig tumor cells. J Androl 1994;15:100–109.
93. Pandey KN. Atrial natriuretic factor inhibits autophosphorylation of protein kinase C and a 240 kDa protein in plasma membranes of bovine adrenal glomerulosa cells: involvement of cGMP-dependent and independent signal transduction mechanisms. Mol Cell Biochem 1994;141:103–111.
94. Schumacher H, Muller D, Mukhopadhyay AK. Stimulation of testosterone production by atrial natriuretic peptide in isolated mouse Leydig cells results from a promiscuous activation of cyclic AMP-dependent protein kinase by cyclic AMP. Mol Cell Endocrinol 1992;90:47–52.
95. Pandey KN, Kovacs WJ, Inagami T. Inhibition of progesterone secretion and regulation of cyclic nucleotides by atrial natriuretic factor in gonadotropin-responsive murine Leydig tumor cells. Biochem Biophys Res Commun 1985;133:800–806.
96. Anand-Srivastava MB, Sairam MR, Cantin M. Ring-deleted analogs of atrial natriuretic factor adenylate cyclase/cAMP system. Possible coupling of clearance atrial natriuretic factor receptors to adenylyl cyclase/cAMP signal transductions system. J Biol Chem 1990;265:8566–8572.
97. Anand-Srivastava MB, Trachte GJ. Atrial natriuretic factor receptor and signal transduction mechanisms. Pharmacol Rev 1993;45:455–497.
98. Goldberg ND, Haddox MK, Nicol SE, Glass DB, Sanford CH, Kuehl FA Jr, Estensen R. Biologic regulation through opposing influences of cyclic GMP and cyclic AMP. The Yin Yang hypothesis. Advances Cycl Nucl Res 1975;5:307–330.
99. Nakamura T, Gold GH. A cyclic nucleotide-gated conductance in olfactory receptor cilia. Nature 1987;325:442–444.
100. Ho AK, Chick CL, Klein DC. Protein kinase C is involved in adrenergic stimulation of pineal cGMP accumulation. J Biol Chem 1987;262:10,059–10,064.
101. Furman RE, Tanaka JC. Photoreceptor channel activation: interaction between cAMP and cGMP. Biochemistry 1989;28:2785–2788.
102. Hipkin RW, Moger WH. Type I and II isoenzymes of cAMP-dependent protein kinase in Leydig cell steroidogenesis. Mol Cell Endocrinol 1991;77:91–96.
103. Hipkin RW, Moger WH. Interaction between cyclic nucleotide second messenger systems in murine Leydig cells. Mol Cell Endocrinol 1991;82:251–257.
104. Khurana ML, Pandoy KN. Modulation of guanylate cyclase-coupled atrial natriuretic factor receptor activity by mastoparan and ANF in murine Leydig tumor cells: role of G-proteins. Biochem Biophys Res Commun 1994;1224:61–67.
105. Khurana ML, Pandey KN. Atrial natriuretic peptide inhibits the phosphoinositide hydrolysis in murine Leydig tumor cells. Mol Cell Biochem 1996;158:97–105.
106. Steegers EAP, Hollanders JMG, Jongsma HW, Hein PR. Atrial natriuretic peptide and progesterone in ovarian follicular fluid. Gynecol Obstet Invest 1990;29:185–187.
107. Mandich A, Isola G, Massari A. Atrial natriuretic peptide in trout ovarian follicles. Gen Comp Endocrinol 1991;84:419–425.
108. Kim SH, Cho KW, Lim SH, Hwang YH, Ryu H, Oh SH, Seul KH, Jeang GB, Yoan S. Presence and release of immunoreactive atrial natriuretic peptide in granulosa cells of the pig ovarian follicle. Regul Pept 1992; 42:153–162.
109. Kim SH, Cho KW, Oh SW, Hwang YH, Lim SH, Ryu H, Seul KH, Jeang GB, Yoan S. Immunoreactive atrial natriuretic peptides in the oocyte. Comp Biochem Physiol 1993;104A:219–223.
110. Kim SH, Cho KW, Hwang YH, Oh SH, Seul KH, Koh GY, Kim SJ. Ovarian atrial natriuretic peptide during the rat estrous cycle. Life Sci 1992;51:1291–1299.
111. Sundsfjord JA, Forsdahl F, Thibault G. Physiological levels of immunoreactive ANH-like peptides in human follicular fluid. Acta Endocrinol 1989;121:578–580.

112. Saheki T, Shimonaka M, Uchida K, Mizuno T, Hirose S. Immunochemical and biochemical distinction of subtypes of atrial natriuretic peptide receptor. J Biochem 1989;106:627–632.
113. Budnik LT, Brunswig B, Mukhopadhyay AK. Atrial natriuretic peptide stimulates luteal guanylate cyclase. Regul Pept 1987;19:23–34.
114. Brunswig B, Budnik LT, Mukhopadhyay AK. Atrial natriuretic peptide induced stimulation of cyclic GMP formation by isolated bovine luteal cells. J Reprod Fert 1989;86:665–669.
115. Fowler RE, Edwards RG, Walters DE, Chan ST, Steptoe PC. Steroidogenesis in preovulatory follicles of patients given human menopausal and chorionic gonadotropins as judged by the radioimmunoassay of steroids in follicular fluid. J Endocrinol 1978;77:161–169.
116. Johnson KM, Hughes FM, Fong YY, Mathur RS, Williamson HO, Gorospe WC. Effects of atrial natriuretic peptide on rat ovarian granulosa cell steroidogenesis *in vitro*. Am J Reproduct Immunol 1994;31:163–168.
117. Musah AI, Black JF, Baker SL, Schrank GD. Human atrial natriuretic peptide infusion and ovarian venous progesterone secretion in Yucatan micropigs. J Reproduct Fert 1994;101:109–113.
118. Usuki S, Saitoh T, Saitoh M, Tanaka J, Kawakura Y, Usuki Y, Shinjo M, Kim SJ. Endothelin-renin-angiotensin-atrial natriuretic peptide system in ovaries: an intraovarian ERAANP system. J Cardiovascul Pharmacol 22 (Suppl.)1993;8:S207–S210.
119. Itskovitz J, Glorioso N, Laragh JH, Rosenwaks Z. The ovarian prorenin-angiotensin system. Ann NY Acad Sci 1988;541:179–188.
120. Glorioso N, Atlas SA, Laragh JH, Jewelewicz R. Sealey JE. Prorenin in high concentrations in human ovarian follicular fluid. Science 1986;233:1422–1424.
121. Tornell J, Carlsson B, Billing H. Atrial natriuretic peptide inhibits spontaneous rat oocyte maturation. Endocrinology 1990;126:1504–1508.
122. Hubbard C, Terranova P. Inhibitory action of cyclic guanosine 5'-phosphoric acid (cGMP) on oocyte maturation: dependence on an intact cumulus. Biol Reprod 1982;26:628–632.
123. Tornell J, Brannstrom M, Hillensjo T. Different effects of cyclic nucleotide derivatives upon the fat oocyte cumulus complex *in vitro*. Acta Physiol Scand 1984;122:507–513.
124. Cho W, Stern S, Biggers J. Inhibitory effects of dibutyryl cAMP on mouse oocyte maturation *in vitro*. J Exp Zool 1974;187:383–386.
125. Magnusson C, Hillensjo T. Inhibition of maturation and metabolism in rat oocytes by cAMP. J Exp Zool 1977;201:139–147.
126. Bornslarger E, Mattei P, Schulz R. Involvement of cAMP-dependent protein kinase and protein phosphorylation in regulation of mouse oocyte maturation. Dev Biol 1986;114:453–462.

# 11 Physiology of the Natriuretic Peptides

*Pituitary Function*

*Zachary Resch, Tonya C. Murphy, and Willis K. Samson*

**CONTENTS**



## INTRODUCTION

The pituitary gland at first seemed an illogical target for the newly described natriuretic peptides (NPs); however, soon after the discovery of ANP, the presence of ANP-like immunoreactivity in both the neurointermediate and anterior lobes, as well as in the neuroendocrine hypothalamus, was described. Even before the presence of distinct NP binding sites was established in those tissues, many investigators correctly reasoned that central nervous system (CNS) and/or pituitary actions of this new class of vasoactive hormones may exist that complement their actions in the periphery. Numerous reports in the 1980s detailed those pharmacologic effects and the actions of at least ANP were extremely predictable. Indeed, the renal actions were paralleled by CNS effects on water intake, salt appetite, and vasopressin secretion. Neuroendocrine actions, particularly those related to the control of ACTH secretion, seemed to mirror those exerted in adrenal gland and kidney. Those initial, descriptive studies have been summarized in several prior reviews *(1,2)*. It is the goal of this chapter not to catalog those numerous effects, but instead to critique our level of understanding of the neuroendocrine actions of the NPs now that more is known about the potentially important role(s) played by the most recently discovered member of the family, C-type natriuretic peptide. Just how far have we come in understanding the physiological relevance of these numerous pharmacologic

From: *Contemporary Endocrinology: Natriuretic Peptides in Health and Disease*
Edited by: W. K. Samson and E. R. Levin Humana Press Inc., Totowa, NJ

effects? Is there clinical relevance to the actions of the NPs in CNS and pituitary gland? Finally, evidence will be reviewed that reveals very different actions of two of the members of the family, ANP and CNP. Those data suggest not only unique effects of these two distinct gene products, but also a potential functional antagonism of the two peptides under physiologic conditions.

## PREDICTORS OF HYPOTHALAMIC AND PITUITARY ACTIONS OF THE NATRIURETIC PEPTIDES

The presence of NP-like immunoreactivity, NP binding sites, and mRNA encoding the NP receptors in the neuroendocrine hypothalamus and pituitary gland all predict local effects of the peptides in those tissues. Moreover, based on those localization studies, and what has been known for years about the functional anatomy of those regions, a framework for studying central effects of the NPs that mirrored those exerted in the periphery has been established. However, there is not always a complementary peripheral action that predicts the CNS or pituitary effect. How to make sense of those unique effects remains a major challenge.

### *Natriuretic Peptide Localization and Gene Expression in the Neuroendocrine Hypothalamus and Pituitary Gland*

Early in the development of the field, immunoreactive ANP was localized to discrete hypothalamic nuclei, regions known to control anterior lobe function. Neurons containing the peptide also innervated the magnocellular elements of the paraventricular nucleus, implying regulatory control over vasopressin (AVP) secretion in the posterior lobe. More recently, those initial radioimmunoassay (RIA)- and immunocytochemistry (ICC)-based peptide maps have been refined by more detailed mapping of the cells expressing the NP genes. Gundlach and Knobe utilized *in situ* hybridization histochemistry to identify the regional distribution of preproANP in the rat brain *(3)*. Cells expressing preproANP mRNA were prominently distributed in the medial preoptic hypothalamic nucleus, a region corresponding to the anteroventral wall of the third cerebroventricle. Earlier reports had identified this region to be the area of highest density of ANP-like immunoreactivity in the diencephalon *(4)*. PreproANP mRNA was also detected in the arcuate nucleus and limbic structures, such as the central amygdala. This initial mapping study agreed closely with an earlier report on the presence of preproANP mRNA in hypothalamus as detected by solution hybridization-nuclease protection assays *(5)*, a study that also identified transcript in anterior pituitary gland. The identification of cells containing preproANP mRNA indicated the local production of the peptide. Together with the *in situ* hybridization histochemistry results, detailed ICC mapping studies identified major terminal fields of ANP-containing neurons, such as paraventricular nucleus, the bed nucleus of the stria terminalis, and the median eminence. The innervation of the median eminence had a physiologic correlate since it was subsequently demonstrated by Fink's group *(6)* that ANP was released into the hypophysial portal blood and therefore could act as a classical neuroendocrine factor. Indeed, levels of ANP in rat hypophysial portal plasma exceeded those in peripheral circulation by two- to fourfold and could be elevated significantly by estradiol treatment, suggesting physiologic regulation of secretion *(7)*. Thus, the anatomic foundation for hypothalamic and pituitary actions of ANP had been established. Subsequent studies identified a similar distribution of BNP-like immunoreactivity and for the purposes of this chapter, since many if not all of BNP's neuroendocrine actions parallel those of ANP, only two members of the NP family will be discussed.

As pharmacologic studies progressed, identifying in rapid succession numerous neuroendocrine and pituitary actions of ANP, the field was expanded by the realization that the other member of the family of hormones, CNP, was present in pituitary in levels that exceeded those of ANP *(8)*. Furthermore, significant amounts of CNP were detected in hypothalamic extracts. When it was recognized that peripheral tissue levels of CNP were greatly eclipsed by those of ANP (and BNP), investigators began to consider CNP to be the potentially more important neuroendocrine factor. Again, *in situ* hybridization histochemistry revealed distinct populations of CNP mRNA-containing neurons in the endocrine hypothalamus. This included the same two areas where ANP expression was the highest, the anteroventral periventricular nucleus and the arcuate nucleus *(9)*. Although paralleling the distribution of CNP message, ANP mRNA was found by these investigators to be more diffuse, widely distributed throughout the hypothalamus.

The presence of peptide and/or mRNA predicted neuromodulatory effects of ANP and CNP in the hypothalamus and neuroendocrine or possibly paracrine actions within the anterior lobe of the pituitary gland. A major advance was the identification of NP receptor subtypes in diencephalon and pituitary gland, if for no other reason than to draw further attention to the potential for distinct actions of CNP vs ANP. Whereas NPR-A receptor mRNA was prominently displayed in cells surrounding the circumventricular organs, such as the subfornical organ and area postrema *(10)*, the NPR-B receptor message was widely distributed throughout the CNS, but of particular note was the abundant presence of NPR-B mRNA in hypothalamus and pituitary gland *(11)*. Indeed, very little NPR-A message was detected in hypothalamic sites where, on the basis of in vivo studies, ANP was thought to exert neuromodulatory effects *(1)*. Thus, more attention was mandated on the actions of CNP. With only few exceptions, however, ANP remains the most studied NP in terms of hypothalamic effects and pituitary actions. Our group would argue that the NPR-A receptors in rostral diencephalon reflect the circumventricular actions of the NPs, such as inhibition of water and salt intakes, whereas the NPR-B receptors transduce a more physiologically relevant action on neuroendocrine and endocrine function. The physiologic function of the NPR-C receptor, thought to be widely distributed in CNS, perhaps more localized to glial elements *(11)*, may be not merely as a means to remove NPs from the interstitium *(12)*, but also to exert significant antigrowth effects on glial cells, possibly via inhibition of phospholipase C *(13,14)*.

### *Activation of Cellular Signal Transduction Mechanisms Suggest Physiological Roles for the Natriuretic Peptides in the Neuroendocrine Regulation of Pituitary Function*

Earliest reports of a pituitary action of ANP were based on the observation by Cantin's group that the peptide lowered cAMP levels in cultured pituitary cells *(15)* and subsequent work verified the prediction that the NPR-C receptor was the transducer of this signal *(16)*. However, the predominant action of the NPs in pituitary gland is the elevation of cGMP levels via an action on the NPR-B receptor, and to some degree possibly through activation of NPR-A receptors *(17,18)*. These effects on cyclic nucleotide signaling pathways are not always reflected by changes in hormone secretion *(19–21)*; however, those biologic effects established a potential for the NPs to act as either neuroendocrine or paracrine agents regulating hormone secretion. Similarly, effects of the NPs on sec-

Table 1
Effects of icv Administration of Natriuretic Peptides on Plasma Pituitary Hormone Levels

| | *CNP* | *ANP* |
|---|---|---|
| Posterior lobe hormones | | |
| Vasopressin | Decrease *(25)* | Decrease *(26,27)* |
| Oxytocin | Not reported | No effect *(27)* |
| Anterior lobe hormones | | |
| Luteinizing hormone | Decrease *(28)* | Decrease *(29)* |
| | | Increase *(30)* |
| Prolactin | Increase *(31)* | Decrease *(32,33)* |
| Growth hormone | No effect *(31)* | Increase *(34)* |
| | | No effect *(29)* |
| Adrenocorticotropin | Increase/decrease *(35)* | No effect *(36)* |
| | | Decrease *(35,37)* |

[a]References are indicated in parentheses.

ond-messenger accumulation in intermediate lobe *(22)*, immortalized hypothalamic neurons *(23)*, and glia *(14)* were demonstrated, again providing biochemical evidence for a physiological role in neuroendocrine control mechanisms. Thus, the stage was set to answer the question of which cell types are affected and what is the physiological relevance of those actions?

## SIMILAR AND CONTRASTING NEUROENDOCRINE ACTIONS OF ANP AND CNP

The natriuretic peptides exert pharmacologic effects on pituitary function at two anatomic sites: the hypothalamus and pituitary gland. Local production of the peptides in those tissues obviates the necessity of circulating hormone being the physiologically relevant agent that might exert those pharmacologically demonstrated actions. The concept of neuromodulatory effects of the NPs in hypothalamus and paracrine or autocrine effects in pituitary gland has been championed by several groups *(1,4,8,11,14,23,24)*. However, mere pharmacologic effects seldom generate longstanding enthusiasm. Only experimental manipulations addressing physiologic and pathologic relevance provide the bridge to potential clinical relevance. As detailed below, those experiments have been difficult to carry out. Only in the case of the potential roles of the NPs in the hypothalamic control of gonadotropin and corticotropin secretion as well as osmoregulation have they to some degree been successful.

### *Hypothalamic Actions of the Natriuretic Peptides to Alter Neuroendocrine Function*

Cerebroventricular administrations of pharmacologic doses of both ANP and CNP result in changes in circulating levels of several pituitary hormones (Table 1) in conscious laboratory animals. With the exception of one study using estrogen-primed, ovariectomized rats *(30)*, it has been consistently reported *(28,29)* that icv administration of ANP or CNP results in decreased circulating levels of luteinizing hormone (LH). The effect of ANP was suggested to be via a direct effect of the peptide on gonadotropin-releasing hormone (GnRH) release since GnRH secretion from median eminence explants was inhibited by ANP *(29)*. Furthermore, iv infusion of ANP resulted in physiologically relevant elevation of the hormone and inhibited the afternoon surge in LH evoked by estrogen and progesterone treatment in ovariectomized rats. This suggested an effect of ANP on GnRH release *(38)*, since the peptide did not affect LH secretion from cultured pituitary cells *(29)*. Additionally, similar infusion of ANP failed to prevent the LH response to bolus iv injection of synthetic GnRH *(39)*. Since the median eminence is positioned outside of the blood–brain barrier, the GnRH axon terminals in the external layer would be exposed to peripherally infused peptide. Finally, the ability of iv-infused ANP to inhibit the pulsatile pattern of LH secretion *(38)* again suggested an effect in median eminence on GnRH release, either directly *(29)* or indirectly *(40)*.

Is this pharmacologic effect physiologically relevant? Because no selective antagonist was available, passive immunoneutralization studies were conducted. Infusion of anti-ANP antiserum into the third cerebroventricle resulted in a significant elevation in plasma LH levels *(38)*, but the effect was transient, lasting only 30 min. A novel experimental approach was developed to conclusively demonstrate a physiologically relevant role for ANP in the hypothalamic control of gonadotropin secretion as a result of our failure using antiserum. We targeted for destruction cellular elements in hypothalamus that expressed the NP receptor by means of a cytotoxin conjugate constructed of ANP crosslinked to the toxic A chain of the plant lectin Ricin *(41,42)*. Binding of the crosslinked peptide to its receptor is followed by internalization of the ligand–receptor complex, bringing the toxic A chain into the cell *(43,44)*. Ricin A chain is shuttled via a trans-Golgi pathway *(44,45)* to the 60 S ribosomal subunit, where it inactivates the elongation factor-2 binding site and inhibits protein synthesis, eventually resulting in cell death *(46)*. This technique has been employed by us *(47)* and others *(24,48)* to selectively target unique subpopulations of pituitary cells in culture. We chose, however, in these studies to employ the technology in vivo.

After establishing the lack of nonspecific toxicity, we hypothesized that central administration of the ANP-ricin A-chain conjugate would destroy the endogenous GnRH system (if ANP's inhibitory effect on LH secretion was directly expressed on the GnRH neuron). This would result in a depletion of GnRH content in median eminence, the absence of a steroid-induced surge of LH, and an interruption of estrous cyclicity in intact female rats. As a control, we monitored water drinking since ANP had been shown to exert antidipsogenic effects when administered centrally *(49)*. Indeed, ANP-ricin A-chain-treated rats drank more water than controls *(41)*, verifying the effectiveness of the conjugate. As expected, estrous cyclicity was interrupted and the steroid- induced LH surge was all but absent *(41)*. However, GnRH content of the median eminence was not significantly altered, indicating that the inhibitory effect of ANP itself was not expressed directly on the GnRH neuron. We previously showed that the pharmacologic effect of ANP on LH secretion was mediated via an inhibitory δ opioid interneuron *(40)*, but baseline levels of LH were not altered in our ANP-ricin A-chain-pretreated animals. This indicated a lack of effect of endogenous ANP under basal conditions (because if it did play a role under those conditions, LH secretion would have been released from the inhibitory influence of the endogenous enkephalins). At first we were unable to explain our results in light of our existing knowledge of the effects of ANP alone.

The subsequent description of CNP and the elucidation of the three NP receptor subtypes provided us our explanation. Infusion of CNP into the cerebroventricle also inhibited LH secretion, an effect that, like that of ANP, was blocked by naltrindole, the δ-opioid antagonist *(28)*. This meant that the NPs could be inhibiting the hypothalamic mechanisms controlling LH secretion via any one of three receptors. At the time we did not know which receptor subtype was the most likely candidate, since the mapping studies had not yet been published. Therefore, we constructed ricin A-chain conjugates to ANP and to CNP and treated rats as in previous studies. In control rats, animals pretreated with the nontoxic, unconjugated ricin A-chain alone both icv ANP and CNP inhibited LH secretion *(42)*. In rats treated with the ANP-ricin A-chain conjugate 1 wk previously, ANP injection failed to lower LH levels in plasma; however, LH levels did fall in response to a subsequent injection of CNP. Thus, the ANP-ricin A-chain conjugate had compromised cells bearing a subtype of receptor that recognized ANP (NPR-A), or ANP and CNP (NPR-C), but not those bearing the receptor subtype selective for CNP (NPR-B). On the other hand, in rats pretreated with the CNP-ricin A-chain conjugate, neither ANP nor CNP injection significantly lowered LH levels in plasma. Importantly, basal LH levels were significantly lower in CNP-ricin A-chain pretreated animals compared to controls before any injections were made. Those data suggested the following hypothalamic mechanism of the NPs to inhibit GnRH release and, as a result, the secretion of LH from the anterior pituitary gland (Fig. 1).

Endogenous CNP binds under normal conditions to a stimulatory interneuron that expresses the NPR-B receptor. The effect of CNP on that interneuron is inhibitory, explaining the LH-lowering effect of icv-administered CNP. Destruction of those stimulatory interneurons results in attentuation of basal LH secretion, resulting from decreased drive for GnRH release into the portal vessels. Furthermore, since CNP exerts its LH-inhibiting effect even in animals targeted with the ANP-ricin A-chain conjugate, the role of the NPR-C receptor is not directly related to the LH-inhibiting action of CNP. Instead, when pharmacologically applied, ANP displaces endogenous CNP from the clearance receptor, making more CNP available to the NPR-B receptor on the interneuron. If under normal in vivo conditions, ANP acts to alter neuroendocrine regulation of LH secretion, that effect would be caused by such a mechanism and could be expressed by either centrally or peripherally derived peptide. Using a negative, mirror-image approach, we hope in the future to identify the phenotype of the hypothesized stimulatory interneuron by determining what cell type is missing following CNP-ricin A-chain treatment.

Do these cytotoxin-based data establish CNP as a physiologically relevant neuromodulatory agent controlling GnRH secretion? They do not. They merely represent further information on the neuronal hierarchy responsible for the control of GnRH release. They do indicate that the cells responsive to CNP (i.e., bearing the NPR-B receptor) are in the chain of excitable elements that drive GnRH secretion. However, with the well-established biologic redundancy controlling GnRH secretion and, therefore, reproduction *(50)*, loss of any one neuromodulatory agent may not be reflected in significant consequences. Thus, merely removing that agent or cells responsive to that agent from the chain may not provide insight into their importance under physiologic conditions.

Hypothalamic actions of ANP and CNP to alter prolactin (PRL) secretion have also been reported. There is agreement that ANP can act centrally to inhibit PRL secretion *(30,32,33)*, while not affecting release of the hormone from pituitary cells in culture *(32)*. Pharmacologic interventions suggest that the effect of ANP is mediated via a stimulation of the release of the dominant PRL release-inhibiting factor, dopamine, into the portal vessels *(33)*. Evidence suggesting the ability of ANP to stimulate DA release centrally has been presented *(51)*. In the same experimental paradigm in which ANP inhibits PRL

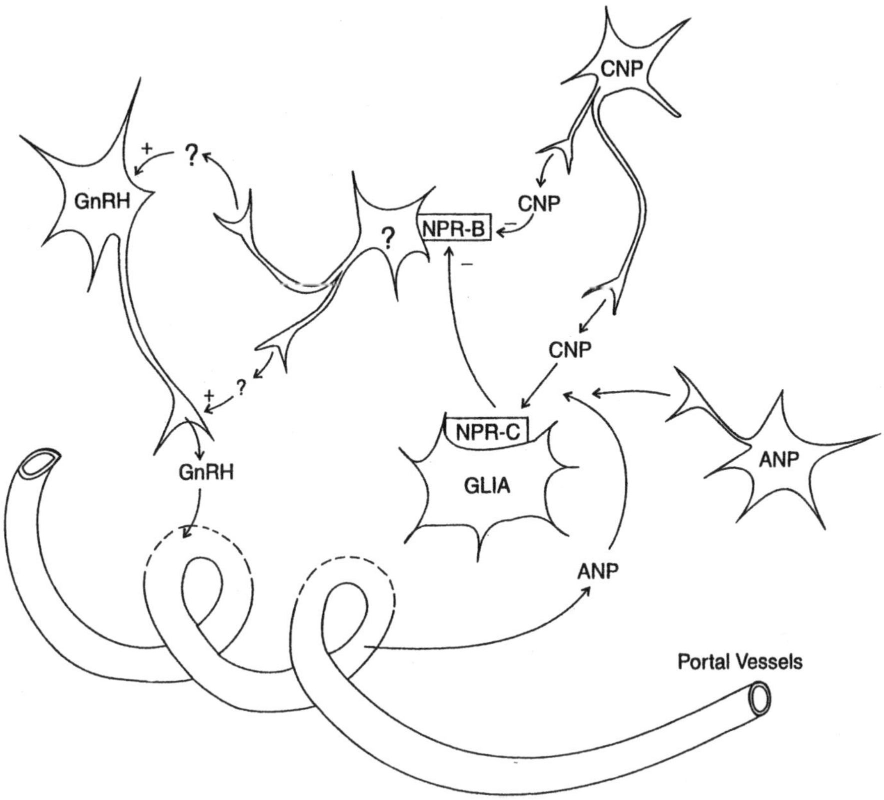


**Fig. 1.** Schematic representation of the hypothesized mechanisms of action of the natriuretic peptides to inhibit gonadotropin-releasing hormone (GnRH) release. C-type natriuretic peptide (CNP) binds to the NPR-B receptor and exerts inhibitory control over a stimulatory interneuron that normally drives GnRH release into the vicinity of the fenestrated capillaries of the hypophysial portal vessels. The CNP also binds to the NPR-C receptor on glia. Circulating atrial natriuretic peptide (ANP), which gains access to the median eminence because of the lack of a blood–brain barrier in the tissue, or ANP of central origin can displace endogenous CNP from the NPR-C receptor, making more CNP available to bind the NPR-B receptor on the neurochemically unidentified stimulatory interneuron. In this manner, both CNP and ANP, acting via displacement of CNP, can inhibit GnRH release into the portal vessels and lower plasma levels of luteinizing hormone (*see* refs. *28,29,41,42*).

secretion, however, CNP exerts potent stimulatory effects. The uniqueness of the two peptides' mechanisms of action was established again using the cytotoxin methodologies. In ANP-ricin A-chain-pretreated rats, CNP still stimulated PRL secretion when infused icv; however, ANP was without effect. On the other hand, in rats pretreated with the CNP-ricin A-chain conjugate, ANP infusion icv significantly lowered PRL levels, but the stimulatory effect of icv-infused CNP was lost. Basal PRL levels were not affected by either conjugate. Thus, the inhibitory effect of ANP on PRL secretion, although expressed via the recruitment of DA, is not exerted directly on the tuberoinfundibular dopamine-containing neurons. If it were, targeting for destruction the DA-containing neurons would remove the lactotroph from inhibitory control and basal PRL levels would be elevated. Additionally, the cells targeted for compromise by either conjugate were not cells expressing the NPR-C receptor. Instead, both guanylyl cyclase-linked NP receptors are involved, the NPR-A mediating the inhibitory effect of ANP in one neuronal population and the NPR-B receptor transducing the stimulatory effect of CNP in another. Unique and opposing effects of ANP and CNP are also expressed on water drinking where ANP inhibits and CNP stimulates the behavior, again suggesting the unique involvement of the NPR-A and NPR-B receptors on differing cell populations *(49,52)*.

The physiological relevance of the hypothalamic effects of the NPs on gonadotropin secretion have not yet been established, nor has that for the effects on PRL secretion. The antireproductive effect of CNP, and possibly ANP, do not seem to mirror the major peripheral actions of the natriuretic peptides. In the case of the hypothalamic effects on PRL secretion, however, there may be some similarity between these hypothalamic actions and the natriuretic effects of the peptides in the kidney. Although controversial for humans, in lower species, PRL exerts profound effects on the renal and extrarenal handling of salt *(53)*. Prolactin receptors are abundantly present in the mammalian kidney *(54)*, and preliminary studies in our laboratory indicate a renal action of PRL to increase sodium excretion in the rat. Therefore, there may be a physiologically relevant reason for the natriuretic peptides to affect PRL secretion. Alternatively, the effects of the natriuretic peptides on PRL secretion may reflect the confluent actions of PRL and the NPs on immune cell function (*55, see also* Chapter 17).

Most well-established and universally accepted is the effect of the NPs on corticotropin release. Although controversy remains regarding the physiological significance of direct pituitary effects on ACTH seen in vitro *(vida infra)*, there is consensus that the peptides can act centrally to reduce the hypothalamic drive for ACTH secretion. Both the releases of corticotropin-releasing hormone (CRH) and vasopressin (AVP) from hypothalamic explants are inhibited by ANP *(39,56,57)*. The effect of CNP on the release of these peptides in vitro has not yet been reported. However, in vivo results *(25)* would predict similar in vitro findings. As does ANP *(35,37)*, infusion of CNP into the ventricular system of conscious animals results in a reduction in stimulated levels of ACTH under a variety of experimental circumstances *(35)*. A physiological role for the neuroendocrine action of ANP at either the hypothalamic or pituitary level is supported experimentally. Stress-induced ACTH is augmented in rats pretreated with anti-ANP antiserum, and glucocorticoid-negative feedback on ACTH is also reduced in the anti-ANP-treated animals *(58)*. This central inhibitory effect of the NPs on ACTH secretion does seem relevant in a teleologic sense, since a reduction of cortisol and to some degree aldosterone secretion at any level would potentiate the renal effects of the peptides. Here also a complementary effect exerted directly at the level of the corticotroph in the pituitary gland has been reported by several groups *(vida infra)*.

One group has reported a significant stimulatory effect of ANP in the hypothalamic control of GH secretion *(34)* when experiments were conducted just after the onset of the light cycle in their colony of rats. Although in our experimental paradigm, conducted later in the day, this stimulatory effect was not observed *(32)*, we did report the ability of anti-ANP, when administered icv, to significantly reduce GH secretion *(38)*. The potential for a significant hypothalamic action of the NPs on GH-releasing hormone or somatostatin release therefore exists; however, the reproducibility of the effect is yet to be established and the significance remains unclear.

### *Direct Pituitary Actions of the Natriuretic Peptides*

Already by 1985, it was apparent that gonadotrophs contained NP-like immunoreactivity, suggesting a paracrine or autocrine action of the peptides in the regulation of anterior-lobe function *(59)*. Those findings were confirmed and extended to include a description of the presence of NP- like immunoreactivity in lactotrophs and corticotrophs, but not somatotrophs or thyrotrophs *(60)*. Not only could these three cell types apparently produce ANP, but at least the gonadotrophs could internalize radiolabel from iodinated ANP-infused iv *(61)*. Most recently, in a series of truly innovative experiments, McArdle and coworkers *(24)* have demonstrated the abundant presence of CNP, but not ANP or BNP, immunoreactivity in mouse and rat pituitary cells, colocalized with LH immunoreactivity. Furthermore, they demonstrated the presence of CNP, but not ANP or BNP, mRNA in pituitary extracts. Although only the mRNA for the NPR-B receptor could be detected by Northern blotting, RT-PCR revealed transcripts for both the NPR-B and NPR-A receptors. Using the αT3-1 line, which is a gonadotropin-derived clonal cell type, McArdle demonstrated that targeting with a GnRH-ricin A-chain conjugate not only inhibited proliferation and GnRH-stimulated LH secretion, it abrogated the ability of CNP to elevate cGMP levels in those cells as well, suggesting again a paracrine or autocrine effect of CNP in the pituitary gland. Morel's group *(18)* then demonstrated the presence of mRNA encoding all three NP receptor subtypes with the NPR-B and NPR-A subtypes present in at least twofold higher amounts than the NPR-C receptor message. Importantly, all three receptor subtypes were found to be expressed in gonadotrophs, lactotrophs, and corticotrophs.

In order for paracrine or autocrine effects of endogenously produced CNP in anterior lobe to be established, several criteria must be experimentally satisfied. First, can significant (i.e., physiologically relevant) amounts of CNP be released from cells in the gland? Second, do established releasing factors, such as GnRH, TRH, or CRH, cause release of the peptides? Third, are levels of peptide in the gland physiologically regulated? For instance, will varying gonadal steroid regimens alter levels of the peptides in or their release from gonadotrophs? Fourth, can endogenous peptide once released avoid enzymatic degradation long enough to exert its proposed actions? In the case of the effects of CNP or even ANP on cyclic nucleotide levels in gonadotrophs, is there a secretion correlate to these effects? To date, all reports indicate that the NPs fail to significantly alter basal or GnRH-stimulated LH secretion from cells in culture *(17,19,21,28,29)*. Similarly, although the lactotroph appears to produce NP and express NP receptors, investigators have failed to demonstrate significant effects on PRL secretion from cultured pituitary cells *(19,21,31,32)*. A cell type that has not been reported to produce NPs or bear NP receptors, the somatotroph, may respond to NPs with a secretion event *(62,63)*. Although the majority of the literature reported a failure of ANP to alter basal or GHRH-stimulated GH release from mixed cell populations in vitro *(19,21,31,32)*, one group did report an effect on basal and stimulated GH secretion *(64)*. In the clonal GH3 cell line, both ANP and CNP stimulated cGMP accumulation, but only CNP stimulated GH secretion *(62)*. The secretion event depended on activation of a cGMP-dependent protein kinase (cGK), but was not affected by activation of either a cAMP-dependent protein kinase or protein kinase C (PKC). In dispersed rat pituitary cells, agents that elevate either soluble guanylyl cyclase (nitric oxide) or particulate guanylyl cyclase (CNP) increase GH release *(63)*. Those authors maintained that any manipulation that increased intracellular cGMP also stimulated GH secretion. That hypothesis is difficult to reconcile with work conducted by others and by our group indicating that nitric oxide production is stimulated by GHRH and that when NO production is blocked using a false substrate for NO-synthase, GHRH-stimulated GH secretion is actually enhanced, thus suggesting that activation of soluble guanylyl cyclase results in a diminution of GH secretion *(65)*. It is possible that when studying mixed cell populations, unlike clonal cell lines, the effects of CNP or ANP on hormone secretion are not exerted directly on the hormone secretor itself, but instead via recruitment of additional, yet to be described, paracrine factors. This is indeed the case for the action of calcitonin on PRL release in vivo and in vitro *(66)*.

Whether secretion is an appropriate index of biologic action is still debatable. The paracrine or autocrine effects of the NPs in the gland may not be related to hormone secretion, but instead to the regulation of hormone–gene transcription. When transfected into GH3 cells, the CNP promoter-CAT reporter construct was functional *(67)*. Tam and coworkers *(68)* furthermore have demonstrated the ability of ANP to reduce the amount of POMC mRNA in AtT-20 cells, suggesting that the action of the NPs in the gland may be more related to production than secretion of a given hormone.

The most controversy in this field has been generated by varying results of the NPs on ACTH secretion. Initially, the ability of ANP to stimulate cGMP accumulation in normal, dispersed pituitary cells and in AtT-20 cells was not associated with significant effects on ACTH secretion *(19,21,69,70)*. However, that same year, Shibasaki and coworkers *(64)* did demonstrate a significant inhibitory effect of ANP on the secretion of POMC-derived peptides. Under certain experimental conditions, an inhibitory effect could be observed; however, the length of exposure, dose levels, and interaction with the right ACTH-releasing hormone mixtures seemed critical to any one investigator's ability to observe this effect *(64,71–74)*. In fact, it had been demonstrated that any biologic agent that elevated intracellular cGMP was capable of inhibiting secretagog-stimulated ACTH secretion *(71)*. Perhaps the most inventive protocol used to examine a potential physiologic effect at the pituitary level was that of Kovacs and Antoni *(75)*, who demonstrated in paraventricular nucleus-lesioned rats that ACTH release in response to CRH/AVP challenge was reduced by iv infusion of ANP.

Virtually all of the work on NP effects on ACTH secretion was conducted using ANP. Since CNP is the major NP in pituitary and since the NPR-B receptor at least in some hands appears to be the major receptor subtype present, attention should be focused on it as a potentially physiologic modulator of corticotroph function. Indeed, Gilkes et al. *(20)* attempted to demonstrate an effect of CNP on ACTH secretion; however, they chose as their model the AtT-20 cell line, which in earlier studies responded to ANP with an increase in cGMP levels but no changes in hormone secretion *(76)*. Their results with CNP were similarly negative, as were those of Shimekake and coworkers *(62)*. Clearly the AtT-20 cell line may not be the optimal choice for these studies.

We have attempted to demonstrate effects of CNP in a wide dose range on ACTH release from dispersed rat pituitary cells under basal and CRH-stimulated conditions. In static incubations CNP was ineffective, failing to alter basal (Fig. 2) and CRH-stimulated (Fig. 3) ACTH secretion. In perifusion incubations, CNP again failed to alter basal ACTH release, and did not significantly alter the secretion of hormone in the presence of CRH (Fig. 4). We did observe, however, in these perifusion experiments a significant inhibition of CRH-stimulated ACTH secretion in the presence of ANP (Fig. 5). The doses tested approximated those employed by Antoni's group, as did the times of exposure *(72)*; however, we did not include in our secretagog challenge AVP along with the CRH. King and Baertschi *(71)* were able to demonstrate significant effects of ANP, in a narrow dose range, on CRH-stimulated ACTH release, but only after 3-h exposure to ANP. We have not yet conducted experiments with the same doses as King and Baertschi *(71)* or with

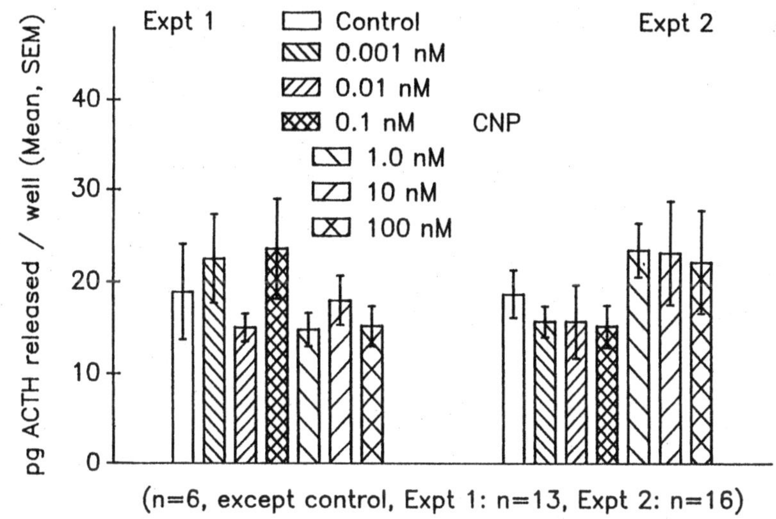


**Fig. 2.** Failure of C-type natriuretic peptide (CNP) to significantly alter basal adrenocorticotropin (ACTH) secretion from dispersed, anterior pituitary cells in static incubation. Data from two separate experiments are presented.

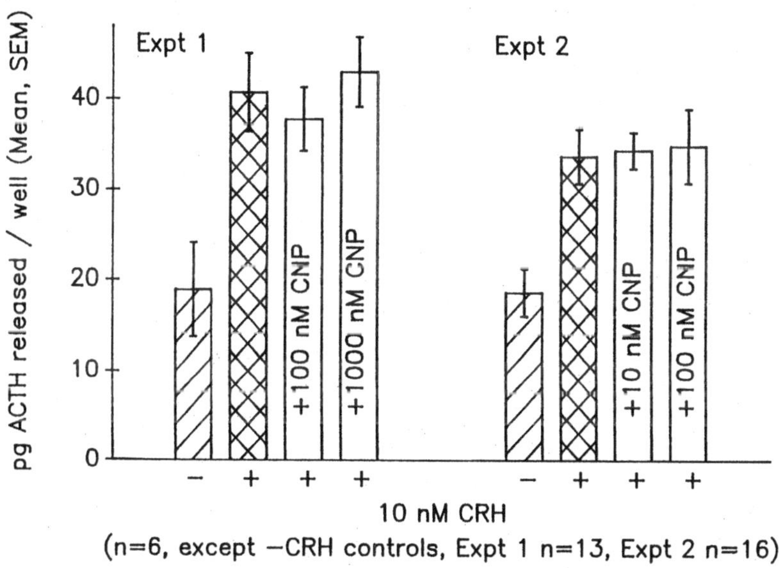


**Fig. 3.** Failure of coadministration of C-type natriuretic peptide (CNP) to alter the adrenocorticotropin (ACTH) response to corticotropin-releasing hormone (CRH) in dispersed, anterior pituitary cells in static incubation. Data from two separate experiments are presented.

similarly long exposure periods to CNP. One caution that should be extended is that although King and Baertschi did employ quite low doses of ANP (10–1000 p*M*, or 30–3000 pg/mL), these may not be physiologically relevant. The whole pituitary gland contains <30 pg ANP-like immunoreactivity, and circulating levels may not be this high *(6)*. Granted, Fink's group has reported three- to fourfold higher levels of ANP in portal blood compared to peripheral; however, since we now know that natriuretic peptides are produced in blood vessels themselves, the method of portal blood collection (i.e., stalk

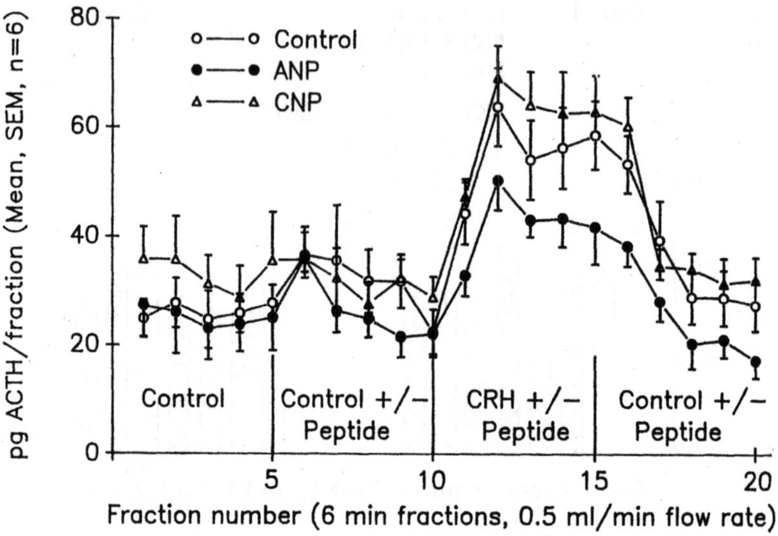


**Fig. 4.** Adrenocorticotropin (ACTH) release from dispersed, anterior pituitary cells in dynamic incubation (0.5 mL/min flow rate, 6 min fractions). Four exposure periods were included. During period 1, cells were perifused with control medium alone. During period 2, cells were perifused with medium alone or medium containing either atrial (ANP) or C-type (CNP) natriuretic peptide at a dose of 100 n*M*. In period 3, corticotropin-releasing hormone (CRH, 1 n*M*) was added to the medium employed during period 2. During the final period of perifusion, cells were exposed to the same medium as in period 2. Cells were loaded on perifusion columns as described previously *(28)* and perifused in parallel. Data presented was generated in seven separate cell dispersions, generating 18-cell columns ($n$ = 6/group).

transection) may have resulted in artifactually elevated levels of the hormone in the samples collected. Just the same, Fink's group did demonstrate physiological regulation of ANP levels in portal blood *(7)*. To be sure, significant effects at doses as low as 10 p*M* for any peptide are impressive and suggest a physiologically relevant event.

## CAN THESE DATA OBTAINED IN ANIMAL MODELS BE EXTRAPOLATED TO HUMANS?

Several biologic effects of the NPs first characterized in animal models have been witnessed or implied in human studies. The ability of ANP to inhibit AVP secretion and water drinking in the rat *(49)* is mirrored by effects of the peptide in humans. Osmotically stimulated thirst was significantly blunted during ANP infusions in healthy male volunteers *(77)*. Another group demonstrated a significant reduction in basal AVP levels during ANP infusion even in the face of the attendant drop in blood pressure *(78)*. A significant reduction in the AVP response to hypertonic saline infusion was observed in young and elderly human volunteers in another study *(79)*, thus establishing a physiologic correlate in humans to the initial observations made in the rat *(26)*. Do the anterior lobe effects reported in animal studies have human correlates? Foresta et al. *(80)* reported that iv infusion of ANP stimulated urine output and sodium excretion, and also reduced basal PRL secretion in male volunteers. The effect was similar in magnitude to that observed following icv injection of ANP in rats *(32)*. In another study, the same group failed to demonstrate any significant action of iv infusion of ANP on LH

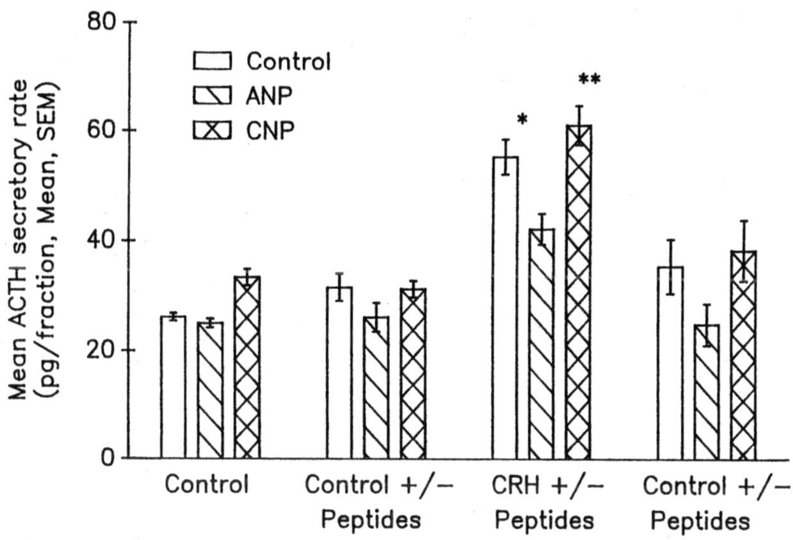


**Fig. 5.** Same data as presented in Fig. 4, calculated on the basis of mean ACTH secretory rate per exposure period; $*p < 0.05$, $**p < 0.01$ vs ANP-exposed cells (ANOVA and Student Newman Keuls multiple comparison test).

secretion *(81)*, as had been demonstrated in the rat *(29,38,39)*; however, they did demonstrate for the first time a direct Leydig cell effect resulting in elevation of plasma testosterone levels *(81)*.

It is not surprising that the controversy experienced in animal studies over the potential interaction of the NPs with the hypothalamo-pituitary-adrenal axis has spilled over into the human literature. Ur and coworkers failed to observe any significant effects of physiologic doses of ANP on CRH-stimulated ACTH secretion *(82)*, suggesting that in humans, ANP is not a major factor regulating the corticotroph. However, a temporal, inverse correlation between changes in plasma ACTH and ANP in critically ill patients has been reported *(83)*. In another human trial, iv infusion of ANP reduced basal ACTH levels and, in women, decreased TRH-stimulated PRL and TSH levels *(84)*. Similarly, CRH-stimulated ACTH secretion was significantly blunted by iv infusion of ANP in the studies of Kellner and coworkers *(85)*.

### *Does the Literature on the Pituitary Effects of the Natriuretic Peptides Demonstrate Physiologically Relevant Actions in the Control of Pituitary Function?*

Using selective pharmacologic antagonists, the effects of blockade of the pituitary actions of endogenous natriuretic peptides have not been reported. Passive immunoneutralization of endogenous peptide levels has been reported, and in experimental animals a physiologically relevant role for ANP in the regulation of LH *(38)*, GH *(38)*, and ACTH *(58)* secretion has been suggested. Numerous other experimental approaches have provided supporting evidence for a physiologically relevant role for ANP and CNP in hypothalamic and pituitary function; however, in the absence of data on the pituitary effects of peptide overexpression or frank absence, it is difficult to conclude that these peptides play an important role in the day-to-day regulation of even posterior lobe function. Correlative data exist, but conclusive evidence is lacking. As a result, there does not at this time appear to be potential for the use of these peptides or their mimetics in clinical settings of aberrant pituitary function. Just the same, a knowledge of their potential actions on posterior lobe hormone secretion or the neuroendocrine regulation of anterior lobe function is important clinically, when their use is entertained under other conditions.

## ACKNOWLEDGMENTS

The authors wish to thank the Max Baer Heart Fund and the Dakota State Aerie of the Fraternal Order of Eagles for their continued support. Studies on the pituitary actions of the natriuretic peptides in the senior author's laboratory were funded by a grant from the National Institutes of Health (HD 25373) and grants from the American Heart Association.

## REFERENCES

1. Samson WK. Natriuretic peptides: a family of hormones. Trends Endocrinol Metabol 1992;3:86–90.
2. Espiner EA. Physiology of the natriuretic peptides. J Int Med 1994;235:527–541.
3. Gundlach AL, Knobe KE. Distribution of preproatrial natriuretic peptide mRNA in rat brain detected by in situ hybridization of DNA oligonucleotides: enrichment in hypothalamic and limbic areas. J Neurochem 1992;59:758–761.
4. Standaert DG, Needleman P, Saper CB. Organization of atriopeptin-like immunoreactive neurons in the central nervous system of the rat. J Comp Neurol 1986;253:315–341.
5. Standaert DG, Needleman P, Day ML, Weigand R, Krause JE. Expression of the gene for preproatriopeptin in the central nervous system of the rat. Mol Brain Res 1988;4:7–13.
6. Lim AT, Sheward WJ, Copolov D, Windmill D, Fink G. Atrial natriuretic factor is released into hypophysial portal blood: direct evidence that atrial natriuretic factor may be a neurohormone involved in hypothalamic pituitary control. J Neuroendocrinol 1990;2:15–18.
7. Sheward WJ, Lim A, Alder B, Copolov D, Dow RC, Fink G. Hypothalamic release of atrial natriuertic factor and beta-endorphin into rat hypophysial portal plasma: relationship to oestrous cycle and effects of hypophysectomy. J Endocrinol 1991;131:113–125.
8. Minamino N, Aburaya M, Kojima M, Miyamoto K, Kangawa K, Matsuo H. Distribution of C-type natriuretic peptide and its messenger RNA in rat central nervous system and peripheral tissues. Biochem Biophys Res Commun 1993;197:326–335.
9. Herman JP, Langub MC, Watson RE. Localization of C-type natriuretic peptide mRNA in rat hypothalamus. Endocrinology 1993;133:1903–1907.
10. Langub MC, Dolgas CM, Watson RE, Herman JP. The C-type natriuretic peptide receptor is the predominant natriuretic peptide receptor mRNA expressed in rat hypothalamus. J Neuroendocrinol 1995;7:305–309.
11. Herman JP, Dolgas CM, Rucker D, Langub MC. Localization of natriuretic peptide-activated guanylate cyclase mRNAs in the rat brain. J Comp Neurol 1996;369:165–187.
12. Maack T, Suzuki M, Almeida FA, Nussenzveig D, Scarborough RM, McEnroe GA, Lewicki JA. Physiological role for silent receptors for atrial natriuretic factor. Science 1987;238:675–678.
13. Levin ER. Natriuretic peptide C-receptor: more than a clearance receptor. Am J Physiol 1993;264: E484–489.
14. Levin ER, Frank HJL. Natriuretic peptide inhibits rat astroglial proliferation: mediation by C receptor. Am J Physiol 1991;261:R453–457.
15. Anand-Srivastava MB, Cantin M, Genest J. Inhibition of pituitary adenylate cyclase by atrial natriuretic factor. Life Sci 1985;36:1873–1879.
16. Anand-Srivastava MB, Sairam MR, Cantin M. Ring-deleted analogs of atrial natriuretic factor inhibit adenylate cyclase/cAMP system. Possible coupling of the clearance atrial natriuretic factor receptors to adenylate cyclase/cAMP signal transduction system. J Biol Chem 1990;265:8566–8572.
17. McArdle CA, Poch A, Kaeppler K. Cyclic guanosine monophosphate production in the pituitary: stimulation by C-type natriuretic peptide and inhibition by gonadotropin-releasing hormone in alphaT3-1 cells. Endocrinology 1993;132:2065–2072.
18. Grandclement B, Brisson C, Bayard F, Tremblay J, Gossard F, Morel G. Localization of mRNA coding for the three subtypes of rat anterior pituitary gland cells. J Neuroendocrinol 1995;7:939–948.

19. Simard J, Hubert JF, Labrie F, Isreal-Assayag E, Heisler S. Atrial natriuretic factor-induced cGMP accumulation in rat anterior pituitary cells in culture is not coupled to hormonal secretion. Regul Peptides 1986;15:269–278.
20. Gilkes AF, Ogden PH, Gould SB, Cramb G. Characterization of natriuretic peptide receptor subtypes in the AtT-20 pituitary tumour cell line. Biochem J 1994;299:481–487.
21. Abou-Samra AB, Catt KJ, Auilera G. Synthetic atrial natriuretic factors (ANFs) stimulate guanine 3',5'-monophosphate production but not hormone release in rat pituitary cells: peptide contamination with a gonadotropin-releasing hormone agonist explains luteinizing hormone-releasing hormone activity of certain ANFs. Endocrinology 1987;120:18–24.
22. Lutz-Bucher B, Schimchowitsch S, Felix JM, Stoeckel ME, Koch B. Stimulation by atrial natriuretic factor of cyclic GMP production in cultured anterior and intermediate pituitary tissues: evidence for a major contribution of proliferating nonendocrine cells. Mol Cell Endocrinol 1989;64:257–266.
23. Olcese J, Middendorf R, Muenker M, Schmidt C, McArdle CA. Natriuretic peptides stimulate cyclic GMP production in an immortalized LHRH neuronal cell line. J Neuroendocrinol 1994;6:127–130.
24. McArdle CA, Olcese J, Schmidt C, Poch A, Kratzmeier M, Middendorff R. C-type natriuretic peptide (CNP) in the pituitary: Is CNP an autocrine regulator of gonadotropes? Endocrinology 1994;135:2794–2801.
25. Shirakami G, Itoh H, Suga S, Komatsu Y, Hama N, Mori K, Nakao K. Central action of C-type natriuretic peptide on vasopressin secretion in conscious rats. Neurosci Lett 1993;159:25–29.
26. Samson WK. Atrial natriuretic factor inhibits dehydration and hemorrhage induced vasopressin secretion. Neuroendocrinology 1985;40:277–279.
27. Samson WK, Aguila MC, Martinovic J, Antunes-Rodrigues, Norris M. Hypothalamic action of atrial natriuretic factor to inhibit vasopressin secretion. Peptides 1987;8:449–454.
28. Huang FLS, Skala KD, Samson WK. Hypothalamic effects of C-type natriuretic peptide (CNP) on luteinizing hormone secretion. J Neuroendocrinol 1992;4:325–330.
29. Samson WK, Aguila MC, Bianchi R. Atrial natriuretic factor inhibits gonadotropin secretion in the rat: evidence for a hypothalamic site of action. Endocrinology 1988;122:1573–1582.
30. Steele MK. Additive effects of atrial natriuretic peptide and angiotensin II on luteinizing hormone and prolactin release in female rats. Neuroendocrinology 1990;51:345–350.
31. Huang FLS, Skala KD, Samson WK. C-type natriuretic peptide stimulates prolactin secretion by a hypothalamic site of action. J Neuroendocrinol 1992;4:593–597.
32. Samson WK, Bianchi R, Mogg RJ. Further evidence for a hypothalamic site of action of atrial natriuretic factor: inhibition of prolactin secretion in conscious rats. Can J Physiol Pharmacol 1988;66:301–305.
33. Samson W, Bianchi R, Mogg RJ. Evidence for a dopaminergic mechanism for the prolactin inhibitory effect of atrial natriuretic factor. Neuroendocrinology 1988;47:268–270.
34. Murakami Y, Kato Y, Tojo K, Inoue T, Yanaihara N, Imura H. Stimulation of growth hormone secretion by central administration of atrial natriuretic polypeptide. Endocrinology 1988;122:2103–2108.
35. Charles CJ, Espiner EA, Richards M, Nicholls MG, Yandle TG. Comparative bioactivity of atrial, brain, and C-type natriuretic peptides in conscious sheep. Am J Physiol 1996;270:R1324–1331.
36. Parkes DG, Coghlan JP, Weisinger RS, Scoggins BA. The effects of intracerebroventricular infusion of atrial natriuretic factor in conscious sheep. Peptides 1988;9:509–513.
37. Itoh H, Nakao K, Katsuura G, Morii N, Yamada T, Shiono S, Sakamoto M, Sugawara A, Saito Y, Eigyo M, Matsushita A, Imura H. Possible involvement of central atrial natriuretic polypeptide in regulation of hypothalamo-pituitary-adrenal axis in conscious rat. Neurosci Lett 1986;69:254–258.
38. Samson WK, Bianchi R, Mogg RJ. Atrial natriuretic peptide modulates hypothalamic control of pituitary function. In: Imura H, Matsuo H, eds. Atrial and Brain Natriuretic Peptides. Kodansha Scientific Publishers, Tokyo, 1991, pp. 219–227.
39. Samson WK. Hypothalamic actions of the atrial factors to alter hormone secretion from both the anterior and posterior pituitary. In: Needleman P, ed. Biological and Molecular Aspects of Atrial Factors. Liss, New York, 1988, pp. 217–230.
40. Huang FLS, Samson WK. An enkephalinergic mechanism for the hypothalamic effect of atrial natriuretic peptide to inhibit luteinizing hormone secretion. J Neuroendocrinol 1991;3:127–132.
41. Samson WK, Skala KD, Huang FLS, Fulton RJ. Central peptidergic mechanisms controlling reproductive hormone secretion: a novel approach uncovers the role of the natriuretic peptides. Can J Physiol Pharmacol 1992;70:773–778.
42. Samson WK, Skala KD, Huang FLS, Fulton RJ. C-type natriuretic peptide mediates the hypothalamic actions of the natriuretic peptides to inhibit luteinizing hormone secretion. Endocrinology 1993;132:504–509.

43. Fulton RJ, Tucker TF, May RD, Till M, Uhr JW. Pharmacokinetics of tumor-reactive immunotoxins in tumor bearing mice: effect of antibody valency and deglycosylation of the ricin A chain on clearance and tumor localization. Cancer Res 1988;48:2618–2624.
44. Vitetta ES, Fulton RJ, May RD, Till M, Uhr JW. Redesigning nature's poisons to create antitumor agents. Science 1987;238:1098–1104.
45. Van Deurs B, Tonneson TI, Petersen OW, Sandvig K, Olsnes S. Routing of internalized ricin conjugates to the Golgi complex. J Biol Chem 1986;102:37–47.
46. Wellner RB, Hewetson JF, Poli MA. Ricin: mechanism of action, detection, and intoxication. J Toxicol Toxin Res 1995;14:483–522.
47. Samson WK, Alexander BA, Skala KD, Huang FLS, Fulton RJ. Ricin-cytotoxin conjugate administration reveals a physiologically relevant role for oxytocin in the control of gonadotropin secretion. Ann NY Acad Sci 1992;652:773–778.
48. Schwartz J, Penke B, Rivier J, Vale W. A new cytotoxin for the target cells of corticotropin releasing factor. Endocrinology 1987;121:1454–1460.
49. Antunes-Rodrigues J, McCann SM, Rogers LC, Samson WK. Atrial natriuretic factor inhibits water intake in conscious rats. Proc Natl Acad Sci USA 1985;82:8720–8724.
50. Kordon C, Drouva SV, Martinez de la Escalera G, Weiner RI. Role of classic and peptide neuromediators in the neuroendocrine regulation of luteinizing hormone and prolactin. In: Knobil E, Neill JD, eds. The Physiology of Reproduction, 2nd ed. Raven, New York, 1994, pp. 1621–1681.
51. Nakao K, Katsuura G, Morii N, Itoh H, Shiono S, Yamada T, Sugawara A, Sakamoto M, Saito Y, Eigyo M, Matsushita A, Imura H. Inhibitory effect of centrally administered atrial natriuretic polypeptide on the brain dopaminergic system in rats. Eur J Pharmacol 1986;131:171–177.
52. Samson WK, Skala KD, Huang FLS. CNP-22 stimulates, rather than inhibits, water drinking in the rat: evidence for a unique biological action of the C-type natriuretic peptides. Brain Res 1991;568:285–288.
53. Loretz CA, Bern HA. Prolactin and osmoregulation in vertebrates. Neuroendocrinology 1982;35:292–304.
54. Royster M, Driscoll P, Kelly PA, Freemark M. The prolactin receptor in the fetal rat: cellular localization of messenger ribonucleic acid, immunoreactive protein, ligand binding activity and induction of expression in late gestation. Endocrinology 1995;136:3892–3900.
55. Leite-de-Moraes M, Touraine P, Gagnerault MC, Savino W, Kelly PA, Dardenne M. Prolactin receptor and the immune system. Ann Endocrinol Paris 1995;56:567–570.
56. Ibanez-Santos J, Tsagarakis S, Rees LH, Besser GM, Grossman A. Atrial natriuretic peptides inhibit the release of corticotropin-releasing factor-41 from the rat hypothalamus in vitro. J Endocrinol 1990;126:223–228.
57. Takao T, Hashimoto K, Ota Z. Effect of atrial natriuretic peptide on acetylcholine induced release of corticotropin releasing factor from rat hypothalamus. Life Sci 1988;42:1199–1203.
58. Fink G, Dow RC, Casley D, Johnston CI, Bennie J, Carroll S, Dick H. Atrial natriuretic peptide is involved in the ACTH response to stress and glucocorticoid negative feedback in the rat. J Endocrinol 1992;135:37–43.
59. McKenzie JC, Tanaka I, Misono KS, Inagami T. Immunocytochemical localization of atrial natriuretic factor in the kidney, adrenal medulla, pituitary, and atrium of rat. J Histochem Cytochem 1985;33:828–832.
60. Morel G, Chabot JG, Belles-Isles M, Heisler S. Synthesis and internalization of atrial natriuretic factor in anterior pituitary cells. Mol Cell Endocrinol 1988;55:219–231.
61. Morel G, Chabot JG, Gossard F, Heisler S. Is atrial natriuretic peptide synthesized and internalized by gonadotrophs? Endocrinology 1989;124:1703–1710.
62. Shimekake Y, Ohta S, Nagata K. C-type natriuretic peptide stimulates secretion of growth hormone from rat-pituitary-derived GH3 cells via a cyclic-GMP-mediated pathway. Eur J Biochem 1994;222:645–650.
63. Hartt DJ, Ogiwara T, Ho AK, Chik CL. Cyclic GMP stimulates growth hormone release in rat anterior pituitary cells. Biochem Biophys Res Commun 1995;214:918–926.
64. Shibasaki T, Naruse M, Yamauchi N, Masuda A, Imaki T, Naruse K, Demura H, Ling N, Inagami T, Shizume K. Rat atrial natriuretic factor suppresses proopiomelanocortin-derived peptides secretion from both anterior and intermediate lobe cells and growth hormone release from anterior lobe cells of the rat pituitary in vitro. Biochem Biophys Res Commun 1986;135:1035–1041.
65. Kato M. Involvement of nitric oxide in growth hormone (GH)-releasing hormone-induced GH secretion in rat pituitary cells. Endocrinology 1992;131:2133–2138.

66. Shah GV, Pedchenko V, Stanley S, Li Z, Samson WK. Calcitonin is a physiological inhibitor of prolactin secretion in ovariectomized female rats. Endocrinology 1996;137:1814–1822.
67. Ohta S, Shimekake Y, Nagata K. Cell-type-specific function of a C-type natriuretic peptide gene promotor in rat anterior pituitary-derived cultured cell lines. Mol Cell Biol 1993;13:4077–4086.
68. Tan TT, Yang Z, Huang W, Lim AT. ANF (1–28) is a potent suppressor of pro-opiomelanocortin (POMC) mRNA but a weak inhibitor of beta EP-LI release from AtT-20 cells. J Endocrinol 1994;143:R1–4.
69. Heisler S, Simard J, Assayag E, Mehri Y, Labrie F. Atrial natriuretic factor does not affect basal, forskolin- and CRF-stimulated adenylate cyclase activity, cAMP formation or ACTH secretion, but does stimulate cGMP synthesis in anterior pituitary. Mol Cell Endocrinol 1986;44:125–131.
70. Hashimoto K, Hattori T, Suemaru S, Sugawara M, Takao T, Kageyama J, Ota Z. Atrial natriuretic peptide does not affect corticotropin-releasing factor-, arginine vasopressin- and angiotensin II-induced adrenocorticotropic hormone release in vivo or in vitro. Regul Peptides 1987;17:53–60.
71. King MS, Baertschi AJ. Physiological concentrations of atrial natriuretic factors with intact N-terminal sequences inhibit corticotropin-releasing factor-stimulated adrenocorticotropin secretion from cultured anterior pituitary cells. Endocrinology 1989;124:286–292.
72. Antoni FA, Dayanithi G. Guanosine 3',5' cyclic monophosphate and activators of guanylate cyclase inhibit secretagogue-induced corticotropin release by rat anterior pituitary cells. Biochem Biophys Res Commun 1989;158:824–830.
73. Dayanithi G, Antoni FA. Atriopeptins are potent inhibitors of ACTH secretion by rat anterior pituitary cells in vitro: involvement of the atrial natriuretic factor receptor domain of membrane-bound guanylyl cyclase. J Endocrinol 1990;125:39–44.
74. Antoni FA, Dayanithi G. Blockade of K channels reverses the inhibitory action of atriopeptin on secretagogue-stimulated ACTH release by perifused isolated rat anterior pituitary cells. J Endocrinol 1990;126:183–191.
75. Kovacs KJ, Antoni FA. Atriopeptin inhibits stimulated secretion of adrenocorticotropin in rats: evidence for a pituitary site of action. Endocrinology 1990;127:3003–3008.
76. Gilkes AF, MacKay KB, Cramb G, Guild SB. Atrial natriuretic peptide effects in AtT-20 pituitary tumor cells. Mol Cell Endocrinol 1992;89:39–45.
77. Burrell LM, Lambert HJ, Baylis PH. Effect of atrial natriuretic peptide on thirst and arginine vasopressin release in humans. Am J Physiol 1991;260:R475–479.
78. Fujio N, Ohashi M, Nawata H, Kato K, Ibayashi H, Kangawa K, Matsuo H. Alpha human natriuretic polypeptide reduces the plasma arginine vasopressin concentration in human subjects. Clin Endocrinol 1986;25:181–187.
79. Clark BA, Elahi D, Fish L, McAloon-Dyke M, Davis K, Minaker KL, Epstein FH. Atrial natriuretic peptide suppresses osmostimulated vasopressin release in young and elderly humans. Am J Physiol 1991;261:E252–256.
80. Foresta C, Mioni R, Scanelli G, Vettor R, Busnardo B, Scandellari C. Alpha human atrial natriuretic peptide and anterior pituitary hormones secretion in men. Horm Metabol Res 1988;20:313–315.
81. Foresta C, Mioni R, Miotto D, DeCarlo E, Facchin F, Varotto A. Stimulatory effects of alpha-hANP on testosterone secretion in man. J Clin Endocrinol Metab 1991;72:392–395.
82. Ur E, Faria M, Tsagarakis S, Anderson JV, Besser GM, Grossman A. Atrial natriuretic peptide in physiological doses does not inhibit the ACTH or cortisol response to corticotrophin-releasing hormone-41 in normal human subjects. J Endocrinol 1991;131:163–167.
83. Vermes I, Beishuizen A, Hampsink RM, Haanen C. Dissociation of plasma adrenocorticotropin and cortisol levels in critically ill patients: possible role of endothelin and atrial natriuretic peptide. J Clin Endocrinol Metab 1995;80:1238–1242.
84. Kentsch M, Lawrenz R. Ball P, Gerzer R, Muller-Esch G. Effects of atrial natriuretic factor on anterior pituitary hormone secretion in normal man. Clin Invest 1992;70:549–555.
85. Kellner M, Wiedemann K, Holshner F. Atrial natriuretic factor inhibits the CRH-stimulated secretion of ACTH and cortisol in man. Life Sci 1992;50:1835–1842.

# 12 Natriuretic Peptide Actions in the Brain

*Alastair V. Ferguson*

***Contents***



## INTRODUCTION

The discovery of atrial natriuretic peptide (ANP) in the heart some 15 yr ago *(1)* prompted an explosion of research directed toward understanding both the integrated physiological roles and the potential sites of action of this vasoactive peptide. These studies pointed to the central nervous system (CNS), and in particular to regions of the brain with recognized regulatory roles in the control of body fluid balance, as locations at which ANP may play significant roles in normal physiological control mechanisms. The early description of binding sites for ANP within localized regions of the brain *(2)*, combined with reports of ANP-like immunoreactivity within some of these areas *(3)*, argued that, in addition to its hormonal actions in the periphery, ANP may act as a nonconventional neurotransmitter within the CNS *(4,5)*. Many studies have since attempted to define functional roles for ANP within the brain using intracerebroventricular (icv) and direct tissue administration techniques. In addition, a small number of electrophysiological studies have attempted to define the precise actions of ANP on specific groups of neurons in the brain. Such studies have been complicated by the subsequent identification first of brain natriuretic peptide (BNP), which has since been demonstrated to be produced in the periphery rather than the brain, and finally CNP (the "C" prefix represents alphabetical order of discovery), which has now been confirmed to be the primary form of the peptide found within the brain *(6)*. Furthermore, the identification of three different receptors for natriuretic peptides (NPs) and the confusion associated with their nomenclature as NPR-A, NPR-B, and NPR-C receptors, which has nothing to do with their relative affinities for ANP, BNP, and CNP, has not aided the emergence of

From: *Contemporary Endocrinology: Natriuretic Peptides in Health and Disease*
Edited by: W. K. Samson and E. R. Levin Humana Press Inc., Totowa, NJ

a coordinated view of potential actions of these substances within the brain. The focus of this chapter will be to first review the anatomical data describing the distribution of NPs and their receptors within the CNS. In particular I will highlight the recent information that has begun to define these localizations according to specific receptor and peptide types. The essential relationship between structure and function will then be highlighted by a consideration of the available evidence illustrating potential physiological functions of NPs in regions of the brain that contain the peptide and/or its receptors. Finally, I will consider the sometimes contradictory electrophysiological evidence that directs us toward the single-neuron level of understanding the mechanisms of natriuretic peptide actions within the CNS.

## ANATOMICAL LOCALIZATION

### *Receptors*

Soon after the initial discovery of the NPs, binding sites for these substances were described within the CNS that appeared to be preferentially located in regions with recognized regulatory roles in the control of body fluid balance. In accordance with this perspective, forebrain binding sites for NPs have been identified in the subfornical organ (SFO), paraventricular nucleus (PVN), supraoptic nucleus (SON), organum vasculosum of the lamina terminalis (OVLT), median preoptic nucleus (MPO), and choroid plexus *(2,7,8)*. Medullary sites with recognized roles in the regulation of fluid balance, including the area postrema (AP), dorsal motor nucleus of the vagus (DMV), and nucleus tractus solitarius (NTS), have also been reported to contain high densities of NP-binding sites *(2)*. However, it must also be emphasized that, in addition to these regions, binding sites for NPs have also been described in a number of other CNS locations, including the medial habenular nucleus, cerebellum, hippocampus, and olfactory tract *(2,7)*, observations that suggest multifunctional roles for NPs as chemical messengers within the brain.

The issue of differential distribution of NPR-A, NPR-B, and NPR-C receptors within the brain has been difficult to address using standard radiolabeled ligand-binding techniques. Inferences have, however, been drawn from studies comparing the relative affinities of different NPs for specific binding sites. Specific binding of iodinated ANP to sites on the choroid plexus, SFO, and olfactory bulb that showed no affinity for CNP was interpreted as indicative of the presence of the NPR-A receptor in these locations *(9,10)*. Similarly, binding studies have indicated the presence of NPR-C receptors in the olfactory bulb, cerebral cortex, and choroid plexus *(10,11)*. In contrast, the limited studies using such binding techniques have suggested the absence of NPR-B receptors within the CNS despite the assertion that CNP is the primary CNS-derived NP *(10)*. Fortunately, the development of specific probes for the NPR-A and NPR-B receptor mRNA has provided a more definitive approach to the description of the distribution of these two receptor subtypes using *in situ* hybridization techniques *(12)*. Interestingly, these studies have shown that the vast majority of previously described binding sites for natriuretic peptides within the hypothalamus, which are protected from access to circulating NPs (ANP and BNP) by the blood–brain barrier, express primarily the mRNA for the NPR-B receptor. The NPR-A receptor is, in contrast, primarily expressed in other regions of the forebrain, including the SFO (this structure lacks a normal blood–brain barrier and thus would be directly accessed by circulating ANP and BNP) and the medial habenula *(12)*. Similar studies applied to the previously demonstrated distribution of NP binding sites in the medulla would thus be predicted to show NPR-A receptor mRNA in the AP (this CVO has direct access to circulating ANP and BNP), and NPR-B receptor mRNA localized primarily in the NTS and DMV, most likely in association with the demonstration of CNP-immunoreactive nerve terminals in these regions.

### *Peptide Distribution*

Soon after the original structural characterization of the NPs, an extensive distribution of NP-like immunoreactive neurons was described in the rat CNS *(3)*. Large numbers of immunoreactive cells were found in the OVLT, MPO nucleus, preoptic suprachiasmatic nucleus, parvocellular PVN (but not SON), medial habenula, parabrachial nucleus, NTS, and AP *(3)*. Interestingly, these studies also demonstrated that the distribution of NP-immunoreactive nerve fibers effectively mirrored that of the cell bodies, although the interpretation of such data is difficult. It may be indicative of nerve fibers emanating from the NP-immunoreactive cell bodies or, alternatively, as suggesting that NP-immunoreactive cell bodies also receive NP innervation from other regions of the brain. This issue is perhaps better addressed by recent *in situ* hybridization studies describing the distribution of NP mRNAs within the CNS, which revealed a widespread distribution of both ANP and CNP, but a complete absence of BNP *(13)*. Interestingly, the observed localization of both ANP and CNP in the same regions of a number of distinct nuclei, including the anteroventral periventricular and arcuate regions, also suggests coexpression of these peptides within some regions of the brain *(13)*. Such information only serves to further highlight the potential neuromodulatory actions of this peptide within the brain. These studies clearly indicate the ability of diverse populations of CNS neurons to produce either ANP, CNP, or in some instances both these peptides. Hopefully, future double-labeling studies in which anatomical tracing techniques can be combined with *in situ* hybridization for peptide or receptor mRNAs will provide the essential information regarding the origins and targets of specific NP innervation.

## FUNCTIONAL STUDIES

The primary emphasis in studies investigating physiological roles for NPs in the CNS has been directed toward determining potential functions in three primary areas: the regulation of pituitary hormone secretion, the control of fluid intake, and blood pressure regulation. The majority of these studies, although not showing obligatory primary actions of the NPs, have suggested important modulatory roles for these peptides. In many cases, such effects are associated with the ability of the NP to modify the animal's response to other regulatory substances.

### *Roles in Control of Fluid Intake*

A number of different studies have now demonstrated roles for NPs in the regulation of fluid intake *(14–16)*. Thus, icv injections of ANP have been shown to inhibit the drinking response to both angiotensin (ANG) and dehydration *(17)* although it has no direct effect on basal fluid intake. In contrast, CNP has been reported to have the opposite effects on ANG and dehydration-induced drinking *(16)*. These observations first suggest that not all actions of NPs are directed toward a common physiological goal, and second indicate that such contrasting effects of these different NPs may be the result specific actions in different regions of the CNS. The studies of Erlich and Fitts *(15)*, in which ANP administered directly into SFO inhibited ANG and dehydration-induced drinking, suggest this CVO, in which high densities of NPR-A receptors are found, to be the site at which ANP exerts such inhibitory effects. This conclusion draws support both from the demonstration of increased numbers of ANP binding sites in SFO and choroid plexus following 4 d of water deprivation *(18)*, and from the demonstrated ability of ANP to stimulate guanylyl cyclase activity in these regions *(19)*. The potential sites of action for the stimulatory actions of CNP on water intake *(16)* have yet to be identified, although one would anticipate regions rich in NPR-B receptors with recognized roles in the control of fluid intake to be likely targets.

### *Modulation of Hormone Secretion*

Inhibitory roles within the CNS for NPs in regulating the secretion of both anterior and posterior pituitary hormones have also been reported. A hypothalamic site of action for ANP whereby it inhibits the secretion of luteinizing hormone from the anterior pituitary has been identified *(20)*, although the physiological relevance of such effects have not yet been elucidated. Samson et al. *(21)* have also shown third ventricular administration of atriopeptin III to specifically reduce circulating concentrations of vasopressin (VP; oxytocin concentrations were unaffected) in euvolemic rats. They proposed a hypothalamic site of action in accordance with their in vitro studies showing that such effects were not observed in isolated neural lobe or median eminence preparations whereas small effects were retained in a hypothalamo-neurohypophysial explant including the supraoptic nucleus *(21)*. In addition, recent studies have reported microinjection of ANP into SFO to inhibit ANG-stimulated VP release *(22)*, supporting the concept that this CVO represents a second potential site for the actions of NPs in the regulation of VP secretion.

### *Central Cardiovascular Control*

The studies described above have demonstrated NP actions within the SFO in modulating centrally mediated drinking and hormonal responses to circulating ANG. The fact that such actions of ANG in the SFO also result in cardiovascular responses led very logically to the observations that icv ANP *(23)* and then later BNP *(24)* (at the time it was believed to be the brain form of the NPs), although without effect on baseline blood pressure, reduced the pressor effects of icv ANG *(24)*. Additional support for the concept of potential roles of NPs within the SFO in both the normal and pathological regulation of the cardiovascular system has since been provided by a number of different receptor-binding studies demonstrating reductions in numbers of ANP receptors in SHR rats prior to the development of hypertension as compared to control animals *(25,26)*.

Alternative forebrain sites of action have been suggested by the studies of Levin et al. *(27)*, demonstrating that fourth ventricular administration of ANP causes significant decreases in blood pressure of both spontaneously hypertensive and normotensive rats. These effects were, however, greater in the former group, and were shown to be adrenergically mediated, because they were totally abolished by central administration of the $\alpha_2$-adrenergic antagonist yohimbine. Similar tissue administration of ANP in the preoptic suprachiasmatic nucleus in the anteroventral third ventricle (AV3V) region has been reported to increase blood pressure and heart rate *(28)*. These contrasting cardiovascular actions of the NPs in different regions of the brain serve to emphasize the likely functional role of these peptides as neurotransmitters in anatomically discrete regions that do not serve a collective single physiological function.

The effects of local administration of NPs into the autonomic control centers of the medulla have provided information suggesting a role for ANP in these areas. Microinjection of ANP into the NTS causes decreases in blood pressure and heart rate *(29,30)*. Ermirio et al. *(31)* combined such local administration of ANP with single-unit recordings and showed that at all sites where ANP activated NTS neurons it also elicited decreases in mean arterial blood pressure and heart rate. Interestingly, these ANP-activated neurons were also shown to receive baroreceptor-afferent information, thus implicating them as important integrators in central cardiovascular control. Such data clearly suggest that modulation of the activity of efferent connections of these ANP-sensitive NTS neurons results in the observed cardiovascular changes. Similar depressor effects of ANP administration have been reported following microinjection into the rostral ventrolateral medulla, whereas ANP has been shown to elicit opposite effects, i.e., it causes increases in blood pressure, when locally applied in the AP *(30)*. Finally, medullary roles for the NPs in salt-sensitive hypertensive rats have been suggested by experiments showing that NTS administration of monoclonal antibodies (MAbs) to ANP causes significant increases in blood pressure in hypertensive animals, but is without effect in normotensive controls *(32)*.

## CELLULAR BASIS OF CNS ACTIONS

Despite the considerable anatomical and physiological evidence indicative of specific actions of the NPs within the CNS, there is to date only a limited amount of electrophysiological evidence describing the effects or these peptides on single neurons within the brain. Almost all of these data have been obtained using extracellular recording techniques, and its interpretation has, in addition, been limited in many cases by the fact that such recordings have been obtained from unidentified (either by function or connectivity) neurons. The initial description of peripheral physiological roles for the NPs in the regulation of body fluid balance served to focus attention on similar potential roles for these peptides in the CNS. The early demonstration of binding sites and immunoreactivity for ANP in hypothalamic and medullary regions intricately involved in fluid homeostasis further reinforced this view and focused electrophysiological studies on understanding the effects of natriuretic peptides on the activity of neurons in these regions.

### *Forebrain Neurons*

One of the first physiological actions of ANP described was its ability to antagonize ANG actions within the SFO to induce drinking *(15)*. A number of different in vitro single-unit recording studies have attempted to determine the effects of NPs on SFO neurons, but unfortunately have yielded contradictory results. Yamashita and Kannan *(33)* reported exclusively inhibitory actions of NPs on a small proportion of SFO neurons (14%), with the remaining cells unaffected. In addition, they observed that a much larger proportion of SFO neurons activated by ANG and unaffected by ANP administered in isolation showed blunted excitatory effects of ANG when this peptide was administered in combination with ANP *(33)*, an observation also reported by Schmid and Simon *(34)*. In contrast, similar in vitro extracellular single-unit recording studies in the rat *(35)* have demonstrated profound dose-dependent excitatory effects of nanomolar ANP in nearly 50% of SFO neurons tested. Finally, the observation that centrally administered CNP enhances water intake in a manner similar to ANG *(16)* suggests that this substance may also have similar effects on SFO neurons, a hypothesis supported by our preliminary

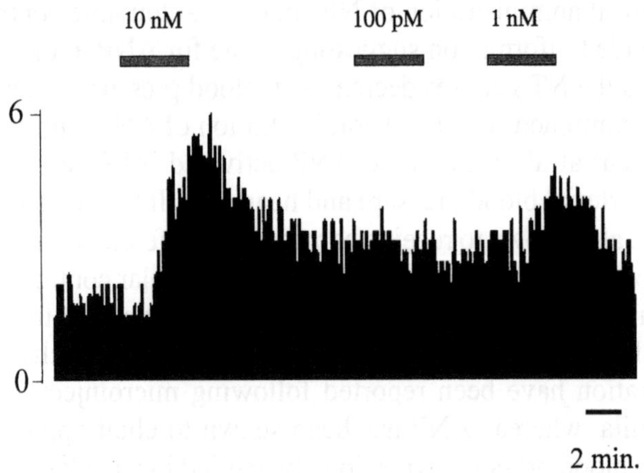


**Fig. 1.** This ratemeter record shows the effects of bath administration of CNP on the spontaneous activity of an SFO neuron recorded in vitro. The peptide was administered for the time periods indicated by the shaded bars at the doses indicated. The horizontal axis indicates the number of action potentials occurring per 10-s period.

single-unit recordings, showing excitatory effects of this peptide on SFO neurons as illustrated in Fig. 1 (Li and Ferguson, unpublished observations). The functional consequences of such actions of CNP on SFO neurons have yet to be established. However, additional preliminary in vivo single-unit recordings demonstrating clear excitatory effects of systemic CNP (which will access SFO neurons) on PVN neurons antidromically identified as projecting to the intermediolateral cell column of the spinal cord (Bains and Ferguson, unpublished observations, Fig. 2), suggest a potential role in controlling sympathetic output. Interestingly these actions are also differentiated from those of circulating ANG because the same population of cells were unaffected by this peptide.

Similar extracellular recordings have been obtained from putative vasopressinergic neurons in the hypothalamic SON and PVN with the purpose of understanding the cellular changes underlying the inhibitory effects of NPs of the secretion of VP from the posterior pituitary. Okuya and Yamashita *(36)* again found primarily inhibitory effects of ANP on hypothalamic neurons although it should be emphasized that only 2 of 25 putative neurosecretory (i.e., oxytocin- or VP-secreting) cells were affected in this way. A larger proportion (22%) of unidentified PVN cells were, however, inhibited, suggesting a potential preferential action of ANP on parvocellular neurons of this nucleus. Inhibitory effects following pressure ejection of NPs on the activity of putative vasopressinergic *(37)* and unidentified neurons *(38)* in PVN have also been observed in vivo. The long latency to the effects observed in the former study *(37)* (up to 200 s), as well as the prolonged duration (over 20 min) raise interesting questions about the mechanism of action. In contrast to the studies of Okuya and Yamashita *(36)*, Standaert et al. *(37)* observed no effects on putative parvocellular neurons of the PVN. Again, although these studies have suggested the expected inhibitory actions of NPs on hypothalamic vasopressinergic neurons, it should be emphasized that the data are not compelling. In fact, the majority of the available data suggest that NPs may not exert direct effects on VP-secreting neurons, although I believe that further experiments are essential to clearly address this question. In such studies it will be important to evaluate effects of the peptide most likely to act on these neurons (presence of NPR-B receptors suggest this to be CNP),

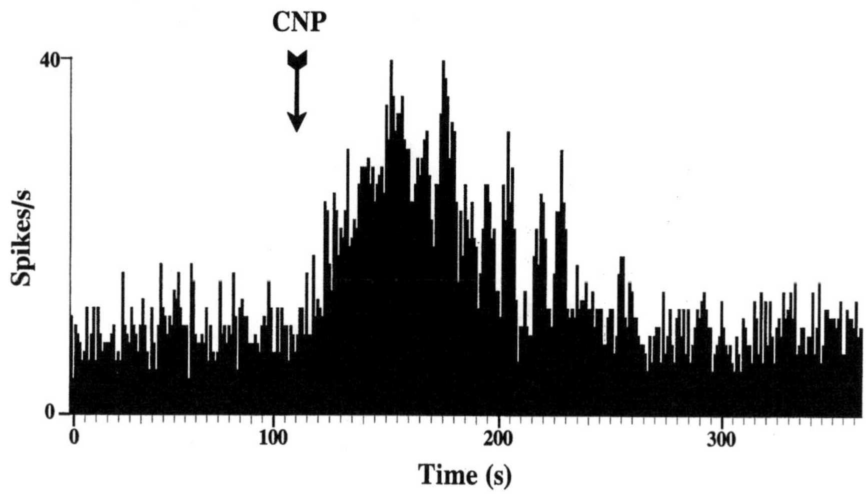


**Fig. 2.** The ratemeter record shown here illustrates the firing frequency of a PVN neuron recorded in vivo that was antidromically identified as projecting to the intermediolateral cell column of the spinal cord. This neuron responded to a single bolus systemic administration of 0.5 μg CNP (injection time indicated by the arrow), with a clear increase in action potential frequency, followed by a return to baseline firing. Although such experiments do not absolutely define the site of action for CNP in eliciting these effects, we have proposed the SFO as a potential location in view of its situation outside the blood–brain barrier.

and also to use the more sophisticated, whole-cell patch-clamp techniques that have recently been applied to single-cell recordings in hypothalamic nuclei *(39,40)*. Our own preliminary studies have, in fact, suggested that CNP administration to PVN slices results in depolarizing effects on both Type 1 (magnocellular) and Type 2 (parvocellular) neurons, effects that are accompanied by small changes in input resistance (Fig. 3, Kusmun and Ferguson, unpublished observations), although firm conclusions regarding these effects await analysis of a larger population of neurons.

Finally, the recognized involvement of the septum and preoptic area of the rat in the regulation of fluid intake when considered in association with the demonstration of ANP-like immunoreactivity in these areas *(3)*, provided a clear rationale for the electrophysiological studies of Wong et al. *(41)*, demonstrating primarily inhibitory effects of iontophoretically administered ANF on the activity of unidentified neurons in these two regions.

### *Medullary Neurons*

Evidence from peptide *(3,13)* and receptor *(12)* localization studies as well whole animal experiments *(30)* have suggested functional roles for the NPs in the regulation of cardiovascular control within medullary autonomic centers. Extracellular single-unit recordings have been obtained from neurons in both the NTS and DMV. In the latter experiments, DMV neurons were antidromically identified by stimulation of the vagus nerve, and found to be primarily excited by local administration of NPs *(42)*. Similar recordings from unidentified neurons within the NTS reported similar excitatory effects on approx 40% of neurons recorded, whereas an additional 20% were inhibited by the peptide *(29)*. Despite the high densities of NP receptors identified in the AP, and its recognized role in cardiovascular regulation, there is to date no direct evidence examining the effects of NPs on the activity of this population of neurons.

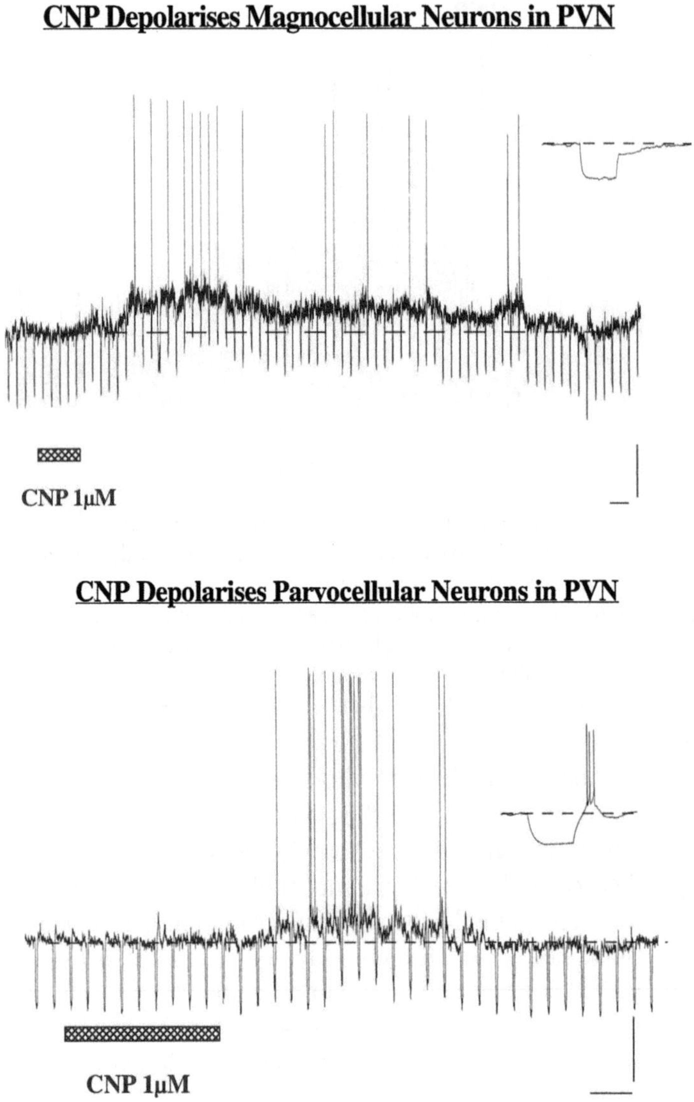


**Fig. 3.** Whole-cell patch recordings from a type I (magnocellular) and a type II (parvocellular) PVN neuron in a hypothalamic slice preparation. These neurons are defined according to their voltage responses to hyperpolarizing current pulses (*see* insets) whereby type I cells show a delayed return to baseline indicative of the presence of an A-current, whereas type II cells show a low threshold calcium current. These records demonstrate that both of these cell types respond to bath administration of CNP with small depolarizations and decreases in input resistance (indicated by the reduced voltage response to the repetitively applied current pulse). The horizontal and vertical scale bars represent 10 s and 10 mV, respectively.

## *Ion Channels*

Perhaps the most likely limitation to our present understanding of the specific roles of NPs within the CNS lies with the complete lack of knowledge of the specific membrane events underlying the observed effects of these substances of the excitability of individual neurons. Although it is well accepted that NP actions at the NPR-A and NPR-B receptor are guanylyl cyclase-mediated *(19,43,44)*, the specific ion channels influenced by such intracellular changes remain to be elucidated. Clues derived from other cell systems do, however, suggest potential actions or NPs in modulating potassium channels *(45,46)*. Future studies to define such effects in **identified** cell groups should considerably enhance our understanding of the mechanisms whereby NPs exert their physiological actions in the CNS.

## INTEGRATIVE PERSPECTIVES

### *Hormone, Neurotransmitter, or Neuromodulator?*

The demonstration of natriuretic peptide (primarily NPR-A) receptors in SFO and AP *(12,13,18,25)* suggests that systemic ANP (preferential actions at NPR-A) plays physiologically relevant roles in controlling the excitability of neurons in these CVOs. Electrophysiological studies have provided conflicting data showing either inhibitory *(47)* or excitatory *(35)* responses of SFO neurons to ANP, which is clearly an area in need of clarification. Such conflicting data may be associated with the additional presence of smaller numbers of NPR-B receptors in these structures at which neurally derived CNP may exert alternative actions. Within such a framework, it would seem logical to suggest that hormonal actions of NPs in these CVOs will complement the physiological roles of these peptides in the periphery, i.e., be primarily directed toward the regulation of body fluid balance.

The localization of all three NP receptors as well as ANP and CNP mRNAs in anatomically diverse regions of the brain that lie behind the blood–brain barrier make it difficult to propose a unified physiological role for the NPs within the CNS. It would appear more likely that diverse populations of neurons have, through the course of evolution, adopted these peptides as chemical messengers used for communication with postsynaptic sites. The primary challenge within this area thus becomes to gain a clearer understanding of the cellular mechanisms through which both ANP and CNP act at postsynaptic sites to control the excitability of their target neurons.

## ACKNOWLEDGMENTS

Supported by the Medical Research Council of Canada and the Heart and Stroke Foundation of Ontario.

## REFERENCES

1. de Bold AJ. Atrial natriuretic factor a hormone produced by the heart. Science 1985;230:767–770.
2. Saavedra JM, Kurihara M. Autoradiography of atrial natriuretic peptide (ANP) receptors in the rat brain. Can J Physiol Pharmacol 1991;69:1567–1575.
3. Skofitsch G, Jacobowitz DM, Eskay RL, Zamir N. Distribution of atrial natriuretic factor-like immunoreactive neurons in the rat brain. Neuroscience 1985;16:917–948.
4. Levin ER. Atrial natriuretic factor neuromodulator of the central nervous system regulation of blood pressure. Am J Hypertens 1989;2:594–598.
5. Debinski W, Kuchel O, Buu NT. Atrial natriuretic factor is a new neuromodulatory peptide. Neuroscience 1990;36:15–20.
6. Sudoh T, Minamino N, Kangawa K, Matsuo H. C-type natriuretic peptide (CNP) a new member of the natriuretic peptide family identified in porcine brain. Biochem Biophys Res Commun 1990;168, No. 2:863–870.
7. Quirion R, Dalpe M, Dam T. Characterization and distribution of receptors for the atrial natriuretic peptides in mammalian brain. Proc Natl Acad Sci USA 1986;83:174–178.

8. Niwa M, Shigematsu K, Kurihara M., Kataoka Y, Maeda T, Nakao K, Imura H, Matsuo H, Tsuchiyama H, Ozaki M. Receptor autoradiographic evidence of specific brain natriuretic peptide binding sites in the porcine subfornical organ. Neurosci Lett 1988;95:113–118.
9. Himeno A, Niwa M, Nakao K, Suga S, Yamashita K, Kataoka Y, Nakane Y, Imura H, Taniyama K. C-type natriuretic peptide-22 differentiates between natriuretic peptide receptors in rat choroid plexus and subfornical organ. Eur J Pharmacol 1992;215:337–440.
10. Brown J, Zuo Z. C-type natriuretic peptide and atrial natriuretic peptide receptors of rat brain. Am J Physiol 1993;264:R513–R523.
11. Konrad E, Thibault MG, Schiffrin EL, Cantin M. Atrial natriuretic factor receptor subtypes in the rat central nervous system. Hypertension 1991;17:1144–1151.
12. Langub MCJ, Dolgas CM, Watson REJ, Herman JP. The C-type natriuretic peptide receptor is the predominant natriuretic peptide mRNA expressed in rat hypothalamus. J Neuroendocrinol 1995;7:305–309.
13. Langub MC, Watson RE, Herman JP. Distribution of natriuretic peptide precursor messenger-RNAs in the rat-brain. J Comp Neurol 1995;356:183–199.
14. Samson WK. Central nervous system actions of atrial natriuretic factor. Brain Res Buff 1988;20:831–837.
15. Ehrlich KJ, Fitts DA. Atrial natriuretic peptide in the subfornical organ reduces drinking induced by angiotensin or in response to water deprivation. Behav Neurosci 1990;104:365–372.
16. Samson WK, Skala KD, Huang FL. CNP-22 stimulates, rather than inhibits, water drinking in the rat evidence for a unique biological action of the C-type natriuretic peptides. Brain Res 1991;568:285–288.
17. Antunes-Rodrigues J, McCann SM, Rogers LC, Samson WK. Atrial natriuretic factor inhibits dehydration and angiotensin II—induced water intake in the conscious unrestrained rat. Proc Natl Acad Sci USA 1985;82:8720–8723.
18. Saavedra JM, Israel A, Kurihara M. Increased atrial natriuretic peptide binding sites in the rat subfornical organ after water deprivation. Endocrinology 1987;120:426,427.
19. Isreal A, Garrido MR, Barbella Y, Becemberg I. Rat atrial natriuretic peptide (99–126) stimulates guanylate cyclase activity in rat subfornical organ and choroid plexus. Brain Res Bull 1988;20:253–256.
20. Samson WK, Aguila MC, Bianchi R. Atrial natriuretic factor inhibits luteinizing hormone secretion in the rat evidence for a hypothalamic site of action. Endocrinology 1988;12:1573–1582.
21. Samson WK, Aguila MC, Martinovic J, Antunes-Rodrigues J, Norris M. Hypothalamic action of atrial natriuretic factor to inhibit vasopressin secretion. Peptides 1987;8:449–454.
22. Steardo L, Steardo MD, Testa N, Attanasio A, Persichella M, Cuomo V. Atrial natriuretic factor antagonizes angiotensin II-induced vasopressin release in rat subfornical organ. Acta Neurol 1994;16:229–234.
23. Shimizu T, Katsuura G, Nakamura M, Nakao K, Morii N, Itoh Y, Shiono S, Imura H. Effect of intracerebroventricular atrial natriuretic polypeptide on blood pressure and urine production in rats. Life Sci 1986;39:1263–1270.
24. Shirakami G, Nakao K, Yamada T, Itoh H, Mori K, Kangawa K, Minamino N, Matsuo H, Imura H. Inhibitory effect of brain natriuretic peptide on central angiotensin II-stimulated presser response in conscious rats. Neurosci Lett 1988;91:77–83.
25. Saavedra J, Israel MA, Kurihara M, Fuchs E. Decreased number and affinity of rat atrial natriuretic peptide (6–33) binding sites in the subfornical organ of spontaneously hypertensive rats. Circ Res 1986;58:389–392.
26. Saavedra JM, Correa FMA, Plunkett LM, Israel A, Kurihara M, and Shigematsu K. Binding of angiotensin and atrial natriuretic peptide in brain of hypertensive rats. Nature 1986;320:758–760.
27. Levin ER, Weber MA, Mills S. Atrial natriuretic factor-induced vasodepression occurs though central nervous system. Am J Physiol 1988;255:H616–H622.
28. Sills MA, Nguyen KQ, Jacobowitz DM. Increases in heart rate and blood pressure produced by microinjections of atrial natriuretic factor into the AV3V region of rat brain. Peptides 1985;6:1037–1042.
29. Ermirio R, Ruggeri P, Cogo CE, Molinari C, Calaresu FR. Neuronal and cardiovascular responses to ANF microinjected into the solitary nucleus. Am J Physiol 1989;256:R577–R582.
30. Okada J, Takayama K, Xiong Y, Miura M. Influence of humoral control peptides on medullary vasomotor control neurons microstimulation and double labeling studies using SHR and WKY rats. J Auton Nerv Syst 1994;49:171–182.
31. Smith GP, Jerome C, Norgren R. Afferent axons in abdominal vagus mediate satiety effect of cholecystokinin in rats. Am J Physiol 1986;249:R638–R641.
32. Yang R, Jin H, Wyss JM, Chen Y, Oparil S. Salt supplementation does not alter the presser effect of blocking atrial natriuretic peptide in nucleus tractus solitarii. Hypertension 1992;20:242–246.

33. Yamashita H, Kannan H. Inhibition of hypothalamic neurons by the atrial natriuretic peptide family. NIPS 1992;72:75–79.
34. Schmid HA, Simon E. Effect of angiotensin II and atrial natriuretic factor on neurons in the subfornical organ of ducks and rats in vitro. Brain Res 1992;588:324–328.
35. Buranarugsa P, Hubbard JI. Excitatory effects of atrial natriuretic peptide on rat subfornical organ neurons in vitro. Brain Res Bull 1988;20:627–631.
36. Okuya S, Yamashita H. Effects of atrial natriuretic polypeptide on rat hypothalamic neurones in vitro. J Physiol (Lond.) 1986;389:717–728.
37. Standaert DG, Cechetto DF, Needleman P, Saper CB. Inhibition of the firing of vasopressin neurons by atriopeptin. Nature 1987;329:151–153.
38. Haskins JT, Zingaro GJ, Lappe RW. Rat atriopeptin III alters hypothalamic neuronal activity. Neurosci Lett 1986;67:279–286.
39. Li Z, Ferguson AV. Electrophysiological properties of paraventricular magnocellular neurons in rat brain slices: modulation of $I_A$ by angiotensin II. Neuroscience 1996;71:133–145.
40. Cui LN, Inenaga K, Nagatomo T, Yamashita H. Sodium nitroprusside modulates NMDA response in the rat supraoptic neurons in vitro. Brain Res Bull 1994;35:253–260.
41. Wong M, Samson WK, Dudley CA, Moss RL. Direct, neuronal action or atrial natriuretic factor in the rat brain. Neuroendocrinology 1986;44:49–53.
42. McCann MJ, Nice-Lepard K, Rogers RC. Dorsal medullary injection of atrial natriuretic rector (ANF) excites vagal efferents and inhibits gastric motility. Brain Res 1991;549(2):247–252.
43. Suga S, Nakao K, Hosoda K, Mukoyama M, Ogawa Y, Shirakami G, Arai H, Saito Y, Kambayashi Y, Inouye K, Imura H. Receptor selectivity of natriuretic peptide family, atrial natriuretic peptide, brain natriuretic peptide and c-type natriuretic peptide. Endocrinology 1992;130:229–239.
44. Kohno M, Horio T, Yokokawa K, Kurihara N, Takeda T. C-type natriuretic peptide inhibits thrombin- and angiotensin II-stimulated endothelin release via cyclic guanosine 3',5'-monophosphate. Hypertension 1992;19:320–325.
45. White RE, Lee AB, Shcherbatko AD, Lincoln TM, Schonbrunn A, Armstrong DL. Potassium channel stimulation by natriuretic peptides through cGMP-dependent dephosphorylation. Nature 1993;361:263–266.
46. Miledi R, Woodward RM. Membrane currents elicited by prostaglandins, atrial natriuretic factor and oxytocin in follicle-enclosed Xenopus oocytes. J Physiol (Lond.) 1989;416:623–643.
47. Hattori Y, Kasai M, Uesugi S, Kawata M, Yamashita H. Atrial natriuretic polypeptide depresses angiotensin II induced excitation or neurons in the rat subfornical organ in vitro. Brain Res. 1988;443:355–359.

# 13 Natriuretic Peptides as Antigrowth Factors

*Ellis R. Levin*

*CONTENTS*



## NATRIURETIC PEPTIDES AS ANTIGROWTH FACTORS

The natriuretic peptides act as antigrowth factors for a variety of cells *(1)*, including mesangial *(2)*, vascular endothelial *(3)* and smooth muscle cells *(4)*, and astrocytes *(5)*. All three members of this family of peptides (ANP, BNP, and CNP) have been shown to inhibit cell proliferation, at least in vitro. This is particularly evident when cells are stimulated to divide by growth factors, such as platelet-derived growth factor (PDGF) or basic fibroblast growth factor (bFGF). CNP has been shown to also inhibit arterial intimal thickening in vivo, probably through inhibiting vascular smooth muscle proliferation induced by vascular injury *(6)*. Therefore, understanding the mechanisms whereby the natriuretic peptides inhibit cell growth may lead to therapeutic applications to prevent the development of disease resulting from deregulated cell proliferation. This chapter explores the mechanisms of antiproliferation and the specific actions of the natriuretic peptides to limit cell growth.

From: *Contemporary Endocrinology: Natriuretic Peptides in Health and Disease*
Edited by: W. K. Samson and E. R. Levin Humana Press Inc., Totowa, NJ

## NATRIURETIC PEPTIDES AND DEVELOPMENT

The natriuretic peptides and their receptors can often be detected temporally at the time of organ development. Regarding the central nervous system (CNS), we have found that atrial natriuretic peptide can be secreted from fetal rat diencephalic neurons in primary culture. These cells, comprised 85% by hypothalamic cells, express ANP mRNA as early as embryonic d 14, a time when the hypothalamus is just being formed *(7)*. We have also been able to detect the binding of ANP and the expression of natriuretic peptide clearance receptors (NPCR) and guanylyl cyclase receptors (undetermined subclass) *(8)*, as early as embryonic d 15. These data strongly suggest that the natriuretic peptide system plays a developmental role, perhaps to regulate proliferation or to modulate the actions of other neurotransmitters on astroglia or other cells within the CNS. There may be an important interaction with other neuropeptide systems at this time. For instance, we found that the expression of the endothelin system in the brain coincides with the hypothalamic expression of the natriuretic peptides, and that endothelin can limit the actions of ANP to generate cGMP in astrocytes *(9)*. In turn, the natriuretic peptides can inhibit endothelin signaling and action on astroglia (*see* Natriuretic Peptides as Antigrowth Factors: Signaling to the Nucleus). It has already been reviewed in Chapter 11 that ANP can interact with the angiotensin system and other CNS neurotransmitter systems. Thus, the temporal and anatomical expression and interaction with other important neurotransmitter systems strongly suggests important roles for the natriuretic peptides in the CNS.

A very analogous situation to the brain is the postnatal vasculature in which the natriuretic peptides limit the actions of other important vasoactive proteins. Therefore, it is likely that the details of growth and developmental regulation by the natriuretic peptides in the CNS will be similar in other organ systems. Although an in-depth examination of the steps by which the natriuretic peptides limit growth will be given only for the astrocyte in culture, many of the details have been confirmed in other systems, most notably cultured vascular smooth muscle cells, and can therefore be used as a general guide to the role and actions of the natriuretic peptides in other organ systems.

## ASTROCYTE PROLIFERATION: WHICH NATRIURETIC PEPTIDE AND WHICH RECEPTOR?

It is well recognized that all three natriuretic peptides (NPs) and their receptors are expressed in the CNS. There are some differences in the anatomical location of the peptides and certainly of the receptors *(10,11)*, as determined from immunocytochemical and ligand-binding localization in brain slices. This diversity may indicate unique functions of the peptides within the CNS. The neurophysiology of the NP has already been reviewed in depth earlier in this text. Regarding astrocyte proliferation, we showed that both ANP and BNP have a roughly equipotent capacity to inhibit the nuclear incorporation of thymidine into cultured diencephalic astrocytes, an index of DNA synthesis and cell proliferation *(5)*. The receptors expressed on the cell membrane probably dictate the specificity of NP effects, since all three family members appear to have the potential to inhibit growth.

In various models of cell proliferation, the NPs have been reported to work through the production of cGMP (Fig. 1). In cultured mesangial cells, it has been shown that the NPs stimulate cGMP and inhibit cell proliferation, and that exogenous cGMP mimics the effects of the NP *(2)*. Supporting a guanylyl cyclase mechanism, several vasoactive

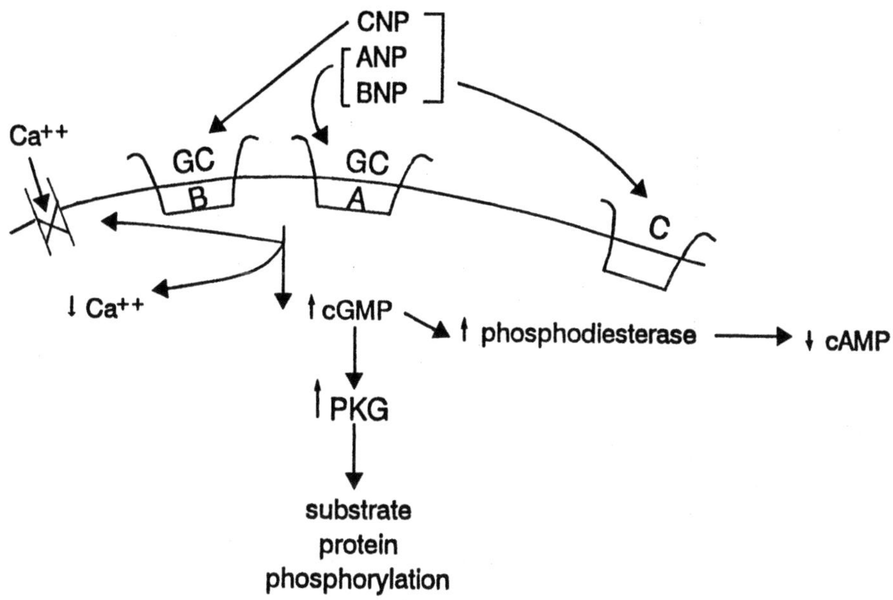


**Fig. 1.** Binding of natriuretic peptide receptors, cGMP generation via guanylyl cyclase receptors, and downstream signaling events. PKG is cyclic GMP-dependent protein kinase.

substances can inhibit the proliferation of vascular smooth muscle cells (VSMC) in culture through the production of cGMP. Garg and Hassid showed that nitric oxide is an antiproliferative factor for VSMC and that these actions probably occur through the generation of cGMP *(12)*. The problem with many of these types of studies is that it has not been shown that a reversal of these antigrowth actions results from inhibiting cGMP formation or related downstream biochemical events, such as cGMP-dependent protein kinase (PKG) activation. Using this approach, we recently showed that prostacyclin and PGE2 can inhibit endothelin-stimulated VSMC proliferation, and that the inhibition can be reversed by LY53853, a compound that inhibits cGMP generation *(13)*. Porter et al. took a slightly different approach and compared relatively selective ligands for the NPCR to the NP. Since the NPs bind both GC and NPCR receptors with equal affinity, this limits determination of which receptor mediates the antiproliferative effects by this approach unless selective agonists are used *(14)*. The authors found that CNP was a potent antigrowth factor for cultured VSMC, whereas ANP (4–23) (clearance receptor ligand) had no effect on cell proliferation. The cells most likely expressed the CNP-selective GC-B receptor, although this was not shown in this study. Therefore, it is most likely that cGMP is the initial signal leading to antigrowth in VSMC, as caused by several vasoactive substances.

In contrast, we investigated the effects of the NP on cultured astrocytes and found that ANP (4–23) was equipotent to ANP (or BNP) in inhibiting cell proliferation, assessed by nuclear thymidine incorporation and cell number after 6 d in culture *(5)*. ANP (4–23) did not stimulate cGMP generation, whereas ANP caused a nearly 12-fold increase. Since both peptides bind to the NPCR, this suggested that the NPCR is involved in this in vitro model.

What is the signal through which the C-receptor can potentially effect antigrowth? Anand-Srivistava and others have previously shown that the NPCR can signal through inhibiting cAMP formation *(15–17)*. We have not been able to show this in astrocytes;

furthermore, it is well known that in many cells the *generation* of cAMP leads to growth inhibition *(18)*. Therefore, it is unlikely that the proximal signal in response to NPCR binding by NP, which then triggers the antiproliferation program, is the inhibition of cAMP. The proximal signal by which the NPCR might inhibit cell proliferation is currently undefined.

What may be the proximal signal by which the NP can inhibit cell proliferation through the GC receptor? Many, but certainly not all, actions of cGMP occur through the activation of PKG (Fig. 1). Cation channel proteins and phosphodiesterases can also be activated by this cyclic nucleotide *(19)*. The generation of cGMP by the NP leads to inhibition of phosphatidylinositol hydrolysis *(20)* and decreased intracellular mobilization of $Ca^+$ *(21)*, thereby inhibiting the actions of a variety of vasoactive effectors in the vasculature. This latter action of the NP is probably mediated through PKG *(22)*. L-type Ca channels can also be inhibited by cGMP, after their activation by cAMP. Since calcium is often necessary for cell proliferation, this modulation provides one potential pathway of growth inhibition via the GC receptor. There is also evidence that cGMP can activate cAMP-dependent protein kinase (PKA) in some cells *(23)*. This would be expected to inhibit signaling systems that are necessary for cell proliferation, such as the MAP kinase cascade, and provides an additional possible mechanism of action.

## NATRIURETIC PEPTIDES AS ANTIGROWTH FACTORS: SIGNALING TO THE NUCLEUS

After NPs bind both the GC and NPCR, the proximal signal for antigrowth is transmitted to effector molecules, ultimately propagating the signal to the cell nucleus. Little is known about the series of events that underlie growth regulation by the NP. We have investigated the NP-induced transmembrane to nuclear events that result in the restraint of growth in astrocytes. First, it appears, as mentioned above, that the clearance receptor is likely to be involved *(5)*. Our model examined the NP inhibition of the mitogenic actions of endothelin, but comparable results were found for several of the steps elucidated that were stimulated by other astrocytic mitogens, such as phorbol ester or epidermal growth factor (EGF) *(24)*.

To understand how the NP restrain growth, the precise steps that are critical for mitogen-activated growth were investigated. ET is known to activate the growth program in mesangial cells, via both PKC-dependent and independent mechanisms *(25)*. This probably results from the binding of the ET ligand to either of the two known ET receptors, effecting a G protein-coupled hydrolysis of phospholipase C, and generation of diacylglycerol (reviewed in ref. *26*). In addition, PLC hydrolysis leads to IP3 generation, which results in the mobilization of intracellular calcium. Both of these steps can activate/cause the translocation of PKC to the cell membrane. We found a similar event in astrocytes, where ET stimulates the phosphorylation of the specific substrate for PKC, $p^{80}$ in these cells. This indicated that ET activates PKC in the astrocyte. We found that ANP or ANP (4–23) can equipotently block this effect *(27)*. Interestingly, the direct activation of PKC by phorbol ester could not be antagonized by the NP, indicating that ANP is probably acting upstream of diacylglycerol action, particularly when opposing the actions of growth factors, such as endothelin. Recent data suggests that the NP inhibit the G protein-related hydrolysis of PLC, which would be expected to reduce diacylglycerol and other PKC-activator substances stimulated by ET *(26)*. The PKC-independent effects of mito-

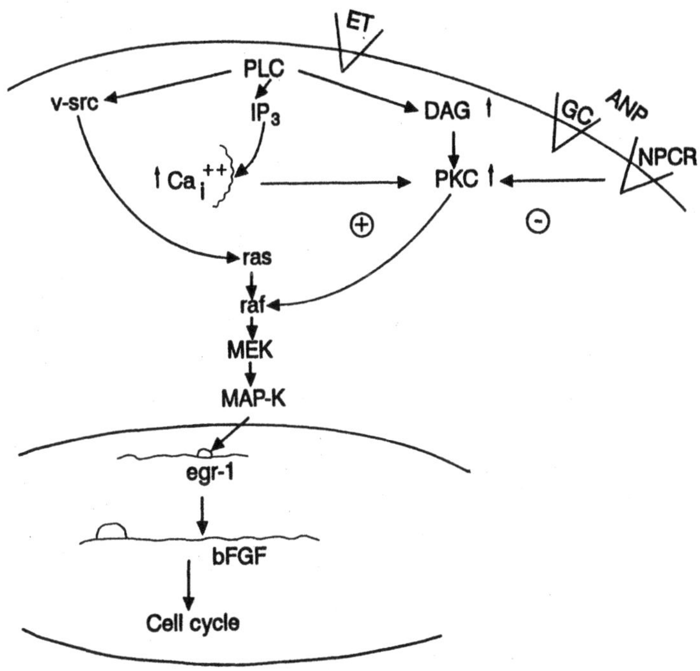


**Fig. 2.** Signaling by growth factor (ET) to the nuclear growth program in astrocytes and inhibition by ANP. PLC is phospholipase C, DAG is diacylglycerol, PKC is protein kinase C, IP3 is inositol tris-phosphate.

gens like ET are not nearly as well studied, but may result from the G-protein-coupled activation of distinct cytosolic tyrosine kinases, like *v-scr (28)*. A cartoon of some of these proximal signaling interactions is seen in Fig. 2.

As a result of activating proximal protein kinase signaling pathways, a series of predominantly cytoplasmic kinase cascades could be enacted to transmit the growth signal through the cyctoplasm to the nucleus. There are currently four well-defined cascades. Recent work suggests that there is crosstalk among several of these systems, and a detailed review is beyond the scope of this chapter. One particular pathway, however, that could be involved in growth factor action is the MAP-kinase pathway *(29)*. As seen in Fig. 2, propagation of this cascade of dual-specificity kinases occurs through each member phosphorylating its substrate and immediate downstream effector protein/enzyme on distinct threonine/serine and tyrosine residues. The signal that links G-protein-coupled receptors (like ET) or classical growth factor receptors that are tyrosine kinases (like PDGF) to cell proliferation is often transmitted through the activation of the *ras* protooncogene *(30)*. This molecule is translocated to the cell membrane by docking proteins, such as grb2 or sos, through physical associations on phosphorylated sarc homology domains (SH2 or SH3) *(31)*. The docking proteins themselves are activated through phosphorylation by the growth factor receptor or nonreceptor tyrosine kinases, such as Src, and Shc can be activated through phosphorylation by Src or through physical association with the growth factor receptor tyrosine kinase domain. This series of phosphorylations ultimately results in the alteration of ras, first translocated, then myristolated at the cell membrane, and then activated in its GTP-bound state. Ras, through association with the family of 14-3-3 proteins and an unidentified kinase activity, can then phosphorylate and activate *c-raf* (MAP kinase kinase kinase) *(32)*. This latter step marks the transition from tyrosine kinases to dual specificity serine/threonine and tyrosine kinase action. Alternatively, it has been shown that PKC can directly phosphorylate and activate c-raf *(33)*, thereby either initiating the cascade independently of ras, or augmenting a ras-mediated signal. Perhaps simply understood, raf then activates MAP kinase kinase (MEK), subsequently leading to MAP kinase activation *(34)*. MAP kinase is the critical transition protein, because it translocates to the nucleus, where it phosphorylates substrate proteins and promotes the transcription of genes that are critical to the growth program *(29)*.

We investigated the potential role of the MAP (erk) kinase signaling system in astrocytes. Several laboratories have published data showing that ET can activate the MAP kinase system in a variety of cell types, including astrocytes *(35–37)*. As mentioned above, this could result from both PKC-dependent and independent pathways. We showed that the NP can significantly inhibit the activation of PKC by ET, thus potentially limiting activation of the MAP kinase cascade through this pathway. The critical role of MAP-kinase as mediating the mitogenic actions of ET has not been shown, however. Recently, we used a dominant-negative MAP-kinase construct to prove this. The Y185F construct has been previously shown to inhibit wild-type MAP-K activity *(38)*, perhaps through titrating off either wild-type MEK or MAP-K substrate proteins in the cell. In astrocytes (and in vascular smooth muscle cells), we showed that this construct caused an approx 60% decrease in ET-stimulated MAP-kinase activity. Furthermore, the dominant-negative MAP-K limited the ability of ET to stimulate DNA synthesis by 65–70%, linking these two events. Based on this stoichiometry, we propose that the MAP-kinase system is the predominant kinase cascade that is responsible for transmitting the ET-stimulated growth signal to the nucleus.

We than assessed the effects of the NP. We found that both ANP and ANP (4–23) caused a significant and equipotent inhibition of MAP-kinase activation by ET and other glial mitogens. The inhibition was 50–75% for the stimulation of MAP-kinase by PDGF, ET, or phorbol ester *(39)*. This finding confirmed more limited and preliminary results from Kasuya et al. *(37)*. Both NPs also potently inhibited the activation of the immediately upstream kinase, MEK, providing evidence that the NPs act at a more proximal level to effect MAP-kinase inhibition. This may result from the previously mentioned inhibition of PLC hydrolysis and PKC activation of raf activity, caused by ET (Fig. 2). It is also possible that the NP stimulated the activity of a MAP-kinase-specific phosphatase, like CL-100/MKP-1 *(40)*, which would result in a decrease of MAP-kinase activity. Recently, Sugimoto et al. showed that ANP (but not C-ANP [4–23]) stimulated the gene expression for CL-100 in mesangial cells *(41)*. However, it has been appreciated that the production of MAP-kinase related phosphatases (even at the protein level) often does not reflect the *activity* of the enzyme. Furthermore, other protein phosphatases, such as 2A (PP2A), and not MKP-1 may be the relevant MAP-K phosphatase, even when MKP-1 gene expression is stimulated by growth factors *(42)*. Therefore, the participation of this enzyme in the NP inhibition of MAP-kinase activity is undetermined at present.

## NATRIURETIC PEPTIDES AS ANTIGROWTH FACTORS: NUCLEAR TARGETS

The next question involved which targets of MAP-kinase in the nucleus play a role in the growth-modulating effects of the vasoactive peptides. When an external growth stimulus triggers a proliferation program in various cells, this results in the production of immediate-early proteins, such as Fos and Jun, which are DNA-binding proteins and transcription factors. In many cases, the immediate-early proteins then stimulate the transcription of autologous growth factors, whose activity is critical to enact the proliferation program. We screened the expression of a variety of immediate-early genes (IEGs) and found that ET stimulated the production of several candidate proteins, including Fos, several Jun family members, Myc, and Egr-1. ANP provided the selectivity, in that only the basal or mitogen-stimulated expression of Egr-1 (also known as Tis 8 in astrocytes *[24]*) was inhibited by ANP. This pointed to Egr-1 as the prime candidate, although other immediate-early proteins could play a role.

From the literature on astrocyte and glioma proliferation, several candidate autologous growth factors were considered. Again, ET stimulated the expression of several of these genes and the production of the related proteins, including PDGF, IGF-1, and bFGF. ANP was again selective in only inhibiting bFGF transcription/protein production. It was then necessary to prove that these proteins were important for the growth program.

Using antisense oligonucleotides (ASO) directed against the translation initiation site for Egr-1 (Tis 8), we showed that ET could not stimulate nuclear thymidine incorporation in the astrocyte when the ASO for Egr-1 was present. In contrast, sense constructs for egr-1, or a c-myc ASO, had no effect on the ET-stimulated proliferative response. The egr-1 ASO also prevented ET from stimulating bFGF transcription, thereby linking the expression of egr-1 to the transcription of the growth factor. In contrast, the ability of ANP to inhibit ET-stimulated egr-1 expression or bFGF expression was augmented by the egr-1 ASO, increased from 35 to 55% inhibition. The egr-1 ASO also comparably augmented ANP inhibition of thymidine incorporation in the glia. This result indicates a mechanism by which ANP restrains growth in these cells (Fig. 3A). To prove a critical role for bFGF, we also used a specific ASO, which inhibited bFGF protein production in response to ET-1 by 65%. The ASO, but not a sense oligonucleotide or lower concentrations of the ASO to bFGF, inhibited ET-stimulated thymidine incorporation by 80%. The ASO also augmented the ANP-inhibition of ET-stimulated DNA synthesis from 40 to 60% (Fig. 3B). An ASO to c-myc had no effect on thymidine incorporation in these cells, and adding exogenous bFGF to the cells overcame the inhibitory action of the ASO. These data provide clear evidence that Egr-1 and b-FGF are critical to the ET-stimulated cell proliferation program in these cells, and that ANP probably works through inhibiting the production of these proteins. Similar sequential steps probably underlie the growth-regulatory events as modulated by the vasoactive peptides in other cells.

We then set out to link the regulation of MAP kinase to the modulation of *egr*-1. It has been recently shown that 1,25 (OH)2 Vit D acts as a mitogen in bone cells to stimulate the expression of egr-1 through a MAP-kinase related pathway *(43)*. Astrocytes were cotransfected with an egr-1/CAT reporter construct, and a MAP-kinase protein expression vector. This resulted in a threefold increase in egr-1/CAT activity, an effect that was inhibited by transfection with the dominant-negative MAP-kinase. This indicates that the *egr*-1 promoter is a substrate for MAP-kinase, which probably acts at SRE proteins like elk-1, and which can potentially mediate the transactivation of the *egr*-1 gene by binding to distinct motifs on the promoter *(44)*. This effect was inhibited by ANP and stimulated by ET-3 (Fig. 4). We recently found that ET-3 activation of *egr*-1 through the MAPK pathway in astrocytes is mediated by SRE elements (E. Biesiada et al., unpublished data), consistent with these data. We then asked how the Egr-1 protein specifically stimulates

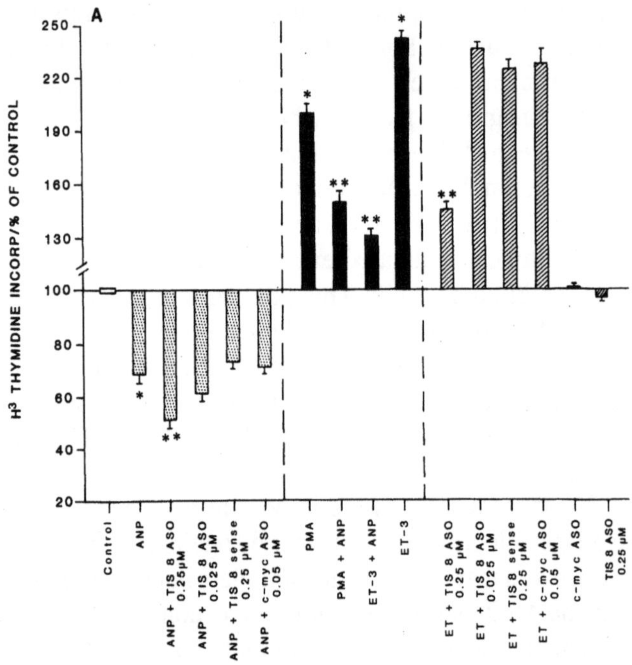


**Fig. 3. (A)** The Tis 8 antisense oligonucleotide augments ANP inhibition and reverses ET stimulation of nuclear thymidine incorporation in astrocytes. *$p < 0.05$ for control vs ANP or ET alone by ANOVA plus Scheffe's test, **$p < 0.05$ for neuropeptide plus Tis 8 ASO vs neuropeptide alone, PMA vs PMA plus ANP, or ET-3 vs ET plus ANP. Reproduced from Hu R-M and Levin ER. Astrocyte growth is regulated by neuropeptides through Tis 8 basic fibroblast growth factor. J Clin Invest 1994,93:1820–1827 by copyright permission of the American Society for Clinical Investigation.

bFGF transcription. We found that there are three noncanonical sites on the bFGF promoter to which Egr-1 can bind. Furthermore, ET stimulated the binding of astrocyte nuclear extract and specifically egr-1 protein, determined by electrophoretic mobility supershifting. ANP inhibited the binding of egr-1 at these sites. By deletional analysis studies, we found that only the site at –160 nucleotides was critical for Egr-1 transactivation of the bFGF promoter (Fig. 5). This provides the molecular frame work to understand how the neuropeptides modulate these important events in astrocyte proliferation.

## NATRIURETIC PEPTIDES AS ANTIGROWTH FACTORS: CELL CYCLE

As a result of MAP kinase transmitting the growth signal to the nucleus, the astrocyte begins the process of cell replication. The coordinated progression of a cell through G1, S (DNA replication), G2, and M (mitosis) phases of the cell cycle results in cell division. In the initial stages, a quiescent cell (Go) begins to replicate the necessary proteins (G1) prior to DNA synthesis. The presence of a growth signal is critical up to a time called the restriction point *(45)*. Prior to this latter phase of G1, the removal of a growth stimulus (growth factor) causes the cell to revert to its quiescent state. After reaching this point, removal of the growth factor has no effect on the irrevocable commitment to cell division, although a second progression factor may be required. Additional time is spent in late G1 on repair of mutated DNA, extending into the early phases of S.

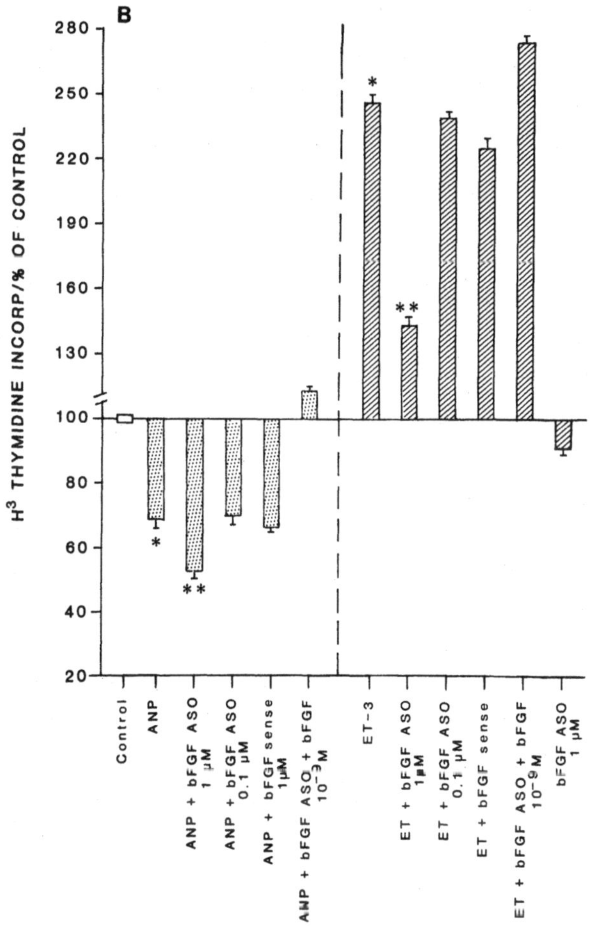

**Fig. 3.** *(continued)* **(B)** An antisense oligonucleotide to bFGF augments ANP-inhibition and inhibits ET stimulation of astrocytic DNA synthesis. $^*p < 0.05$ for control vs ANP or ET alone by ANOVA plus Scheffe's test, $^{**}p < 0.05$ for neuropeptide plus bFGF ASO vs neuropeptide alone. Reproduced from the J Clin Invest 1994,93:1820–1827 by copyright permission of the American Society for Clinical Investigation.

The progression of a cell through the G1 phase is governed by an extraordinary series of regulatory events, involving kinase stimulatory and restraining proteins, modulated in turn by both activating and inactivating phosphorylations (Fig. 6). In a temporal fashion, a family of G1 regulatory proteins, the cyclins, bind and modulate the kinase activity of a family of cell cycle kinases, the cyclin-dependent kinases (CDKs) *(45)*. The D class of cyclins, notably D1 and D3, are first expressed in mid-G1 phase. These proteins bind the Cdk4 and Cdk6 kinases and change the conformation of Cdk, allowing activating phosphorylation by CAKS *(see below)*. Cyclin D1 was identified as the *Prad 1* oncogene, whose activity correlates with the development of some parathyroid and other neoplasms *(47)*. A second important cyclin, cyclin E, is increasingly expressed in late G1, binds to Cdk2, and plays a prominent role in the G1 to S phase transition *(48)*.

Another family of activating kinases, directed against the CDKs, and known as CAKS or cyclin-activating kinases, provides activating phosphorylation of the CDKs on Thr160 *(49)*. However, the activity of the CAKs does not seem to vary much through G1, so it

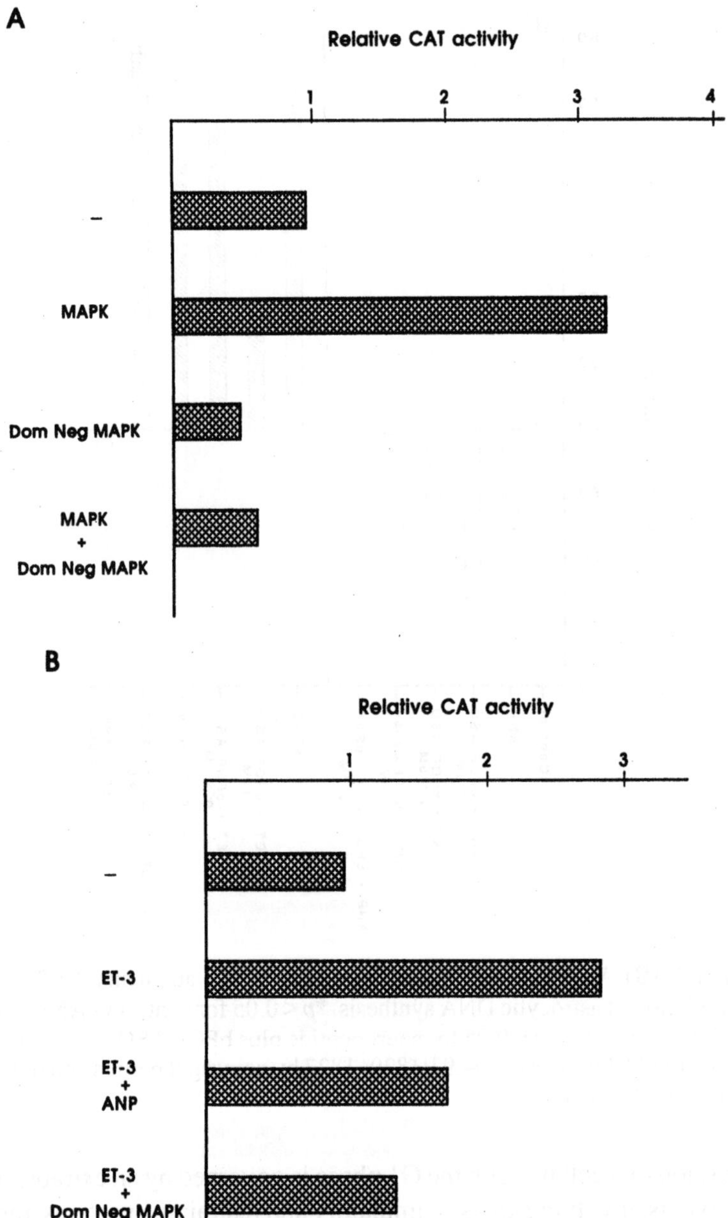


**Fig. 4.** Mitogen-activated protein kinase (MAPK) expressed in astrocytes by transient transfection stimulates egr-1/CAT transcription. The dominant-negative MAPK construct *(Dom Neg MAPK)* and ANP inhibit ET or MAPK-induced egr-/CAT transcription. Each experimental condition bar represents the mean of three experiments, triplicate determinations per experiment.

is not clear how they are in turn regulated to provide activating regulation of CDK activity. The CDKs are also the target for activating dephosphorylation by the Cdc25 family of phosphatases, as a secondary but important event *(50)*.

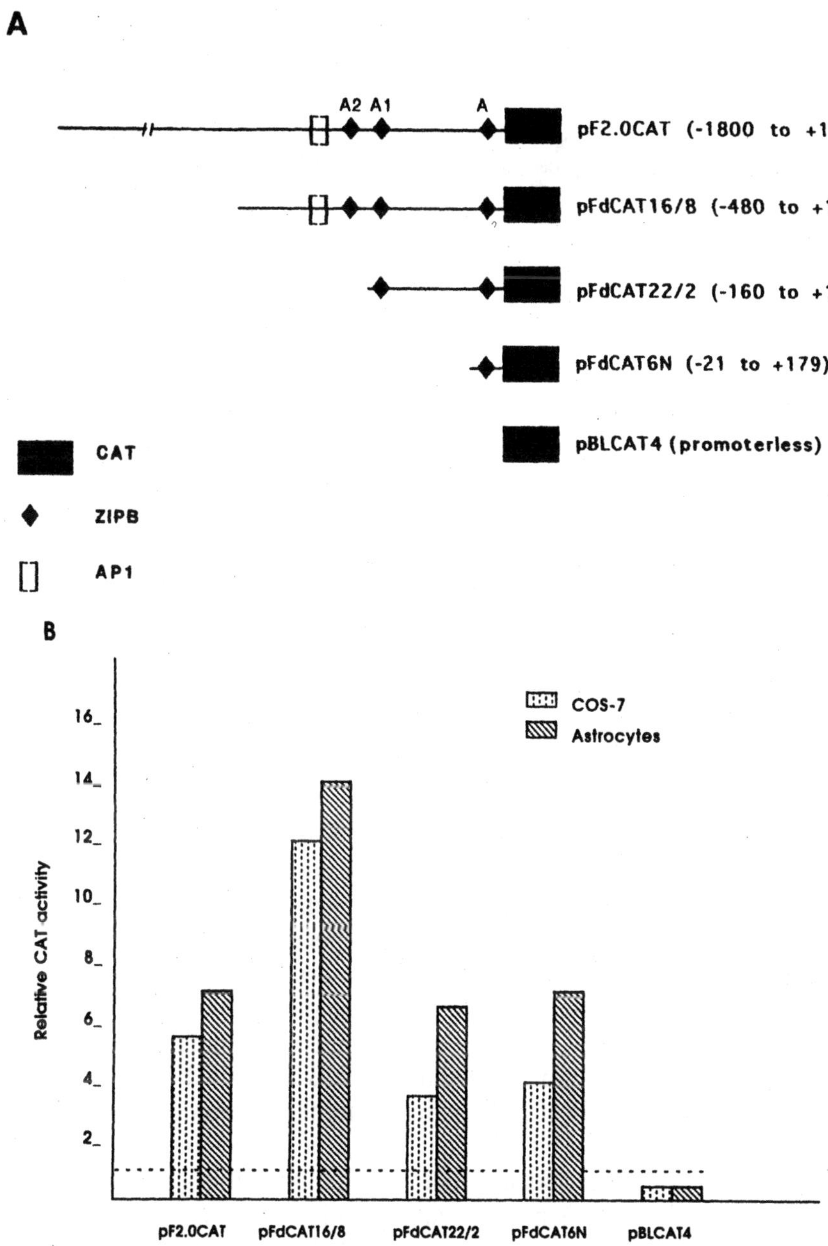


**Fig. 5.** Egr-1 induction of the human bFGF promoter. **(A)** Representation of promoter sequence of basic FGF/CAT and related 5' deletional constructs. The AP1 and ZiPB regions (egr-1 and overlapping Sp1 binding sites) are indicated. **(B)** Relative CAT activity in astrocytes and COS-7 cells. Cells were transfected with the plasmids in the presence or absence of an Egr-1 expression vector (pCMV5-Egr1). Each bar represents the average of triplicate transfections per experiment, repeated three times.

Finally, a family of restraining proteins, the cyclin-dependent kinase inhibitors (or CDKI) have been isolated *(51)*. These proteins include $p^{15}$ and $p^{16}$, which are directed against the Cyclin D/Cdk4 or Cdk6 complexes, and $p^{27}$, which inhibits all cyclin/cdk activations, including the activation of Cyclin E/Cdk2, which occurs late in G1 and is

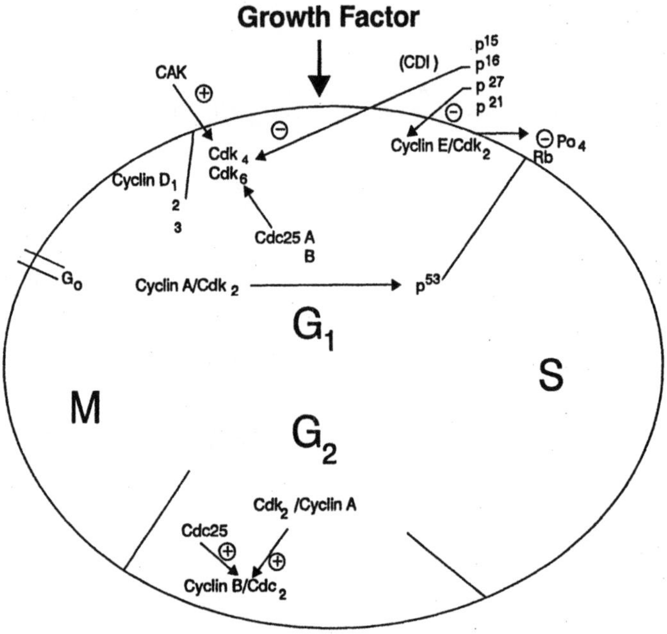


**Fig. 6.** Key proteins responsible for the transition through the phases of the cell cycle. CAK is cyclin-activating kinase, CDI is cyclin-dependent inhibitory protein, Cdk is cyclin-dependent kinase, Rb is retinoblastoma protein.

important for cell-cycle progression *(52,53)*. In addition, $p^{21}$ is an inhibitor of the activity of all known cyclin/Cdk complexes *(54)*. This latter protein is transcriptionally induced by the $p^{53}$ oncogene, and thereby mediates the cell-cycle restraining effects of $p^{53}$. However, $p^{21}$ can also be regulated through p53-independent mechanisms. When $p^{53}$ is mutated, as occurs in many cancers, the ability to transcribe $p^{21}$ is lost, and passage of such cells through G1 to S is increased *(55)*. $p^{21}$ may also play a role in cell senescence and apoptosis. A second protein, $p^{57}$, which is homologous to $p^{21}$, is expressed mainly in the CNS and likely contributes to growth restraint in cells from the brain *(56)*.

We first investigated the role of endothelin to alter the expression or activity of these various proteins. We found that ET stimulated the gene or protein expression of Cyclin D1 and Cycle E over a 24-h period, and that ANP inhibited this stimulated production. This indicates that one mechanism of growth regulation by these vasoactive peptides is at the level of cyclin production. Independently, ANP stimulated the production of the $p^{16}$, $p^{27}$, and $p^{57}$ CDKIs. ET inhibited both the basal and ANP-stimulated production of several of these cell-cycle inhibitory proteins. This led us to hypothesize that ANP might act as an antigrowth factor through the production of these cell-cycle restraining proteins. To prove this, we determined whether ANP inhibited ET-stimulated thymidine incorporation in cultured astrocytes through these proteins. ANP caused a 45% inhibition of ET-stimulated thymidine incorporation, which was reversed 85% by an antisense oligonucleotide (ASO) to $p^{57}$, 63% by an ASO to $p^{16}$, and 33% by an ASO to $p^{27}$. In contrast, lower ASO concentrations for the three proteins or missense oligonucleotides had no effect on ANP actions. These data provide evidence that a second mechanism by which the natriuretic peptides can inhibit cell-cycle progression is through the restraint of Cyclin/Cdk activity, resulting from increased CDKI production. ET may relieve this restraint, thereby promoting cell proliferation. It is established that growth inhibitors, such as TGF-β, inhibit cell-cycle progression by preventing the inactivating phosphorylation of the retinoblastoma protein (Rb) *(53)*, thereby preventing the G1-S transition. This occurs by TGF-β stimulating $p^{27}$ production, which then inhibits Cdk2 E/Cdk2 kinase and cyclin D/Cdk4 activity. Rb in a hypophosphorylated state sequesters transcriptional factors like E2F, preventing the transcription of undefined genes that are critical to cell cycle progression. We propose that ANP acts in part through comparable mechanisms to TGF-β to restrain astrocyte proliferation.

## SUMMARY

The pathways through which the natriuretic peptides can act as antiproliferation factors for a variety of cells often depends on the cell-membrane receptor expressed. Both the GC receptors and NPCR have the capacity to mediate the antigrowth action of the NP, and the signal transduction pathways that lead to antigrowth appear to be initiated at the membrane and extend to the nucleus. A precise understanding of the steps that result in antigrowth will serve as a useful model to compare other growth inhibitors, to perhaps uncover novel mechanisms of growth restraint, and to suggest therapeutic targets for clinical intervention. Finally, these investigations may also provide us with new insights regarding the signaling pathways for the NPCR and GC receptors, pointing to new physiological roles for the NP as mediated through the various classes of transmembrane receptors.

## ACKNOWLEDGEMENT

This work was supported by grants from the Medical Research Service of the Veteran's Administration and the NIH (HL-50161 and NS-30521).

## REFERENCES

1. Appel RG. Growth-regulatory properties of atrial natriuretic factor. Am J Physiol 1992;262: F911–F918.
2. Appel RG. Mechanism of atrial natriuretic factor-induced inhibition of rat mesangial cell mitogenesis. Am J Physiol 1990;259:E312–E318.
3. Itoh H, Pratt RE, Dzau VJ. Atrial natriuretic peptide as a novel antigrowth factor of endothelial cells. Hypertension 1992;19(6):758–761.
4. Itoh H, Pratt RE, Dzau VJ. Atrial natriuretic polypeptide inhibits hypertrophy of vascular smooth muscle cells. J Clin Invest 1990;86(5):1690–1697.
5. Levin ER, Frank HJL. Natriuretic peptides inhibit astroglial proliferation: mediation by C receptor. Am J Physiol 1990;261:R453–R457.
6. Furuya M, Aisaka K, Miyazaki T, Honbou N, Kawishima K, Ohno T, Tanaka S, Minamino M, Kangawa K, Matsuo H. C-type natriuretic peptide inhibits intimal thickening after vascular injury. Biochem Biophys Res Commun 1993;193(1):248–253.
7. Levin ER, Loughlin SE, Kaplan G. Atrial natriuretic peptide secretion from fetal rat diencephalon in culture. J Neuroendocrinol 1990;2(3):317–321.
8. Levin ER, Frank HJL, Gelfand R, Loughlin S, Kaplan G. Natriuretic peptide receptors in cultured rat diencephalon. J Biol Chem 1990;265(17):10,019–10,024.
9. Levin ER, Frank HJL, Pedram A. Endothelin receptor expression in fetal rat diencephalon. J Neurochem 1992;58:659–666.
10. Brown J, Czarnecki A. Distribution of atrial natriuretic peptide receptor subtypes in rat brain. Am J Physiol 1990;258:R1078–R1083.
11. Konrad EM, Thibault G, Schriffin EL, Cantin M. Atrial natriuretic factor receptor subtypes in the rat central nervous system. Hypertension 1991;17:1144–1151.

12. Garg UC, Hassid A. Nitric oxide-generating vasodilators and 8-bromo-cyclic guanosine monophosphate inhibit mitogenesis and proliferation of cultured rat vascular smooth muscle cells. J Clin Invest 1989;83:1774–1777.
13. Prins B, Hu R-M, Nazario B, Pedram A, Frank HJL, Weber M, Levin ER. Prostaglandin $E_2$ and prostacyclin inhibit the production and secretion of endothelin from cultured endothelial cells. J Biol Chem 1994;269(16):11,938–11,944.
14. Porter JG, Catalano R, McEnroe G, Lewicki JA, Protter AA. C-type natriuretic peptide inhibits growth factor-dependent DNA synthesis in smooth muscle cells. Am J Physiol 1992;263:C1001–C1006.
15. Anand-Srivastava MB, Srivastava AK, Cantin M. Pertussis toxin attenuates atrial natriuretic factor-mediated inhibition of adenyl cyclase. J Biol Chem 1987;262(11):4931–4934.
16. Anand-Srivastava MB, Sairam MR, Cantin M. Ring-deleted analogs of atrial natriuretic factor inhibit adenylate cyclase/cAMP system. J Biol Chem 1990;265(15):8526–8572.
17. Drewett JG, Ziegler RJ, Trachte GJ. Neuromodulatory effects of atrial natriuretic peptides correlate with an inhibition of adenylate cyclase but not an activation of guanylate cyclase. J Pharm Exp Ther 1992;260(2):689–696.
18. Wu J, Dent P, Jelinek T, Wolfman A, Weber MJ, Sturgill TW. Inhibition of the EGF-activated MAP kinase signalling pathway by adenosine 3',5'-monophosphate. Science 1993;262:1065–1069.
19. Francis SH, Corbin JD. Progress in understanding the mechanism and function of cyclic GMP-dependent protein kinase. Adv Pharmacol 1994;26:115–170.
20. Hirata M, Kohse KP, Chang CH, Ikebe T, Murad F. Mechanism of cyclic GMP inhibition of insositol phosphate formation in rat aortic segments and cultured bovine aortic smooth muscle cells. J Biol Chem 1990;265:1268–1273.
21. Lincoln TM. Cyclic GMP and mechanisms of vasodilation. Pharmacol Ther 1989;41:479–502.
22. Cornwell TL, Lincoln TM. Regulation of intracellular calcium levels in cultured vascular smooth muscle cells. Reduction of calcium by atriopeptin and 8-bromo-cyclic GMP is mediated by cyclic-GMP-dependent protein kinase. J Biol Chem 1989;264:1146–1155.
23. Cornwell TL, Arnold E, Boerth NJ, Lincoln TM. Inhibition of smooth muscle cell growth by nitric oxide and activation of cAMP-dependent protein kinase by cGMP. Am J Physiol 1994;267:C1405–C1413.
24. Arenander AT, Lim RW, Varum BC, Cole R, de Vellis J, Herschman HR. TIS gene expression in cultured astrocytes: multiple pathways of induction by mitogens. J Neurosci Res 1989;23(3):257–265.
25. Simonson MS, Herman WH. Protein kinase C and protein tyrosine kinase activity contribute to mitogenic signalling by endothelin-1. J Biol Chem 1993;268:9347–9357.
26. Levin ER. Endothelins. New Engl J Med 1995;333(6):356–363.
27. Prins B, Hu R-M, Levin ER. Unpublished observations.
28. Simonson MS, Wang, Y, Herman WH. Nuclear signaling by endothelin-1 requires Src protein tyrosine kinases. J Biol Chem 1996;271(1):77–82.
29. Davis RJ. The mitogen-activated protein kinase signal transduction pathway. J Biol Chem 1993; 268(20):14,553–14,556.
30. Knall C, Young S, Nick JA, Buhl AM, Worthen GS, Johnson GL. Interleukin-8 regulation of the ras/raf/mitogen-activated protein kinase pathway in human neutrophils. J Biol Chem 1996;271(5):2832–2838.
31. Yajnik V, Blaikie P, Bork P, Margolis B. Identification of residues within the SHC phosphotyrosine binding/phosphotyrosine interaction domain crucial for phosphopeptide interaction. J Biol Chem 1996;271(4):1813–1816.
32. Marais R, Light Y, Paterson HF, Marshall CJ. Ras recruits raf-1 to the plasma membrane for activation by tyrosine phosphorylation. EMBO J 1995;14(13):3136–3145.
33. Kolch W, Heidecker G, Kochs G, Hummel R, Vahidi H, Mischak H, Finkenzeller G, Marme D, Rapp UR. Protein kinase C alpha activates RAF-1 by direct phosphorylation. Nature 1993;364(6434):249–252.
34. Marshall CJ. Map kinase kinase kinase, MAP kinase kinase and MAP kinase. Curr Opin Genet Dev 1994;4:82–89.
35. Koide M, Kawahara Y, Tsuda T, Ishida Y, Shii K, Yokoyama M. Endothelin-1 stimulates tyrosine phosphorylation and the activities of two mitogen-activated protein kinases in cultured vascular smooth muscle cells. J Hyperten 1992;10:1173–1182.
36. Wang Y, Rose PM, Webb ML, Dunn MJ. Endothelins stimulate mitogen-activated protein kinase cascade through either ETA or ETB. Am J Physiol 1994;267:C1130–C1135.
37. Kasuya Y, Abe Y, Hama H, Takeshi S, Asada S, Masaki T, Goto K. Endothelin-1 activates mitogen-activated protein kinase through two independent signalling pathways in rat astrocytes. Biochem Biophys Res Commun 1994;204(3):1325–1333.

38. Robbins, DJ, Zhen E, Owaki H, Vanderbilt CA, Ebert D, Geppert TD, Cobb MH. Regulation and properties of extracellular signal-regulated protein kinases 1 and 2 in vitro. J Biol Chem 1993; 268(7):5097–5106.
39. Prins B, Weber MJ, Hu R-M, Pedram A, Daniels M, Levin, ER. Atrial natriuretic peptide inhibits mitogen-activated protein kinase through the clearance receptor: potential role in the inhibition of astrocyte proliferation. J Biol Chem 1996;271(24):14,156–14,162.
40. Sun H, Charles CH, Lau LF, Tonks NK. MKP-1, an immediate early gene product, is a dual specificity phosphatase that dephosphorylates MAP kinase in vivo. Cell 1993;75:487–493.
41. Sugimoto T, Haneda M, Togawa M, Isono M, Shikano T, Araki S-I, Nagakawa T, Kashiwagi A, Guan K-L, Kikkawa R. Atrial natriuretic peptide induces the expression of MKP-1, a mitogen-activated protein kinase phosphatase, in glomerular mesangial cells. J Biol Chem 1996;271(1):544–547.
42. Alessi DR, Gomez N, Moorhead G, Lewis T, Keyse SM, Cohen P. Inactivation of p42 MAP kinase by protein phosphatase 2A and a protein tyrosine phosphatase, but not CL 100, in various cell lines. Curr Biol 1995;5(3):283–295.
43. Beno DW, Brady LM, Bissonnette M, Davis BB. Protein kinase C and mitogen-activated protein kinase are required for 1,25-dihydroxyvitamin D3-stimulated Egr induction. J Biol Chem 1995;270:3642–3647.
44. Sharrocks AD. ERK2/p42 MAP kinase stimulates both autonomous and SRF-dependent DNA binding by Elk-1. FEBS Lett 1995;368(1):77–80.
45. Dou QP, Levin AH, Zhao S, Pardee AB. Cyclin E and cyclin A as candidates for the restriction point protein. Can Res 1993;53(7):1493–1497.
46. Sherr CJ. G1 phase progression: cycling on cue. Cell 1994;79:551–555.
47. Arnold A. The cyclin D1/Prad 1 oncogene in human neoplasia. J Invest Med 1995;43(6):543–549.
48. Dulic V, Drullinger LF, Lees E, Reed SI. Association of human cyclin E with a periodic G1-S phase protein kinase. Science 1992;257:1958–1961.
49. Yee A, Wu L, Liu L, Kobayashi R, Xiong Y, Hall FL. Biochemical characterization of the human cyclin-dependent protein kinase activating kinase. J Biol Chem 1996;271(1):471–477.
50. Jinno S, Suto K, Nagata A, Igarashi M, Kanaoka Y, Nojima H, Okayama H. Cdc25A is a novel phosphatase functioning early in the cell cycle. EMBO J 1994;13(7):1549–1556.
51. Harper JW, Adami GR, Wei N, Keyomarsi K, Elledge SJ. The $p^{21}$ cdk-interacting protein Cip 1 is a potent inhibitor of G1 cyclin-dependent kinases. Cell 1993;75:805–816.
52. Hannon GJ, Beach D. $p^{15INK4B}$ is a potential effector of TGF-b-induced cell cycle arrest. Nature (Lond) 1993;371:257–261.
53. Massague J, Polyak K. Mammalian antiproliferative signals and their targets. Curr Opin Genet Devel 1995;5(1):91–96.
54. Xiong Y, Hannon GJ, Zhang H, Casso D, Kobayashi R, Beach D. p21 is a universal inhibitor of cyclin kinases. Nature 1993;366:701–704.
55. Yin Y, Tainsky MA, Bischoff FZ, Strong LC, Wahl GM. Wild-type p53 restores cell cycle control and inhibits gene amplification in cells with mutant p53 alleles. Cell 1992;70:937–948.
56. Lee MH, Reynisdottir I, Massague J. Cloning of p57KIP2, a cyclin-dependent kinase inhibitor with unique domain structure and tissue distribution. Gen Devel 1995;9(6):639–649.



# 14 Natriuretic Peptides and Salt Sensitivity

*Daniel Villarreal and Ronald H. Freeman*

*CONTENTS*



## HISTORICAL CONCEPTS OF NATRIURETIC PEPTIDES

Normal humans and animals excrete appropriate quantities of salt in their urine to maintain homeostasis of body fluid volume and blood pressure. Several complex physiological systems converge on the kidneys to alter renal salt excretion, including one that is mediated by a family of cardiac natriuretic peptides *(1)*. During sodium surfeit and attendant volume expansion, secretion of these cardiac natriuretic peptides is enhanced to interact with receptors in the kidney, the vasculature, and the adrenal glands to promote salt excretion and lower blood pressure. The direct renal actions of the cardiac peptides to promote natriuresis and diuresis are effected through increments of glomerular filtration and reductions in tubular sodium reabsorption *(1–3)*.

In 1956, Kisch *(4)* published the first description of granules located within the cardiac atria. In 1964, Jamieson and Palade *(5)* confirmed the observations made by Kisch and suggested that the atrial granules had all the characteristics of secretory granules found in neuroendocrine cells. Fifteen years later, deBold *(6)* showed that the granularity of atrial tissue was altered by changes in water and electrolyte balance. This observation led directly to studies in 1981 by deBold and colleagues *(7)* that demonstrated a profound natriuresis and diuresis elicited by an iv injection into rats of atrial homogenates despite a marked reduction in blood pressure. Shortly thereafter, atrial natriuretic peptides had been localized to the atrial granules, purified, sequenced, and synthesized *(8,9)*.

From: *Contemporary Endocrinology: Natriuretic Peptides in Health and Disease*
Edited by: W. K. Samson and E. R. Levin Humana Press Inc., Totowa, NJ

## ENDOCRINE CARDIO-RENAL INTEGRATION FOR SODIUM BALANCE

Atrial natriuretic peptide (ANP) functions as a classic hormone, integrating cardio-renal function for sodium homeostasis. In response to an increase in atrial stretch *(10)*, ANP is released into the circulation and delivered to the kidney to exert its biological effects appropriate to the stimulus, i.e., to increase excretion of salt and fluids, thereby reducing intravascular volume and atrial distention.

The renal effects of ANP are varied and include inhibition of tubular sodium reabsorption in the cortical and inner medullary collecting duct system, an increase in the glomerular filtration rate, and inhibition of renin release *(1–3)*. The most unique renal effect of ANP is the marked increments of glomerular filtration rate (GFR) that can occur even when renal perfusion pressure is reduced *(2,11)*. The apparent mechanism for ANP-induced increases in GFR appears to involve efferent arteriolar vasoconstriction, afferent arteriolar vasodilation, and increased glomerular capillary permeability *(2,11)*. The direct tubular effects of ANP appear to involve inhibition of luminal amiloride-sensitive sodium channels located in the inner medullary collection duct *(3,11)*. This results in reduced cellular entry of sodium and limits its access to be actively transported across the basolateral membrane of these cells. These two renal effects of ANP to increase GFR and to inhibit sodium reabsorption in the inner medullary collecting duct appear to be the major mechanisms by which ANP can enhance renal natriuresis and diuresis during states of sodium surfeit.

Pathophysiological derangements of this cardio-renal axis may contribute to several salt-sensitive cardiovascular pathologies, including hypertension *(12)* and decompensated congestive heart failure *(1)*. This concept is illustrated by recent studies by John and colleagues *(12)* in which mice with a disruption of the pro-ANP gene were generated. Homozygous mutants had no circulating or atrial ANP, and their blood pressures were elevated when they were fed standard (0.5% NaCl) and intermediate (2% NaCl) salt diets. On standard salt diets, the heterozygotes had normal levels of plasma ANP and were normotensive, but they became hypertensive on high (8% NaCl) salt intake. Thus, it appears that animals with genetically reduced production of ANP develop salt-sensitive arterial hypertension *(12)*.

The syndrome of congestive heart failure is characterized by extreme salt sensitivity *(1)*. In the early compensated phase of experimental heart failure, increased atrial distention stimulates secretion of ANP, which then serves an important compensatory function to attenuate sodium retention. As cardiac systolic performance continues to deteriorate over time, potent sodium-retaining mechanisms are activated and begin to dominate the compensatory actions of ANP, with a consequent enhancement of systemic congestion and edema formation. Indeed, with the progression of heart failure, underperfusion of the renal vasculature leads to activation of the renin–angiotensin–aldosterone axis (RAAS) *(1)*, and also attenuates the renal natriuretic actions of ANP. The previously balanced interaction between ANP and renin–aldosterone axis becomes disturbed, and sodium homeostasis becomes pathologically altered to produce a marked salt-sensitive pathological state.

## ROLE OF ANP AND THE RAAS IN BODY FLUID HOMEOSTASIS

### *Introduction*

The normal regulation of sodium balance, and hence extracellular fluid volume, involves the complex interaction of extrinsic neurohumoral and intrinsic renal mechanisms that function to produce appropriate adjustments of salt and water excretion commensurate with dietary intake. One major feedback endocrine system known to contribute significantly to sodium homeostasis is the renin–angiotensin–aldosterone axis. This system is activated to promote sodium conservation under conditions of sodium deficit, volume contraction, or reductions in renal perfusion pressure *(13,14)*. However, the relevance of this mechanism may be diminished during sodium surfeit and volume expansion, when other systems are recruited to enhance sodium excretion and restore external balance *(7,15–19)*. In the last 15 yr, evidence has accumulated to suggest that an important compensatory adjustment to increase the renal elimination of sodium involves the secretion of ANP *(7,15,16,18,19)*. The primary stimulus for the release of this peptide hormone is cardiac distention *(7,9,10,20,21)*, and it has direct and potent renal natriuretic and diuretic actions that are effected through reductions in tubular sodium reabsorption and increases in glomerular filtration *(2,3)*. Additionally, ANP indirectly accentuates natriuresis and diuresis via inhibition of the renin–aldosterone axis *(9,20–26)*. The biologic effects of ANP are exerted through receptor-mediated processes coupled positively to particulate guanylyl cyclase *(20,21,28–30)*, and in the kidney, it decreases tubular sodium reabsorption by inhibiting a cation channel in the apical membrane via phosphorylation-dependent and phosphorylation-independent mechanisms *(28)*. Thus, the available information is supportive of a substantial role of ANP for the control of sodium excretion in a physiologic setting of body-fluid volume expansion appropriate to stimulate its secretion and facilitate its natriuretic expression.

### *ANP and the RAAS in Physiological and Pathophysiological Conditions*

The normal daily control of sodium balance can be considered to consist of two phases: a postprandial phase involved with the elimination of excess ingested sodium, and a postabsorptive phase concerned with the tubular reabsorption of the filtered load of sodium necessary to prevent the dissipation of sodium stores *(31)*. In this context, the characteristics of ANP and the renin–aldosterone axis suggest a temporal integration of the two systems to exert their sodium-regulatory actions differentially in the postprandial and postabsorptive phases, respectively. Studies to examine the relevance of these concepts have been conducted during the normal physiology of intact dogs and in the pathophysiology of dogs with compensated high-output heart failure (HOHF) produced by a large infrarenal aorto-caval (AV) fistula *(31)*. Similar to patients with early asymptomatic ventricular dysfunction *(32)*, the compensated state of this experimental model is characterized by normal sodium balance in the presence of sustained elevations of cardiac filling pressures and circulating ANP, and the concomitant normalization of the plasma levels of renin and aldosterone *(31,33,34)*.

Figures 1 and 2 illustrate the postprandial responses to an identical oral load of sodium in normal dogs and in dogs with an AV fistula and HOHF of approx 4 wk duration *(31,35)*. The two groups of dogs exhibited temporally related increases in atrial filling pressures, plasma ANP, and urinary sodium excretion, without modifications of plasma renin and aldosterone *(31,35)*. It is of interest to observe that although postprandial sodium excretion was qualitatively and quantitatively very similar in the two group of dogs (Figs. 1 and 2), this renal excretory response was achieved at absolute plasma ANP elevations 10–15-fold greater in the AV fistula dogs than in the normal animals (Fig. 3). This augmented ANP release in the dogs with compensated HOHF is probably reflective of its heightened cardiac biosynthetic and secretory activity during chronic elevations in cardiac filling

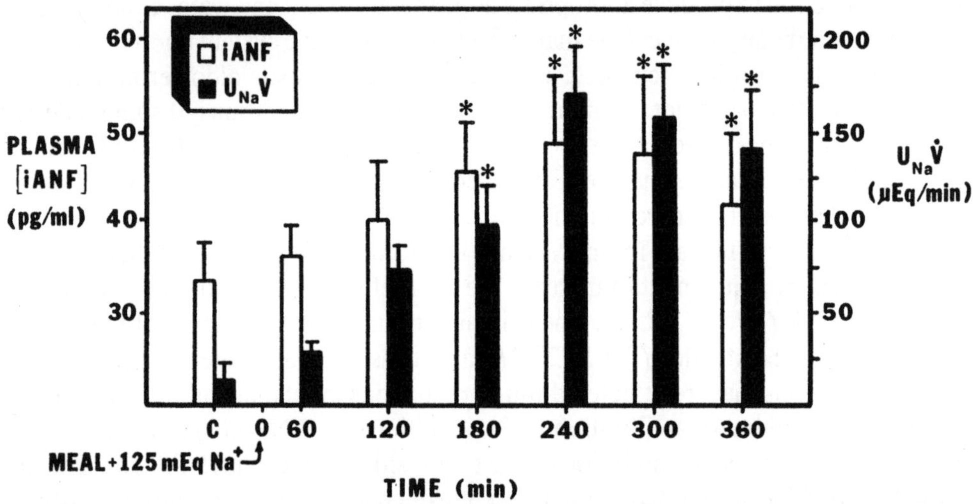


**Fig. 1.** Effects of a high-sodium meal in normal conscious dogs. Values are expressed as mean ± SEM; $n = 5$ dogs. *C*, Control period before meal; [iANF], concentration of immunoreactive atrial natriuretic factor; $U_{Na}V$, urinary sodium excretion. *$p < 0.05$ vs C. (From Verburg KM, Freeman RH, Davis JO, Villarreal D, Vari RC. Am J Physiol 1986;251:R947–R956, reproduced by permission of the American Physiological Society.)

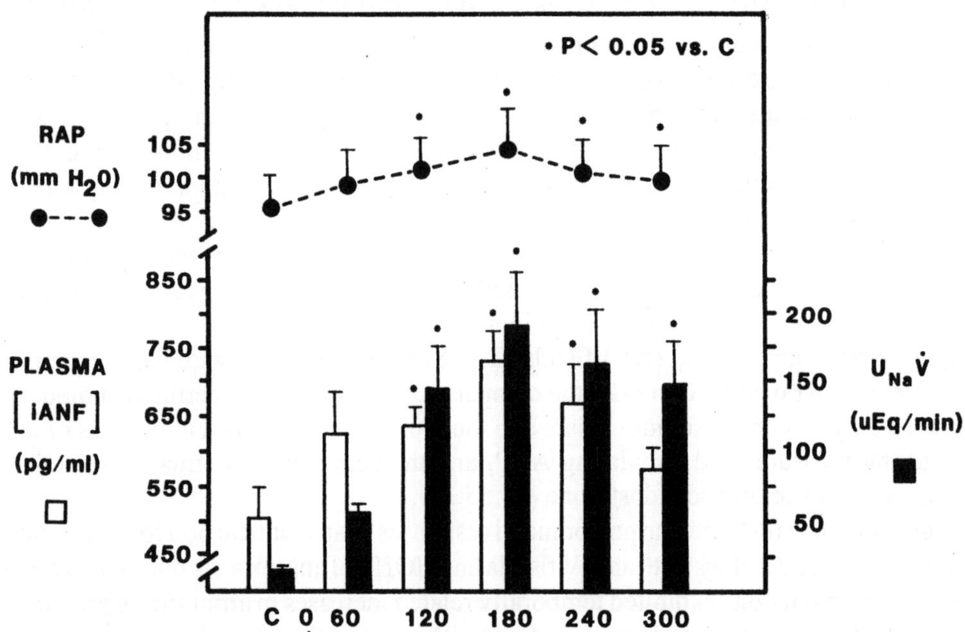


**Fig. 2.** Effects of a high-sodium meal in conscious dogs with an AV fistula and chronic compensated HOHF. Values are expressed as mean ± SEM; $n = 5$ dogs. Abbreviations as in Fig. 1, except for *RAP*, right atrial pressure. *$p < 0.05$ vs C. (From Villarreal D, Freeman RH, Brands MW. Am J Physiol 1990;258:R232–R239, reproduced by permission of the American Physiological Society.)

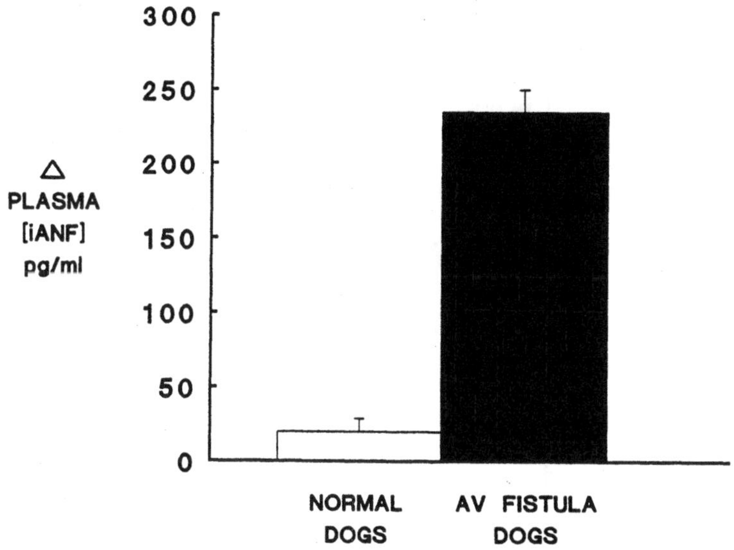


**Fig. 3.** Peak responses in plasma [iANF] after a high-salt meal in normal dogs ($n = 5$) and in dogs with an AV fistula and chronic compensated HOHF ($n = 5$). Values are expressed as mean ± SEM. Abbreviations as in Fig. 1. (From Villarreal D and Freeman RH. J Lab Clin Med 1991;118:515–522, reproduced by permission of Mosby-Year Book, Inc.)

pressures *(31)*. Under these conditions, it is therefore possible to suggest that the ANP endocrine system is sodium-volume sensitive, and can reset to perform at a high level of activity that is physiologically appropriate to achieve a postprandial natriuresis and maintain sodium balance in compensated heart failure when the renin–aldosterone axis is not activated. It is relevant to point out that the postprandial volume expansion and the increased renal hemodynamic function that occur concurrently with a high-salt protein meal *(36–38)* are important modulatory factors for the natriuretic actions of the cardiac peptides. Studies in rats *(39)* and dogs *(40)* have indicated that ANP acts synergistically with volume expansion to exert marked natriuretic effects. Indeed, ANP infusion studies in groups of normal and AV-fistula dogs have demonstrated that increases in ANP levels without a concomitant intravascular volume expansion produced a substantially lower natriuresis than after the ingestion of a high-salt meal *(31,35)*. Thus, the importance of ANP as a regulatory system should not be considered simply from absolute changes in its circulating levels. Rather, it appears that the full natriuretic effects of the cardiac peptides are achieved in an appropriate physiologic environment involving intravascular volume expansion and augmented renal hemodynamic function *(19,31,35,39,40)*.

In the postabsorptive phase, the major physiological renal requirement becomes the necessary reabsorption of the filtered load of sodium to minimize its loss. The hormonal adjustments that are needed to maintain sodium homeostasis under postabsorptive conditions were examined in normal dogs and in compensated AV-fistula dogs of 3–4 wk duration that were fed normal and low-salt diets (Fig. 4). During the normal-sodium diet of 40 mEq/d, both groups of animals exhibited normal postabsorptive levels of renin and aldosterone, but sodium balance was achieved in the AV-fistula dogs in the presence of a fourfold elevation in plasma ANP compared to normal animals *(31)*. After 2 wk of sodium restriction to < 6 mEq/d, the reduced excretion of sodium and the marked stimulation of renin and aldosterone were similar in the two subsets of dogs, but plasma ANP

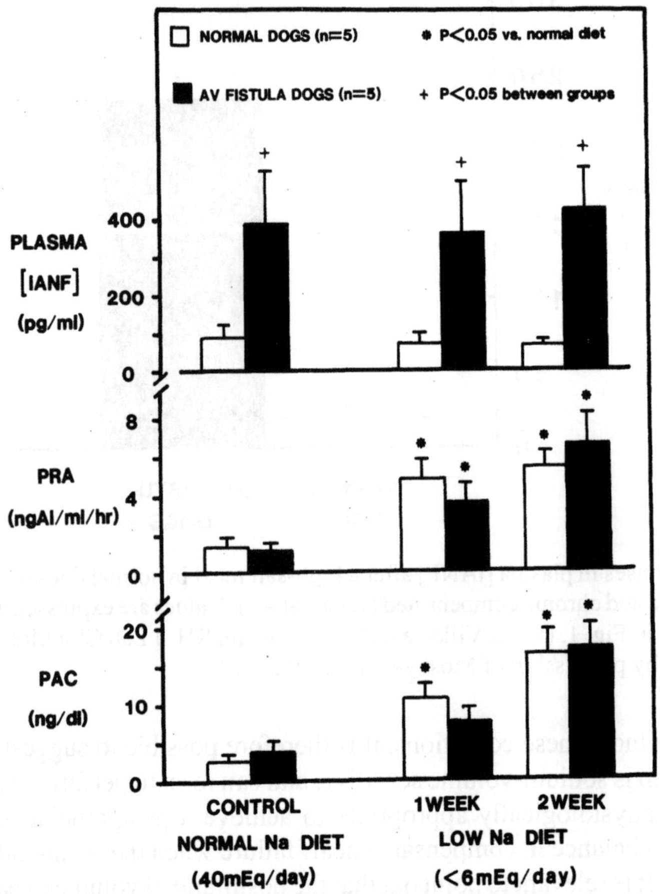


**Fig. 4.** Effects of normal and low sodium *(Na)* diets in normal dogs and in dogs with an AV fistula and chronic compensated HOHF. Values are expressed as mean ( SEM; normal dogs, $n = 5$; AV-fistula dogs, $n = 5$. *PRA*, Plasma renin activity; *PAC*, plasma aldosterone concentration. Other abbreviations as in Fig. 1. *$p < 0.05$ vs control (sodium replete); +$p < 0.05$ vs corresponding values in normal dogs. (From Villarreal D, Freeman RH. J Lab Clin Med 1991;118:515–522, reproduced by permission of Mosby-Year Book, Inc.)

did not change significantly in either group (Fig. 4). These results were interpreted to suggest that the postabsorptive levels of renin and aldosterone, but not ANP, were closely related to chronic levels of sodium intake and sodium conservation. Taken together, the observations of the acute feeding and chronic sodium restriction studies *(31,35)* are consistent with the concept of a temporal hormonal integration for sodium balance, in which ANP exerts its actions to eliminate sodium in the postprandial volume-expanded phase, whereas the renin–aldosterone system prevents the dissipation of sodium stores in the postabsorptive retaining phase. It is of particular interest that this hormonal integration continues to be operational in compensated HOHF, but in this case the ANP endocrine system functions at a high level that appears appropriate for maintaining normal postprandial sodium excretion *(31)*.

In human subjects, the relevance of these considerations is not clear and requires further investigation. Studies in patients with heart failure that would examine the postprandial ANP responses to high sodium intake are difficult to perform; however, the experiments in normal humans that have evaluated the postprandial changes in ANP are conflictive. Although most of the results by different investigators have demonstrated elevations in plasma ANP and sodium excretion after acute or chronic increases in dietary sodium *(39–43)*, a natriuresis without detectable increments in circulating cardiac peptides has also been reported *(44–46)*. The reasons for these discrepant findings are unclear, but may be related, at least in part, to differences in methods and protocols used.

### *Longitudinal Interactions of ANP and the RAAS in Heart Failure*

Independent of their primary renal effects, ANP and the renin–aldosterone axis can have modulatory interactions *(20,21)* that may become most important during pathophysiological situations that activate both systems *(33,34,47,48)*. These interactions, which can occur at several levels of organization, ultimately have an impact on the chronic regulation of sodium balance and systemic hemodynamics. The mechanisms by which ANP can acutely inhibit renin secretion are likely related to its physiological interaction with both the intrarenal vascular baroreceptor and the macula densa mechanisms *(25)*. The suppression of aldosterone results partly from the decrease in plasma renin and hence angiotensin II *(20,24,27)*, as well as from a direct inhibitory action of ANP on the adrenal glomerulosa cells *(20,27)*.

The importance of the long-term inhibitory actions of ANP on the renin–aldosterone system has been the subject of intense research since early studies in humans indicated that plasma ANP levels are markedly elevated in patients with cardiac dysfunction *(49,50)*. Experimental animal preparations have been extremely useful to help characterize the longitudinal alterations of the two hormonal systems during the initiation and progression of heart failure. Studies in the AV-fistula dog model that evaluated the temporal pattern in the secretion of ANP and the renin–aldosterone axis, as well as their relationship to cardiovascular hemodynamics and sodium balance, are shown in Fig. 5 *(33)*. In the initial 5 d postfistula, the animals retained sodium markedly in association with significant increases in plasma renin and aldosterone, but only modest elevations occurred in right atrial pressure and plasma ANP. From d 6–14, the dogs increased their sodium excretion and approached sodium balance. The natriuretic response was accompanied by progressive and significant increases in plasma ANP and with the return of renin and aldosterone levels to baseline, despite a sustained 10 mmHg reduction in mean arterial pressure. Subsequent studies have demonstrated that this renal, hormonal, and circulatory stabilization persists for at least 1 mo after creation of the AV fistula *(34)*. Interestingly, initial data in patients with minimally symptomatic or asymptomatic heart failure suggests a hormonal response pattern similar to that seen in compensated AV-fistula dogs *(32)*. Thus, the longitudinal observations in this canine experimental model suggest that during the early stages of mild heart failure, chronic compensatory hormonal adjustments include the increased secretion of ANP, which functions to restore sodium balance, presumably via the tonic suppression of renin and aldosterone and through its direct renal effects for the promotion of a natriuresis. This proposed tonic inhibition of ANP on the renin–aldosterone axis has also been suggested to occur during early and mild heart failure of rats with ventricular infarction *(48)* and of dogs with rapid ventricular pacing *(47)*. In this latter preparation, these concepts have been recently confirmed during the acute administration of specific ANP receptor antagonists *(51)*. Moreover, long-term studies in the dog model of thoracic inferior vena cava constriction have documented parallel decreases in central venous pressure, plasma ANP, and urinary

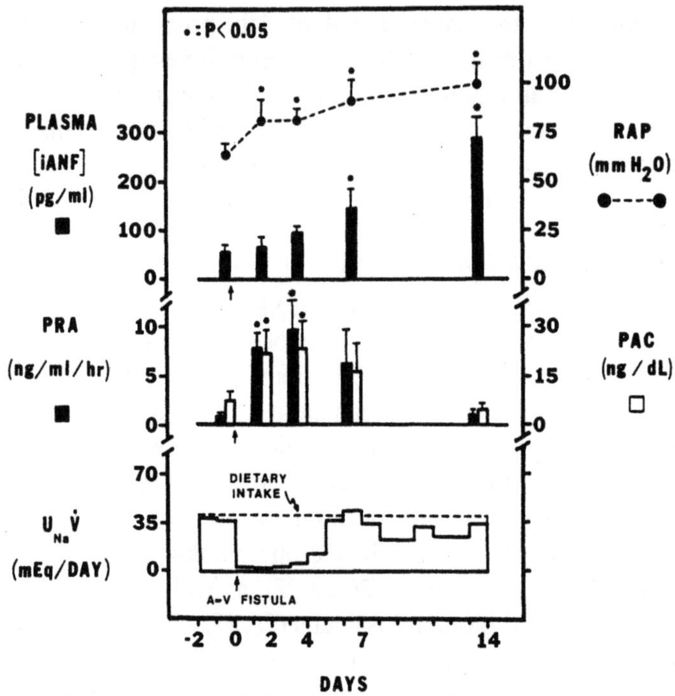


**Fig. 5.** Effects of an AV fistula in conscious dogs. Values are expressed as mean ± SEM; $n = 6$ dogs, except d 8–14, $n = 5$ dogs. Abbreviations as in Figs. 1, 2, and 4. $^*p < 0.05$ vs control values before AV fistula. (From Villarreal D, Freeman RH, Davis JO, Verburg KM, Vari RC. Am J Physiol 1987;252:h692–h696, reproduced by permission of the American Physiological Society.)

sodium excretion in association with the stimulation of plasma renin and aldosterone *(52)*. These observations have been interpreted to suggest that a sustained deficit in ANP may facilitate the chronic activation of the renin–aldosterone axis and sodium retention *(52)*. Indeed, acute infusions of synthetic ANP in the caval dogs have been shown to produce significant decreases in both plasma renin and aldosterone levels *(26)*.

With the progression of heart failure, however, the balanced interaction between ANP and the renin–angiotensin–aldosterone system becomes disturbed, and sodium hemostasis becomes pathologically altered. Figure 6 shows the observations in one AV-fistula dog that developed a rapid cardiac decompensation and is highly illustrative of the hormonal responses during progressive heart failure *(34)*. After the fourth week following the AV fistula in this animal, right atrial pressure rapidly increased over a 2-wk period concurrent with a 10–20 mmHg reduction in mean arterial pressure. The marked elevation in cardiac preload was associated with striking further elevations in plasma ANP to values >1000 pg/mL. Despite of this enhanced ANP secretion, renin and aldosterone also increased >20-fold, and marked sodium retention occurred with formation of peripheral edema and ascites. Previous clinical and animal studies have indicated that a sufficient deterioration in cardiac performance leading to large and chronic reduction in renal perfusion pressure can produce marked activation of the renin–angiotensin–aldosterone axis *(13,53)*. The data from the AV-fistula dogs with decompensated HOHF and similar long-term observations in the rapid ventricular pacing model *(54,55)* suggest that under these conditions, the excessive stimulation of the renal–adrenal hormonal axis can

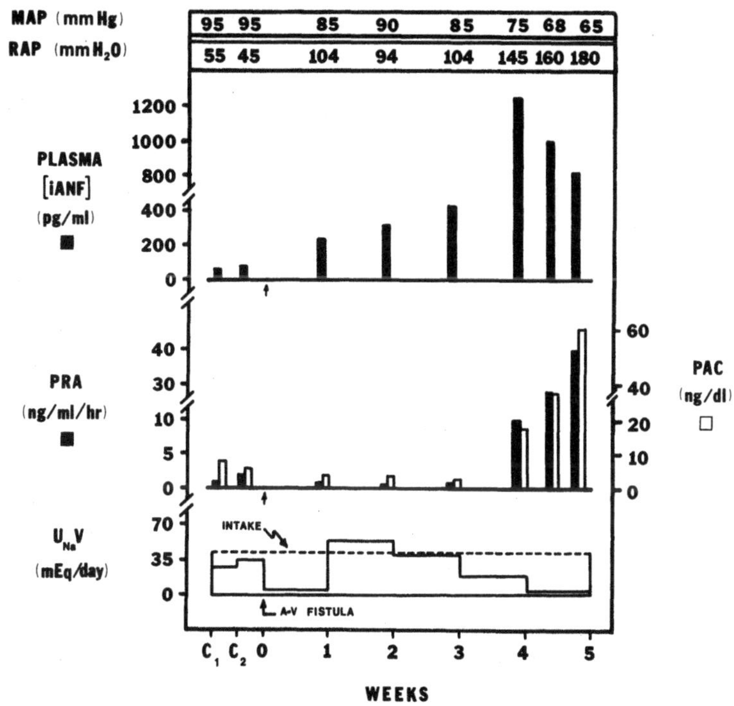


**Fig. 6.** Hemodynamic, hormonal, and sodium balance responses in one dog with an AV fistula and HOHF during cardiac compensation (wk 1–3) and decompensation (wk 3–5). Abbreviations as in Figs. 1–4, except for *MAP*, mean arterial pressure. (From Villarreal D, Freeman RH. J Lab Clin Med 1991;118:515-522, reproduced by permission of Mosby-Year Book, Inc.)

effectively blunt the natriuretic actions of ANP. Thus, the composite information suggests that the role of the ANP endocrine system as a compensatory mechanism should be most apparent during the early and mild stages of heart failure, before marked reduction in cardiac function occurs with the attendant excessive stimulation of the renin–aldosterone axis. However, this interpretation of an important antagonism of the renin–angiotensin–aldosterone system on the natriuretic actions of ANP may be overestimated, because reductions of renal perfusion pressure *per se*, which are characteristic of heart failure, can also blunt the sodium excretory effects of ANP *(56–58)*. To further investigate these hormonal interactions in the absence of a concurrent deterioration in renal hemodynamic function, additional studies *(34)* were performed in AV-fistula dogs with compensated HOHF of 4–5 wk duration during chronic exogenous administration of the mineralocorticoid DOCA (Fig. 7). Three weeks of excess DOCA produced virtually complete sodium retention despite a chronic twofold elevation in plasma ANP to a peak level of approx 600 pg/mL. Plasma renin was significantly decreased during DOCA-induced sodium retention and volume expansion. During the DOCA regimen, mean arterial pressure increased by 10–12 mmHg above baseline, but still the high plasma ANP levels were unable to counterbalance the sodium-retaining actions of excess mineralocorticoid. In the DOCA-washout period (Fig. 7), the animals exhibited a marked natriuresis, and all parameters returned to pre-DOCA control values. These data are consistent with the concept that a hormonal imbalance between the renin–aldosterone axis and the ANP endocrine system may contribute substantially to the resultant sodium retention in decompensated heart failure.

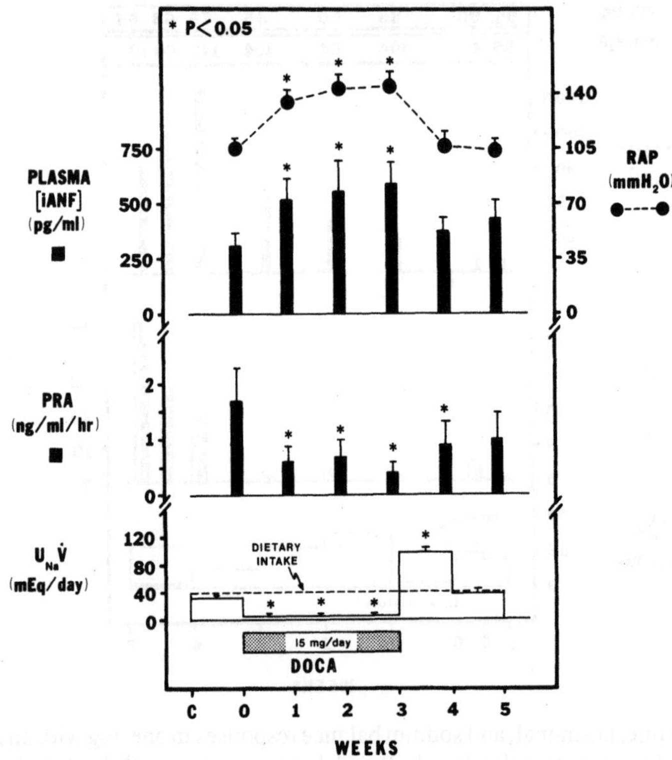


**Fig. 7.** Effects of DOCA in dogs with a chronic AV fistula and compensated HOHF. Values are expressed as mean ± SEM for five dogs, except for wk 4 and 5, when $n = 3$ dogs. Abbreviations as in Figs. 1–4. $*p < 0.05$ vs control. (From Villarreal D, Freeman RH, Brands MW. Am J Physiol 1989;257:h739–h745, reproduced by permission of the American Physiological Society.)

Other lines of evidence are also supportive of this hypothesis. Investigations in patients and animal models of heart failure have demonstrated attenuation of the natriuretic effect of ANP during infusions of synthetic peptide *(31,50,54)*, and it has been suggested that this phenomenon is related, at least in part, to the antinatriuretic actions of the renin–aldosterone axis *(31,50,54)*. Indeed, studies in rats and dogs with an AV fistula and HOHF have demonstrated restoration of the natriuretic effects of synthetic ANF during chronic blockade of the systemic and/or the intrarenal renin–angiotensin system *(59,60)*. However, it is relevant to recognize that in heart failure, mechanisms other than the renin–aldosterone system may have a negative impact on the renal excretory response to synthetic ANP. These potential factors include decreases in renal perfusion pressure *(31,50,54)*, downregulation of renal ANP receptors *(61)*, activation of the sympathetic nervous system *(62,63)*, and excessive degradation of ANP by renal endopeptidases *(64)*.

## PHYSIOLOGICAL FUNCTIONS OF RELATED NATRIURETIC PEPTIDES IN SODIUM-VOLUME BALANCE

### *Introduction*

In the past decade, the family of natriuretic peptides has expanded considerably with the discovery of brain natriuretic peptide (BNP), C-type natriuretic peptide (CNP), and urodilatin. Similar to ANP, these peptide hormones have a 17-amino-acid ring structure formed by a disulfide bond and possess natriuretic and/or vasorelaxant actions that are linked to guanylyl cyclase receptors *(29,65)*. The source of circulating BNP is mainly cardiac, whereas urodilatin does not circulate and is only found in the kidney *(65–67)*. CNP is localized in the heart and is highly expressed in vascular smooth muscle *(65,68)*, and its major role appears to be paracrine and related to antimitogenesis and remodeling *(68)*. In addition, more recent information suggests that several circulating peptide fragments of the N-terminal end of the pro-ANP hormone, in particular pro-ANP (31–67), are also vasodilatory and produce salt and water excretion *(69)*. Although the physiologic significance of these newer peptide hormones remains to be firmly established in health and disease, the biological properties of BNP, urodilatin, and pro ANP (31–67) suggest their important involvement in the control of cardiorenal hemodynamics and body-fluid homeostasis.

The scope of the following section is limited to a consideration of the role of pro-ANP (31–67) in the integrative regulation of sodium balance in normal physiology and during the pathophysiology of experimental heart failure. A number of excellent reviews are available on different aspects of BNP, CNP, and urodilatin *(65,66,68,70–74)*. The interested reader is referred to these reviews and other chapters of this book for more detailed knowledge.

### *ANP Prohormone Fragment 31–67*

In the past 5–8 yr, studies utilizing high-performance gel-permeation chromatography followed by specific RIAs have demonstrated that following cleavage of the C-terminal 28-aa ANP, presumably by nonmyocyte cells *(75)*, the N-terminal portion of the ANP prohormone is further proteolytically cleaved in the heart and plasma to render distinct peptide fragments consisting of aa 1–30, 31–67, and 79–98 *(76,77)*. These peptides circulate in several animal species, and in healthy humans exhibit circadian variation that is inversely related to the levels of arterial blood pressure *(78)*. In addition, initial studies have suggested that these peptides have physiological actions similar to ANP (99–126) *(79–81)*. The three pro-ANP fragments produce vasodilation *(79)*, and pro-ANP (1–30) and (31–67) are natriuretic and diuretic *(81)*, with the latter reportedly possessing similar effects to ANP for the enhancement of urinary sodium excretion in the normal rat *(81)*.

Studies in normal humans *(82)* and dogs *(69)* have demonstrated that pro-ANP (31–67) circulates at levels 10–20-fold higher compared to ANP. The reasons for this phenomenon are not well determined, but may be related, in part, to the different metabolic clearance rates of the two peptides. Indeed, it is now documented that the plasma half-life of the pro-ANP (31–67) is approx 45 min compared to 3–5 min for ANP *(83)*. Acute stimuli, such as head-out water immersion in normal humans, releases both pro-ANP (31–67) and ANP, although the peak plasma elevation and return to baseline of pro ANP (31–67) are delayed for at least 1 h compared to ANP *(84)*. The potential physiological implications of these findings are unclear, but it is possible to suggest that under situations of acute cardiac stretch/distention, such as produced by hypervolemia secondary to a high-salt meal, the initial increases in urinary sodium excretion may be related to elevation of both peptides, whereas the commonly observed delayed natriuretic response that persists despite normalization of plasma ANP levels may involve the continued effects of pro-ANP (31–67).

Table 1
Hormonal, Hemodynamic, and Renal Effects in Dogs with a Chronic Arteriovenous Fistula and High-Output Heart Failure

| *Days* | *pro-ANF (31–67) pg/mL* | *ANF (99–126) pg/mL* | $U_{Na}V$ *mEq/d* | *PRA ng of AI/mL/h* | *MAP mmHg* | *RAP mm* $H_2O$ |
|---|---|---|---|---|---|---|
| Control | 936 ± 51 | 53 ± 8 | 36 ± 1 | 0.5 ± 0.2 | 105 ± 1 | 52 ± 2 |
| Arteriovenous fistula | | | | | | |
| 3 | 997 ± 82 | 62 ± 5 | 2 ± 0.5[a] | 12 ± 2[a] | 88 ± 3[a] | 69 ± 6[a] |
| 5 | 1209 ± 66[a] | 106 ± 23 | 2 ± 0.7[a] | 11 ± 3[a] | 88 ± 3[a] | 86 ± 3[a] |
| 7 | 1322 ± 71[a] | 176 ± 26[a] | 19 ± 8[a] | 2.7 ± 1.6 | 87 ± 2[a] | 91 ± 3[a] |
| 14 | 1557 ± 78[a] | 303 ± 120[a] | 52 ± 8 | 0.8 ± 0.2 | 88 ± 2[a] | 102 ± 2[a] |
| 28 | 1613 ± 80[a] | 303 ± 97[a] | 34 ± 3 | 0.4 ± 0.1 | 85 ± 2[a] | 108 ± 5[a] |
| DOCA 15 mg/d | | | | | | |
| 38 | 1938 ± 98[a,b] | 591 ± 140[a,b] | 5 ± 0.6[a,b] | 0.2 ± 0.1 | 94 ± 2[a,b] | 143 ± 11[a,b] |

Values are means ± SE; $n = 5$ dogs.
[a] $p < 0.05$ vs control.
[b] $p < 0.05$ vs d 28.
pro-ANF (31–67), plasma concentration of atrial natriuretic factor 31–67 prohormone fragment.
ANF 99–126, plasma concentration atrial natriuretic factor (C-terminal peptide).
Other abbreviations as in Figs. 1–7. (From Habibullah AA, Villarreal D, Freeman RH, Dietz JR, Vesley DL, Simmons JC. Clin Exp Pharmacol Physiol 1995;22,130, reproduced by permission of Blackwell Science).

## *Role of pro-ANP (31–67) in Experimental Heart Failure: Temporal Observations*

In pathophysiological conditions of chronic cardiac volume overload, including heart failure, the potential significance of pro-ANP (31–67) in the regulation of sodium-volume balance has been examined only recently *(69)*. In an initial study in humans with congestive heart failure, it was suggested that plasma pro-ANP (31–67) was increased according to the clinical severity of the disease *(82)*. This study, however, consisted of only single measurements obtained in different patients, and therefore, the alterations in the circulating levels of the peptide during the initiation and progression of heart failure could not be longitudinally determined in the same subjects. Moreover, the physiological role of pro-ANP (31–67) in the natural history of heart failure was not characterized in this investigation *(82)*.

Studies in the AV-fistula dog model of HOHF that evaluated the temporal pattern in the secretion of pro-ANP (31–67), ANP, the renin–aldosterone axis, and their relationship to cardiovascular hemodynamics and sodium balance are shown in Table 1 *(69)*. In the initial 5–6 d after the fistula, the dogs retained sodium markedly in association with a 20–25-fold elevation in PRA. On d 3 and 5 postfistula, plasma ANP exhibited parallel changes to right atrial pressure and appeared to increase from a control of 53 ± 8 pg/mL to 62 ± 5 and 106 ± 23 pg/mL, although these increments were not significant from control. In contrast, plasma pro-ANP (31–67) became significantly elevated from baseline earlier, from a control level of 936 ± 51 to 1209 ± 66 pg/mL on the fifth day postfistula. Beginning on d 7 postsurgery and throughout the subsequent 3 wk, the animals approached and maintained sodium balance by exhibiting urinary sodium excretion that approximated sodium intake. Concurrently, PRA returned to baseline, and both circulating peptides stabilized at high levels in the second to fourth weeks after the fistula (Table 1).

At the end of the fourth week of observations, while the animals were in compensated stage, a regimen of deoxycorticosterone acetate (DOCA; 15 mg/d im) was given for 10 consecutive days to evaluate the changes in circulating ANP and pro-ANP (31–67) during progressive elevations in cardiac filling pressures (Table 1). This period of avid sodium retention and consequent intravascular volume expansion was characterized by additional significant elevations of right atrial pressure, mean arterial pressure, and the plasma levels of the two peptides. These pre- and post-DOCA longitudinal results suggest that the cardiac reserve for the release of the N-terminal prohormone and the C-terminal peptide is well maintained in experimental HOHF. In addition, data from human studies and the AV-fistula dogs appear to indicate that plasma pro-ANP (31–67) is more sensitive than ANP to determine mild increases in atrial filling pressures, and this peptide fragment could be utilized as a useful noninvasive marker to follow the clinical course of heart failure *(69,82)*.

### *Renal, Hormonal, and Hemodynamic Effects of Synthetic pro-ANP (31–67) in Normal and Compensated AV-Fistula Dogs*

The initial longitudinal observations obtained in the AV fistula dogs suggested that pro-ANP (31–67) could serve as one compensatory mechanism to achieve body-fluid homeostasis in heart failure. This postulate was subsequently evaluated by examining the biological effects of the synthetic peptide in groups of normal dogs and compensated AV fistula dogs at 4 wk postfistula *(69)*. Figure 8 demonstrates that during progressive infusion-doses of pro-ANP (31–67), similar directional responses in renal excretory function were obtained in the two subsets of animals (Fig. 8, top panel). In both situations, a progressive natriuresis occurred, with peak three- to fourfold elevations from control during the high-infusion dose of 100 ng/kg/min. This natriuretic effect was not accompanied by changes in glomerular filtration rate or renal plasma flow in the two groups of animals. However, the fractional sodium excretion paralleled the increases in urinary sodium excretion for both experiments, and interestingly, this tubular effects of pro-ANP (31–67) for the promotion of a natriuresis was not associated with any changes in the rate of excretion of urinary cyclic GMP (UcCMP; Fig. 8, upper middle panel).

The systemic hemodynamic responses to synthetic pro-ANP (31–67) were also comparable in the two groups of animals, with dose-related reductions in right atrial pressure (Fig. 8, middle lower panel) and mean arterial pressure (data not shown). Similar to the effects on the kidney, these hemodynamic responses were not accompanied by changes in cGMP, which remained unaltered from control in both groups of animals throughout the pro-ANP (31–67) infusion protocol *(69)*.

Figure 8 (bottom panel) illustrates the plasma pro-ANP (31–67) levels during the administration of the synthetic peptide. In the two studies, plasma pro-ANP (31–67) increased in a dose-related fashion, achieving peak levels of 10,000–11,000 pg/mL during the infusion dose of 100 ng/kg/min. Finally, in the two groups of dogs, the baseline normal levels of PRA remained unmodified throughout the experimental protocols despite significant reductions in mean arterial pressure *(69)*.

Previous investigations in the canine model of HOHF have suggested that compensatory adjustments to overcome the sodium retention induced by the opening of the AV fistula may involve the increased circulating levels of ANP as well as the decreases in renin and aldosterone secretion (33,34). In view of earlier studies indicating natriuretic properties of pro-ANP (31–67) similar to ANP in normal experimental animals *(81)*,

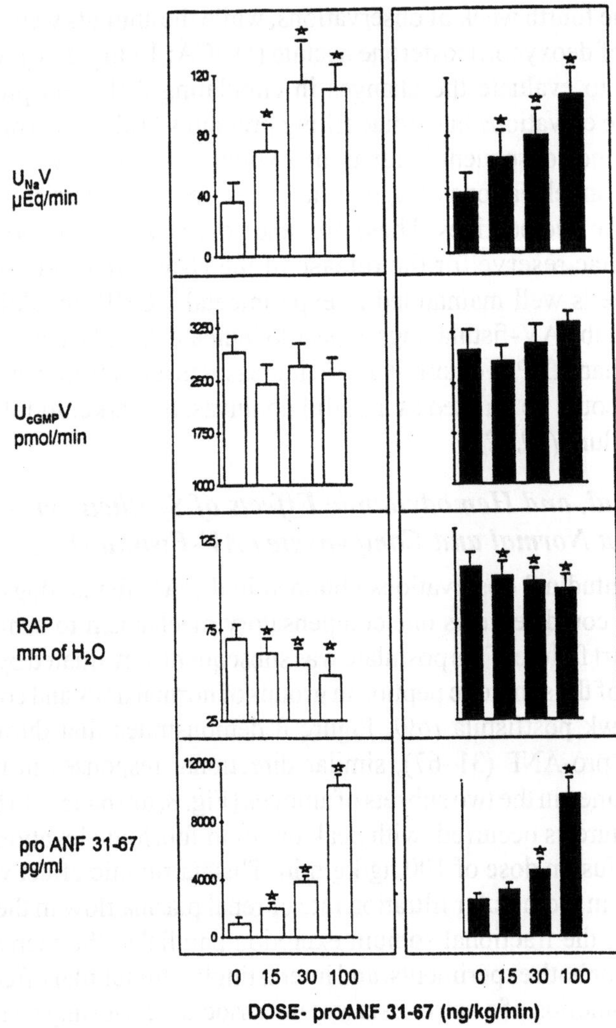


**Fig. 8.** Effects of synthetic pro-ANF (31–67) in conscious dogs. Values are mean ± SEM, $n = 4$ normal dogs (□) and 5 dogs with an AV fistula and compensated HOHF (■). $U_{cGMP}V$ = urinary excretion rate of guanosine 3',5'-cyclic monophosphate. Other abbreviations as in Figs. 1–4 and Table 1. *$P < 0.05$ compared with controls (0 dose). Each dose was infused for 60 min periods. (From Habibullah AA, Villarreal D, Freeman RH, Dietz JR, Vesley DL, Simmons JC. Clin Exp Pharmacol Physiol 1995;22:130–135, reproduced by permission of Blackwell Science.)

coupled with the results of synthetic pro-ANP (31–67) infusions in the intact and the AV-fistula dogs, it is possible to suggest that in this model of experimental heart failure, the chronically increased levels of pro-ANP (31–67) could have contributed to the reestablishment and maintenance of sodium balance. Although the mechanisms by which pro-ANP (31–67) may influence sodium excretion are not well defined *(85,86)*, the data derived from the present studies suggest that its precise renal actions are different from ANP. In both subsets of dogs, synthetic pro-ANP (31–67) stimulated sodium excretion by a tubular effect that, in contrast to ANP, did not appear to be dependent on the renal generation of cGMP. This information is consistent with earlier in vitro studies in medullary collecting duct cells, suggesting that pro-ANP (31–67) increases prostaglandin $E_2$

production, which, in turn, reduces tubular sodium transport by inhibiting $Na^+$, $K^+$ ATPase *(85)*. Moreover, recent investigations with indomethacin administration in normal conscious dogs have clearly indicated that the hemodynamic and renal excretory effects of pro-ANP (31–67) are dependent, at least in part, on the local synthesis of prostaglandins *(86)*. In this regard, it is worthwhile to emphasize that in contrast to these results with pro-ANP (31–67), numerous studies have demonstrated that prostaglandin synthesis blockade with indomethacin or other cyclooxygenase inhibitors does not alter the natriuretic effect of the C-terminal ANP *(87–91)*.

At variance with other previously reported biological effects of ANP *(60,67)*, the results with synthetic pro-ANP (31–67) infusions in the intact and AV-fistula dogs did not suggest that the natriuresis was related to either the enhancement of glomerular filtration rate or inhibition of the renin–aldosterone system. It is relevant to point out, however, that the associated reductions in arterial blood pressure could have masked any potential effects of the peptide on renal hemodynamics and/or renin secretion. Indeed, preliminary observations from this laboratory in sodium-depleted anesthetized dogs with a single filtering kidney suggest that with intrarenal infusions of synthetic pro-ANP (31–67) that avoided systemic vascular effects, renin secretion was strikingly reduced by 66% from baseline, and this response was associated with significant increases in glomerular filtration rate and sodium excretion *(92)*. Thus, it is possible that pro-ANP (31–67) may play an important role in the modulation of renin secretion, and it is likely that this inhibitory effect is mediated, at least in part, via the macula densa.

The mechanisms involved in pro-ANP (31–67)-induced vasodilation are not well elucidated, and further investigations are required for its better definition. In both the intact and AV-fistula dogs, the reductions in arterial blood pressure and central venous pressure were not associated with increases in plasma cGMP, and these findings are in contrast to previous in vitro studies suggesting that the dilation of porcine aortic rings produced by synthetic pro-ANP (31–67) was cGMP-mediated *(79)*. The reasons for these conflicting results are not readily apparent, but may be related to the different experimental conditions. Also, it is possible that the plasma levels of cGMP may not have been a sensitive enough marker to adequately reflect the vascular generation cGMP. As an alternative cGMP-independent mechanism, it is reasonable to suggest that, similar to the kidney, pro-ANP (31–67) could have enhanced the production of vasodilatary protaglandins, including $PGE_2$. Indirect support for this concept is provided by recent studies in normal conscious dogs in which cyclooxygenase inhibition with indomethacin completely blocked the synthetic pro-ANP (31–67)-induced reductions of mean arterial pressure and right atrial pressure *(86)*.

Of major physiological and pharmacological interest derived from the current studies in experimental heart failure are the significant renal excretory responses to synthetic pro-ANP (31–67) observed in the AV-fistula dogs. The natriuresis in these animals with compensated HOHF is at variance with earlier synthetic ANP infusion studies in this canine model, in which a marked subsensitivity to the natriuretic effects of the C-terminal peptide has been consistently demonstrated *(60,67)*. The reasons for this latter phenomenon are unclear, but may include, among others, downregulation of renal ANP receptors *(60)*. Since, as discussed earlier, pro-ANP (31–67)-induced sodium excretion appears to involve mechanism(s) other than cGMP-linked receptors, it is not surprising that the natriuresis was quantitatively and qualitatively similar in both normal and AV-fistula dogs. However, it is important to observe that in the two subsets of dogs, the peak three-to fourfold elevation in sodium excretion was attained at plasma pro-ANP (31–67) levels ranging between 10,000 and 11,000 pg/mL. Even though it is unclear if these circulating peptide values are within the physiological range, it should be recognized that previous studies in humans with congestive heart failure have documented endogenous plasma pro-ANP (31–67) levels of up to 45,000–50,000 pg/mL *(82)*.

The relevance of the data in experimental models in the context of congestive heart failure in humans remains undetermined. Although an initial report in patients with congestive heart failure has documented increased plasma levels of pro-ANP (31–67) *(82)*, longitudinal studies during progressive cardiac decompensation with measurements of circulating pro-ANP (31–67) and its relationship to renal hemodynamics, sodium excretion, and the renin–aldosterone axis have not been conducted. In addition, it would be of interest to examine if humans with heart failure exhibit renal and vascular responses to synthetic pro-ANP (31–67) comparable to the AV-fistula dogs. Nevertheless, with the available information it is possible to suggest that via potentially different mechanisms of action, ANP and pro-ANP (31–67) may coordinate and contribute to the regulation of systemic hemodynamic and renal function during physiological and pathophysiological elevations in cardiac filling pressures.

## SUMMARY AND CONCLUSIONS

The history of ANP research spans a period of over four decades, beginning with the identification of atrial granules using electron microscopy *(4)*. Initially, progress was slow, but then was advanced rapidly by the application of modern molecular biological methodology. Today, research in this field has slowed again because the difficult challenge now is to clarify the physiology and pathophysiology of this salt-excreting peptide hormone. There are clear outlines to guide future endeavors, however.

First, it is clear that both the secretion of ANP and its consequent biological actions on the kidneys are enhanced in the physiological context of postprandial volume expansion and associated increments in renal blood flow and glomerular filtration rate. Second, other peptide fragments derived from the pro-ANP hormone also have natriuretic and diuretic properties with a potency rivaling those of ANP. The interrelationships among ANP and these other peptide fragments may have important implications for regulating the renal excretion of salt surfeit both in normal and pathophysiological states. Third, animal models of salt-sensitive cardiovascular pathologies, including hypertension *(12)* and decompensated heart failure *(1)*, should continue to be utilized extensively to evaluate the genetic/environmental basis for regulation of the cardiac natriuretic peptide system in an appropriate setting. Finally, it is clear that other, more powerful salt-retaining hormonal systems, i.e., the renin–angiotensin–aldosterone axis, can attenuate the salt-excreting actions of ANP under specific conditions. Indeed, the renal tubules demonstrate an enhanced sensitivity to the sodium-retaining effects of aldosterone in animals and humans with heart failure *(1)*. Thus, the physiological and pathophysiological integration of these two potent salt-regulating hormonal systems must be the ultimate goal to better understand the basis for salt sensitivity in biological systems.

## ACKNOWLEDGMENTS

The authors wish to acknowledge the expert technical assistance of Tamara Day, Mary Flood, and Charles Gay.

## REFERENCES


1. Villarreal D, Freeman RH. ANF and the renin-angiotensin system in the regulation of sodium balance: longitudinal studies in experimental heart failure. J Lab Clin Med 1991;118:515–522.
2. Cogan MG. Renal effects of atrial natriuretic factor. Annu Rev Physiol 1990;52:699–708.
3. Zeidel ML. Renal actions of atrial natriuretic peptide: regulation of collecting duct sodium and water transport. Annu Rev Physiol 1990;52:747–759.
4. Kisch B. Electron microscopy of the atrium of the heart. Exp Med Surg 1956;14:99–112.
5. Jamieson JD, Palade GE. Specific granules in atrial muscle cells. J Cell Biol 1964;23:151–172.
6. DeBold AJ. Heart granularity effects of changes in water-electrolyte balance. Proc Soc Exptl Biol Med 1979;161:508–511.
7. DeBold AJ, Borenstein HB, Veress AT, Somberg H. A rapid and potent natriuretic response to intravenous injection of atrial myocardial extracts in rats. Life Sci 1981;28:89–94.
8. Flynn TG, DeBold ML, DeBold AJ. The amino acid sequence of an atrial peptide with potent diuretic and natriuretic properties. Biochem Biophys Res Commun 1983;117:859–865.
9. DeBold AJ. Atrial natriuretic factor: a hormone produced by the heart. Science 1985;230:767–770.
10. Dietz JR. Release of natriuretic factor from heart-lung preparation by atrial distention. Am J Physiol 1989;247:R1093–1096.
11. Brenner BM, Ballermann BJ, Gunning ME, Zeidel ML. Diverse biological actions of ANF. Physiol Rev 1990;70:666–695.
12. John SW, Krege JH, Oliver PM, Hagaman JR, Hodgin JB, Pang SC, Flynn TG, Smithies O. Genetic decreases in atrial natriuretic peptide and salt sensitive hypertension. Science 1995;267:679–681.
13. Davis JO, Freeman RH. Mechanisms regulating renin release. Physiol Rev 1976;56:1–56.
14. Laragh JH. Atrial natriuretic hormone: the renin-aldosterone axis and blood pressure-electrolyte homeostasis. N Engl J Med 1985;313:1330–1340.
15. Gonzalez-Campoy JM, Romero JC, Knox FG. Escape from the sodium retaining effects of mineralocorticoids: role of ANF and intrarenal hormone systems. Kidney Int 1989;35:767–777.
16. Metzler CH, Gardner DG, Keil LC, Baxter JD, Ramsay JD. Increased synthesis and release of atrial peptide during DOCA escape in conscious dogs. Am J Physiol 1987;252:R188–192.
17. Hall JE, Granger JP, Smith MJ Jr, Premen AJ. Role of renal hemodynamics and arterial pressure in aldosterone "escape." Hypertension 1984;6(Suppl I):183–192.
18. Granger JP, Burnett JC Jr, Romero JC, Opgenorth TJ, Salazar J, Joyce M. Elevated levels of atrial natriuretic peptide during aldosterone escape. Am J Physiol 1987;252:R878-882.
19. Blaine EH. Atrial natriuretic factor plays a significant role in body fluid homeostasis. Hypertension 1990;15:2–8.
20. Goetz KT. Physiology and pathophysiology of atrial peptides. Am J Physiol 1988;254:E1–15.
21. de Zeew D, Janssen WMT, de Jong PE. Atrial natriuretic factor: its (patho) physiological significance in humans. Kidney Int 1992;41:1115–1133.
22. Needleman P, Adams SP, Cole BR, et al. Atriopeptins as cardiac hormones. Hypertension 1985;7:469–482.
23. Winquist RJ. Possible mechanisms underlying the vasorelaxant response to atrial natriuretic factor. Fed Proc 1986;45:2371–2375.
24. Burnett JC Jr, Granger JP, Opgenorth TJ. Effects of synthetic atrial natriuretic factor on renal function and renin release. Am J Physiol 1984;247:F863–868.
25. Villarreal D, Freeman RH, Davis JO, Verburg KM, Vari RC. Renal mechanisms for suppression of renin secretion by atrial natriuretic factor. Hypertension 1986;8(Suppl II):28–35.
26. Freeman RH, Davis JO, Vari RC. Renal response to atrial natriuretic factor in conscious dogs with caval constriction. Am J Physiol 1985;248:R495–500.
27. Brands MW, Freeman RH. Aldosterone and renin inhibition by physiological levels of atrial natriuretic factor. Am J Physiol 1988;254:R1011–1016.
28. Light DB, Corbin JD, Stanton BA. Dual ion-channel regulation by cyclic GMP and cyclic GMP-dependent protein kinase. Nature 1990;344:336–339.
29. Rosenzweig A, Seidman CE. Atrial natriuretic factor and related peptide hormones. Annu Rev Biochem 1991;60:223–255.
30. Inagami T. Atrial natriuretic factor. J Biol Chem 1989;264:3043–3046.
31. Villarreal D, Freeman RH, Brands MW. ANF and postprandial control of sodium excretion in dogs with compensated heart failure. Am J Physiol 1990;258:R232–239.

32. Francis GS, Benedict C, Johnstone DE, Kirlin PC, Nicklas J, Chang-seng L, Kubo SH, Rudin-Toretsky E, Yusuf S. Comparison of neuroendocrine activation in patients with left ventricular dysfunction with and without congestive heart failure. Circulation 1990;82:1724–1729.
33. Villarreal D, Freeman RH, Davis JO, Verburg KM, Vari RC. Atrial natriuretic factor secretion in dogs with experimental high-output heart failure. Am J Physiol 1987;252:H692–696.
34. Villarreal D, Freeman RH, Brands MW. DOCA administration and atrial natriuretic factor in dogs with chronic heart failure. Am J Physiol 1989;257:H739–745.
35. Verburg KM, Freeman RH, Davis JO, Villarreal D, Vari RC. Control of atrial natriuretic factor release in the conscious dog. Am J Physiol 1986;251:947–967.
36. Kaczmarczyk G, Schimmrich B, Mohnhaupt R, Reinhardt HW. Atrial pressure and postprandial volume regulation in conscious dogs. Pfluegers Arch 1979;381:143–150.
37. Kaczmarczyk G, Drake A, Eisele R, et al. The role of the cardiac nerves in the regulation of sodium excretion in conscious dogs. Pfluegers Arch 1981;390:125–130.
38. Davis JO, Howell DS. Mechanisms of fluid and electrolyte retention in experimental preparations in dogs. Circ Res 1953;1:171–178.
39. Cuneo RC, Espiner EA, Nicholls MG, Yandle TG, Joyce SL, Gilchrist NL. Renal, hemodynamic, and hormonal responses to atrial natriuretic peptide infusions in normal man, and effects of sodium intake. J Clin Endocrinol Metab 1986;63:946–953.
40. Gutkowska J, Schiffrin EL, Cantin M, Genest J. Effect of dietary sodium on plasma concentration of immunoreactive atrial natriuretic factor in normal humans. Clin Invest Med 1986;9:222–224.
41. Sagnella GA, Markandu ND, Shore AC, MacGregor GA. Plasma immunoreactive atrial natriuretic peptide and changes in dietary sodium intake in man. Life Sci 1987;40:139–143.
42. Shirataka M, Marumo F, Ando K, Sato T. Response of atrial natriuretic peptide in plasma and urine to changes in dietary sodium chloride in man. Jpn J Physiol 1988;38:677–687.
43. Sagnella GA, Markandu ND, Buckley MG, Miller MA, Singer DRJ, MacGregor GA. Plasma atrial natriuretic peptide, aldosterone, and plasma renin activity responses to gradual changes in dietary sodium intake. Am J Hypertens 1990;3:863–865.
44. Gnadinger MP, Weidmann P, Rascher W, Lang RE, Hellmuller B, Uchlinger DE. Plasma arginine-vasopressin levels during infusion of synthetic atrial natriuretic peptide on different sodium intakes in man. J Hypertens 1986;4:623–629.
45. Saville MA, Geer PG, Wang BC, Leadly RJ Jr, Goetz KL. A high salt meal produces natriuresis in humans without elevating plasma atriopeptin. Proc Soc Exp Biol Med 1988;188:387–393.
46. Drummer C, Franck W, Heer M, Forssmann WG, Gerzer R, Goetz K. Postprandial natriuresis in humans: further evidence that urodilatin, not ANP, modulates sodium excretion. Am J Physiol 1996;270:F301–310.
47. Holmer SR, Riegger AJG, Notheis WF, Kromer EP, Kochsick K. Hemodynamic changes and renal plasma flow in early heart failure: implications for renin, aldosterone, norepinephrine, atrial natriuretic peptide and prostacyclin. Basic Res Cardiol. 1987;82:101–108.
48. Hodsman GP, Kohzuki M, Howes LG, Sumithran E, Tsunoda K, Johnston CI. Neurohumoral responses to chronic myocardial infarction in rats. Circulation 1988;78:376–381.
49. Burnett JC Jr, Kao PC, Hud C, et al. Atrial natriuretic peptide elevation in congestive heart failure in the human. Science 1986;231:1145–1147.
50. Cody RJ, Atlas SA, Laragh JH, et al. Atrial natriuretic factor in normal subjects and heart failure patients. Plasma levels and renal, hormonal and hemodynamic responses to peptide infusion. J Clin Invest 1986;78:1362–1374.
51. Redfield MM, Stevens TL, Sanderg SM, Aarhus LL, Kinoshita M, Burnett JC Jr. Effects of acute and chronic antagonism of atrial natriuretic peptide on renal and humoral function in a canine model of early ventricular dysfunction. Circulation 1994;90:I-547.
52. Freeman RH, Villarreal D, Vari RC, Verburg KM. Endogenous atrial natriuretic factor in conscious dogs with caval constriction. Circ. Res. 1987;61(Suppl I):96–99.
53. Dzau VH. Renal and circulatory mechanisms in congestive heart failure. Kidney Int 1987;31:1402–1415.
54. Riegger GAJ, Elsner D, Kromer EP, et al. Atrial natriuretic peptide in congestive heart failure in the dog: plasma levels, cyclic guanosine monophosphate, ultrastructure of atrial myoedocrine cells, and hemodynamic, hormonal, and renal effects. Circulation 1988;77:398–406.
55. Armstrong PW, Stopps TP, Ford SE, DeBold AJ. Rapid ventricular pacing in the dog: pathophysiologic studies in heart failure. Circulation 1986;74:1075–1084.
56. Blaine EH, Heinel LA, Schorn TW, Marsh EA, Whinnery MA. The character of the atrial natriuretic factor response: pressure and volume effects. J Hypertens 1986;4(Suppl 2):517–524.

57. Seymour AA, Smith SG III, Mazack EK. Effects of renal perfusion pressure on the natriuresis induced by atrial natriuretic factor. Am J Physiol 1987;253:F234–238.
58. Mizelle HL, Hildebrandt DA, Gaillard CA, Brands MW, Montani JP, Smith MJ Jr, Hall JE. Atrial natriuretic peptide induced sustained natriuresis in conscious dogs. Am J Physiol 1990;258:R1445–1452.
59. Abassi Z, Haramati A, Hoffman A, Burnett JC Jr, Winaver J. Effect of convertin-enzyme inhibition on renal response to ANF in rats with experimental heart failure. Am J Physiol 1990;259:R84–89.
60. Villarreal D, Freeman RH, Johnson RA. Captopril enchances renal responsiveness to ANF in dogs with compensated high-output heart failure. Am J Physiol 1992;262:R509–516.
61. Johnston CI, Hodsman GP, Harrison RW, Medelsohn FAO, Tsunoda K. Regulation of cardiac preload by atrial natriuretic peptide in congestive heart failure. Am J Med 1988;84(Suppl 3A):105–111.
62. Koepke JP, Sones S, DiBona GF. Blunted natriuresis to atrial natriuretic peptide in chronic sodium-retaining disorders. Am J Physiol 1987;252:F865–871.
63. Villarreal D, Freeman RH, Johnson RA, Simmons JC. Effects of renal denervation on postprandial sodium excretion in experimental heart failure. Am J Physiol 1994;266:R1599–1604.
64. Cavero PG, Margulies KB, Winaver J, Seymour AA, Delaney NG, Burnett JC Jr. Cardiorenal actions of neutral endopeptidase inhibition in experimental congestive heart failure. Circulation 1990;8:2196–2201.
65. Wei CM, Heublein DM, Perella MA, Lerman A, Rodeheffer RJ, McGregor CGA, Edwards WD, Schaff HV, Burnett JC Jr. Natriuretic peptide system in human heart failure. Circulation 1993;88:1004–1009.
66. Goetz KL. Evidence that atriopeptin is not a physiological regulator of sodium excretion. Hypertension 1990;15:9–19.
67. Villarreal D, Freeman RH, Johnson RA. Renal effects of ANF (95–126), a new atrial peptide analogue, in dogs with experimental heart failure. Am J Hypertens 1991;4:508–515.
68. Espiner EA. Physiology of natriuretic peptides. J Int Med 1994;235:527–541.
69. Habibullah AA, Villarreal D, Freeman RH, Dietz JR, Vesley DL, Simmons JC. Atrial natriuretic peptide fragments in dogs with experimental heart failure. Clin Exp Pharmacol Physiol 1995;22:130–135.
70. Nicholls MG. The natriuretic peptides in heart failure. J Int Med 1994;235:515–526.
71. Richards AM. The natriuretic peptides in hypertension. J Int Med 1994;235:543–600.
72. Yandle TG. Biochemistry of natriuretic peptides. J Int Med 1994;235:561–576.
73. Gunning ME, Brenner BM. Natriuretic peptides and the kidney: current concepts. Kidney Int 1994;42(Suppl 38):S127–133.
74. Ogawa Y, Itoh H, Nakao K. Molecular biology and biochemistry of natriuretic peptide family. Clin Exp Pharmacol Physiol 1995;22:49–53.
75. Greenwald JE, Needleman P, Siegel N, Tetens E, Biel B, Ritter D. Processing of atriopeptin prohormone by nonmyocytic atrial cells. Biochem Biophys Res Comm 1992;188:644–654.
76. Winters CJ, Sallman AL, Meadows J, Rico DM, Vesely DL. Two new prohormones: prohormone atrial natriuretic peptides 1–30 and 31–67 circulate in man. Biochem Biophys Res Comm 1988;150:231–236.
77. Vesely DL. Atrial natriuretic hormones originating from the N-terminus of the atrial natriuretic factor prohormone. Clin Exp Pharmacol Physiol 1995;22:108–114.
78. Winters CJ, Sallman AL, Vesely DL. Circadian rhythm of prohormone atrial natriuretic peptides 1–30, 31–67 and 99–126 in man. Chronobiol Int 1988;5:403–409.
79. Vesely DL, Norris JS, Walters JM, Jespersen RR, Baeyens DA. Atrial natriuretic factor prohormone peptides 1–30, 31–67 and 79–98 vasodilate the aorta. Biochem Biophys Res Comm 1987;148:1540–1548.
80. Ngo L, Wyeth RP, Bissett JK, Hester WL, Newton MT, Sallman AL, Winters CJ, Vesely DL. Prohormone atrial natriuretic peptides 1–30, 31–67, and 99–126 increase in proportion to right ventricular pacing. Am Heart J 1989;117:385–390.
81. Martin DR, Pevahous JR, Trigg DJ, Vesely DL, Buerkart JE. Three peptides from the ANF prohormone $NH_2$-terminus are natriuretic and/or kaliuretic. Am J Physiol 1990;258:F1401–1408.
82. Winters CJ, Sallman AL, Baker BJ, Meadows J, Rico DM, Vesely DL. The *N*-terminus and a 4,000-MW peptide from the midportion of the *N*-terminus of the atrial natriuretic factor prohormone each circulate in humans and increase in congestive heart failure. Circulation 1989;80:438–449.
83. Ackerman BH, Wyeth RP, Vesely DL, Ngo WL, Bissett JK, Winters CJ, Sallman AL. Pharmacokinetic characterization of the postdistribution phase of prohormone atrial natriuretic peptides amino acids 1–98, 31–67 and atrial natriuretic factor during and after rapid ventricular pacing in dogs. J Clin Pharmacol 1992;32:415–421.
84. Vesely DL, Norsk D, Winters CJ, Rico DM, Sallman AL, Epstein M. Increased release of the *N*-terminal and *C*-terminal portions of the prohormone of atrial natriuretic factor during immersion-induced central hypervolemia in normal humans. Proc Soc Exptl Biol Med 1989;192:230–235.

85. Gunning ME, Brady HR, Otuchera G, Brenner BM, Zeidel ML. Atrial natriuretic peptide (31–67) inhibits $Na^+$ transport in rabbit inner medullary collecting duct cells. J Clin Invest 1992;89:1411–1417.
86. Habibullah AA, Villarreal D, Freeman RH, Knoblich PR. Effects of indomethacin on the renal actions of atrial natriuretic prohormone fragment 31–67 in normal dogs. J Invest Med 1995;43:452A.
87. Vesley DL. Peptides from *N*-terminus of the atrial natriuretic factor prohormone enhance guanylate cyclase activity and increase cyclic GMP levels in a wide variety of tissues. Mol Cell Biochem 1992;109:43–50.
88. Gaillard CA, Koomans AA, Rabelinik AH, Mees EJ. Effects of indomethacin on renal response to atrial natriuretic peptide. Am J Physiol 1987;253:F868–873.
89. Miyamori M, Ikeda T, Matsubara S. The renal, cardiovascular and hormonal actions of human atrial natriuretic peptide in man: effects of indomethacin. Br J Clin Pharmacol 1987;23:425–431.
90. Salazar FJ, Bollerman R, Riksen-Olsen MJ, Quesada T, Romero JC. Role of prostaglandins in mediating the renal effects of atrial natriuretic factor. Hypertension 1988;12:274–278.
91. Keeler R. Atrial natriuretic factor has a direct, prostaglandin independent action on kidneys. Can J Physiol Pharmacol 1982;60:1078–1082.
92. Taraben A, Villarreal D, Freeman RH, Habibullah AA, Knoblich PR. Modulation of renin secretion by atrial natriuretic factor prohormone fragment 31–67. FASEB J 1996;10:A268.

# 15 Physiology of the Clearance Receptor

*George J. Trachte*

*CONTENTS*



## INTRODUCTION

### *General Features of Clearance Receptors*

INTRODUCTION

Natriuretic peptide receptors are divided into two general categories containing three well-recognized receptors, two particulate guanylyl cyclases and the clearance receptor. Most of the widely recognized actions of natriuretic peptides have been attributed to activation of particulate guanylyl cyclases. Two emerging concepts in the natriuretic peptide field include the clearance receptor as a mediator of certain natriuretic peptide activities *(1,2)* and the existence of multiple forms of the clearance receptor *(3)*. Furthermore, heterogeneous forms of the clearance receptor have separate and distinct biological activities *(4)*. Thus, the classical view of natriuretic peptides interacting solely with guanylyl cyclases to mediate biological/physiological effects is challenged by the accumulating evidence depicting the clearance receptor as a signal-transducing entity. This chapter will evaluate the data favoring a physiological role for the clearance receptor in various biological actions of natriuretic peptides. After an initial discussion of the synthesis, structure, and location of clearance receptors, the known signal-transducing pathways coupled to clearance receptors will be discussed. The contribution of these pathways to biological actions of natriuretic peptides will then be discussed. Finally, the involvement of the clearance receptor in known physiological activities of natriuretic peptides will be assessed. The role of clearance receptors in mediating cellular internalization of natriuretic peptides will be avoided because this subject was presented in Chapter 5.

From: *Contemporary Endocrinology: Natriuretic Peptides in Health and Disease*
Edited by: W. K. Samson and E. R. Levin Humana Press Inc., Totowa, NJ

## Synthesis

Clearance receptors are synthesized by most cells. The gene is encoded with eight exons, six coding for the extracellular region of the receptor, one coding for the transmembrane domain, and one coding for the intracellular portion of the clearance receptor *(5)*. Alternative splicing at the interface between the sixth and seventh exons causes the formation of a clearance receptor containing an additional cysteine residue but a net loss of one amino acid *(6)*. This smaller clearance receptor exists as a monomer, rather than the predominant dimeric form of clearance receptors. A related, but completely distinct, form of the clearance receptor has been identified in the eel *(3)*. This eel receptor possesses seven cysteine residues, in contradistinction from the two forms listed above, which possess five or six. All of these receptors have predicted molecular weights of approx 60 kDa and contain 500–541 amino acids. The monomeric forms of these clearance receptors exhibit a relative size of 66 kDa during sodium dodecyl sulfate-polyacrylamide gel electrophoresis (SDS-PAGE). A larger version of the clearance receptor, migrating at 77 kDa after SDS-PAGE, also exists but has not been cloned at this point *(4)*. This larger clearance receptor binds the clearance receptor ligand, cANP, but has a low affinity for C-type natriuretic peptide *(4)*. Thus, multiple forms of the clearance receptor exist as distinct proteins that vary in their affinities for natriuretic peptides. Evidence for distinct functional roles of these receptors is also accumulating. For instance, the clearance receptor binding C-type natriuretic peptide in the kidney fails to internalize the peptide *(4,7)*, in marked contrast to the postulated physiological role of clearance receptors to internalize natriuretic peptides *(8)*.

## Structure

The clearance receptor represents a group of related proteins possessing a large extracellular domain, a single transmembrane-spanning domain, and a small intracellular domain *(9)*. The receptor is normally present in a dimerized form in renal and cardiovascular tissues, although the monomeric form predominates in platelets *(10)*. The extracellular region is responsible for binding natriuretic peptides and is 33% homologous to the extracellular regions of the particulate guanylyl cyclases responding to natriuretic peptides. The transmembrane regions of these receptors also exhibit significant homology. The intracellular region of the clearance receptor is only 37 amino acids long and is enriched in basic amino acids containing a motif defined by Nishimoto *(11)* as coupling to heterotrimeric GTP-binding proteins. However, the clearance receptor structure dramatically differs from most receptors coupled to GTP-binding proteins in that it possesses only one, instead of seven, transmembrane-spanning domains.

## Location

Natriuretic peptide clearance receptors represent the predominant natriuretic peptide receptor in the plasma membrane of the renal cortex, the vasculature, the heart, and the lung *(12)*. They are the less abundant natriuretic peptide receptor in most neuronal tissue *(13)* and in the adrenal cortex. They have been reported in liver *(14)* and ovary *(15)*, but are absent from rat Leydig cells *(12)*, intestine, and spleen *(16)*. Many brain regions lack the clearance receptor *(16)*, although the most recently cloned clearance receptor is primarily expressed in the brain *(3)*. They are also present in most endocrine organs (e.g., ref. *17*). The general distribution of these receptors appears to be consistent with a clearance function inasmuch as they are often positioned to contact blood, thus potentially serving as regulators of plasma natriuretic peptide concentrations. However, clearance receptors also are displaced from the bloodstream, such as in podocytes of renal epithelial cells *(18)* and chondrocytes of cartilage *(19)*. In summary, this receptor is widely distributed throughout the body. The distribution is generally consistent with a clearance function, although some clearance receptor locations are totally inconsistent with a physiological role in clearing natriuretic peptides from plasma.

## *Dichotomy of Functions*

### Introduction

The natriuretic peptide clearance receptor has been proposed to function in either the removal of natriuretic peptides from plasma or in the transduction of natriuretic peptide signals. These two potential functions are not mutually exclusive and could be occurring simultaneously. However, the more likely scenario involves heterogeneous clearance receptors functioning to perform only one function. This possibility is best illustrated by the recent study of Brown and Zuo *(4)*, in which C-type natriuretic peptide reduced cAMP concentrations by an influence on a 66-kDa clearance receptor but was not internalized by the same receptor in the kidney. This section will provide a broad overview of the evidence supporting clearance or signal-transducing roles for the natriuretic peptide clearance receptor.

### Clearance of Natriuretic Peptides

The initial study identifying the natriuretic peptide clearance receptor *(8)* demonstrated a metabolic function for the receptor but no other biological action in the kidney. Subsequent studies confirmed this assessment with the clearance receptor ligand, cANP, markedly reducing the metabolic clearance of infused natriuretic peptides *(20)*. However, the observation that renal clearance receptors fail to internalize C-type natriuretic peptide also suggests other roles for at least a subset of these receptors *(4)*. Thus, although a clearance function of these receptors is evident, other types of clearance receptors could perform other functions as well.

### Influence on Signal-Transduction Pathways

Natriuretic peptide fragments or synthetic peptides that bind solely to natriuretic peptide clearance receptors often exhibit biological activity in isolated systems. These probes serve as tools to isolate natriuretic peptide activities attributable to the clearance receptor rather than guanylyl cyclase receptors. The first observation of this phenomenon involved an activation of phospholipase C by truncated natriuretic peptides that failed to influence guanylyl cyclase activity *(21,22)*. Subsequently, Anand-Srivastava et al. *(23)* and Drewett et al. *(24)* observed the synthetic clearance receptor ligand, cANP, to suppress adenylyl cyclase activity. Finally, another ligand selective for the clearance receptor, ANP (7–23), augmented calcium currents through L-type channels in bovine adrenal glomerulosa cells *(25)*. Thus, at least three signal-transduction pathways are influenced by agents whose only known receptor is the clearance receptor. Moreover, these agents failed to alter cyclic guanosine monophosphate concentrations and none of them have been reported to alter tyrosine kinase activity. Thus, the clearance receptor appears to couple to select signal-transduction pathways. This coupling will be considered in greater detail in the next section.

## *Biological Activity of Clearance Receptor Ligands*

### Introduction

The clearance receptor binds all natriuretic peptides and most truncated analogs of natriuretic peptides. Clearance receptor ligands include agents that preferentially bind to clearance receptors relative to other natriuretic peptide receptors, such as particulate guanylyl cyclases. Most clearance receptor ligands are truncated versions of atrial natriuretic peptide (ANP), a peptide composed of 28 amino acids. The truncated derivatives are designated by the amino acids they share with atrial natriuretic peptide. These agents include the classic clearance receptor ligand, cANP *(8)*; the truncated natriuretic peptide, atriopeptin I *(22)*; other truncated natriuretic peptides, ANP (7–23) *(25)*, ANP (18–28) *(26)*, des ($Cys^7$,$Cys^{23}$)-ANP (5–28) *(27)*, and AP-811 *(28)*; and, in tissues devoid of guanylyl cyclase-B, C-type natriuretic peptide *(4,29)*.

### cANP (des-[$Gln^{18}$, $Ser^{19}$, $Gly^{20}$, $Leu^{21}$, $Gly^{22}$]-ANP [4–23])

The clearance receptor ligand, cANP, was initially identified by Maack et al. *(8)*. This agent reduced metabolism of natriuretic peptides but failed to reproduce their renal effects in isolated kidneys; therefore, cANP was proposed to interfere with cellular internalization of natriuretic peptides. More recent studies found this selective clearance receptor ligand to suppress adenylyl cyclase activity *(23,24)*, adrenergic neurotransmitter efflux from nerves *(24)*, and mitogenesis *(30)*. The cANP also antagonized natriuretic peptide effects in some studies *(27,29)*. Thus, this compound is useful in assessing clearance receptor functions with the caveat that it must be examined both for agonist and antagonist activity. Most reports of clearance receptor biological activity involve cANP as the clearance receptor ligand.

### Atriopeptin I

Atriopeptin I (ANP [5–25]) contains four fewer amino acids on the amino terminus and three fewer on the carboxy terminus than atrial natriuretic peptide but has the full complement of 17 amino acids in the disulfide ring formed by two cysteine residues. Atriopeptin I is a weak guanylyl cyclase stimulant in most tissues, indicating its relative selectivity for the clearance receptor. It was observed to stimulate inositol phosphate generation, an indicator of phospholipase C activity, more potently than atrial natriuretic peptide *(22)*. The mild activation of guanylyl cyclase limits its use as a selective clearance receptor agonist in most systems.

### Other Truncated Atrial Natriuretic Peptide Analogs

These shortened forms of atrial natriuretic peptide usually retain affinities for clearance receptors but have a reduced affinity for guanylyl cyclases. ANP (7–23) has six fewer amino-terminal amino acids and five fewer carboxy-terminal amino acids than atrial natriuretic peptide. It exhibits a higher affinity for the clearance receptor relative to particulate guanylyl cyclases *(25)*. ANP (18–28) contains the carboxyl terminal 11 amino acids of atrial natriuretic peptide and differs from both cANP and ANP (7–23) in that it lacks an amino-acid ring formed by two cysteine residues. The ANP (18–28) has biological activity in the central nervous system (CNS) but does not activate guanylyl cyclase *(26)*. The des[$Cys^7$, $Cys^{23}$]ANP (7–28) analog of atrial natriuretic peptide also lacks the 17-amino-acid ring. It reacts with the clearance receptor, but not guanylyl cyclases, and suppresses growth in vascular tissues *(27)*. AP–811 is one of the smallest active fragments of natriuretic peptides, consisting of a mere four amino acids with nonpeptide amino and carboxy termini *(28)*. This molecule is a linear analog devoid of the disulfide ring characteristic of most active natriuretic peptides and derivatives. AP-811 suppressed adenylyl cyclase activity in one study but failed to stimulate guanylyl cyclase *(28)*.

### C-Type Natriuretic Peptide

C-type natriuretic peptide is considered to be the third major member of the natriuretic peptide family. It binds to both the clearance receptor and the guanylyl cyclase-B (NPR-B) receptor. The C-type natriuretic peptide associates only with clearance receptors in some tissues, as indicated by complete displacement of C-type natriuretic peptide binding by cANP in glomerular, pheochromocytoma, and select brain cells *(4,29,31)*. The C-type natriuretic peptide reduced adenylyl cyclase activities or suppressed adrenergic neurotransmitter efflux in the absence of alterations in guanylyl cyclase activities *(29,31)*. Thus, C-type natriuretic peptide appears to be a selective clearance receptor agonist in tissues lacking guanylyl cyclase B (NPR-B).

### Summary of Clearance Receptor Ligands

The seven agents listed above can be utilized to assess clearance receptor activity; but most require simultaneous measurement of guanylyl cyclase activity to verify a selective effect on clearance receptors. Some of the agents, such as cANP, ANP (18–28), and des[Cys$^7$, Cys$^{23}$]ANP (7–28), fail to activate guanylyl cyclase directly; thus, they have an inherent advantage over the others. The C-type natriuretic peptide represents an endogenous peptide; therefore, its actions probably have the most physiological relevance. The other three agents represent useful compounds but their utility has been limited to few studies thus far.

## SIGNAL TRANSDUCTION PATHWAYS ASSOCIATED WITH CLEARANCE RECEPTORS

### *Adenylyl Cyclase*

#### Inhibitory Effects

The first clear demonstration of a clearance receptor influence on signal-transduction pathways in a cell-free system involved an inhibitory effect of cANP to suppress adenylyl cyclase activity in plasma membranes obtained from the anterior pituitary, aorta, brain, and adrenal tissues *(23)*. The cANP failed to alter guanylyl cyclase activity, thus confirming its selectivity for the clearance receptor.

#### Involvement of GTP-Binding Proteins

The inhibitory effect of cANP on adenylyl cyclase activity was ablated by prior treatment with pertussis toxin to abrogate responses dependent on inhibitory GTP-binding proteins *(23)*. The latter result suggested the existence of a process involving a coupling of the clearance receptor to an inhibitory GTP-binding protein to suppress adenylyl cyclase activity. A prior study also found atrial natriuretic peptide to suppress adenylyl cyclase activity by a process both requiring guanine nucleotides and exhibiting sensitivity to pertussis toxin or guanosine 5'-O-(2-thiodiphosphate) (GDP$\beta$S), another inhibitor of GTP-binding proteins *(32)*.

### Tissues Exhibiting this Response

An inhibition of adenylyl cyclase activity was the first reported action of natriuretic peptides on cyclic nucleotide generation *(33)*. This inhibitory action has been confirmed in almost every cell type examined, including kidney *(4)*, adrenal *(34)*, vasculature *(35)*, heart *(36)*, lung *(37)*, endocrine *(38)*, neural *(24)*, platelet *(10)*, liver *(14)*, oocyte *(39)*, fibroblast *(40)*, and parotid cells *(41)*. Natriuretic peptide inhibition of adenylyl cyclase has not been reported in the intestine *(42)*. The role of the clearance receptor in mediating this inhibition of adenylyl cyclase was established in many of these cells by the use of cANP to mimic the effect of the naturally occurring natriuretic peptides *(23,24)*. The inhibitory effect of natriuretic peptides on adenylyl cyclase activity is usually inhibited by pertussis toxin *(23)*, but has also been attributed to a stimulation of phosphodiesterases by cyclic guanosine monophosphate *(43)*. The net conclusion from these studies is that clearance receptors attenuate adenylyl cyclase activity, probably by affecting GTP-binding proteins, although cAMP concentrations also can be influenced by phosphodiesterases activated by cyclic guanosine monophosphate.

## Phospholipase C

### Biphasic Effects

Phospholipase C stimulation by natriuretic peptides has not been reported in cell-free systems except in abstract form. We have observed clearance receptor stimulants and intracellular clearance receptor fragments to augment phospholipase C activity in quiescent membranes of pheochromocytoma (PC12) cells. Additionally, phospholipase C activity is attenuated after stimulation with either calcium (10 μ*M*) or muscarinic receptor agonists *(44)*. These data suggest a coupling between clearance receptors and phospholipase C, but the details of the coupling have not been defined yet.

### Tissues Exhibiting Response

The first report indicating clearance receptor coupling to signal-transduction events involved both an activation of phospholipase C and an inhibition of adenylyl cyclase activity by natriuretic peptides binding to a receptor that did not activate guanylyl cyclase *(21)*. The stimulation of phospholipase C activity by a clearance receptor agonist was confirmed by Hirata et al. *(22)* in aortic smooth muscle. Berl et al. *(45)* observed a similar activation of phospholipase C by natriuretic peptides, and this stimulation was potentiated by pharmacological antagonists of the cyclic guanosine monophosphate-sensitive protein kinase. The last study suggests that the potentiative effect of natriuretic peptides on phospholipase C was mediated by the clearance receptor and not via an elevation in cyclic guanosine monophosphate concentrations coupled to activation of protein kinases.

## L-Type Calcium Channels

### Stimulatory Effects

A third signal-transducing activity assigned to clearance receptors involves alterations in calcium conductance. A clearance receptor ligand potentiated depolarization-induced increases in calcium concentrations of bovine adrenal cells and suppressed adenylyl cyclase activity at concentrations that did not alter guanylyl cyclase activity *(25)*. Clearance receptor agonists also potentiate calcium currents in porcine renal cells *(46)*. Le Grand et al. *(47)* found a similar effect in cardiac cells grown in the absence of guanosine triphosphate to prevent cyclic guanosine monophosphate formation. These data suggest a functional relationship between the clearance receptor and calcium channels but the ultimate physiological relevance of this interaction is unknown.

#### Inhibitory Effects in Heart

McCall and Fried *(36)* demonstrated an inhibitory effect of atriopeptin II (ANP [5–27]) on calcium currents, adenylyl cyclase, and contractility in rat isolated myocardial cells. These effects occurred independently of changes in cyclic guanosine monophosphate concentrations, thereby suggesting a clearance receptor involvement. Pertussis toxin eliminated the inhibitory effects of atriopeptin II on all three parameters. These results are consistent with a putative role for GTP-binding proteins in coupling clearance receptors to the ultimate effector enzymes or channels mediating responses to atriopeptin II. An inhibitory effect of atriopeptin III (ANP [5–28]) on cardiac calcium currents also was reported in frog ventricular cells *(48)*; the inhibition was eliminated by prior intracellular injection of cAMP. Thus, both studies suggest a pathway involving a reduction in adenylyl cyclase activity that ultimately results in a reduction in calcium currents. Inasmuch as the clearance receptor is believed to couple to adenylyl cyclase in an inhibitory manner, the most likely explanation of these results involves clearance receptors mediating the inhibitory effects of natriuretic peptides on cardiac contractility and calcium currents. However, contradictory conclusions by LeGrand et al. *(47)* attribute these inhibitory effects to the generation of cyclic guanosine monophosphate. Currently there is no consensus of opinion regarding the influence of clearance receptors on calcium currents in cardiac cells. Both potentiative and inhibitory effects have been hypothesized.

### *Hypothesized Scheme Involving Clearance Receptor Signal-Transduction*

Figure 1 depicts the proposed coupling of clearance receptors to adenylyl cyclase, phospholipase C, and calcium channels. The clearance receptor probably suppresses adenylyl cyclase activity by interacting with an inhibitory GTP-binding protein. These actions presumably result in biological effects, including a suppression of exocytosis, growth, and cardiac contractility, in addition to a potentiation of platelet aggregation.

## POTENTIAL PHYSIOLOGICAL ROLES OF CLEARANCE RECEPTORS

### *Neuromodulation*

#### Inhibition of Sympathetic Neurotransmission

Natriuretic peptides initially were reported to attenuate sympathetic neurotransmission by Nakamaru and Inagami *(49)*. The mechanism of action was attributed to clearance-receptor activation because cANP mimicked the effect of atrial natriuretic peptide *(24)*. Atrial natriuretic peptide suppressed catecholamine efflux from pheochromocytoma cells at a physiologically relevant concentration, with the half-maximal effect ($EC_{50}$) averaging 75 p*M*. Babinski et al. *(50)* also found both C-type natriuretic peptide and cANP to suppress catecholamine efflux from bovine adrenal glands with an $EC_{50}$ of 10 p*M*, confirming that clearance receptors mediate the effect and that the effect occurs at physiological concentrations. Natriuretic peptides are synthesized in sympathetic nerve synapses *(51)* and adrenal glands *(52)*, further emphasizing the potential physiological relevance of this observation.

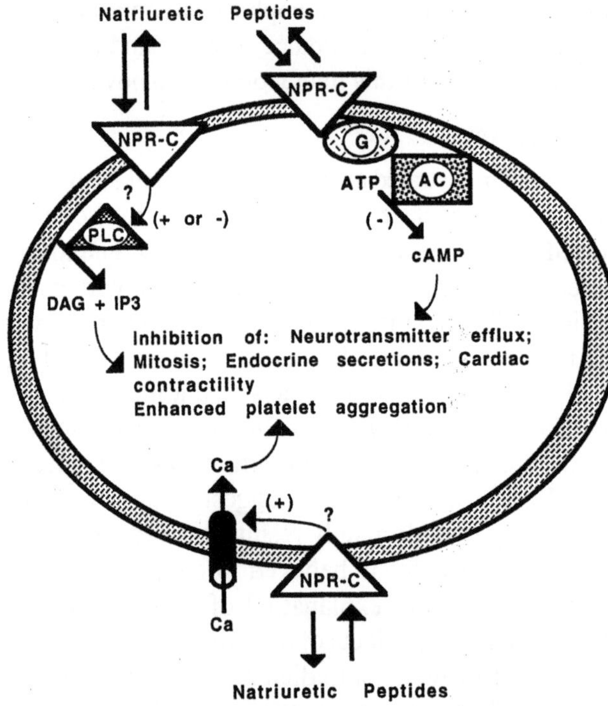


**Fig. 1.** Natriuretic peptides are depicted as interacting with clearance receptors (NPR-C) resulting in influences on three signal-transducing systems, phospholipase C (PLC), adenylyl cyclase (AC), and calcium (Ca) channels. The coupling to phospholipase C results in a stimulation or inhibition (represented by + or –) of the generation of diacylglycerol (DAG) and inositol trisphosphate (IP3). The coupling to adenylyl cyclase appears to involve an inhibitory GTP-binding protein (G) to suppress cAMP concentrations. The coupling to calcium channels involves a potentiative effect (represented by +) on calcium conductances, at least in the adrenal glomerulosa. Undefined coupling mechanisms between clearance receptors and effector enzymes or channels are indicated by ?. Ultimately these effects on signal-transduction pathways are presumed to result in inhibition of exocytotic processes, cell division and cardiac contractility, and modulation of platelet aggregation.

The inhibitory action of natriuretic peptides on sympathetic neurotransmission persists in isolated organs and in vivo, as well as in the isolated cellular systems listed in the previous paragraph. Atrial natriuretic peptide and cANP suppressed both adrenergic neurotransmission and evoked catecholamine efflux from rabbit isolated vas deferens by a mechanism sensitive to pertussis toxin *(53)*. Furthermore, Ebert et al. *(54)* observed atrial natriuretic peptide to suppress plasma norepinephrine concentrations when cardiac filling pressure was maintained constant, indicating that the attenuation of sympathetic nerve activity occurs in vivo. However, the receptor responsible for the reduction in norepinephrine efflux from the adrenergic nervous system was not ascertained in this study.

The signal-transduction pathway mediating these inhibitory neuromodulatory effects of natriuretic peptides have not been identified definitively. We have correlated changes in dopamine efflux from pheochromocytoma cells with changes in either adenylyl cyclase or phospholipase C activity. As shown in Fig. 2, either of these signal transducing systems could participate in intracellular processes, resulting in a reduction in dopamine efflux because alterations in dopamine efflux correlate with changes in diacylglycerol or cAMP

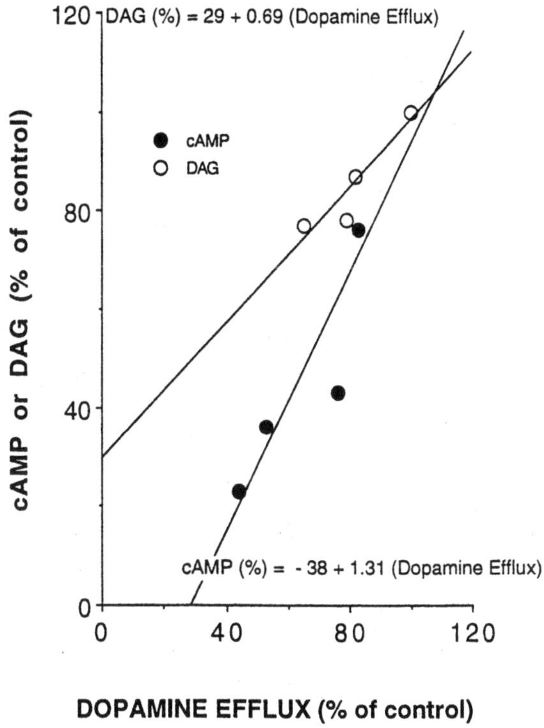


**Fig. 2.** Correlation among dopamine efflux, cyclic adenosine monophosphate (cAMP), and diacylglycerol (DAG) concentrations in depolarized pheochromocytoma cells. Data points represent average responses to varying concentrations of clearance receptor agonists (i.e., either cANP or CNP; 10–10,000 p*M*). The responses correlate in either comparison.

concentrations. The potentiative effect of natriuretic peptides on calcium conductance in these cells should oppose the inhibitory effect on catecholamine efflux; therefore, it is unlikely to account for the inhibitory effect of natriuretic peptides.

### Effects in the CNS

Natriuretic peptides injected into the CNS suppress sympathetic nerve traffic *(55)* and lower blood pressure *(56)*. These actions are consistent with the peripheral sympatho-inhibitory effects discussed in the previous section. Injected natriuretic peptides also suppress nerve firing of hypothalamic *(57)* and paraventricular nucleus *(58)* neurons. Some additional effects of natriuretic peptides injected centrally include a reduction in both drinking *(59)* and parasympathetic nervous system effects to constrict bronchioles *(26)*. The latter two effects were reproduced by injections of the truncated atrial natriuretic peptide analogs, ANP (18–28) *(26)* or cANP *(59)*. Thus, at least some of the central effects of natriuretic peptides are mediated by clearance receptors.

These central actions of natriuretic peptides lower blood pressure. Interestingly, natriuretic peptide receptors are more scarce in brains of hypertensive rats in comparison to normotensive rats *(60)*. Hypertensive rats also exhibit an increased efflux of catecholamines from their adrenergic nerves and attenuated responses to natriuretic peptides in comparison to normotensive rats *(61)*. Thus, the tempering effect of natriuretic peptides on the sympathetic nervous system could be physiologically relevant in the control of blood pressure.

## *Antimitogenic Actions*

### Central Nervous System

Antimitogenic effects of natriuretic peptides were reviewed in Chapter 14 and in a recent review *(2)*; therefore, the review of the data on this topic will be cursory. Atrial natriuretic peptide or cANP initially were reported to suppress thymidine incorporation and growth of astroglial cell cultures *(30)*. The clearance receptor presumably regulates this effect because cANP mimics the action of other natriuretic peptides.

### Cardiovascular System

Natriuretic peptides suppress thymidine incorporation into, and growth of, rat cultured aortic cells; these effects are blocked by cANP, indicative of a clearance receptor involvement *(27)*. Furthermore, a linear analog of atrial natriuretic peptide lacking the disulfide ring structure mimicked the effect of naturally occurring natriuretic peptides *(27)*. This linear natriuretic peptide analog did not activate guanylyl cyclase, thereby indicating its selectivity for clearance receptors. This study indicates the potential of cANP to act as an antagonist of clearance receptor responses in contrast to most other reports, indicating it to be an agonist at clearance receptors.

Cardiac fibroblast growth and incorporation of thymidine were reduced by cANP, indicative of clearance receptor mediation *(62)*. As in other studies, the cANP lacked influences on guanylyl cyclase. A further indication of clearance receptor involvement in growth is indicated by the increased expression of clearance receptors in vascular tissue following injury *(63)*. The studies presented in this section indicate that clearance receptor ligands suppress growth in cardiovascular tissues.

## *Inhibition of Endocrine Secretions*

### Thyroglobulin Secretion

Atrial natriuretic peptide suppresses thyroglobulin secretion at physiological concentrations (i.e., 10 p*M*) and higher *(17)*. This inhibition correlates with an inhibition of adenylyl cyclase activity but occurs at concentrations failing to activate guanylyl cyclase. Additionally, human thyroid tissue contained only low-mol-wt natriuretic peptide receptors migrating at 66 kDa after reduction in SDS-PAGE. These data are consistent with clearance receptors mediating inhibitory actions on thyroglobulin secretion. No specific clearance receptor agonists or antagonists were utilized to evaluate this conclusion definitively.

### Others

Natriuretic peptides suppress the secretion of many other endocrine factors, such as aldosterone, renin, antidiuretic hormone, adrenocorticotropin, progesterone, oxytocin, prolactin, melanocyte-stimulating hormone, and luteinizing hormone. Some of these natriuretic peptide effects have been associated with both an attenuation of adenylyl cyclase activity and an absence of guanylyl cyclase-stimulating activity. Examples of this scenario include suppression of progesterone secretion from the testes and renin release from juxtaglomerular cells. The mechanism of inhibitory natriuretic peptide actions on hormone secretion is unresolved in most of the other examples.

### Progesterone

Natriuretic peptides suppress both adenylyl cyclase activity in, and progesterone secretion from, mouse Leydig cells in a manner reminiscent of natriuretic peptide effects in the thyroid *(38)*. Natriuretic peptides suppressed adenylyl cyclase activities and progesterone secretion more potently than guanylyl cyclase activity. This scenario suggests a mediation by clearance receptors although no other definitive experiments were performed to verify or negate this conclusion.

### Renin

Atrial natriuretic peptide suppresses renin release from juxtaglomerular cells with an $EC_{50}$ of 10 p*M*, whereas the stimulation of guanylyl cyclase has an $EC_{50}$ of 100,000 p*M* *(64)*. Furthermore, the threshold effect of atrial natriuretic peptide on guanylyl cyclase occurs at a concentration of 100 p*M*, an order of magnitude higher than the half-maximal effect on renin secretion. These data appear to differentiate the suppression of renin release from an activation of guanylyl cyclase. Natriuretic peptides suppress adenylyl cyclase activities in the renal glomerulus with an $EC_{50}$ of 1000 p*M* and a threshold of 1 p*M* *(4)*. Thus, the reduction in renin release correlates better with a reduction in adenylyl cyclase activity than an activation of guanylyl cyclase. Furthermore, pertussis toxin eliminated the attenuation of renin secretion by atrial natriuretic peptide *(65)*. The conclusion from these studies is that natriuretic peptides suppress renin release and cAMP concentrations at physiological levels. The role of clearance receptors in mediating these actions was not addressed. However, cANP inhibits cardiac generation of the message for renin *(66)*. This observation emphasizes the possibility that clearance receptors could mediate inhibitory effects on renin release.

### Summary

Natriuretic peptides suppress endocrine secretions similarly to their effects on neurotransmitter exocytosis. The effects occur at physiological concentrations and appear to be independent of guanylyl cyclase activity. Furthermore, the natriuretic peptide action is inhibited by pertussis toxin in the case of thyroglobulin and renin release. These data are consistent with the hypothesis that clearance receptors mediate these effects by coupling to GTP-binding proteins.

## *Negative Cardiac Inotropic Effects*

Natriuretic peptides suppress calcium conductances in cardiac tissue *(36,47,48)*. This action should suppress cardiac contractility because calcium influx is necessary to promote cardiac contractions. Only one study has definitively established a suppressant effect of natriuretic peptides on contractility. McCall and Fried *(36)* found ANP (5–27) to reduce adenylyl cyclase activity, calcium conductance, and contractility in rat isolated heart cells. The reduction in contractility correlated with both the reduction in adenylyl cyclase activity and calcium conductance, but failed to correlate with guanylyl cyclase activation. Furthermore, pertussis toxin pretreatment eliminated the inhibitory effects. These findings suggest the involvement of a guanylyl cyclase-independent receptor, such as the clearance receptor, in mediating effects of natriuretic peptides on cardiac contractility. Moreover, the data are consistent with the involvement of a GTP-binding protein to suppress adenylyl cyclase activity and calcium conductance.

## *Modulation of Platelet Aggregation*

Platelets are an interesting model for investigating natriuretic peptide effects because they lack guanylyl cyclase receptors for natriuretic peptides but possess clearance recep-

Table 1
Natriuretic Peptide Potencies for Biological Responses or Signal Transduction Pathways[a]

| *Process affected* | $EC_{50}$ | *Adenylyl cyclase* $EC_{50}$ | *Guanylyl cyclase* $EC_{50}$ |
|---|---|---|---|
| Neuromodulation | 75*(23)* | 21*(23)* | 10,000*(51)* |
| Antimitogenesis | 100*(30)* | —*(30)* | 30,000*(31)* |
| Thyroglobulin release | 100*(17)* | 250*(17)* | >10,000*(17)* |
| Progesterone release | 100*(38)* | 100*(38)* | 10,000*(38)* |
| Renin release | 10*(64)* | 1000*(4)* | 100,000*(64)* |
| Platelet aggregation | 10*(10)* | 250*(10)* | —*(10)* |

[a]All values indicate $EC_{50}$ (in p*M*). —: absence of a response. These biological activities of natriuretic peptides generally correlate better with a suppression of adenylyl cyclase than an activation of guanylyl cyclase. The references for the data are given in parentheses.

tors *(10)*. Therefore, biological activities in these cells are attributable only to clearance receptor activation. Atrial natriuretic peptide suppressed adenylyl cyclase activity *(10)*, an action consistent with a potentiative effect on platelet aggregation. This potentiative effect of atrial natriuretic peptide was observed *(67)*. The maximal potentiative effect occurred at a concentration of 10 p*M*, whereas the $EC_{50}$ for suppression of adenylyl cyclase occurred at a concentration of 250 p*M*. Neither clearance receptor ligands nor pertussis toxin have been investigated to ascertain the role of clearance receptors or GTP-binding proteins in the modulatory effects of natriuretic peptides on platelet aggregation.

## CONCLUSIONS

### *Internalization*

The studies listed above indicate overwhelming evidence favoring an internalization role for the clearance receptor, as initially proposed by Maack et al. *(8)*. This clearance function is almost certainly physiological because blockade of the clearance receptor with selective ligands results in an elevation of natriuretic peptide concentrations in plasma. Therefore, there is little controversy regarding the physiological relevance of the internalization function of clearance receptors.

### *Signal Transduction*

The other obvious conclusion from the studies cited in this chapter involves clearance receptors coupling to adenylyl cyclase, phospholipase C, and calcium conductances. The physiological relevance of these actions is controversial at present. However, the potencies of natriuretic peptides certainly are consistent with a physiological role in modulating these processes, because physiological concentrations of natriuretic peptides are approx 30 p*M* *(68)*. Natriuretic peptide potencies in modulating cellular functions, adenylyl cyclase activity, and guanylyl cyclase activity are depicted in Table 1.

### *Undefined Mechanisms of Clearance Receptor Coupling*

The coupling of clearance receptors to adenylyl cyclase presumably involves GTP-binding proteins because these responses are eliminated by pertussis toxin. The coupling of clearance receptors to phospholipase C or calcium channels is undefined. Furthermore, there is no report definitively linking the clearance receptor physically to GTP-binding proteins. Thus, defining the coupling of clearance receptors with signal transduction systems represents a major future challenge.

### *Physiological Roles*

The ultimate physiological relevance of the clearance receptor to natriuretic peptide actions has not yet been defined. As indicated in Table 1, natriuretic peptide actions in certain tissues are too potent to be explained by an activation of guanylyl cyclases. Instead, these actions roughly correlate with an inhibition of adenylyl cyclase, an action attributed to clearance receptor activation by numerous investigators. Therefore, the clearance receptor probably functions as a physiological regulator of the following: adrenergic neurotransmission, growth, thyroglobulin, renin, and progesterone secretion; and platelet aggregation. These conclusions are based on the potency of natriuretic peptides influencing these processes. Definitive tests of the physiologic relevance of clearance receptors require more selective antagonists or transgenic models deficient in these receptors. In the absence of these developments, the suggestive data cited in this review support a physiological role for the clearance receptors in multiple processes involving exocytosis, growth, and cardiac contractility.

## ACKNOWLEDGMENTS

This work was supported by grant RO1 HL42525 from the United States Public Health Service. The excellent secretarial assistance of S. Kurki is gratefully appreciated, as are the helpful comments and collaborations with B. Elmquist, S. Kanwal, J. Ross, and B. Clarke.

## REFERENCES

1. Anand-Srivastava MB, Trachte GJ. Atrial natriuretic factor receptors and signal transduction mechanisms. Pharmacol Rev 1993;45:455–497.
2. Levin ER. Natriuretic peptide C-receptor: more than a clearance receptor. Am J Physiol 1993;264: E483–489.
3. Kashiwagi M, Katafuchi T, Kato A, Inuyama H, Ito T, Hagiwara H, Takei Y, Hirose S. Cloning and properties of a novel natriuretic peptide receptor, NPR-D. Eur J Biochem 1995;233:102–109.
4. Brown J, Zuo Z. Receptor proteins and biological effects of C-type natriuretic peptides in the renal glomerulus of the rat. Am J Physiol 1994;266:R1383–R1394.
5. Saheki T, Mizuno T, Iwata T, Saito Y, Nagasawa T, Mizuno KU, Ito F, Ito T, Hagiwara H, Hirose S. Structure of the bovine atrial natriuretic peptide receptor (type C) gene. J Biol Chem 1991;266:11,122–11,125.
6. Mizuno T, Iwashina M, Itakura M, Hagiwara H, Hirose S. A variant form of the type C atrial natriuretic peptide receptor generated by alternative RNA splicing. J Biol Chem 1993;268:5162–5167.
7. Brandt RR, Heublein DM, Aarhus LL, Lewicki JA, Burnett JC. Role of natriuretic peptide clearance receptor in *in vivo* control of C-type natriuretic peptide. Am J Physiol 1995;269:H326–331.
8. Maack T, Suzuki M, Almeida FA, Nussenzveig D, Scarborough RM, McEnroe GA, Lewicki JA. Physiological role of silent receptors of atrial natriuretic factor. Science 1987;238:675–678.
9. Porter JG, Arfsten A, Fuller F, Miller JA, Gregory LC, Lewicki JA. Isolation and functional expression of the human atrial natriuretic peptide clearance receptor cDNA. Biochem Biophys Res Comm 1990;171:796–803.
10. Anand-Srivastava MB, Gutkowska J, Cantin M. The presence of atrial-natriuretic-factor receptors of ANF-R2 subtype in rat platelets. Biochem J 1991;278:211–217.
11. Nishimoto I. The IGF-II receptor system: a G protein-linked mechanism. Mol Reprod Devel 1993;35:398–407.
12. Leitman DC, Andresen JW, Catalano RM, Waldman SA, Tuan JJ, Murad F. Atrial natriuretic peptide binding, cross-linking, and stimulation of cyclic GMP accumulation and particulate guanylate cyclase activity in cultured cells. J Biol Chem 1988;263:3720–3728.

13. Rathinavelu A, Isom GE. Differential internalization and processing of atrial-natriuretic-factor B and C receptors in PC12 cells. Biochem J 1991;276:493–497.
14. Rashed HM, Sun H, Patel TB. Atrial natriuretic peptide inhibits growth of hepatoblastoma (HEP G2) cells by means of activation of clearance receptors. Hepatol 1993;17:677–684.
15. Saheki T, Shimonaka M, Uchida K, Mizuno T, Hirose S. Immunochemical and biochemical distinction of subtypes of atrial natriuretic peptide receptor. J Biochem 1989;106:627–632.
16. Lin X, Hanze J, Heese F, Sodmann R, Lang RE. Gene expression of natriuretic peptide receptors in myocardial cells. Circ Res 1995;77:750–758.
17. Tseng Y-CL, Lahiri S, Sellitti DF, Burman KD, D'avis JC, Wartofsky L. Characterization by affinity cross-linking of a receptor for atrial natriuretic peptide in cultured human thyroid cells associated with reductions in both adenosine 3',5'-monophosphate production and thyroglobulin secretion. J Clin Endocrinol Metabol 1990;70:528–533.
18. Shimonaka M, Saheki T, Hagiwara H, Hagiwara Y, Sono H, Hirose S. Visualization of ANP receptor on glomeruli of bovine kidney by use of a specific antiserum. Am J Physiol 1987;253:F1058–1062.
19. Sakeguchi H, Katafuchi T, Hagiwara H, Takei Y, Hirose S. High-density localization of ANP receptors in chondrocytes of eel gill cartilage. Am J Physiol 1993;265:R474–479.
20. Chiu PJS, Tetzloff G, Romano MT, Foster CJ, Sybertz EJ. Influence of C-ANF receptor and neutral endopeptidase on pharmacokinetics of ANF in rats. Am J Physiol 1991;260:R208–216.
21. Resink TJ, Scott-Burden T, Baur U, Jones CR, Buhler FR. Atrial natriuretic peptide induces breakdown of phosphatidylinositol phosphates in cultured vascular smooth-muscle cells. Eur J Biochem 1988;172:499–505.
22. Hirata M, Chang C-H, Murad F. Stimulatory effects of atrial natriuretic factor on phosphoinositide hydrolysis in cultured bovine aortic smooth muscle cells. Biochim Biophys Acta 1989;1010:346–351.
23. Anand-Srivastava MB, Sairam MR, Cantin M. Ring-deleted analogs of atrial natriuretic factor inhibit adenylate cyclase/cAMP system. J Biol Chem 1990;265:8566–8572.
24. Drewett JG, Ziegler RJ, Trachte GJ. Neuromodulatory effects of atrial natriuretic factor are independent of guanylate cyclase in adrenergic neuronal pheochromocytoma cells. J Pharmacol Exp Ther 1990;255:497–503.
25. Isales CM, Lewicki JA, Nee JJ, Barrett PQ. ANP-(7–23) stimulates a DHP-sensitive $Ca^{2+}$ conductance and reduces cellular cAMP via a cGMP-independent mechanism. Am J Physiol 1992;263:C334–342.
26. Haxhiu MA, Deal EC, Lunteren EV, Cherniack NS. Central effects of somatostatin and atrial natriuretic peptide on tracheal tone. J Am Applied Physiol 1993;75:2353–2359.
27. Cahill PA, Hassid A. Clearance receptor-binding atrial natriuretic peptides inhibit mitogenesis and proliferation of rat aortic smooth muscle cells. Biochem Biophys Res Comm 1991;179:1606–1613.
28. Koyama S, Inoue T, Terai T., Takimoto K, Kato M, Ito K, Neya M, Seki J, Kobayashi Y, Kyogoku Y, Yoshida K. AP-811, a novel ANP-C receptor selective agonist. Int J Peptide Protein Res 1994;43: 332–336.
29. Trachte GJ, Kanwal S, Elmquist BJ, Ziegler RJ. C-type natriuretic peptide neuromodulates via "clearance" receptors. Am J Physiol 1995;268:C978–984.
30. Levin ER, Frank HJL. Natriuretic peptides inhibit rat astroglial proliferation: mediation by C receptor. Am J Physiol 1991;261:R453–457.
31. Brown J, Zuo Z. C-type natriuretic peptide and atrial natriuretic peptide receptors of rat brain. Am J Physiol 1993;264:R513–523.
32. Resink TJ, Panchenko MP, Tkachuk VA, Buhler FR. Involvement of $N_i$ protein in the functional coupling of the atrial natriuretic factor (ANF) receptor to adenylate cyclase in rat lung plasma membranes. Eur J Biochem 1988;174:531–535.
33. Anand-Srivastava MB, Franks DJ, Cantin M, Genest J. Atrial natriuretic factor inhibits adenylate cyclase activity. Biochem Biophys Res Comm 1984;121:855–862.
34. Barrett PQ, Isales CM. The role of cyclic nucleotides in atrial natriuretic peptide-mediated inhibition of aldosterone secretion. Endocrinology 1988;122:799–808.
35. Kishi Y, Ashikaga T, Watanabe R, Numano F. Atrial natriuretic peptide reduces cyclic AMP by activating cyclic GMP-stimulated phosphodiesterase in vascular endothelial cells. J Cardiovasc Pharmacol 1994;24:351–357.
36. McCall D, Fried TA. Effect of atriopeptin II on Ca influx, contractile behavior and cyclic nucleotide content of cultured neonatal rat myocardial cells. J Mol Cell Cardiol 1990;22:201–212.
37. Tkachuk VA, Hoffenberg SI, Starikova MG, Panchenko MP. Signal-transducing GTP-binding proteins of mammalian heart and lungs. J Mol Cell Cardiol 1989;21(Suppl I):91–95.

38. Pandey KN, Kovacs WJ, Inagami T. The inhibition of progesterone secretion and the regulation of cyclic nucleotides by atrial natriuretic factor in gonadotropin responsive murine leydig tumor cells. Biochem Biophys Res Comm 1985;133:800–806.
39. Sandberg K, Bor M, Carvallo PM, Catt KJ. Atrial natriuretic factor activates cyclic adenosine 3',5'-monophosphate phosphodiesterase in Xenopus laevis oocytes and potentiates progesterone-induced maturation via cyclic guanosine 5'-monophosphate accumulation. Biol Reprod 1993;49:1074–1082.
40. Lee MA, West RE, Moss J. Atrial natriuretic factor reduces cyclic adenosine monophosphate content of human fibroblasts by enhancing phosphodiesterase activity. J Clin Invest 1988;82:388–393.
41. Shimomura H, Nashida T, Imai A. Effect of PT-treatment on ANP-mediated inhibition of adenylate cyclase and amylase release in rat parotid gland. Mol Cell Biochem 1994;139:53–58.
42. Semrad CE, Cragoe EJ, Chang EB. Inhibition of Na/H exchange in avian intestine by atrial natriuretic factor. J Clin Invest 1990;86:585–591.
43. MacFarland RT, Zelus BD, Beavo JA. High concentrations of a cGMP-stimulated phosphodiesterase mediate ANP-induced decreases in cAMP and steroidogenesis in adrenal glomerulosa cells. J Biol Chem 1991;266:136–142.
44. Kanwal S, Trachte GJ. Mechanism of C-type natriuretic peptide neuromodulation: involvement of ANP-C receptor, phospholipase C and protein kinase C. FASEB J 1996;10:A159.
45. Berl T, Mansour J, Teitelbaum I. ANP stimulates phospholipase c in cultured RIMCT cells: roles of protein kinases and G protein. Am J Physiol 1991;260:F590–595.
46. Dai L-J, Quamme GA. Atrial natriuretic peptide initiates $Ca^{2+}$ transients in isolated renal cortical thick ascending limb cells. Am J Physiol 1993;265:F592–597.
47. Le Grand B, Deroubaix E, Couetil J-P, Coraboeuf E. Effects of atrionatriuretic factor on $Ca^{2+}$ current and $Ca_i$-independent transient outward $K^+$ current in human atrial cells. Pflugers Arch 1992;421: 486–491.
48. Gisbert M-P, Fischmeister R. Atrial natriuretic factor regulates the calcium current in frog isolated cardiac cells. Circ Res 1988;62:660–667.
49. Nakamaru M, Inagami T. Atrial natriuretic factor inhibits norepinephrine release evoked by sympathetic nerve stimulation in isolated perfused rat mesenteric arteries. Eur J Pharmacol 1986;123:459–461.
50. Babinski K, Haddad P, Vallerand D, McNicoll, De Lean A, Ong H. Natriuretic peptides inhibit nicotine-induced whole-cell currents and catecholamine secretion in bovine chromaffin cells: evidence for the involvement of the atrial natriuretic factor $R_2$ receptors. J Neurochem 1995;64:1080–1087.
51. Kuchel O, Debinski W, Racz K, Nguyen TB, Garcia R, Cusson JR, Larochelle P, Cantin M, Genest J. An emerging relationship between peripheral sympathetic nervous activity and atrial natriuretic factor. Life Sci 1987;40:1545–1551.
52. Morel G, Chabot J-G, Garcia-Caballero T, Gossard F, Dihl F, Belles-Isles M, Heisler S. Synthesis, internalization, and localization of atrial natriuretic peptide in rat adrenal medulla. Endocrinology 1988;123:149–158.
53. Johnson BG, Trachte GJ, Drewett JG. Neuromodulatory effect of the atrial natriuretic factor clearance receptor binding peptide, cANF(4–23)-$NH_2$ in rabbit isolated vasa deferentia. J Pharmacol Exp Ther 1991;257:720–726.
54. Ebert TJ, Groban L, Muzi M, Hanson M, Cowley AW. ANP-mediated volume depletion attenuates renal responses in humans. Am J Physiol 1992;263:R1303–1308.
55. Schultz HD, Steele MK, Gardner DG. Central administration of atrial peptide decreases sympathetic outflow in rats. Am J Physiol 1990;258:R1250–1256.
56. Levin ER, Weber MA, Mills S. Atrial natriuretic factor-induced vasodepression occurs through central nervous system. Am J Physiol 1988;255:H616–622.
57. Wong M, Samson WK, Dudley CA, Moss RL. Direct, neuronal action of atrial natriuretic factor in the rat brain. Neuroendocrinology 1986;44:49–53.
58. Haskins JT, Zingaro GJ, Lappe RW. Rat atriopeptin III alters hypothalamic neuronal activity. Neurosci Letts 1986;67:279–284.
59. Samson WK, Skala KD, Huang F-LS. CNP-22 stimulates, rather than inhibits, water drinking in the rat: evidence for a unique biological action of the C-type natriuretic peptides. Brain Res 1991;568:285–288.
60. Tang W, Paulding W, Summers C. ANP receptors in neurons and astrocytes from spontaneously hypertensive rat brain. Am J Physiol 1993;265:C106–112.
61. Tsuda K, Shima H, Nishio I, Masuyama Y. Inhibitory action on alpha-human atrial natriuretic polypeptide on vascular adrenergic neurotransmission is attenuated in spontaneously hypertensive rats. Jap Circ J 1987;51:589–593.

62. Cao L, Gardner DG. Natriuretic peptides inhibit DNA synthesis in cardiac fibroblasts. Hypertension 1995;25:227–234.
63. Brown J, Chen Q. Regional expression of natriuretic peptide receptors during the formation of arterial neointima in the rabbit. Circ Res 1995;77:906–918.
64. Kurtz A, Bruna RD, Preilschifter J, Taugner R, Bauer C. Atrial natriuretic peptide inhibits renin release from juxtaglomerular cells by a cGMP-mediated process. Proc Natl Acad Sci USA 1986;83:4769–4773.
65. Devlin AM, Leckie BJ. Second messengers involved in the control of renin secretion in cultured human nephroblastoma cells. Am J Physiol 1994;266:F667–673.
66. Dostal D. (1996) (personal communication).
67. Loeb AL, Gear RL. Potentiation of platelet aggregation by atrial natriuretic peptide. Life Sci 1988;43:731–738.
68. Hirata Y, Matsuoka H, Hayakawa H, Sugimoto T, Suzuki E, Sugimoto T, Kangawa K, Matsuo H. Role of endogenous atrial natriuretic peptide in regulating sodium excretion in spontaneously hypertensive rats. Hypertension 1991;17:1025–1032.

# 16 Natriuretic Peptides and Immune Function

*Angelika M. Vollmar*

**CONTENTS**



## INTRODUCTION

The natriuretic peptides (NP) are important regulators in the renal and vascular systems. The idea of an additional involvement of natriuretic peptides in immune function initially emerged by reports on the presence of the peptides in various immune organs. Data on regulation of NP expression and NP receptors in immune cells followed and supported the initial hypothesis. Finally, various effects of NP on immune cells could be demonstrated.

## LINK BETWEEN NATRIURETIC PEPTIDES AND THE IMMUNE SYSTEM

The atrial natriuretic peptide (ANP) was the first natriuretic peptide discovered (for review, *see* Chapters 1 and 2). Therefore, most of the early information regarding a link of the NP to the immune system focused on ANP. ANP was originally isolated from the heart atria and it is commonly accepted that the heart is the major site of synthesis for this peptide. However, shortly after its discovery evidence for an extra-atrial existence of ANP was given *(1)*. The demonstration that ANP was present in organs not related to cardiovascular physiology drew attention to other possible biological roles of ANP in addition to its effects on volume homeostasis. In this respect, it is worth mentioning that ANP was found in the digestive system *(2–4)*. ANP was localized in cells of the lamina propria, in enterochromaffin cells, and surprisingly also within lymphoid follicles of the

From: *Contemporary Endocrinology: Natriuretic Peptides in Health and Disease*
Edited by: W. K. Samson and E. R. Levin Humana Press Inc., Totowa, NJ

intestine *(2,4)*, findings that have recently been confirmed *(5)*. Based on this observation, the hypothesis of a link between ANP and the immune system emerged.

### *Discovery of Atrial Natriuretic Peptide in Various Immune Organs*

To support this hypothesis, organs of the immune system have been searched for ANP: The thymus of newborn and adult rats as well as of newborn humans was shown to synthesize ANP as ANP-immunoreactivity and mRNA coding for ANP were detected *(6–9)*. ANP-like material in newborn rats was characterized predominately as the prohormone ANP (1–126) accompanied by smaller amounts of ANP (99–126) *(6)*. In thymus of adult rats, additionally the N-terminal truncated form of the peptide, ANP (5–28) has been identified *(7)*. The amount of ANP in the thymus is about four orders of magnitude lower than in the atria. Furthermore, the expression of ANP in the thymus seems to be highly dependent on the age of the animals (*see* Regulation of Expression of Natriuretic Peptides in the Immune System). The age seems to be also an important factor regarding the localization of ANP in the thymus. Examining newborn rats we found ANP-positive cells mostly in the outer cortex *(6)*, presumably in thymocytes. In adult organs, ANP has been demonstrated in macrophages *(7,10)*.

In addition to the thymus, ANP was found in the spleen of guinea pigs *(11)* and rats *(8,12)*, in lymph nodes of various animal species, and in human tonsils *(13)*. Recently, ANP has also been identified in isolated immune cells, e.g., in macrophages of various origin *(14)*.

### *Expression of Brain- and C-Type Natriuretic Peptides in the Immune System*

Not only ANP, but also the other two natriuretic peptides seem to be expressed by the immune system. The extent of information on this issue, though, is rather scarce. We have detected specific mRNAs for BNP and CNP as well as their translational products coexpressed in the rat thymus *(15,16)* (Fig. 1). However, levels of both BNP and CNP-immunoreactivity were about one-third lower than that of ANP *(15)*. Minamino et al. reported similar data *(8)*. Coexpression of the three NPs is also evident in cultured macrophages *(17)*. In this respect, special attention should be paid to CNP. Ishizaka et al. have demonstrated that CNP was produced by two human leukemia cells lines, THP-1 and HL-60 cells *(18)*. Moreover, CNP production was drastically augmented on stimulation with a phorbolester (TPA) known to differentiate these cells into macrophages. Data of our work show that CNP is the predominate NP in macrophages *(17)*. Furthermore, CNP-immunoreactivity has been found in bone marrow in concentrations considerably higher than in other peripheral tissues *(8)*. Taken together, the widespread occurrence of NP in the immune system suggests a functional role of these peptides in the immune system.

## REGULATION OF EXPRESSION OF NATRIURETIC PEPTIDES IN THE IMMUNE SYSTEM

The fact that NP expression underlies certain regulatory mechanisms may provide a first hint regarding the functional significance of natriuretic peptides in immune organs.

### *Thymus*

ANP expression in the thymus was shown to be influenced by the animal's age as well as by certain pharmacological interventions. It is well known that the morphology and the

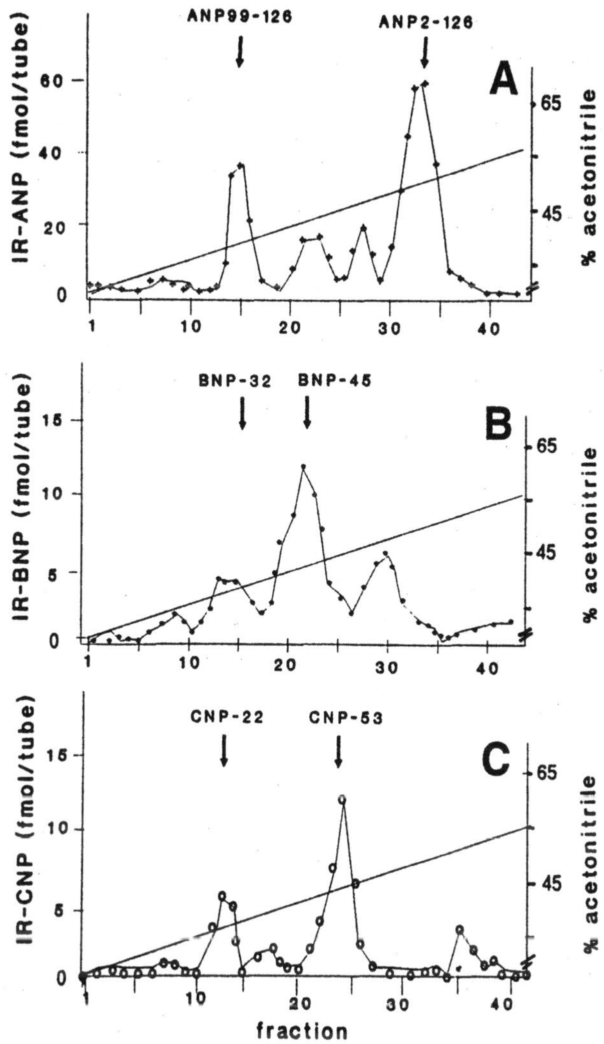


**Fig. 1.** NP immunoreactivity present in adult rat thymus. **(A)** Reverse-phase HPLC profiles of separated ANP-IR. Arrows indicate the elution positions of synthetic ANP (99–126) and prohormone ANP (2–126). **(B)** Fractions are assayed for BNP-IR. Column was calibrated with rat BNP-32 and rat BNP-45. **(C)** Elution profile of thymic CNP-IR compared with the positions of synthetic porcine CNP-22 and CNP-53.

functional status of the thymus is highly age-dependent *(19)*. Thus, it might be meaningful that the functional active thymus of newborn rats expresses significantly higher levels of ANP mRNA (up to 100-fold) and ANP-immunoreactivity (up to sixfold) as compared to thymus tissue of adult rats *(9)*. Decreasing thymic concentrations of ANP as a function of animal age was also reported by Throsby et al. *(20)*. Moreover, there is speculation on age-related differences in the posttranslational processing of the peptides *(20)* since ANP (1–126) dominates in neonatal tissue, whereas in adult animals considerable amounts of ANP (99–126) are found *(15,20,22)*.

Pharmacological tools known to have a powerful effect particularly on the thymus, such as administration of glucocorticoids or exposure to X-rays, profoundly alter ANP production *(10,21,22)*. Both agents dose-dependently induce apoptosis of thymocytes, peaking at d 3–4 after single exposure. As a function of time, however, involution of the thymus is reversible. Interestingly, on dexamethasone administration, the acutely involuted thymus expresses a strikingly increased ANP mRNA (up to 30-fold). During regeneration of the tissue, ANP mRNA decreases again and reaches control levels after about 3 wk. Data on ANP immunoreactivity parallel the mRNA data; i.e., a fivefold increase of ANP (1–126) immunoreactivity was found in the highly involuted thymus, which decreases to control ANP concentrations during degeneration *(21)*. BNP as well as CNP expression were not affected by thymic involution induced by dexamethasone *(15)*. Exposure of animals to X-rays employed as another mechanism to cause acute thymus involution resulted in a similar stimulation of ANP production *(22)*. Therefore, stimuli associated with induction of acute involution rather than direct effects of either the glucocorticoid or X-ray seem to be responsible for augmented ANP production. The enormous cell death evokes increased migration of macrophages into the thymus that display increased phagocytotic activity as judged by morphological and histochemical analysis *(10)*. These activated macrophages, however seem to be the thymic cells responsible for stimulated ANP expression in the acute involuted thymus. Immunohistochemistry revealed numerous heavily stained ANP-immunoreactive cells in the cortical area identified as macrophages (Fig. 2). Control thymus contains a rather low number of macrophages that do not display high phagocytotic activity, and these macrophages stain only weakly for ANP (Fig. 2). Thus, it is hypothesized that increased ANP-gene expression is linked to activation of macrophages. ANP may be released by activated macrophages and serve cytokine functions. In vitro experiments using cultivated macrophages support this idea and are communicated below.

### *Macrophages*

First, expression of ANP in macrophages is not restricted to the thymus tissue. mRNA coding for ANP is present in peritoneal- as well as in bone-marrow-derived macrophages *(14)*. Both types of macrophages synthesize and release ANP. In contrast to the thymic macrophages, ANP immunoreactivity of cultured macrophages mainly consists of the biologically active fragment ANP (99–126) accompanied with a smaller amount of ANP prohormone ANP (1–126). This different molecular pattern might be caused by specific culture conditions *(14)*. In any case, ANP seems to be a common constituent of macrophages.

Second and even more important, ANP expression in macrophages is indeed regulated by compounds affecting the activity of these cells, and thus supports the hypothesis raised above. Incubation of cells with lipopolysaccharides (LPS) and zymosan increased the ANP mRNA up to sixfold as determined by use of competitive PCR (Fig. 3, black bars). Intracellular ANP concentrations and ANP secretion were increased up to three- and sixfold, respectively. Both LPS and zymosan are known to induce a highly activated status of macrophages. They increase, for instance, phagocytic activity and secretion of various cytokines *(23,24)*. In contrast, dexamethasone negatively regulates macrophage functions *(25,26)* and elicits very small stimulatory effects on ANP expression (Fig. 3). Dexamethasone, however, is known to stimulate ANP production in various other cells *(27)*.

Summarizing these observations, the following conclusions can be drawn: A general activation of macrophages rather than a solely dexamethasone-specific effect may account for the stimulated ANP synthesis by thymic macrophages. This finding seems to apply to macrophages in general and the potent effect of zymosan on ANP expression suggests

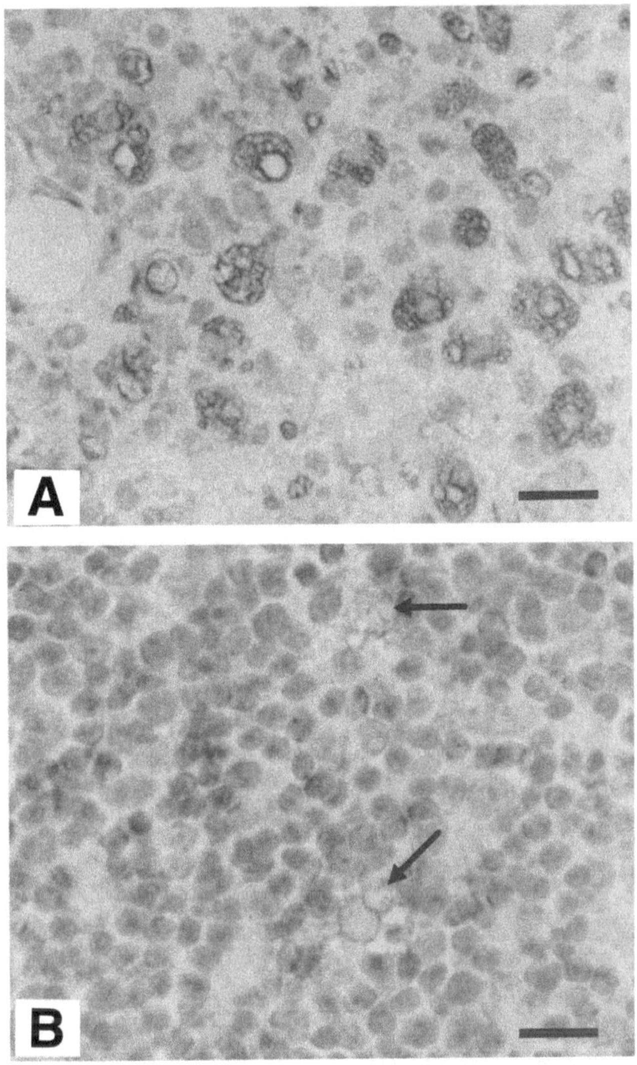


**Fig. 2.** Localization of ANP in the acute involuted thymus and normal thymus. The cell-depleted cortex of a thymus at d 4 after dexamethasone administration **(A)** contains numerous phagocytosing macrophages that strongly stain for ANP-IR. Macrophages of untreated thymus **(B)** show only very weak ANP-immunoreaction (arrows); bars: 17 μm.

that activation of phagocytosis may represent a major mechanism for stimulated ANP production by macrophages. Thus, ANP displays cytokine-like features.

Further studies showed that macrophages are able to synthesize and secrete BNP and CNP in addition to ANP *(17)*. Thus, all three peptides may be involved in modulation of macrophage-mediated immunological functions. As assumed for NP in ventricular tissue *(28–30)*, the physiological selection among the three NPs, however, may occur by way of differential regulation of their expression. Indeed, NP expression is differentially affected by various immunomodulatory substances also shown in Fig. 3. Most notably, a striking increase of mRNA coding for CNP occurs when cells were exposed to LPS (up to 300-fold)

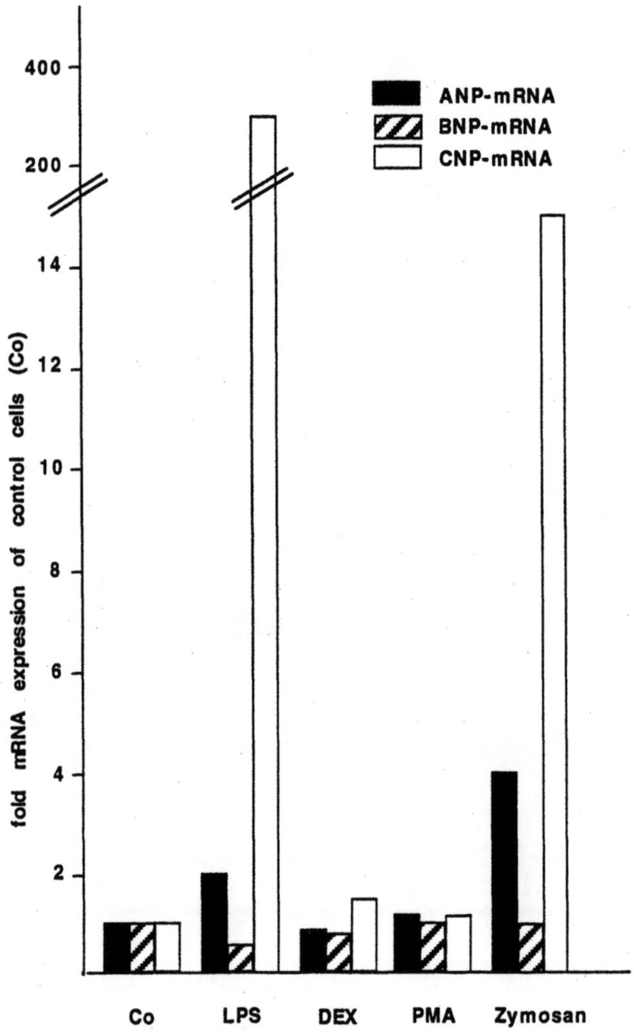


**Fig. 3.** Regulation of natriuretic peptide expression in macrophages by immunostimulants. Relative ANP, BNP, and CNP mRNA levels in unstimulated bone marrow macrophages (Co) and after stimulation with LPS, dexamethasone (Dex), phorbolester (PMA), and zymosan, respectively.

and zymosan (up to 15-fold) (Fig. 3, open bars). In parallel, intracellular as well as secreted CNP immunoreactivity was augmented up to 10-fold. Macrophage CNP-like immunoreactivity is composed of CNP-22 as a major form accompanied by smaller amounts of CNP-53 *(17)*. In contrast to ANP and CNP, BNP mRNA concentrations were shown to be decreased (up to fivefold) in LPS-activated cells (Fig. 3, hatched bars). BNP immunoreactivity was not altered in stimulated vs control cells *(17)*.

In addition to primary macrophages, monocytic leukemia cell lines (THP-1, HL-60) have the potential to synthesize CNP. CNP production is augmented more than 100-fold when cells were differentiated into macrophages by exposure to TPA *(18)*. ANP and BNP seem not to be expressed by these cell lines. These results suggest that differentiation into macrophage-like cells induces an active production of CNP. Furthermore, a different proteolytic processing of the CNP precursor in these cells was suggested since a 29 amino acid CNP (CNP-29) has been identified and neither CNP-22 nor CNP-53 as in the brain.

The striking effect of macrophage activation on CNP expression raises the question about the functional aspect of this observation. Initially, CNP was recognized as a neuropeptide being localized in the brain (*see* Chapters 12 and 13). Meanwhile, the peptide has also been identified in a variety of peripheral tissues *(7,15,31,32)*. Interestingly, in endothelial cells, CNP synthesis is also strongly stimulated by LPS and other cytokines *(31)*. Moreover, CNP production in the endothelium is influenced by ANP and BNP *(33)*. CNP itself blocks the growth of muscle cells induced by cytokines (*34* and *see also* Chapter 14).

Based on this information, one may speculate that CNP secreted from activated macrophages acts as local mediator in the growth regulation of either smooth muscle cells *(34)* or other cells, including immune cells. With respect to possible functions of the NP in the immune system, information on cells or tissues bearing receptors for NP is helpful.

## NATRIURETIC PEPTIDE RECEPTORS IN IMMUNE ORGANS AND CELLS

### *Identification and Localization*

Kurihara et al. were the first to report binding sites for ANP (99–126) in rat thymus and spleen as well as in isolated thymocytes *(35,36)*. They employed quantitative autoradiographic techniques and found a single class of saturable receptors. In the thymus, the number of binding sites was higher in the medulla than in the cortex. The white pulp was the predominant localization of ANP receptors in the spleen. In both thymocytes and spleen cells, addition of ANP (99–126) stimulated the formation of cyclic GMP, indicating the presence of guanylyl cyclase-coupled receptors. At the time of this investigation, the existence of NP-receptor subtypes was not known, and thus, no further distinction was attempted.

At present, NP receptors are categorized into at least three subtypes as thoroughly reviewed in Chapters 3–5: the guanylyl cyclase-coupled receptors NPR-A and NPR-B as well as the NPR-C, which lacks guanylyl cyclase activity. Whole thymus tissue expresses the mRNAs coding for all three NP-receptors shown by PCR analysis *(15)*. Moreover, within this organ, thymocytes seem to represent a cell population bearing NPR. Northern blot as well as PCR analysis demonstrated mRNAs for NPR-A, -B, and -C. Exposure of thymocytes to ANP and CNP increased cGMP accumulation in the medium *(37)*, suggesting the presence of functional NPR-A and -B proteins. Interestingly, mRNA for NPR-B, which is thought to be the specific receptor for CNP, was considerably higher expressed than that of NPR-A and NPR-C *(37)*.

Stroma cells seem to represent another thymic cell type expressing NP receptors, since the three NPRs could be detected in a murine thymic stromal cell line (MRL 104.8a) *(38)*. In addition, NPRs are also localized on cultured stromal cells originating from the bone marrow as well as on two bone-marrow-derived stromal cells lines (A-3, ALC) *(39)*. Binding- and affinity-crosslinking experiments with C-ANF revealed that most of the binding sites correspond to NPR-C. However, in addition, ALC cells contain both NPR-A and NPR-B receptors, whereas A-3 cells express mainly NPR-B.

ANP-specific guanylyl cyclase-coupled receptors seem to be also present on human mononuclear blood cells and neutrophils since ANP evokes a cGMP response on both cell types *(40,41)*. Moreover, expression of NPR-A mRNA was shown for the mononuclear blood cells *(40,41)*.

Finally, ANP receptors are suggested to be also localized on macrophages. Stimulation of guanylyl cyclase after incubation with ANP has been reported for human peritoneal macrophages as well as for the macrophage cell line J774 *(42,43)*. Our own unpublished data obtained from mouse bone-marrow macrophages and peritoneal macrophages support this assumption.

### *Regulation of NPR Expression*

It was again Kurihara et al. *(44)* who reported that expression of NP receptors in the thymus and spleen underlies regulatory mechanisms. Thymocytes and spleen cells of spontaneously hypertensive rats show lower binding capacities. The immune system of these animals is altered as decreased number of immature T-lymphocytes and an abnormal production of thymic hormone have been demonstrated.

Recently, it has been shown that NPR expression in thymocytes is regulated by mitogenic stimuli *(37)* and importantly the expression of NP receptors seems to be differentially modulated on stimulation of cell growth. Exposure of cells with concanavalin A (Con A, 48 h) results in an increase of mRNA coding for NPR-A, but not for NPR-B or NPR-C. In agreement with these data, only ANP, but not CNP, elicits increased cGMP response in Con A-stimulated cells.

Similar observations have been made on human peripheral blood mononuclear cells (PBMCs) *(40)*. Freshly isolated PBMCs showed no sensitivity to ANP, whereas cells which have been cultivated in the presence of phytohemagglutinin (PHA, 48 h) respond to ANP by a 10-fold increased cGMP concentration as compared to unstimulated cells (Fig. 4). CNP had no detectable effect on cGMP accumulation in these cells. The appearance of this ANP-sensitive guanylyl cyclase activity in mitogen-stimulated cells seems to be caused by *de novo* synthesis of the enzyme rather than because of activation of an inactive enzyme: The mRNA for NPR-A in PBMC increases time dependently over the period of incubation with PHA (Fig. 4). Subfractionation of cultured PBMC and *in situ* PCR revealed that the majority of T-cells showed specific PCR-products; the B-cell-enriched fraction and monocytes displayed only background levels of mRNA *(40)*. The data on both thymocytes and PBMC NPR suggest that ANP may play a role as a regulator of T-cell growth and differentiation.

As mentioned before, stromal cells are another type of target cells for NP in the thymus. The growth factor TGF-$\beta_1$ has been shown to increase ANP-binding sites in a murine stromal cell line *(38)*. Interestingly, the augmented binding sites are NPR-C and not NPR-A in these cells. This augmentation of NPR-C expression occurs also at the transcriptional level shown by Northern blot and in vitro nuclear transcription assay.

Expression of NPR-A and NPR-B is downregulated by TGF-$\beta_1$, thus TGF-$\beta_1$ elicits opposite actions on the NPR-genes. TGF-$\beta_1$ and ANP may alter the function of stromal cells, known to play an important role in thymopoesis *(45)*. Taken together, these data further strengthen the idea that ANP is involved in the complex mechanism of T-cell development.

## FUNCTIONAL ASPECTS OF NATRIURETIC PEPTIDES

### *Proliferation of Thymocytes*

As discussed above, stimulation of thymocyte proliferation causes an augmented expression of their ANP-receptor (NPR-A). Consequently, the question arises regarding

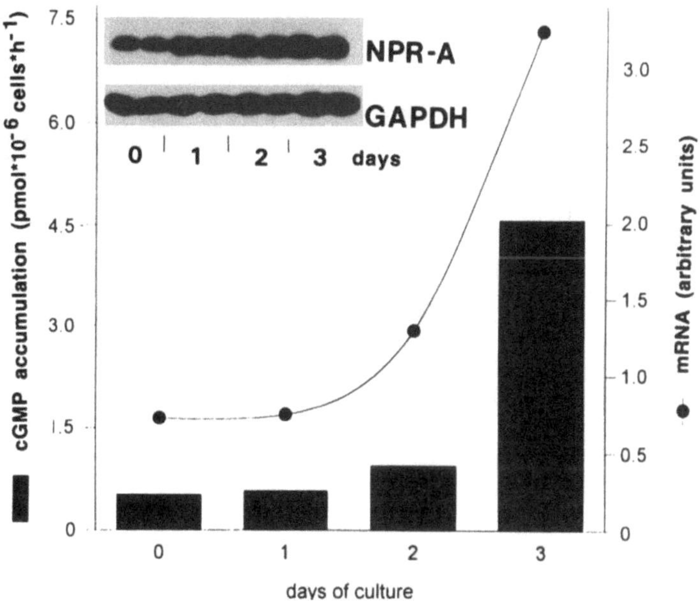


**Fig. 4.** Regulation of NPR-A expression in human PBMCs. Time-dependent increase of cGMP accumulation (■) and NPR-A-specific mRNA (●). Cells were cultured for up to 3 d in the presence of phytohemagglutinin (1 mg/L). ANP-sensitive guanylyl cyclase activity (1 $\mu M$ ANP, 60 min, 37°C) was measured at the indicated timepoints. NPR-A specific mRNA in cells was detected by PCR (insert). NPR-A mRNA was quantified by the amount of $^{32}$P-ATP incorporated in the PCR products, which were equalized to the signals obtained for GAPDH mRNA *(40)*.

whether ANP interferes with the activated cell growth of thymocytes. Proliferation of thymocytes can be stimulated with Con A and monitored by measurement of incorporation of radioactive-labeled thymidine.

ANP at concentrations of $10^{-9}$–$10^{-6}M$ significantly inhibited mitogen-activated thymocyte proliferation *(37)* (Fig. 5). CNP, in contrast, did not share the inhibitory properties of ANP. C-ANF, a specific ligand of the NPR-C, was also unable to mimic the effect of ANP, suggesting that the suppression was mediated via the NPR-A. Thus, one may speculate that an increased expression of NPR-A represents a mechanism rendering the cells more sensitive for ANP action. As reviewed in Chapter 14, NPs interfere with the proliferation of various other cell types, and it is surprising that CNP, known to be a potent inhibitor of growth in other cell systems, did not show any effect on thymocytes. Regarding the fact that receptors specific for CNP (NPR-B) are rather highly expressed in the thymus, the functional significance of CNP in this organ remains to be elucidated. ANP, however, seems to function as a regulator of growth in thymocytes and might indeed influence thymopoesis, as suggested earlier *(6,9)*.

## *NO Synthesis*

Macrophages are another possible target cells for NP effects beside thymocytes. Macrophages play their central role in the regulation of immune function partly by secretion of various organic and inorganic compounds *(23)*. Synthesis of NO by activated macrophages has recently been the subject of numerous investigations (for review, *see*

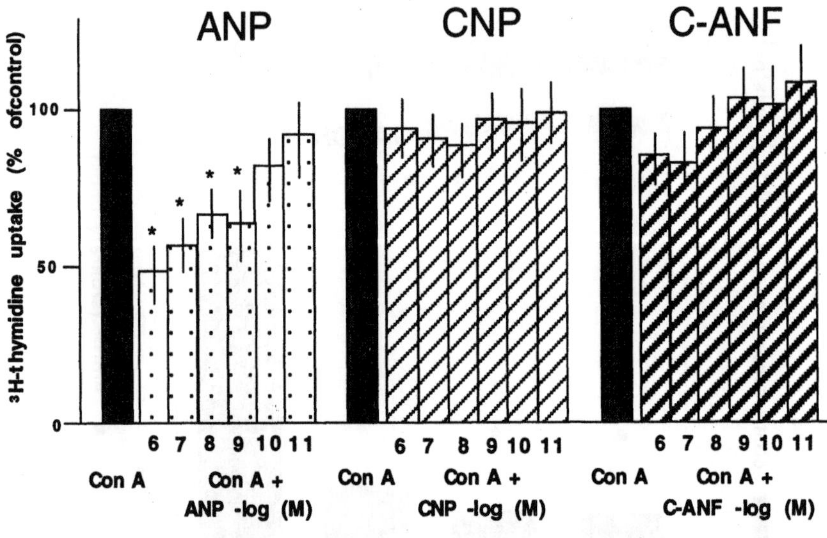


**Fig. 5.** Effect of natriuretic peptides on proliferation of thymocytes. Thymocytes were treated with Con A (1 μg/mL, 48 h) in the presence or absence of decreasing concentrations of ANP, CNP, and (des-$Gln^{18}$, $Ser^{19}$,$Gly^{20}$,$Leu^{21}$,$Gly^{22}$)-ANF (4–23)-$NH_2$ (C-ANF). Rate of proliferation is measured by incorporation of $^3$H-thymidine. Radioactivity taken up by Con A-stimulated cells is set as 100%.

ref. *46*). Importantly, ANP is able to inhibit the production of NO by LPS-stimulated macrophages in a dose- and time-dependent manner *(47)*. CNP as well as NP-related peptides, such as urodilatin, and atriopeptin I did not show this effect (Vollmar et al., manuscript in preparation). The effect can be blocked by the guanylyl cyclase-specific receptor antagonist HS-142-11, and C-ANF does not mimic the effect (Vollmar et al., manuscript in preparation). Thus, the effect is most likely mediated by the NPR-A receptor. Since macrophages express only one type of NO synthase, i.e., the inducible NOS or NOS-II, ANP might interfere with the expression and/or activity of this enzyme. The exact mechanism underlying the inhibition of iNOS by ANP is the subject of our current investigations.

An interference of ANP with the NO system is reported likewise in human proximal tubular cells *(48,49)*. In addition to the expression of iNOS, these cells contain the constitutive NOS-III or eNOS. Exposure of tubular cells to ANP resulted in an increased NO production. This stimulation by ANP might be caused via either inducible or constitutive NOS. Interestingly, NO production that was induced by a mixture of cytokines was inhibited in the presence of ANP. Thus, the negative regulation of NO production by ANP is most likely caused by an effect of the peptide on the iNOS system. The functional significance of both the stimulation and the inhibition of NO production by ANP remains to be investigated.

### *Respiratory Burst Activity and Phagocytosis of Macrophages*

ANP influences further important functions of phagocytes, i.e., their production of reactive oxygen radicals as well as their phagocytic activity. Wiedermann et al. reported that ANP is a potent signal for priming polymorphonuclear neutrophils (PMN) to secrete superoxide anion *(50)*. Cells were briefly exposed to ANP ($10^{-7}$ *M*) and the respiratory burst of cells stimulated with phorbolester, opsonized zymosan, or the chemotractant

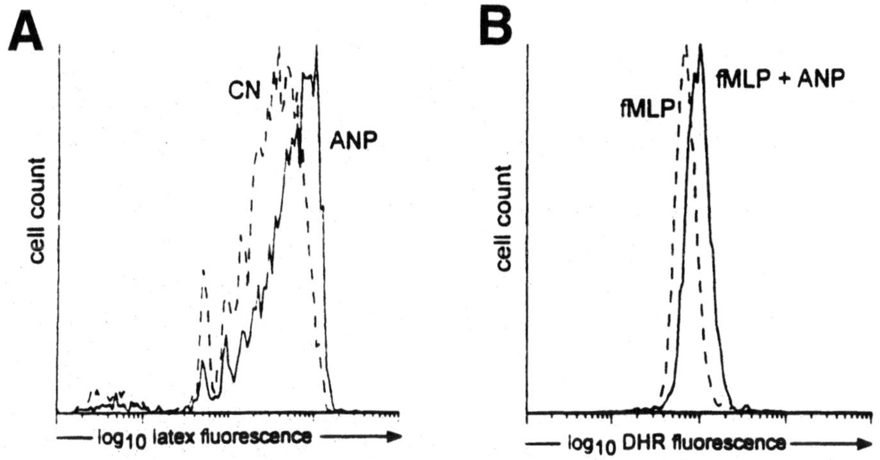


**Fig. 6.** Flowcytometric analysis of ANP effects on phagocytosis and respiratory burst. **(A)** Phagocytosis activity is quantified by the number of ingested particles, which is proportional to the fluorescence intensity. The histogram (fluorescence intensity vs number of cells) shows the stimulatory effect of ANP ($10^{-9}M$, 18 h) on uptake of fluorescent latex particles by J774 cells as a fluorescence shift to the right of ANP-exposed cells (ANP, solid line) in comparison to control cells (CN, dashed line) is evident. **(B)** Histogram showing the stimulatory effect of ANP ($10^{-8}M$) (solid line) on reactive oxygen production of J774 cells induced by fMLP ($10^{-6}M$) (dashed line). Cells were preincubated with ANP for 2 h and respiratory burst activity was monitored by fluorescence of oxidized dihydrorhodamine-123 (DHR-123).

molecule FMLP was markedly enhanced. Because ANP failed to stimulate cGMP accumulation in PMN, the authors suggest that the effect is mediated by the NPR-C. We observed similar effects of ANP on oxygen radical production using the murine macrophage cell line J774 and flow cytometry as a technique to quantify reactive oxygen species in the cell (*51*; Fig. 6B).

Recently these data have been supported and extended by the work of Biselli et al. *(50)*. Coincubation of human PMN with ANP and FMLP or ANP and the eicosanoid chemoattractant leukotriene B4 (LTB4) resulted in a potentiated peroxide production at ANP concentrations as low as $10^{-14}$ and $10^{-12}M$. The discrepancy in ANP concentration necessary for maximal effect might be explained by differences in the experimental designs of the studies.

Phagocytosis of IgG complexes by macrophages (J774) was significantly suppressed when cells were incubated with ANP at high concentrations of $10^{-5}$ and $10^{-6}M$ *(43)*. This inhibitory effect on phagocytic activity can be mimicked by a stable cGMP analog, suggesting that the effect is mediated via cGMP. In contrast, we found that ANP stimulates the uptake of opsonized fluorescent particles by J774 cells employing flow cytometry to measure phagocytosis (Fig. 6A). In this respect, it is important to note that ANP elicited its maximal effect at $10^{-9}M$. The effect progressively decreased with increasing ANP concentrations and reached basal levels at a concentration of $10^{-6}M$ ANP *(51)*. ANP might influence phagocytosis in a bell-shaped dosis curve. Based on the observation that ANP plays a role in three most important functions of macrophages, i.e., NO production, respiratory burst, and phagocytosis an involvement of the peptide in cellular defense mechanisms is suggested.

### *Miscellaneous*

A possible function of ANP in host defense is further supported by the fact that ANP affects migration and chemotaxis of neutrophils *(41,50)*. Wiedermann et al. *(50)* showed that ANP at concentrations of $10^{-7}M$ maximally inhibited the FMLP-stimulated directed migration of PMN into nitrocellulose filters. ANP failed to have an effect on PMN whose locomotion was not stimulated by FMLP. The authors suggest that the observed inhibition of migration may reflect an enhancement of cell adhesiveness induced by ANP as observed for other PMN-priming factors, such TNF or GM-CSF. Lower concentrations of ANP ($2 \times 10^{-8}M$) seem to stimulate migration of neutrophils, however, only in the absence of a chemotactic stimulus, such as FMLP *(41)*. Furthermore, the activating effect of ANP is restricted to a rather small concentration range ($10^{-8} - 4 \times 10^{-8}M$). Higher concentrations of the peptide are also shown to be inhibitory. Furthermore, an involvement of pertussis-toxin-sensitive G proteins was suggested since ANP-activated migration is inhibited by pertussis toxin *(41)*.

ANP is able to augment natural killer cell (NK) activity in vitro *(53)*. The effect seems to be highly dose- and time-dependent. At concentrations of 250 pg/mL and an incubation time of 24 h, ANP evokes the highest response. The authors discuss the fact that the ANP concentrations inducing the greatest augmentation of NK activity were at levels found in patients with cardiac diseases. A short-term increase of NK activity is seen during tachycardia and NK activity is involved in early mild organ rejection. Thus, ANP might influence immunological function in cardiovascular disease *(53)*.

Interestingly, ANP is also able to induce histamine release from mast cells of the peritoneum *(54)*. This original report has recently been extended by others *(55)* since the effect of the other two natriuretic peptides, BNP and CNP, on histamine release has been tested. ANP and BNP were equally effective at concentration of $10^{-6}M$ and higher *(54,55)*. The histamine-releasing activities of ANP and BNP were twice as potent than that of substance P; that of CNP was equal to substance P. Although the mechanism of histamine release by neuropeptides is still uncertain, it is generally agreed that a high degree of arginine rather than lysine residues and spacing of basic residues are important factors. NP are poly-basic peptides and the greater positive charge of ANP and BNP in comparison to that of CNP may explain the difference in effects among the NP. On one hand, the induction of histamine release by NP may limit their usefulness as therapeutic agents. On the other hand, corresponding studies into the possible role of NP as regulators in inflammation might be interesting.

## REFERENCES

1. Gutkowska J, Nemer M. Structure, expression and function of atrial natriuretic factor in extraatrial tissues. Endocr Rev 1989;10:519–537.
2. Vollmar AM, Friedrich A, Sinowatz F, Schulz R. Presence of atrial natriuretic peptide-like material in guinea pig intestine. Peptides 1988;9:965–971.
3. Vuolteenaho O, Arjamaa O, Vakkuri O, Maksniemi T, Nikkilä L, Kangas J, Puurunen J, Ruskoaho H, Leppäluoto J. Atrial natriuretic peptide (ANP) in rat gastrointestinal tract. FEBS Lett 1988;233:79–82.
4. Gerbes AL, Nathrath W, Cantin M, Denecke H. Presence of atrial natriuretic factor prohormone in enterochromaffin cells of the large intestine. Gastroenterology 1991;101:424–429.
5. Hervieu G, Volant K, Grishina O, Descroix-Vagne M, Nahon JL. Similarities in cellular expression and functions of melanin-concentrating hormone and atrial natriuretic factor in the rat digestive tract. Endocrinology 1995;137:561–571.
6. Vollmar AM, Schulz R. Evidence for the presence of ANP-precursor material in the rat thymus. Biochem Biophys Res Commun 1988;155:700–708.

7. Throsby M, Yang Z, Lee D, Huang W, Copolov DL, Lim AT. *In vitro* evidence for atrial natriuretic factor-(5–28) production by macrophages of adult rat thymi. Endocrinology 1993;132:2184–2190.
8. Minamino N, Aburaya M, Kojima M, Miyamoto K, Kangawa K, Matsuo H. Distribution of C-type natriuretic peptide and its messenger RNA in rat central nervous system and peripheral tissue. Biochem Biophys Res Commun 1993;197:326–335.
9. Vollmar AM, Schulz R. Atrial natriuretic peptide is synthesized in the human thymus. Endocrinology 1990;126:2277–2280.
10. Vollmar AM, Colbatzky F, Schulz R. Expression of atrial natriuretic peptide in thymic macrophages after treatment of rats with dexamethasone. Cell Tissue Res 1992;268:397–399.
11. Vollmar AM, Friedrich A, Schulz R. Immunoreactive atrial natriuretic peptide in the guinea pig spleen. Life Sci 1989;45:1293–1297.
12. Throsby M, Lee D, Huang W, Yang Z, Copolov DL, Lim AT. Evidence for atrial natriuretic peptide-(5–28) production by macrophages of the rat spleen: an immunochemical and nonradioactive *in situ* hybridization approach. Endocrinology 1991;129:991–1000.
13. Vollmar AM, Schulz R. Atrial natriuretic peptide in lymphoid organs of various species. Comp Biochem Physiol 1990;96A:459–463.
14. Vollmar AM, Schulz R. Expression and secretion of atrial natriuretic peptide in cultured macrophages. J Clin Invest 1994;94:539–545.
15. Vollmar AM, Wolf R, Schulz R. Co-expression of the natriuretic peptides (ANP, BNP, CNP) and their receptors in normal and acutely involuted thymus. J Neuroimmunol 1995;57:117–127.
16. Vollmar AM, Gerbes AL, Nemer M, Schulz R. Detection of C-type natriuretic peptide (CNP) transcript in the rat heart and immune organs. Endocrinology 1993;132:1872–1874.
17. Vollmar AM, Schulz R. Expression and differential regulation of natriuretic peptides in mouse macrophages. J Clin Invest 1995;95:2442–2450.
18. Ishizaka Y, Kangawa K, Minamino N, Ishii K, Takano S, Eto T, Matsuo H. Isolation and identification of C-type natriuretic peptide in a human monocytic line, THP-1. Biochem Biophys Res Commun 1992;189:697–704.
19. Stutman O. Ontogenity of T-cells. Clin Immunol Allergy 1985;5:191–234.
20. Throsby M, Yang Z, Copolov DL, Lim AT. Colocalization of atrial natriuretic factor and β-endorphin in rat thymic macrophages. Peptides 1994;15:291–296.
21. Vollmar AM, Schulz R. Dexamethasone action on rat thymic atrial natriuretic peptide. Endocrinology 1990;127:3240–3242.
22. Vollmar AM, Colbatzky F, Schulz R. Increased production of atrial natriuretic peptide in the rat thymus after irradiation. Immunopharmacology 1993;26:65–72.
23. Adams, DO, Hamilton TA. The cell biology of macrophage activation. Annu Rev Immunol 1984;2:283–318.
24. Green SP, Hamilton JA, Philips WA. Zymosan-triggered tyrosine phosphorylation in mouse bone marrow-derived macrophages—a possible mechanism for glucocorticoid-mediated suppression of endotoxicity. J Immunol 1992;149:4041–4047.
25. Snyder DS, Unanue ER. Corticoids inhibit murine macrophage la expression and interleukin-1 production. J Immunol 1982;129:1803–1809.
26. Di Rosa M, Radomski R. Carnuccio R, Moncada S. Glucocorticoids inhibit the induction of nitric oxide synthase in macrophages. Biochem Biophys Res Commun 1990;172:1246–1252.
27. Gardner DG, Gertz BJ, Deschepper CF, Kim DY. Gene for the atrial natriuretic peptide is regulated by glucocorticoids in vitro. J Clin Invest 1988;82:1275–1281.
28. Takahashi T, Allen PD, Izumo S. Expression of A-, B-, and C-type natriuretic peptide genes in failing and developing human ventricles. Correlation with expression of the $Ca^{2+}$ ATPase gene. Circ Res 1992;71:9–17.
29. Wei CM, Heublein DM, Perella MA, Lerman A, Rodeheffer CG, McGregor A, Edwards WD, Schaff H, Burnett JC. Natriuretic peptide system in human heart failure. Circulation 1993;88:1004–1009.
30. Gerbes AL, Dagnino L, Nguyen T, Nemer M. Transcription of brain natriuretic peptide and atrial natriuretic peptide genes in human tissues. J Clin Endocrinol Metab 1994;78:1307–1311.
31. Suga SK, Itoh H, Komatsu Y, Ogawa Y, Hama T, Yoshimasa T, Nakao K. Cytokine-induced C-type natriuretic peptide (CNP) secretion from vascular endothelial cells: evidence for CNP as a novel autocrine/paracrine regulator from endothelial cells. Endocrinology 1993;133:3038–3041.
32. Stingo AJ, Clavell AL, Heublein DM, Wei CM, Pittelkow MR, Burnett Jr JC. Presence of C-type natriuretic peptide in cultured human endothelial cells and plasma. Am J Physiol 1992;263: H1318–1321.

33. Nazario B, Hu R-M, Pedram A, Prins B, Levin ER. Atrial and brain natriuretic peptides stimulate the production and secretion of C-type natriuretic peptide from bovine aortic endothelial cells. J Clin Invest 1995;95:1151–1157.
34. Porter JG, Catalano R, McEnroe G, Lewicki JA, Protter AA. C-type natriuretic peptide inhibits growth factor-dependent DNA synthesis in smooth muscle cells. Am J Physiol 1992;263:C1001–C1006.
35. Kurihara M, Sigematsu K, Saavedra JM. Localization of atrial natriuretic peptide, ANP (99–126) binding sites in the ret thymus and spleen with quantitative autoradiography. Regul Peptides 1986;15:341–346.
36. Kurihara M, Katamine S, Saavedra JM. Atrial natriuretic peptide, ANP (99–126), receptors in rat thymocytes and spleen cells. Biochem Biophys Res Commun 1987;145:789–796.
37. Vollmar AM, Schmidt KN, Schulz R. Natriuretic peptide receptors on rat thymocytes: inhibition of proliferation by atrial natriuretic peptide. Endocrinology 1996;137:1706–1713.
38. Agui T, Xin X, Cai Y, Shim G, Muramatsu Y, Yamada T, Fujiwara H, Matsumoto K. Opposite actions of transforming growth factor-β1 on the gene expression of atrial natriuretic peptide biological and clearance receptors in a murine thymic stromal cell line. J Biochem 1995;118:500–507.
39. Agui T, Yamada T, Legros G, Nakajima T, Clark M, Peschel C, Matsumoto K. Expression of receptors for atrial natriuretic peptide on the murine bone marrow-derived stromal cells. Endocrinology 1992;130:2487–2494.
40. Heim JM, Knau B, Sturm C, Fricke H, Hartmann J, Pachmann K, Schmidt KN, Vollmar AM, Gerzer R. Particulate ANP-sensitive guanylyl cyclase: induction in cultured human peripheral CD3$^+$ mononuclear blood cells. J Mol Med 1996;74:629–635.
41. Elferink JGR, De Koster BM. Atrial natriuretic factor stimulates migration by human neutrophils. Eur J Pharmacol (Mol Pharm Sec) 1995;288:335–340.
42. Houdijik AP, Adolfs MJ, Bonta IL, DeLonge HR. Atriopeptins and nitroprusside provoke opposite changes in cGMP and cAMP levels in human macrophages. Eur J Pharmacol 1990;179:413–417.
43. Mattana J, Singhal PC. Effects of atrial natriuretic peptide and cGMP on uptake of IgG complexes by macrophages. Am J Physiol 1993;265:C92–98.
44. Kurihara M, Castren E, Gutkind JS, Saavedra JM. Lower number of atrial natriuretic peptide receptors in thymocytes and spleen cells of spontaneously hypertensive rats. Biochem Biophys Res Commun 1987;149:1132–1140.
45. Dardenne M, Savino W. Control of thymus physiology by peptidic hormones and neuropeptides. Immunol Today 1994;15:518–523.
46. Xie Q, Nathan C. The high output nitric oxide pathway: role and regulation. J Leukoc Biol 1994;56: 576–582.
47. Vollmar AM, Schulz R. Atrial natriuretic peptide inhibits nitric oxide synthesis in mouse macrophages. Life Sci 1995;56,PL:149–155.
48. McLay JS, Chatterjee PK, Jardine AG, Hawsworth GM. Atrial natriuretic factor modulates nitric oxide production: an ANF-C receptor mediated effect. J Hypertens 1995;13:625–630.
49. McLay JS, Chatterjee PK, Mistry SK, Weerakody RP, Jardine AG, McKay NG, Hawksworth GM. Atrial natriuretic factor and angiotensin 11 stimulate nitric oxide release from human proximal tubular cells. Clin Sci 1995;89:527–531.
50. Wiedermann CJ, Niedermühlbichler M, Braunsteiner H. Priming of polymorphonuclear neutrophils by atrial natriuretic peptide in vitro. J Clin Invest 1992;89:1580–1586.
51. Vollmar AH, Förster R, Schulz R. Effects of atrial natriuretic peptide on phagocytosis and respiratory burst in murine macrophages. Eur J Pharm 1987;319:279–285.
52. Biselli R, Farrace S, DeSimone C, Fattorossi A. Potentiation of human polymorphonuclear leukocyte activation by atrial natriuretic peptide. Inhibitory effect of carnitine congeners. Inflammation 1996;20:33–42.
53. Moss RB, Golightly MG. In vitro enhancement of natural cytotoxicity by atrial natriuretic peptide fragment 4–28. Peptides 1991;12:851–854.
54. Opgenorth TJ, Budzik GP, Mollison KW, Davidsen SK, Holst MR, Holleman WH. Atrial peptides induce mast cell histamine release. Peptides 1990;11:1003–1007.
55. Yoshida H, Inagaki Y, Yamaki K, Beppu Y, Kawashima T, Takagi K. Histamine release induced by human natriuretic peptide from rat peritoneal mast cells. Reg Peptides 1996;61:45–49.

# 17 Design and Utilization of Natriuretic Peptide Antagonists

*Yuzuru Matsuda*

*CONTENTS*



## INTRODUCTION

Natriuretic peptides play an important role in the hormonal regulation of blood pressure and sodium balance (for review, *see* refs. *1–3*). Recently, the actions of these hormones have extended beyond vasodepressor and natriuretic functions to embrace antiproliferative effects and a role in vascular remodeling *(3)*. The naturally occurring members of the natriuretic peptide family include atrial (ANP), brain (BNP), and C-type natriuretic peptide (CNP) *(1–3)*. Their tissue distribution differs substantially. Although the intracellular mechanisms through which natriuretic peptides elicit their diverse biological effects are not known, their biological effects are mediated by binding to specific membrane-associated receptors in the target tissues *(4,5)*.

At least two biochemically and functionally distinct classes of receptors are present in a variety of tissues: guanylyl cyclase-linked and guanylyl cyclase-free receptors with molecular weights of 120–140 and 60–70 kDa, respectively *(1–3)*. The guanylyl cyclase-linked receptor, which is presumed to be the biological receptor, contains a guanylyl cyclase domain in its structure. The binding of the peptide hormone to an extracellular ligand-binding domain is directly associated with an increase in intracellular cyclic GMP (cGMP) accumulation by activation of the membrane-bound guanylyl cyclase *(6,7)*. The cloning and expression of cDNA encoding a mammalian guanylyl cyclase-linked type of natriuretic peptide receptor has shown two different subtypes of the receptor: an A-type (ANPR-A or GC-A) and a B-type (ANPR-B or GC-B) *(8,9)*. The other major class of

From: *Contemporary Endocrinology: Natriuretic Peptides in Health and Disease*
Edited by: W. K. Samson and E. R. Levin Humana Press Inc., Totowa, NJ

natriuretic peptide receptors, guanylyl cyclase-free receptor (ANPR-C), is unique among polypeptide receptors because it is thought to be specific for storage or clearance of natriuretic peptides, playing an important role in the removal of natriuretic peptides from circulation *(10,11)*. Recent studies concerning natriuretic peptides and their receptors have revealed the complex nature of the relationships between ligands and multiple receptor subtypes *(12,13)*.

On the other hand, the physiological and pathophysiological functions of natriuretic peptides are still unclear *(14,15)*. A specific antagonist for the functional natriuretic peptide receptor would be highly useful in order to clarify this point *(2)*. HS-142-1 seems to be such an antagonist for functional ANP-A and ANP-B receptors for natriuretic peptide *(16)*. In this chapter, we describe its biochemical and pharmacological properties and utilization of the antagonist.

## DESIGN OF NATRIURETIC PEPTIDE ANTAGONISTS

Several ANP analogs and structurally related peptides were found to have antagonistic activities against guanylyl cyclase-linked and/or guanylyl cyclase-free type receptors *(17–19)*. Von Geldern et al. *(20)* have reported that another structurally related peptide antagonizes exogenous ANP-induced diuresis, natriuresis, and enhancement of urinary cGMP excretion in vivo, but it also has an agonistic effect on vascular guanylyl cyclase-linked receptors.

Furthermore, metabolic instability of these peptides might limit their usefulness for in vivo studies. It is well known that microorganisms are a virtually unlimited source of novel and diverse chemical compounds. Weber et al. *(21)* discovered anantin, a peptide binding to the ANP receptors, from a strain of genus *Streptomyces*. Although anantin was found to bind competitively to ANP receptors from adrenal cortex, it completely displaced [$^{125}$I]rANP binding to the receptors. This fact suggests that anantin is a nonselective ligand for guanylyl cyclase-linked and guanylyl cyclase-free type receptors, since adrenocortical membranes are composed of these two types of receptors. In any case, a true antagonist that specifically blocks the actions of natriuretic peptides in vivo through specific antagonism at the guanylyl cyclase-linked receptor had not yet been available. We discovered a microbial metabolite named HS-142-1 to be such an antagonist *(16)*.

## THE DISCOVERY OF HS-142-1

During our screening program for ANP receptor ligands of microbial origin, a fungal strain of genus *Aureobasidium*, KAC-2383, isolated from a soil sample collected in Japan, was found to produce a novel substance that inhibited [$^{125}$I]rANP [=ANF (99–126)] binding to ANP receptor in rabbit renal cortical membranes. Structural analysis with $^{1}$H and $^{13}$C NMR and fast atom bombardment (FAB-MS) spectrometers showed that HS-142-1 was a novel polysaccharide composed of 20–30 kinds of linear β-1,6-glucan esterified by caproyl groups. By repeated gel-filtration column chromatography, we separated these substances into single active constituents that were almost equally potent as ANP antagonists *(22,23)*. However, the amount of single active constituents was too low to permit further structural and pharmacological characterization. To obtain the biochemical and pharmacological profile of these substances, we used a chromatographically defined mixture of active constituents as shown in Fig. 1. HS-142-1 is also the first example of a polysaccharide ligand for peptide hormone receptors.

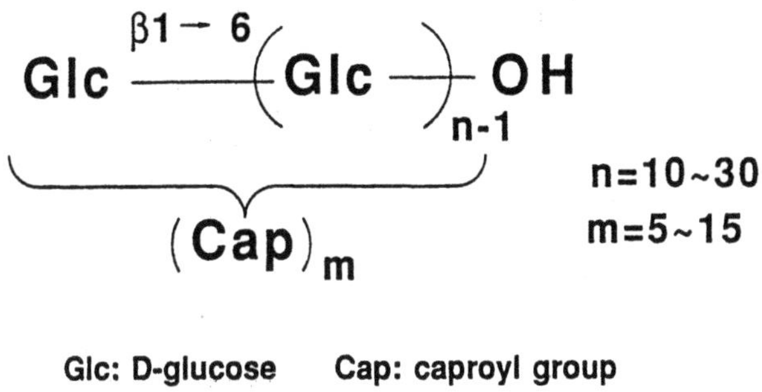


**Fig. 1.** Chemical structure of HS-142-1. Glc, D-glucose; Cap, capronic acid.

# BIOCHEMISTRY OF HS-142-1

## *Receptor Binding*

HS-142-1 inhibited [$^{125}$I]rANP binding to rabbit renal cortical membranes with an $IC_{50}$ value (concentration that produces 50% inhibition of the specifically bound [$^{125}$I]rANP) of 0.3 μg/mL, but at concentrations up to 100 μg/mL did not inhibit specific binding of [$^{125}$I]angiotensin II or [$^{125}$I]endothelin-1 to bovine lung membranes. Scatchard analysis of the effects of HS-142-1 on the binding characteristics of [$^{125}$I]rANP to purified guanylyl cyclase-linked receptor was examined, and the estimated inhibition constant (Ki) of HS-142-1 was 0.3 μ*M*, assuming the mean molecular weight of HS-142-1 to be 4000 *(22)*.

The selectivity of HS-142-1 for inhibition of [$^{125}$I]rANP binding to different ANP receptor subtypes was extensively studied *(22–27)*. In bovine adrenocortical membranes, which have been shown to express predominantly guanylyl cyclase-linked receptors *(28)*, HS-142-1 inhibited [$^{125}$I]rANP binding in a dose-dependent manner with an $IC_{50}$ value of 8 μg/mL, but approx 35% of specific binding was resistant to displacement by HS-142-1. Atriopeptin-1 (AP-1), which binds selectively at rather lower concentrations to guanylyl cyclase-free type receptors *(9,29)*, inhibited [$^{125}$I]rANP binding to adrenocortical membranes by approx 35%, whereas approx 65% of specific binding was resistant to AP-1. The similarity between the amount of [$^{125}$I]rANP binding resistant to displacement by HS-142-1 at 100 μg/mL and that displaced by AP-1 at $10^{-8}$*M* suggests the specificity of HS-142-1 toward guanylyl cyclase-linked receptors. Indeed, when [$^{125}$I]rANP binding to ANP-C receptors was blocked by AP-1 at $10^{-8}$*M*, HS-142-1 at 100 μg/mL completely displaced [$^{125}$I]rANP binding to adrenocortical membranes (Fig. 2A). In contrast, HS-142-1 had little or no effect on the specific binding of [$^{125}$I]-rANP to bovine lung membranes, which have been shown to express predominantly guanylyl cyclase-free receptors (ANPR-C) *(30)*. AP-1 inhibited [$^{125}$I]rANP binding to the lung membranes with an $IC_{50}$ of 0.7 n*M*, whereas 10% of specific binding remained resistant. In the presence of 100 μg/mL of HS-142-1, AP-1 displaced the specific binding completely (Fig. 2B). HS-142-1 also slightly inhibited the specific bindings of [$^{125}$I]rANP to rat glomeruli *(27)* and cultured bovine vascular smooth muscle cells *(24,25)* as well as [$^{125}$I]CNP to human mesangial cells *(31)*, where only a small portion of the binding sites are coupled to guanylyl cyclase. These results indicate that HS-142-1 selectively interacts with guanylyl cyclase-linked receptor.

Recently, Koller et al. *(12)* described ligand specificity of the two different subtypes of ANP functional receptor: A-type (ANPR-A or GC-A) and B-type (ANPR-B or GC-B).

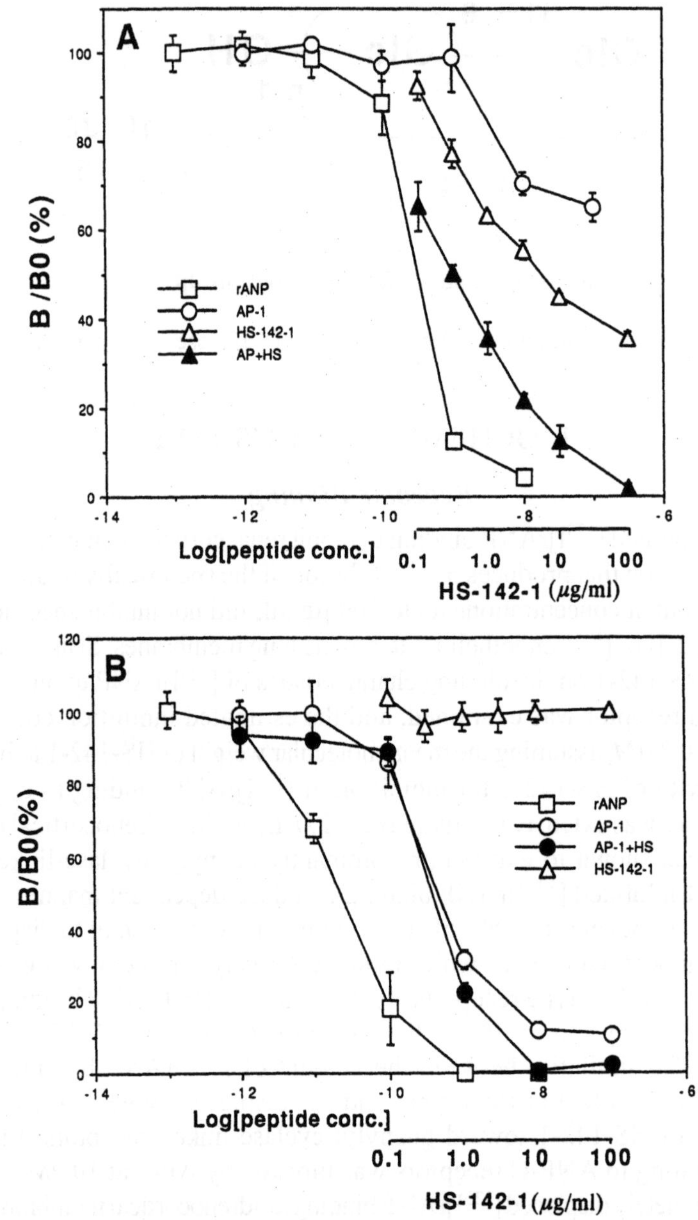


**Fig. 2.** Effects of rANP, atriopeptin I (AP-1), and HS-142-1 on the specific binding of [$^{125}$I]rANP to bovine adrenocortical and lung membranes. **(A)** Competitive inhibition of [$^{125}$I]rANP binding to bovine adrenocortical membranes by rANP (□), AP-1 (○), HS-142-1 (△), or HS-142-1 in combination with 0.1 $\mu M$ AP-1 (▲). Inhibitory effects are expressed as percentages of the specific binding of [$^{125}$I]rANP. Bars represents the mean ±SEM ($n$ = 4–5). **(B)** Competitive inhibition of [$^{125}$I]rANP binding to bovine lung membranes by rANP (□), AP-1 (○), HS-142-1 (△), or AP-1 in combination with 100 μg/mL HS-142-1 (●). Reproduced from ref. *26*, with permission.

ANP and, to a lesser extent, BNP are effective activators of ANPR-A. The ANPR-B receptor is activated most efficiently by CNP. The receptor subtype specificity of HS-142-1 was confirmed by experiments with cells that specifically express each receptor subtype. HS-142-1 competitively blocked specific binding of the natriuretic peptides to ANPR-A

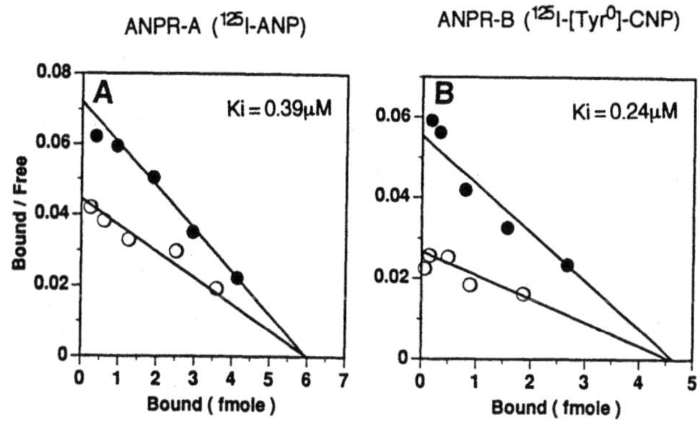


**Fig. 3.** Scatchard analysis of the effects of HS-142-1 on the specific binding of $^{125}$I-labeled natriuretic peptides to cells stably expressing ANPR-A **(A)** or ANPR-B **(B)**. The specific binding of [$^{125}$I]rANP or [$^{125}$I,Tyr$^0$]CNP was measured in the absence of HS-142-1 (●, Kd; 167 and 167 p*M*, respectively) or the presence of 1.0 μg/mL of HS-142-1 (○, Kd; 273 and 340 p*M*, respectively). Reproduced from ref. *32*, with permission.

and ANPR-B with $IC_{50}$ values of 2.2 and 3.3 μg/mL, respectively *(32)*. Inhibition constant (Ki) of HS-142-1 for ANPR-A and ANPR-B were calculated to be 0.39 and 0.24 μ*M*, respectively, indicating that antagonistic ability of HS-142-1 against CNP is similar to that against ANP (Fig. 3). HS-142-1 also interacts with a novel natriuretic peptide receptor NPR-D (ANPR-D) cloned from eel as it does with mammalian ANPR-A and ANPR-B *(33)*. Taken together, these results indicate that HS-142-1 interacts competitively with guanylyl cyclase-linked receptors and recognizes it selectively.

## *Affinity Crosslinking*

To establish more conclusively that HS-142-1 selectively recognizes the guanylyl cyclase-linked ANP receptor, [$^{125}$I]rANP affinity crosslinking studies were performed in the presence of varying concentrations of HS-142-1 or AP-1 with membranes prepared from adrenocortical and lung membranes *(26,34)*. As shown in Fig. 4, the 135- and 60-kDa bands should be derived from the labeling of the guanylyl cyclase-linked and guanylyl cyclase-free (ANPR-C) receptors, respectively *(28)*. AP-1 inhibited the labeling of the 60-kDa band in a concentration-dependent manner but did not interfere with the labeling of the 135-kDa band (Fig. 4A). Under identical conditions, HS-142-1 at concentrations ranging from 1.0 to 1000 μg/mL specifically abolished the labeling of the 135-kDa band in a concentration-dependent manner and had no effect on the labeling of the 60-kDa band (Fig. 4A). In lung membranes, similarly to adrenocortical membranes, AP-1 and HS-142-1 selectively recognized 60- and 135-kDa bands, respectively (Fig. 4B). Furthermore, HS-142-1 also recognized only the 130-kDa ANP receptor in membranes prepared from glomerular and vascular endothelial cells *(24,25,27)*. Similar results were obtained with adrenocortical membranes using the photoaffinity labeling technique *(34)*. On the other hand, in affinity crosslinking studies with membranes prepared from bovine aortic smooth muscle cells, a major portion of the 135-kDa band was unexpectedly HS-142-1-insensitive *(24)*. Although the identity of the HS-142-1-insensitive ANP receptor is not yet known, it is expressed only in bovine aortic smooth muscle cells, and not in bovine aortic endothelium, lung, adrenal cortex, or rat glomeruli. These results suggest the possible existence of a novel subtype of ANP functional receptor that is insensitive to HS-142-1.

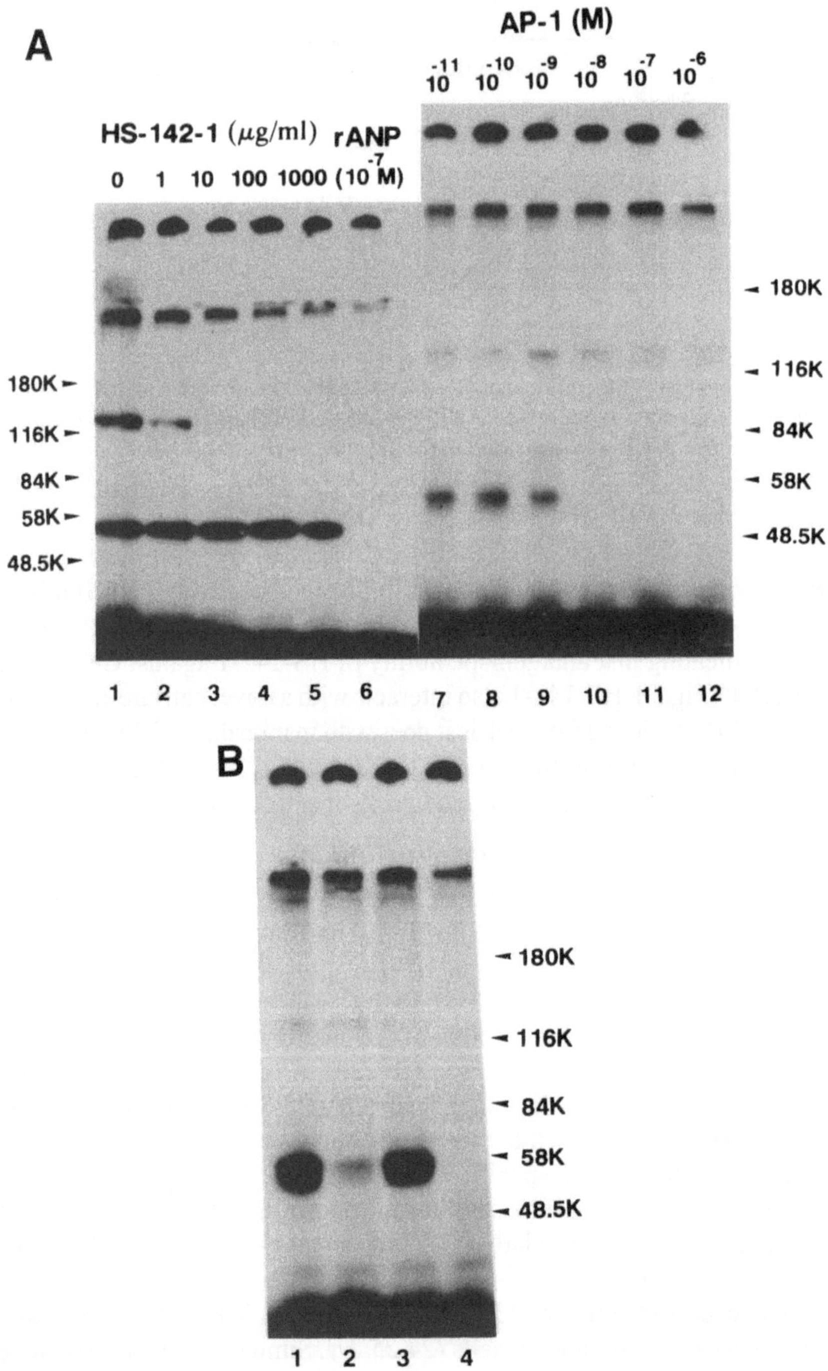


**Fig. 4.** Specific recognition by HS-142-1 of guanylyl cyclase-linked ANP receptor. **(A)** Affinity crosslinking of [$^{125}$I]rANP to bovine adrenocortical membranes in the absence of competing ligand (lane 1), or in the presence of indicated concentrations of HS-142-1 (lanes 2–5), 0.1 μ*M* rANP (lane 6), or indicated concentrations of AP-1 (lanes 7–12). **(B)** Affinity crosslinking of [$^{125}$I]rANP to bovine lung membranes in the absence of competing ligand (lane 1), or in the presence of 0.1 μ*M* AP-1 (lane 2), 100 μg/mL HS-142-1 (lane 3), or 0.1 μ*M* rANP (lane 4). Reproduced from ref. *26*, with permission.

Taken together, the affinity crosslinking studies confirmed the binding specificity of HS-142-1 for guanylyl cyclase-linked ANP receptors.

## PHARMACOLOGY OF HS-142-1

### *Antagonism In Vitro*

#### Cyclic GMP Production

In bovine adrenocortical membrane *(22)*, bovine lung membrane *(23)*, rat glomeruli *(27)*, bovine or rat aortic smooth muscle and endothelial cells *(24,25)*, rat pheochromocytoma PC12 cells *(35)*, mouse neuroblastoma-rat glioma hybridoma NG108-15 cells *(35)*, and porcine kidney epithelial LLC-PK1 cells *(34)*, HS-142-1 significantly attenuated cGMP production stimulated by rANP in a concentration-dependent manner at 1.0–100 μg/mL. $IC_{50}$ estimates ranged from 1.5–30 μg/mL, since the inhibitory potency varies among different types of cells. HS-142-1 alone in the absence of rANP had no effect on the basal cGMP production in these cells.

By activating soluble guanylyl cyclase, sodium nitroprusside elicited marked accumulation of cGMP in many types of cells *(36)*. HS-142-1, even at 100 μg/mL, did not inhibit cGMP accumulation elicited by sodium nitroprusside in PC12 *(35)*, NG108-15 *(35)*, or LLC-PK1 cells *(34)*, indicating that a general effect on cGMP metabolism is not involved in the HS-142-1-elicited inhibition of ANP-mediated cGMP formation *(34)*. No change in intracellular cyclic AMP (cAMP) was observed on treatment of the LLC-PK1 cells with HS-142-1, indicating that a general effect of HS-142-1 on cyclic nucleotide phosphodiesterase is unlikely *(34)*. Furthermore, HS-142-1 exerts neither agonistic nor antagonistic activity for the third mammalian form of membrane-bound guanylyl cyclase (GC-C), a receptor for heat-stable enterotoxins (STa), in T84 cells, showing that HS-142-1 is not a common antagonist for a family of membrane-bound guanylyl cyclase receptors *(37)*. Therefore, these results indicate that HS-142-1 is a specific receptor antagonist of the ANP functional receptor.

As mentioned above, there are two different subtypes of ANP functional receptor: ANPR-A and ANPR-B. The selectivity of HS-142-1 for the two subtypes of ANP functional receptor was first examined in PC12 rat pheochromocytoma cells *(35)*. At concentration range of 1.0–1000 μg/mL, HS-142-1 prevented, with almost equal potency, cGMP production in PC12 cells, stimulated by rANP, porcine BNP (pBNP), and porcine CNP (pCNP), suggesting that HS-142-1 recognizes equally the ANPR-A and ANPR-B in PC12 cells. Similar results were obtained in cultured vascular smooth muscle and endothelial cells with the three natriuretic peptides and HS-142-1 *(24)*. It is worthy of comment that, although the biological action and molecular form of BNP are known to differ substantially among species *(38–40)*, HS-142-1 inhibited completely the stimulant effects of pBNP, rat BNP (rBNP), or human BNP (hBNP) at concentrations similar to those that inhibited rANP-induced cGMP production *(24)*. This receptor subtype specificity of HS-142-1 was confirmed by experiments with cells that specifically express each receptor subtype, ANPR-A or ANPR-B *(32)*. HS-142-1 inhibited cGMP production elicited by ANP or CNP with $IC_{50}$ values of 1.8 and 1.5 μg/mL, respectively (Fig. 5). These results are consistent with the observation seen in the receptor binding studies, as described above.

#### Vasorelaxation

HS-142-1 prevented human ANP (hANP)-induced relaxation of isolated rabbit thoracic aorta rings contracted with phenylephrine. HS-142-1, at a concentration range of

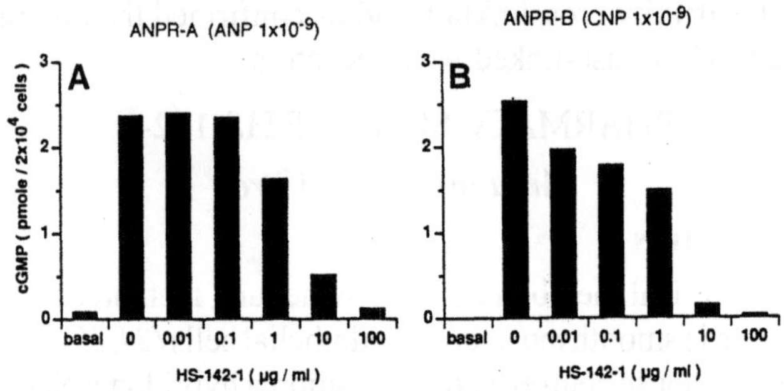


**Fig. 5.** The effect of HS-142-1 on the cGMP production induced by natriuretic peptides in cells stably expressing ANPR-A **(A)** or ANPR-B **(B)**. The cGMP production elicited by $10^{-9}$ *M* rANP (A), or $10^{-9}$ *M* rCNP (B) was measured in the presence of varying concentrations of HS-142-1. Reproduced from ref. *32*, with permission.

30–300 μg/mL, caused a concentration-related parallel rightward shift in the concentration response curve of hANP, without altering the maximal responses to hANP *(24,25)*. It is worthy of note that HS-142-1 is several folds more potent in inhibiting the pBNP-induced relaxation than in blocking the hANP-induced relaxation, although HS-142-1 inhibits equally the rANP- and BNP-induced cGMP production by aortic smooth muscle cells *(24)*. The inhibition by HS-142-1 of hANP or pBNP-induced relaxation of aortic rings was reversible *(24,25)*. The action of HS-142-1 was specific for the relaxation elicited by natriuretic peptides. Pretreatment of aortic rings with HS-142-1 at 100 μg/mL had no effect on relaxation induced by other relaxants, such as sodium nitroprusside, isoproterenol, or papaverine, which cause relaxation through a mechanism different from that of ANP. HS-142-1 alone had no contractile or vasorelaxant effects on rabbit aortic rings. Furthermore, it also had no effect on relaxation of aortic rings induced by forskolin, an adenylyl cyclase activator, or on contractions induced by angiotensin II, norepinephrine, or KCl *(24,25)*. These data suggest that the attenuation by HS-142-1 of the hypotensive response induced by exogenously injected ANP in vivo can be attributed to the specific interaction of HS-142-1 with ANP receptors.

### Isolated Perfused Kidneys

The effect of HS-142-1 on ANP- or BNP-induced reductions in renal vascular resistance (RVR) was studied in isolated perfused kidneys *(41)*. HS-142-1 at 1–5 μg/mL reversed ANP- or BNP-induced decreases in RVR in a concentration-dependent manner. HS-142-1 by itself did not change the perfusion pressure. Furthermore, HS-142-1 did not influence the effects of sodium nitroprusside, endothelin-1, angiotensin II, or norepinephrine on perfusion pressure.

### Antiproliferative Action

Excessive vascular smooth muscle proliferation is considered to be a key event in the initiation and progression of atherosclerotic disease. Recently, CNP has been shown to have a growth-suppressive effect on cultured vascular smooth muscle cells *(42)*. We examined the effects of CNP in the presence or absence of HS-142-1 on serum-induced proliferation of cells stably expressing ANPR-B, a selective receptor for CNP *(32)*. CNP suppressed the growth of cells that specifically express ANPR-B by 22% in a concentration-dependent manner. This antiproliferative action of CNP was completely blocked by

HS-142-1. At similar doses of the antiproliferative action, HS-142-1 blocked the cGMP production elicited by CNP. This fact clearly indicates that the antiproliferative action of CNP is mediated by the increased production of cGMP through the activation of ANPR-B. Similar results were obtained in the studies with glomerular mesangial cells, which are the cells of importance in the regulation of glomerular filtration rates through their contractility *(43)*. HS-142-1 blocked ANP-induced cellular cGMP accumulation and prevented ANP-induced inhibition of both serum-stimulated incorporation of [$^3$H] thymidine and cell proliferation.

### Aldosterone Production

ANP is known to inhibit aldosterone production induced by angiotensin II in adrenal glomerulosa cells *(44,45)*. HS-142-1 restored the rANP-induced inhibition of aldosterone production and, at the same time, reduced the increased intracellular cGMP accumulation in bovine adrenal glomerulosa cells *(46)*. HS-142-1 also reversed the effects of CNP on the basal secretion of cGMP and ACTH-induced aldosterone production from bovine adrenal zona glomerulosa cells in culture *(47)*. These results suggest that ANP or CNP inhibits aldosterone production through the guanylyl cyclase-coupled pathway. This is in contrast with the previous findings that cast doubt on the role of cGMP in the inhibition of aldosterone production *(48,49)*.

### Others

HS-142-1 inhibits ANP or CNP-induced various cellular responses in vitro, such as testosterone production, together with cGMP accumulation in isolated mouse Leydig cells *(50)*, glibenclamide-sensitive $K^+$ currents in follicle-enclosed *Xenopus* oocytes *(51)*, and upregulation of endothelin receptors in cultured rat vascular smooth muscle cells *(52)*. HS-142-1 significantly increased ANP secretion concomitantly with reduction in cGMP production from cultured atrial myocytes, supporting the hypothesis of an autoregulatory mechanism of ANP secretion in atrial myocytes *(53)*.

## *Antagonism In Vivo*

### Blood Pressure

The hypotensive effect of rANP was antagonized by HS-142-1 at a dose of 1.0 mg/rat/iv and only during rANP (10 μg/h) infusion, as shown in Fig. 6 *(54)*. This may be best explained by antagonism of ANP-mediated reductions in cardiac output or arterial vasorelaxation, as suggested in studies that have reported that HS-142-1 inhibited ANP- and BNP-induced vasorelaxation in rabbit aortic rings *(24)*. This, however, is not conclusive because the inhibition by HS-142-1 of the hypotensive action was not as salient as in the case of renal responses, as described below.

### Renal Responses

The renal interactions of HS-142-1 with exogenous natriuretic peptides were examined in anesthetized rats and dogs *(54–57)*. In rats, the increase in urine flow and in the urinary excretion of sodium elicited by hANP (3 μg/kg/iv bolus followed by 0.6 μg/kg/min sustaining infusion for 20 min) was prevented by HS-142-1 at a dose of 1.0 or 0.3 mg/kg/iv (Fig. 7A,B). This was accompanied by inhibition of an increase in urinary cGMP excretion (Fig. 7C). Furthermore, the renal responses to hANP or pBNP were rapidly reversed by HS-142-1, injected during natriuretic peptide infusion. Microcatheterization studies showed significant partial reversal of ANP-induced inhibition of sodium chloride and

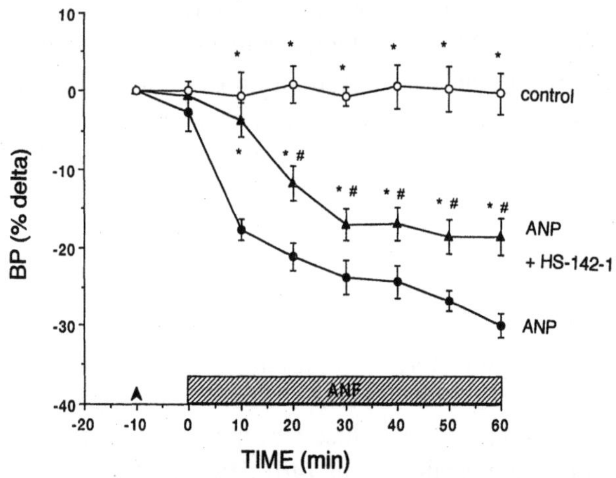


**Fig. 6.** Blockade by pretreatment with HS-142-1 of depressor response caused by rANP in rats. Percentage change in arterial blood pressure from initial value in ANP-infused rats (group 1, solid circles), in ANP-infused, HS-142-1 pretreated rats (group 2, solid triangles), and in control rats (group 3, open circles). The arrow and the shaded bar on the abscissa indicate injection of vehicle or HS-142-1, and the infusion of ANP, respectively, in groups 1 and 2. Statistically significant differences from group 1 (*), and from group 3 (#), are indicated at each time period. Reproduced from ref. *54*, with permission.

water reabsorption in the medullary collecting duct *(54)*. Even at higher doses, HS-142-1 did not alter the increase in urine flow or in the urinary excretion of sodium induced by furosemide or arginine, a precursor of nitric oxide *(55,56)*. HS-142-1 alone, at a 10 mg/kg/iv bolus, had no effect on these parameters *(55)*.

### *Toxicity and Stability*

HS-142-1 does not have a general cytotoxic action in mouse, as judged by the lack of acute toxicity at doses up to 10 mg/kg/iv, nor does it permanently damage the aortic rings, as shown by the ready reversibility of the inhibition by HS-142-1 *(24,26)*. HS-142-1 is stable as a powder or in a solution at 5°C for several months. The stability in vivo was examined using HS-142-1 conjugated with fluorescent probe. When intravenously injected to rats, half-life of HS-142-1 is approx 2 h (unpublished observation).

## UTILIZATION OF HS-142-1

### *Physiological and Pathophysiological Role of Endogenous Natriuretic Peptides*

There has been a strong debate on the role of natriuretic peptides; some pharmacologists say ANP plays a significant role in body fluid homeostasis, whereas some physiologists say evidence that atriopeptin, another name for the natriuretic peptide, is not a physiological regulator of sodium excretion *(14,15)*. Because of the lack of a specific antagonist for ANP functional receptors, the true physiological and pathophysiological functions of ANP were still unclear, and a discovery of ANP antagonist had been long awaited. HS-142-1 was thought to be such an antagonist. Many researchers, including us, therefore used this compound to conclude that endogenous natriuretic peptides are truly involved in some physiological and pathophysiological conditions.

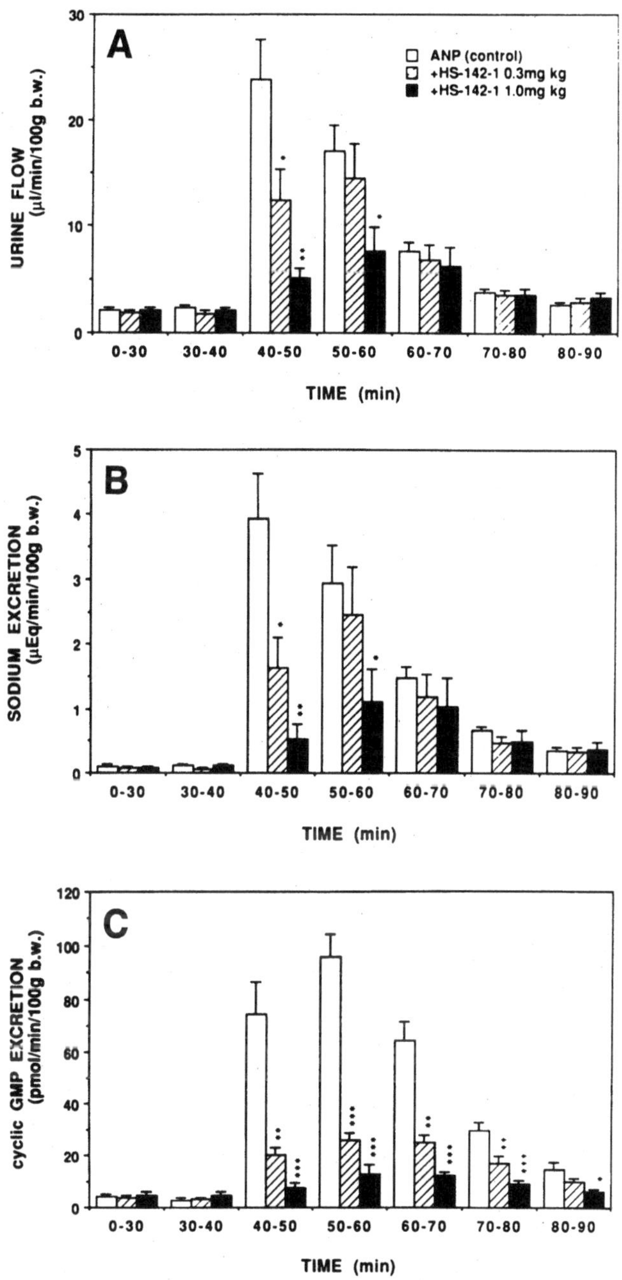


**Fig. 7.** Blockade by pretreatment with HS-142-1 of renal responses caused by hANP in anesthetized rats. **(A)** Urine flow; **(B)** sodium excretion; **(C)** cGMP excretion. Saline and HS-142-1 (0.3 or 1.0 mg/kg/iv) were administered by a bolus injection at 30 min, and 10 min later hANP was infused by a bolus injection (3 µg/kg/iv) followed by a sustaining infusion (0.6 µg/kg/min) for 20 min. Each value represents the mean ±SEM of eight animals. **P*<0.05, ***P*<0.01, and ****P*<0.001, significantly different from the control group (hANP alone). Reproduced from ref. *55*, with permission.

### Acute Volume Expansion

HS-142-1 was used to clarify whether endogenous natriuretic peptides play a true physiological role in volume homeostasis *(58,59)*. Anesthetized male rats received an acute volume expansion by a constant infusion of isotonic saline (4.0 mL/kg/min for 5 min). HS-142-1 at 3.0 mg/kg/iv or saline vehicle (control) was injected 10 min before volume expansion. Acute volume expansion caused a fivefold increase in urine flow and a 22-fold increase in urinary sodium excretion. Pretreatment of rats with HS-142-1 significantly reduced these responses without affecting potassium excretion (Fig. 8). A significant elevation of the urinary cGMP level, as a marker of endogenous ANP activity, was not observed during volume expansion. A slight reduction of the urinary cGMP level was observed after volume loading both in control and HS-142-1-treated groups; this reduction was more significant in the treated group. These results suggest that endogenous natriuretic peptide and its functional receptor participate in the diuresis and natriuresis produced by acute volume expansion under the present experimental conditions, and that endogenous natriuretic peptides play an important role in volume homeostasis under physiological conditions.

### DOCA-Salt Hypertensive Rats

The renal and systemic effects of HS-142-1 in deoxycorticosterone acetate (DOCA)-salt rats were examined *(41)*. DOCA-salt rats showed higher plasma concentrations of ANP and BNP than the controls. Bolus administration of 8 mg/kg of HS-142-1 elevated blood pressure by 8%, which was attributed to an increase in systemic vascular resistance (+14%), whereas urinary sodium excretion (–41%), glomerular filtration rate (–27%), renal blood flow (–18%), and plasma (–77%) and urinary cGMP (–69%) were significantly decreased. These effects were seen in a dose-dependent manner in DOCA-salt rats, but slight or negligible in the controls. These results demonstrate that endogenous ANP and/or BNP may be actually involved in the regulation of blood pressure and body fluid volume in DOCA-salt rats.

It has been known that kidney has the ability to "escape" the continued sodium-retaining effects of mineralocorticoid excess *(60)*. The mechanism underlying the escape phenomenon, however, still remains unclear. To assess the possible contribution of natriuretic peptides to mineralocorticoid escape, Yokota et al. *(61)* examined the effects of chronically administered HS-142-1 on the mineralocorticoid escape phenomenon induced by DOCA injection in conscious rats. HS-142-1 suppressed the increase in urinary volume, urinary sodium excretion, and urinary cGMP excretion observed during DOCA escape in a dose-dependent manner, showing that plasma natriuretic peptides appear to be a major factor initiating mineralocorticoid escape through a guanylyl cyclase-dependent pathway.

### Severe Congestive Heart Failure

To elucidate the compensatory role of endogenous ANP in dogs with severe congestive heart failure induced by chronic rapid ventricular pacing, we examined changes in hemodynamics and neuroendocrine and renal functions by administration of HS-142-1 *(62)*. When injected as a bolus at three doses of 0.3, 1.0, and 3.0 mg/kg at intervals of 30 min into right atrium, HS-142-1 lowered the plasma and urinary cGMP levels in a dose-dependent manner to 32 and 37% of the control levels, respectively. Mean arterial, pulmonary capillary wedge, and right atrial pressures and cardiac output did not change significantly. However, plasma renin activity, aldosterone level, and norepinephrine level increased rapidly to 226, 179, and 252% of the control values, respectively. Urine flow rate (–42%) and urinary sodium excretion (–82%) were significantly inhib-

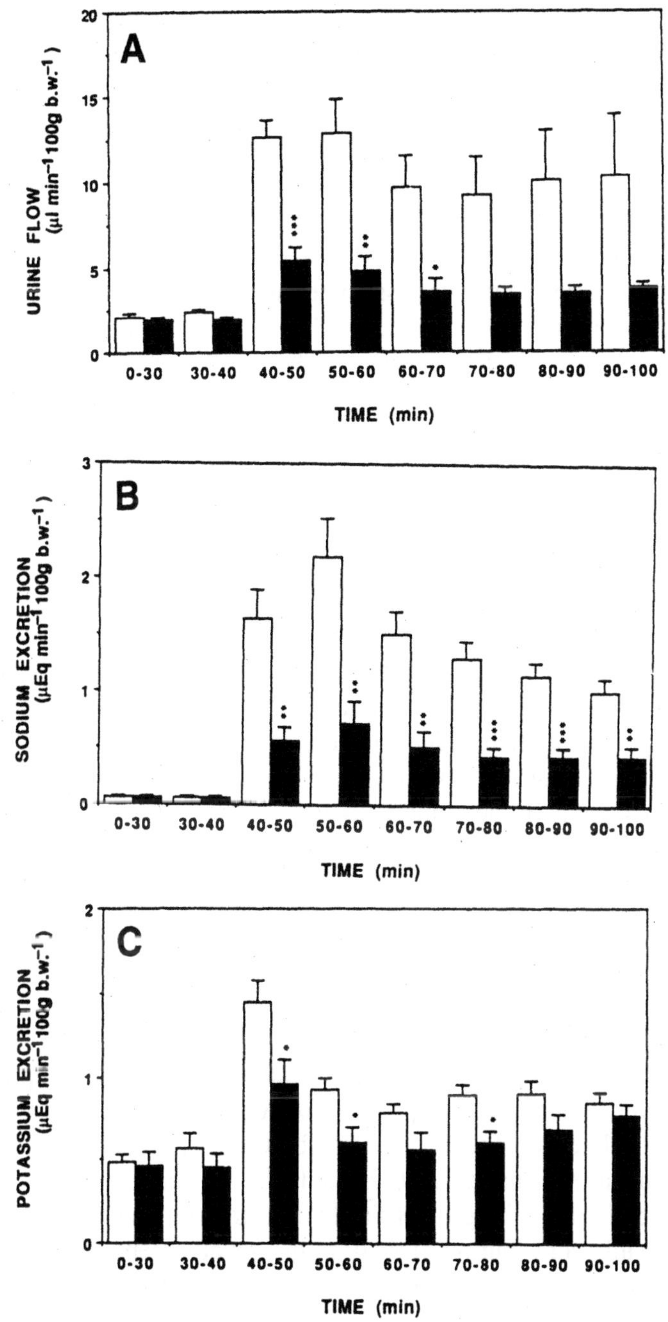


**Fig. 8.** Effects of HS-142-1 on the increase in urine flow **(A)**, urinary excretion of sodium **(B)**, and urinary excretion of potassium **(C)** caused by acute volume expansion (4.0 mL/kg/min for 5 min at = 40). HS-142-1 at 3.0 mg/kg/iv or saline vehicle was injected at 10 min before volume expansion. Open bars represent saline-vehicle-treated rats (control), whereas solid bars represent HS-142-1-treated rats. Values are mean ±SEM of six animals. *$P < 0.05$, **$P < 0.01$, and ***$P < 0.001$ vs respective control. Reproduced from ref. *58*, with permission.

ited, with no concomitant change in glomerular filtration rate or renal plasma flow. These results suggest that the elevated ANP levels in congestive heart failure do not contribute directly to the maintenance of systemic hemodynamics but rather contribute to the suppression of the activation of the renin–aldosterone system, sympathetic nervous activity, and body fluid retention. The results obtained in the studies of rats with heart failure induced by ligation of coronary artery are in general agreement with those seen in dogs *(63,64)*.

### Cirrhosis and Ascites

The effect of HS-142-1 on systemic hemodynamics, renal function, and the renin–aldosterone system in rats with cirrhosis and ascites induced by carbon–tetrachloride was examined *(65)*. The blockade by HS-142-1 of endogenous natriuretic peptides receptors in rats with cirrhosis and ascites was associated with marked reduction of renal plasma flow (RPF) and glomerular filtration rate (GFR), as well as marked increases in plasma renin activity and plasma aldosterone concentrations in the absence of significant changes in systemic hemodynamics, such as mean arterial pressure, heart rate, cardiac output, or total peripheral resistance. In control rats, only a significant reduction of GFR was observed. These results indicate the endogenous natriuretic peptides play a major role in the maintenance of renal perfusion and GFR and in the regulation of the renin–aldosterone system in rats with cirrhosis and ascites.

### Others

Several other studies with HS-142-1 support a functional role for endogenous natriuretic peptides in the regulation of basal sodium excretion in anesthetized normal dogs *(57)*, blood pressure and renal responses induced by a neutral endopeptidase inhibitors in DOCA-salt rats *(66)*, remnant kidney function in rats *(59)*, sodium excretion and renin–angiotensin–aldosterone system in a canine model of early left ventricular dysfunction *(67)*, and chronic anemia in the ovine fetus *(68)*.

## *Possibility as a Therapeutic Agent*

We examined a possibility of HS-142-1 as a therapeutic agent and have chosen two disease targets, diabetic nephropathy and sepsis, in which excessive actions of natriuretic peptides seem to be involved.

### Diabetic Nephropathy

Early diabetes mellitus is associated with elevations of GFR, which appears before the emergence of overt diabetic nephropathy *(69)*. Among several possible mechanisms, ANP has recently been postulated to contribute to the development of glomerular hyperfiltration in diabetic animals *(70)*. Indeed, plasma ANP levels have been shown to be higher in diabetic rats with glomerular hyperfiltration than in control rats *(70)*. To evaluate the possible contribution of ANP to the development of glomerular hyperfiltration, we examined the effect of HS-142-1 on GFR and RPF in streptozotosin-induced diabetic rats *(71,72)*. Blood glucose was maintained between 300 and 400 mg/dL by injection of insulin. Three to four weeks after streptozotosin injection, the plasma concentration of ANP, urinary cGMP excretion rate, GFR, and RPF were significantly higher in diabetic rats than in control rats. The increase in GFR and RPF in diabetic rats was significantly reduced by a single iv injection of HS-142-1 in a dose-dependent manner at doses up to 10 mg/kg body weight. Continuous sc administration of HS-142-1 with an osmotic minipump for 3–4 wk, beginning 2 d after streptozotosin injection, prevented the increases in urinary cGMP excretion rate, GFR, and RPF observed in untreated diabetic rats. Similar results were obtained by Zhang et al. *(73)* and Hirata et al. *(74)*. Then, in diabetic rats with moderate hyperglycemia, plasma glucose may lead to an increase in ANP secretion, which in turn results in the development of diabetic glomerular hyperfiltration, and an ANP antagonist might be useful in the treatment of diabetic nephropathy.

### Sepsis

Human sepsis is now becoming a more and more serious disease as AIDS (acquired immunodeficiency syndrome) patients increase in number. In human sepsis, we see reduced mean arterial pressure (MAP) and the following symptoms: reduced systemic vascular resistance, elevated cardiac output, elevated heart rate, and reduced myocardial contractility. Among them, rapid MAP reduction leading to septic shock is the most serious. The purpose of this study was, therefore, to evaluate the possible contribution of ANP to the ovine hyperdynamic sepsis. Ehardt et al. examined the effect of HS-142-1 on the vasodilation and the fall in MAP in septic sheep model *(75)*. Adult sheep were surgically prepared. Seven days later the sheep were infused with live *Pseudomonas aeruginosa* for 48 h. Experimental model sheep received continuous infusion of HS-142-1 at a dose of 3 mg/kg/h, and control sheep received an equivalent volume of saline. Plasma ANP level was increased in this sheep sepsis model, and MAP was rapidly decreased during sepsis. Under the same conditions, HS-142-1 treatment returned the MAP to untreated levels in this sheep sepsis model. HS-142-1 would support the cardiovascular system during sepsis and therefore could be useful in the treatment of septic humans.

## CONCLUSION

Receptor antagonists have traditionally provided powerful tools toward the understanding of the physiological role of ligands. It is informative to recall arguments concerning the true functions of the renin–angiotensin system. It was only when effective chemical inhibitors of the renin–angiotensin system were discovered, e.g., angiotensin–converting enzyme inhibitors, that the true nature of the renin–angiotensin system was appreciated.

In a similar manner, because of the lack of a specific antagonist, the true physiological and pathophysiological functions of the natriuretic peptides were unclear. The development of a specific nonpeptide antagonist for the functional natriuretic peptide receptor should be very useful. Such an antagonist HS-142-1 was discovered in a microbial culture broth. This novel compound specifically recognizes functional guanylyl cyclase-linked receptors, namely, ANP-A and ANP-B receptors, and antagonizes the actions of the peptide hormones both in vitro and in vivo. The lack of acute toxicity and the reversibility of its inhibitory effect suggest the usefulness of HS-142-1 as a natriuretic peptide antagonist. There has been no previous report on any compound, including peptide ligands, with properties similar to those of HS-142-1. Indeed, HS-142-1 has been used to support a functional role for endogenous natriuretic peptides in a variety of physiological and pathophysiological conditions, but we have to be very careful in choosing experimental animal models, since the relative density, affinity, and proportion of natriuretic peptide receptor subtypes in the kidney exhibit significant species differences *(76)*. Finally, HS-142-1 is a novel chemical entity that could be beneficial in the treatment of diseases involving excessive actions of the natriuretic peptides.

## REFERENCES

1. Inagami, T. Atrial natriuretic factor. J Biol Chem 1989;264:3043–3046.
2. Needleman P, Blaine EH, Greenwald JE, Michener ML, Saper CB, Stockmann PT, Tolunay HE. The biochemical pharmacology of atrial peptides. Ann Rev Pharm Tox 1989;29:23–54.

3. Espiner EA, Richards AM, Yandle TG, Nicholls MG. Natriuretic hormones. Endocrinol Metab Clin North Am 1995;24:481–509.
4. De Lean A, Gutkowska J, McNicoll N, Schiller PW, Cantin M, Genest J. Characterization of specific receptors for atrial natriuretic factor in bovine adrenal zona glomerulosa. Life Sci 1984;35:2311–2318.
5. Napier MA, Vandlen RL, Albers-Schonberg G, Nutt RF, Brady S, Lyle T, Winquist R, Faison EP, Heinel LA, Blaine EH. Specific membrane receptors for atrial natriuretic factor in renal and vascular tissues. Proc Natl Acad Sci USA 1984;81:5946–5950.
6. Chinkers M, Garbers DL, Chang MC, Lowe DG, Chin H, Goeddel DV, Schulz S. A membrane form of guanylate cyclase is an atrial natriuretic peptide receptor. Nature 1989;338:78–83.
7. Waldman SA, Rapoport RM, Murad F. Atrial natriuretic factor selectively activates particulate guanylate cyclase and elevates cyclic GMP in rat tissues. J Biol Chem 1984;259:14,332–14,334.
8. Chang MC, Lowe DG, Lewis M, Hellmiss R, Chen E, Goeddel DV. Differential activation by atrial and brain natriuretic peptides of two different receptor guanylate cyclases. Nature 1989;341:68–72.
9. Schulz S, Singh S, Bellet RA, Singh G, Tubb DJ, Chin H, Garbers DL. The primary structure of a plasma membrane guanylate cyclase demonstrates diversity within this new receptor family. Cell 1989;58:1155–1162.
10. Fuller F, Porter JG, Arfsten AE, Miller J, Schilling JW, Scarborough RM, Lewicki JA, Schenk DB. Atrial natriuretic peptide clearance receptor. Complete sequence and functional expression of cDNA clones. J Biol Chem 1988;263:9395–9401.
11. Nussenzveig DR, Lewicki JA, Maack T. Cellular mechanisms of the clearance function of type C receptors of atrial natriuretic factor. J Biol Chem 1990;265:20,952–20,958.
12. Koller KJ, Lowe DG, Bennett GL, Minamino N, Kangawa K, Matsuo H, Goeddel DV. Selective activation of the B natriuretic peptide receptor by C-type natriuretic peptide (CNP). Science 1991;252:120–123.
13. Suga S, Nakao K, Hosoda K, Mukoyama M, Ogawa Y, Shirakami G, Arai H, Saito Y, Kambayashi Y, Inouye K, Imura H. Receptor selectivity of natriuretic peptide family, atrial natriuretic peptide, brain natriuretic peptide, and C-type natriuretic peptide. Endocrinology 1992;130:229–239.
14. Blaine EH. Atrial natriuretic factor plays a significant role in body fluid homeostasis. Hypertension 1990;15:2–8.
15. Goetz KL. Evidence that atriopeptin is not a physiological regulator of sodium excretion. Hypertension 1990;15:9–19.
16. Matsuda Y, Morishita Y. HS-142-1: a novel nonpeptide atrial natriuretic peptide antagonist of microbial origin. Cardiovasc Drug Rev 1993;11:45–59.
17. Kambayashi Y, Nakajima S, Ueda M, Inouye K. A dicarba analog of β-atrial natriuretic peptide (β-ANP) inhibits guanosine 3',5'-cyclic monophosphate production induced by α-ANP in cultured rat vascular smooth muscle cells. FEBS Lett 1989;248:28–34.
18. Kitajima Y, Minamitake Y, Furuya M, Takehisa M, Katayama T, Tanaka S. Linear α-human atrial natriuretic peptide analogs display receptor binding activity and inhibit α-hANP-induced cGMP accumulation. Biochem Biophys Res Commun 1989;164:1295–1301.
19. Abell TJ, Richard AM, Yandle TG, Espiner EA, Frampton C, Ikram H. Competitive peptide antagonists of ANF-induced cyclic guanosine monophosphate production. Biochem Biophys Res Commun 1989;164:108–113.
20. Von Geldern TW, Budzik GP, Dillon TP, Holleman WH, Holst MA, Kiso Y, Novosad EI, Opgenorth TJ, Rockway TW, Thomas AM, Yeh S. Atrial natriuretic peptide antagonists: biological evaluation and structural correlations. Mol Pharmacol 1990;38:771–778.
21. Weber W, Fischli W, Hochuli E, Kupfer E, Weibel EK. Anantin—a peptide antagonist of the atrial natriuretic factor (ANF). I. Producing organism, fermentation, isolation and biological activity. J Antibiotics 1991;44:164–171.
22. Morishita Y, Sano T, Ando K, Saitoh Y, Kase H, Yamada K, Matsuda Y. Microbial polysaccharide, HS-142-1, competitively and selectively inhibits ANP binding to its guanylyl cyclase-containing receptor. Biochem Biophys Res Commun 1991;176:949–957.
23. Morishita Y, Takahashi M, Sano T, Kawamoto I, Ando K, Sano H, Saitoh Y, Kase H, Matsuda Y. Isolation and purification of HS-142-1, a novel nonpeptide antagonist for the atrial natriuretic peptide receptor, from *Aureobasidium* sp. Agric Biol Chem 1991;55:3017–3025.
24. Imura R, Sano T, Goto J, Yamada K, Matsuda Y. Inhibition by HS-142-1, a novel nonpeptide atrial natriuretic peptide antagonist of microbial origin, of atrial natriuretic peptide-induced relaxation of isolated rabbit aorta through the blockade of guanylyl cyclase-linked receptors. Mol Pharmacol 1992;42:982–990.

25. Matsuda Y, Sano T, Imura R, Goto J, Yamada K. Specific inhibition by a novel atrial natriuretic peptide (ANP) antagonist of ANP-induced relaxation of aorta through the inhibition of cyclic GMP production. J Hypertension 1992;10(Suppl 4):S77.
26. Morishita Y, Sano T, Kase H, Yamada K, Inagami T, Matsuda Y. HS-142-1, a novel nonpeptide atrial natriuretic peptide (ANP) antagonist, blocks ANP-induced renal responses through a specific interaction with guanylyl cyclase-linked receptors. Eur J Pharmacol Mol Pharmacol Sec 1992;225:203–207.
27. Sano T, Imura R, Morishita Y, Matsuda Y, Yamada K. HS-142-1, a novel polysaccharide of microbial origin, specifically recognizes guanylyl cyclase-linked ANP receptor in rat glomeruli. Life Sci 1992;51:1445–1451.
28. Takayanagi R, Inagami T, Snajdar RM, Imada T, Tamura M, Misono KS. Two distinct forms of receptors for atrial natriuretic factor in bovine adrenocortical cells. Purification, ligand binding, and peptide mapping. J Biol Chem 1987;262:12,104–12,113.
29. Leitman DC, Murad F. Comparison of binding and cyclic GMP accumulation by atrial natriuretic peptides in endothelial cells. Biochim Biophys Acta 1986;885:74–79.
30. Uchida K, Mizuno T, Shimonaka M, Sugiura N, Nara K, Ling N, Hagiwara H, Hirose S. Purification and properties of active atrial-natriuretic-peptide receptor (type C) from bovine lung. Biochem J 1989;263:671–678.
31. Zhao J, Ardaillou N, Lu CY, Placier S, Pham P, Badre L, Cambar J, Ardaillou R. Characterization of C-type natriuretic peptide receptors in human mesangial cells. Kidney Int 1994;46:717–725.
32. Ohyama Y, Miyamoto K, Morishita Y, Matsuda Y, Saito Y, Minamino N, Kangawa K, Matsuo H. Stable expression of natriuretic peptide receptors: effect of HS-142-1, a non-peptide ANP antagonist. Biochem Biophys Res Commun 1992;189:336–342.
33. Kashiwagi M, Katafuchi T, Kato A, Inuyama H, Ito T, Hagiwara H, Takei Y, Hirose S. Cloning and properties of a novel natriuretic peptide receptor, NPR-D. Eur J Biochem 1995;233:102–109.
34. Tanaka T, Ichimura M, Nakajo S, Snajdar RM, Morishita Y, Sano T, Yamada K, Inagami T, Matsuda Y. HS-142-1, a novel non-peptide antagonist for atrial natriuretic peptide receptor, selectively inhibits particulate guanylyl cyclase and lowers cyclic GMP in LLC-PK1 cells. Biosci Biotech Biochem 1992;56:1041–1045.
35. Toki S, Morishita Y, Sano T, Matsuda Y. HS-142-1, a novel non-peptide ANP antagonist, blocks the cyclic GMP production elicited by natriuretic peptides in PC12 and NG108-15 cells. Neurosci Lett 1992;135:117–120.
36. Ishii K, Chang B, Kerwin JF Jr, Wagenaar FL, Huang ZJ, Murad F. Formation of endothelium-derived relaxing factor in porcine kidney epithelial LLC-PK1 cells: an intra- and intercellular messenger for activation of soluble guanylate cyclase. J Pharmacol Exp Ther 1991;256:38–43.
37. Ohyama Y, Miyamoto K, Morishita Y, Matsuda Y, Kojima M, Minamino N, Kangawa K, Matsuo H. HS-142-1, a novel antagonist for natriuretic peptides, has no effect on the third member of membrane bound guanylate cyclases (GC-C) in T84 cells. Life Sci 1993;52:PL153–157.
38. Kambayashi Y, Nakao K, Itoh H, Hosoda K, Saito Y, Yamada T, Mukoyama M, Arai H, Shirakami G, Suga S, Ogawa Y, Jougasaki M, Minamino N, Kangawa K, Matsuo H, Inoue K, Imura H. Isolation and sequence determination of rat cardiac natriuretic peptide. Biochem Biophys Res Commun 1989;163:233–240.
39. Kambayashi Y, Nakao K, Mukoyama M, Saito Y, Ogawa Y, Shiono S, Inouye K, Yoshida N, Imura H. Isolation and sequence determination of human brain natriuretic peptide in human atrium. FEBS Lett 1990;259:341–345.
40. Kambayashi Y, Nakao K, Kimura H, Kawabata T, Nakamura M, Inoue K, Yoshida N, Imura H. Biological characterization of human brain natriuretic peptide (BNP) and rat BNP: species-specific actions of BNP. Biochem Biophys Res Commun 1990;173:599–605.
41. Hirata Y, Matsuoka H, Suzuki E, Hayakawa H, Sugimoto T, Matsuda Y, Morishita Y, Kangawa K, Minamino N, Matsuo H, Sugimoto T. Role of endogenous atrial natriuretic peptide in DOCA-salt hypertensive rats. Effects of a novel non-peptide antagonist for atrial natriuretic peptide receptor. Circulation 1993;87:554–561.
42. Furuya M, Yoshida M, Hayashi Y, Ohnuma N, Minamino N, Kangawa K, Matsuo H. C-type natriuretic peptide is a growth inhibitor of rat vascular smooth muscle cells. Biochem Biophys Res Commun 1991;177:927–931.
43. Haneda M, Kikkawa R, Koya D, Sakamoto K, Nakanishi S, Matsuda Y, Shigeta Y. Biological receptors mediate anti-proliferative action of atrial natriuretic peptide in cultured mesangial cells. Biochem Biophys Res Commun 1993;192:642–648.

44. Atarashi K, Mulrow PJ, Franco-Saenz R, Snajdar R, Rapp J. Inhibition of aldosterone production by an atrial extract. Science 1984;224:992–994.
45. Atlas SA, Volpe M, Sosa RE, Laragh JH, Camargo MJF, Maack T. Effect of atrial natriuretic factor on blood pressure and the renin–angiotensin–aldosterone system. Fed Proc 1986;45:2115–2121.
46. Oda S, Sano T, Morishita Y, Matsuda Y. Pharmacological profile of HS-142-1, a novel non-peptide atrial natriuretic peptide (ANP) antagonist of microbial origin. II. Restoration by HS-142-1 of ANP-induced inhibition of aldosterone production in adrenal glomerulosa cells. J Pharmacol Exp Ther 1992;263:241–245.
47. Kawai M, Naruse M, Yoshimoto T, Naruse K, Shionoya K, Tanaka M, Morishita Y, Matsuda Y, Demura R, Demura H. C-type natriuretic peptide as a possible local modulator of aldosterone secretion in bovine adrenal zona glomerulosa. Endocrinology 1996;137:42–46.
48. Ganguly A, Chiou S, West LA, Davis JS. Atrial natriuretic factor inhibits angiotensin-induced aldosterone secretion: not through cGMP or interference with phospholipase C. Biochem Biophys Res Commun 1989;159:148–154.
49. Okamoto M. Effects of α-human atrial natriuretic polypeptide, sodium nitroprusside and dibutyryl cyclic GMP on aldosterone production in bovine zona glomerulosa cells. Acta Endocrinol (Copenh) 1988;119:358–366.
50. Schumacher H, Matsuda Y, Mukhopadhyay AK. HS-142-1 inhibits testosterone production and guanosine-3':5'-cyclic monophosphate accumulation stimulated by atrial natriuretic peptide in isolated mouse Leydig cells. Mol Cell Endocrinol 1993;94:105–110.
51. Sakuta H, Okamoto K, Tandai M. Atrial natriuretic factor potentiates glibenclamide-sensitive $K^+$ currents via the activation of receptor guanylate cyclase in follicle-enclosed Xenopus oocytes. Eur J Pharmacol Mol Pharmacol Sec 1994;267:281–287.
52. Eguchi S, Hirata Y, Imai T, Marumo F. C-type Natriuretic peptide upregulates vascular endothelin type B receptors. Hypertension 1994;23[part 2]:936–940.
53. Nachshon S, Zamir O, Matsuda Y, Zamir N. Effects of ANP receptor antagonists on ANP secretion from adult rat cultured atrial myocytes. Amer J Physiol Endocrinol Metab 1995;31:E428–432.
54. Honrath U, Matsuda Y, Sonnenberg H. Cardiovascular and renal functional effects of an antagonist of the guanylyl cyclase-linked ANF receptor. Regul Peptides 1994;49:211–216.
55. Sano T, Morishita Y, Matsuda Y, Yamada K. Pharmacological profile of HS-142-1, a novel non-peptide atrial natriuretic peptide antagonist of microbial origin. I. Selective inhibition of the actions of natriuretic peptides in anesthetized rats. J Pharmacol Exp Ther 1992;260:825–831.
56. Zhang PL, Jimenez W, Mackenzie HS, Guo J, Troy JL, Ros J, Angeli P, Arroyo V, Brenner BM. HS-142-1, a potent antagonist of natriuretic peptides *in vitro* and *in vivo*. J Am Soc Nephrol 1994;5:1099–1105.
57. Stevens TL, Wei CM, Aahrus LL, Heublein DM, Kinoshita M, Matsuda Y, Burnett JC Jr. Modulation of exogenous and endogenous atrial natriuretic peptide by a receptor inhibitor. Hypertension 1994;23:613–618.
58. Sano T, Morishita Y, Yamada K, Matsuda Y. Effects of HS-142-1, a novel non-peptide ANP antagonist, on diuresis and natriuresis induced by acute volume expansion in anesthetized rats. Biochem Biophys Res Commun 1992;182:824–829.
59. Zhang PL, Mackenzie HS, Troy JL, Brenner BM. Effects of natriuretic peptide receptor inhibition on remnant kidney function in rats. Kidney Int 1994;46:414–420.
60. Higgins JT Jr. Escape from sodium-retaining effects of deoxycorticosterone in hypotensive and hypertensive dogs. Proc Soc Exp Biol Med 1970;134:768–772.
61. Yokota N, Bruneau BG, de Bold MLK, de Bold AJ. Atrial natriuretic factor significantly contributes to the mineralocorticoid escape phenomenon. Evidence for a guanylate cyclase-mediated pathway. J Clin Invest 1994;94:1938–1946.
62. Wada A, Tsutamoto T, Matsuda Y, Kinoshita M. Cardiorenal and neurohumoral effects of endogenous atrial natriuretic peptide in dogs with severe congestive heart failure using a specific antagonist for guanylate cyclase-coupled receptors. Circulation 1994;89:2232–2240.
63. Zhang PL, Mackenzie HS, Totsune K, Troy JL, Brenner BM. Renal effects of high-dose natriuretic peptide receptor blockade in rats with congestive heart failure. Circ Res 1995;77:1240–1245.
64. Nishikimi T, Miura K, Minamino N, Takeuchi K, Takeda T. Role of endogenous atrial natriuretic peptide on systemic and renal hemodynamics in heart failure rats. Am J Physiol Heart Circ Physiol 1994;36:H182–H186.

65. Angeli P, Jimenez W, Arroyo V, Mackenzie HS, Zhang PL, Claria J, Rivera F, Brenner BM, Rodes J. Renal effects of natriuretic peptide receptor blockade in cirrhotic rats with ascites. Hepatology 1994;20:948–954.
66. Hirata Y, Suzuki Y, Suzuki E, Hayakawa H, Kimura K, Goto A, Omata M, Minamino N, Kangawa K, Matsuo H. Mechanisms underlying the augmented responses of deoxycorticosterone acetate-salt hypertensive rats to neutral endopeptidase inhibitors. J Hypertension 1994;12:367–374.
67. Stevens TL, Burnett Jr JC, Kinoshita M, Matsuda Y, Redfield MM. A functional role for endogenous atrial natriuretic peptide in a canine model of early left ventricular dysfunction. J Clin Invest 1995;95:1101–1108.
68. Silberbach M, Woods LL, Hohimer R, Shiota T, Matsuda Y, Davis LE. Role of endogenous atrial natriuretic peptide in chronic anemia in the ovine fetus: effects of a non-peptide antagonist for atrial natriuretic peptide receptor. Pediatric Res 1995;38:722–728.
69. Mogensen CE. Glomerular filtration rate and renal plasma flow in short-term and long-term juvenile diabetes mellitus. Scand J Clin Lab Invest 1971;28:91–100.
70. Ortola F, Ballermann BJ, Anderson S, Mendez RE, Brenner BM. Elevated plasma atrial natriuretic peptide levels in diabetic rats—potential mediator of hyperfiltration. J Clin Invest 1987;80:670–674.
71. Kikkawa R, Haneda M, Sakamoto K, Koya D, Shikano T, Nakanishi S, Matsuda Y, Shigeta Y. Antagonist for atrial natriuretic peptide receptors ameliorates glomerular hyperfiltration in diabetic rats. Biochem Biophys Res Commun 1993;193:700–705.
72. Sakamoto K, Kikkawa R, Haneda M, Shigeta Y. Prevention of glomerular hyperfiltration in rats with streptozotocin-induced diabetes by an atrial natriuretic peptide receptor antagonist. Diabetologia 1995;38:536–542.
73. Zhang PL, Mackenzie HS, Troy JL, Brenner BM. Effects of an atrial natriuretic peptide receptor antagonist on glomerular hyperfiltration in diabetic rats. J Am Soc Nephrol 1994;4:1564–1570.
74. Hirata Y, Suzuki Y, Hayakawa H, Suzuki E, Kimura K, Goto A, Kangawa K, Matsuo H, Omata M. Participation of endogenous atrial natriuretic peptide in the regulation of urinary protein excretion in experimental diabetic rats. Clin Sci 1995;88:413–419.
75. Ehardt JS, Nelson SH, Hinder F, Brooke M, Traber LD, Traber DL. Effects of an atrial natriuretic peptide (ANP) inhibitor, HS-142-1, in isolated blood vessels from septic sheep. FASEB J 1995;9:A888.
76. Rutherford RAD, Matsuda Y, Wilkins MR, Polak JM, Wharton J. Identification of renal natriuretic peptide receptor subpopulations by use of the non-peptide antagonist, HS-142-1. Br J Pharmacol 1994;113:931–939.

# 18 Natriuretic Peptides in Cardiovascular Disease

## *A Focus on Congestive Heart Failure*

*J. Aaron Grantham and John C. Burnett, Jr.*

***Contents***



## INTRODUCTION

It is now well established that the heart and vascular endothelium are endocrine glands that synthesize and release natriuretic and vasoactive peptides. Specifically, the cardiac myocyte produces atrial and brain natriuretic peptides (ANP and BNP), whereas endothelial cells produce C-type natriuretic peptide (CNP). All three peptides share structural similarities, but are genetically distinct and have different roles in cardiovascular, renal, and endocrine regulation under physiological and pathophysiological conditions. Studies support the concept that the natriuretic peptides play a homeostatic role to maintain sodium excretion, promote vasodilatation, inhibit hypertrophic responses, and oppose the synthesis and actions of cardiovascular peptides, such as angiotensin II (ANG II) and endothelin (ET-1). Although a number of pathophysiologic states are characterized by involvement of the natriuretic peptide system, congestive heart failure (CHF) emerges as the paradigm in which the natriuretic peptides are activated to their greatest magnitude in nature. This has provided significant insight into the pathophysiology of this endocrine

From: *Contemporary Endocrinology: Natriuretic Peptides in Health and Disease*
Edited by: W. K. Samson and E. R. Levin Humana Press Inc., Totowa, NJ

Table 1
The Natriuretic Peptides: Cardiorenal Sites of Synthesis and Actions

| |
|---|
| Atrial and brain natriuretic peptides |
| Sites of synthesis: atrial and ventricular myocytes |
| Stimuli for synthesis and release: myocardial cell stretch and humoral mediators |
| Biological actions: natriuresis, arterial vasodilatation, inhibition of the RAAS and ET release, sympatho-inhibition, increase in hydraulic conductivity, and antimitogenic |
| C-type natriuretic peptides |
| Sites of synthesis: vascular endothelial and renal epithelial cells |
| Stimuli for synthesis and release: cytokines and humoral mediators |
| Biological actions: arterial venous vasodilatation, inhibition of ET release, and antimitogenic |

system. After reviewing the physiologic role of this system in cardiorenal homeostasis, this chapter will focus on the pathophysiology of the natriuretic peptides in CHF and explore potential therapeutic and diagnostic implications.

## PHYSIOLOGICAL ROLE OF THE NATRIURETIC PEPTIDES IN CARDIORENAL HOMEOSTASIS

The work of Kirsch *(1)* first suggested an endocrine role for the heart with the observation of membrane-bound storage granules, referred to as specific granules in atrial cardiocytes. The density of these granules was reported to be affected by various experimental procedures, such as changes in fluid and electrolyte balance. The physiological significance of these atrial granules was established by de Bold et al. *(2,3)*, who reported the natriuretic and hypotensive effect of iv injection of atrial extracts into rats. This hallmark investigation led to the characterization of ANP as a 28-residue C-terminal peptide that is derived from a 126-amino-acid precursor pro-ANP, which is the principal storage form *(4)*. Repeated studies have demonstrated that ANP possesses unique biological properties that include natriuretic, vasodilatory, renin, aldosterone, and endothelin-inhibiting and antihypertrophic actions (Table 1) *(5,6)*. Brain natriuretic peptide (BNP) is now the second member of the natriuretic peptide family that complements ANP *(7)*. BNP is also translated as a prohormone that when cleaved results in a 32-amino-acid biologically active form in the human *(8)*. BNP, like ANP, is synthesized in cardiac myocytes and released in response to atrial and ventricular stretch, although ANP emerges as the most responsive to acute increases in atrial pressure. BNP is characterized by greater species specificity than ANP, but shares similar biological actions to ANP and therefore the two peptides function as a dual cardiac natriuretic peptide system (Table 1) *(9)*. CNP consists of 22 amino acids and, unlike ANP and BNP, is an endothelial cell-derived peptide that functions as a vasorelaxing and antigrowth factor (Table 1) *(10–12)*.

While the cellular mechanisms for ANP and BNP synthesis and processing continue to emerge, Edwards et al. have shown in an elegant study creating cardiac tamponade in the dog that atrial stretch is the principal stimulus for ANP secretion *(13)*. Cardiac tamponade produced a balanced increase in intra-atrial and pericardial pressures with no change in atrial transmural pressure or atrial stretch. No change in circulating ANP was observed. In contrast, aortic and pulmonary artery constriction resulted in increased transmural pressure and atrial stretch in association with elevated plasma ANP concentrations. This mechanism explains ANP release evoked by a variety of maneuvers and conditions associated with central volume overload *(14,15)*. The cellular mechanism underlying stretch-mediated ANP release is dependent on increases in cytosolic calcium with activation of the phosphoinositide pathway *(16)*. In contrast to ANP, increases in circulating BNP occur hours or days after increases in cardiac filling pressures, which suggests that BNP may serve as a complementary natriuretic peptide that is recruited only if increases in cardiac filling are sustained *(17)*. An exception is the immediate release of BNP in response to myocardial cell ischemia and cell death *(18)*. CNP functions not as a circulating humoral system, but rather a local autocrine or paracrine factor. The stimulus for CNP synthesis and release remains poorly characterized, but cytokine and mechanical stretch have been implicated *(19,20)*.

Three receptors have been identified that mediate the biological actions of ANP, BNP, and CNP. A biologically active receptor, termed the NPR-A receptor, is linked to the activation of particulate guanylyl cyclase. Binding of ANP and BNP to this receptor results in increases in guanosine 3',5'-cyclic monophosphate (cGMP), which leads to the biological actions of the natriuretic peptides in many tissues. CNP functions via a separate guanylyl cyclase-linked receptor that has been termed the natriuretic peptide-B (NPR-B) receptor. The most abundant receptor, however, is a clearance receptor, termed the NPR-C receptor, that functions to bind and clear all three natriuretic peptides from the circulation *(21)*. Following binding, receptor-mediated endocytosis occurs with intracellular degradation. To complement the clearance action of the NPR-C, the ectoenzyme-neutral endopeptidase 24.11 (NEP) functions to cleave the amino side of hydrophilic residues on all three natriuretic peptides, rendering them inactive *(22,23)*. This enzyme is widely distributed in the kidney, lung, heart, and vasculature, is nonsaturable, colocalized with angiotensin-converting enzyme (ACE), and quickly degrades the natriuretic peptides (CNP>ANP>BNP), thus limiting the magnitude and duration of their biologic actions *(24)*. Although the natriuretic peptides function via the second messenger cGMP, evidence supports an additional mechanism of action that is activation of potassium channels resulting in cell membrane hyperpolarization *(25)*. The physiologic actions and signaling pathways of the natriuretic peptides are listed in Table 1 and Fig. 1.

## ACTIVATION OF THE NATRIURETIC PEPTIDE SYSTEM IN HEART FAILURE

A hallmark of chronic CHF is the elevation of circulating ANP and BNP (Fig. 2) *(26,27)*. This elevation is secondary to enhanced cardiac synthesis and release, which are activated by increased cardiac volume and pressure overload. Recent investigations also support the role of humoral stimulation of natriuretic peptide synthesis and release by other local and circulating humoral factors *(28)*. In early CHF, the increase of circulating ANP is secondary to release of stored atrial ANP, with enhanced cardiac synthesis maintaining elevated levels with more sustained ventricular failure *(29)*. In the cardiomyopathic strain of Syrian hamster, the content of ANP granules within the cardiac atria varies inversely with the circulating plasma level of ANP, consistent with increased atrial synthesis and release *(30)*. It is possible that the synthetic capacity of the atria is overwhelmed relative to the demands of the system, leading to a state of relative deficiency with chronic and severe ventricular dysfunction. Indeed, Redfield et al. *(31)* demonstrated in chronic experimental CHF an impaired capacity to release ANP in

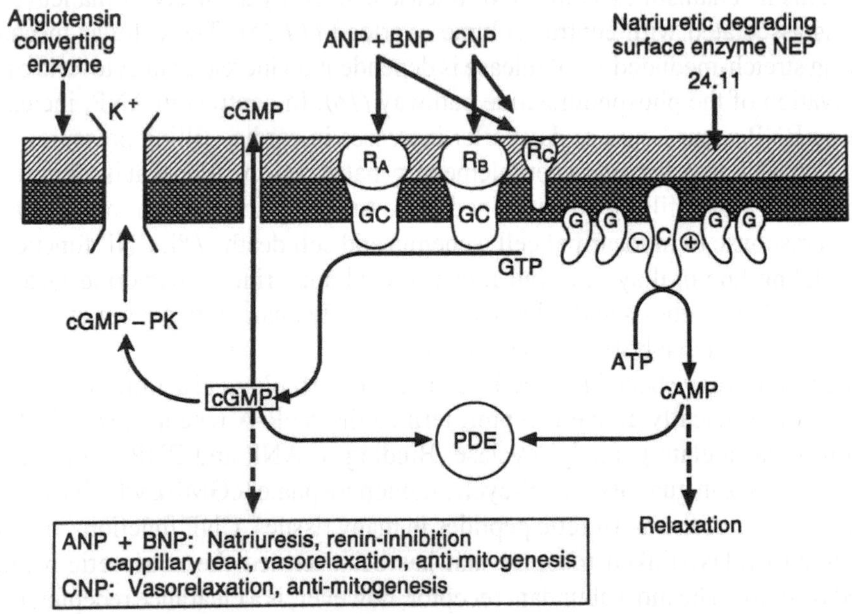


**Fig. 1.** Natriuretic peptide receptor organization and signal transduction mechanisms for ANP, BNP, and CNP with colocalization of NEP and ACE at the cell surface.

response to acute volume expansion with increases in atrial pressures. Recently, Volpe et al. *(32)* have also demonstrated an impaired release of ANP in humans with dilated cardiomyopathy and mild CHF. Thus, a relative deficiency may occur in chronic CHF with biological consequences. Conversely, BNP and CNP are not released in response to acute ventricular overload, such as occurs in early or acute CHF *(17)*. Only with sustained overload does BNP synthesis and release occur. Thus, ANP is the immediately responsive natriuretic peptide and BNP serves as a backup hormone for sustained ventricular overload.

Recent studies have demonstrated increased myocardial CNP concentrations late in the progression of experimental *(33)* and human CHF *(34)*. These studies, together with studies suggesting a role for CNP in the maintenance of vascular tone *(35)* and regulation of cellular growth *(36)*, support the concept that CNP may act as an autocrine or paracrine factor in the regulation of myocardial and vascular growth and function in CHF.

CHF in humans and animals is characterized by the presence of ventricular ANP and BNP, in contrast to normal physiologic conditions, in which these two natriuretic peptides are solely synthesized in atrial myocardium of the human heart (Fig. 3) *(37)*. It should be noted, however, that a role for ventricular BNP synthesis in the normal human heart has been suggested in other studies *(29)*. Since ventricular ANP and BNP are present in primitive organisms, the occurrence in higher species might represent the reactivation of fetal genes. Protooncogenes that regulate the hypertrophic process in cardiomyopathy may therefore also control the recruitment of ventricular ANP and BNP synthesis *(38)*. The importance of ventricular synthesis of the natriuretic peptides in CHF is supported by investigations in cardiomyopathic hamsters that demonstrate that the ventricular myocardium becomes the principal source for circulating natriuretic peptides *(39)*.

A portion of total atrial immunoreactive ANP in the severe failing human heart is composed of β-ANP, an antiparallel dimer with reduced biological activity (Fig. 4). Wei et al. *(40)* have

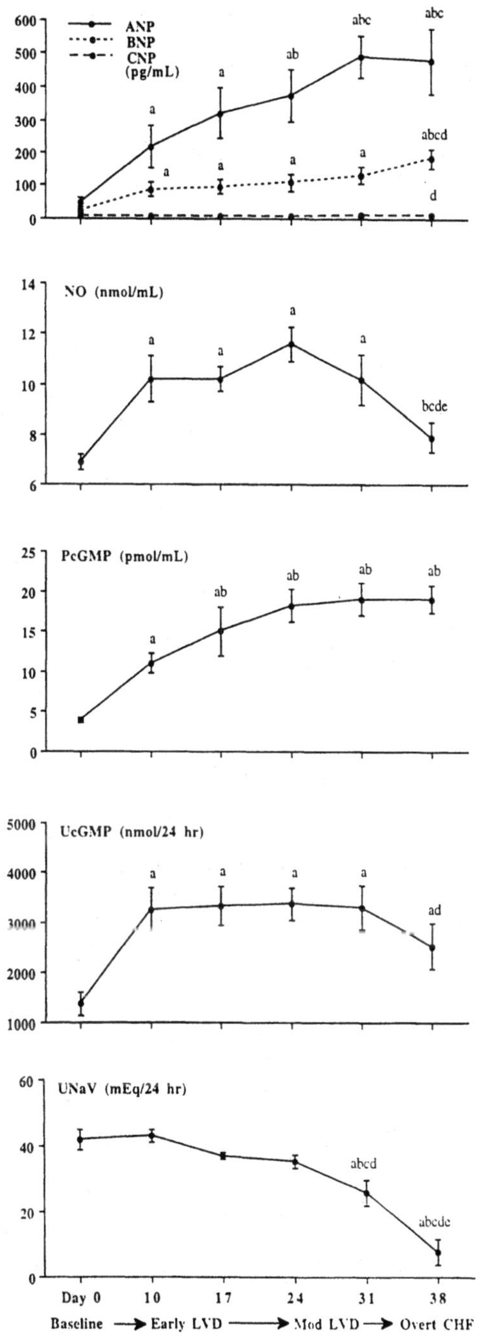


**Fig. 2.** Temporal changes in circulating natriuretic peptides, cGMP, and urinary sodium excretion in experimental progressive left ventricular dysfunction *(33)*.

reported that this altered biological form of ANP is also elevated in the circulation of patients with severe CHF. Thus, plasma β-ANP becomes a major molecular form of ANP in overt heart failure and may have biological significance for this sodium-retaining and vasoconstrictive state. Such a concept is supported by evidence that β-ANP may not effectively bind to the NPR-A receptor, resulting in reduced ability to activate cGMP production *(41)*.

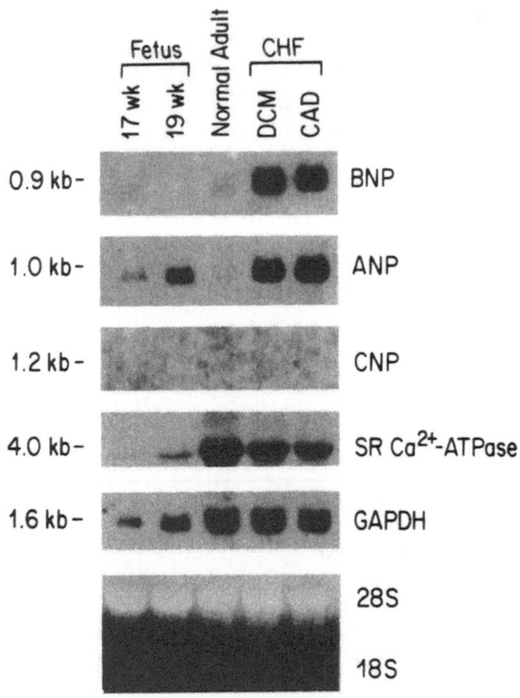


**Fig. 3.** Northern blot analysis of left ventricular natriuretic peptide mRNAs from fetal, normal adult, and failing adult hearts *(37)*.

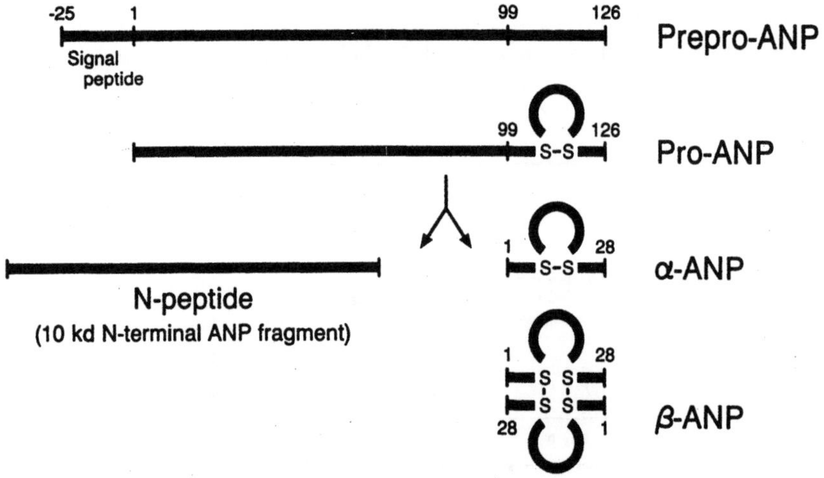


**Fig. 4.** Schematic diagram of the molecular forms of ANP.

Based on its elevation in chronic CHF, circulating ANP has emerged as an important diagnostic and prognostic serum marker in CHF. Repeated studies have demonstrated that elevated plasma ANP correlates with the functional class of symptomatic CHF. Gottlieb et al. *(42)* reported that ANP provides prognostic data on survival, ventricular ectopy, and hemodynamic abnormalities. Davis et al. *(44)* extended these findings and identified ANP as a specific and sensitive test for predicting CHF in elderly subjects.

Table 2
Superiority of BNP as a Hormonal Marker of Ventricular Systolic and Diastolic Dysfunction and Ventricular Hypertrophy at an Abnormal Peptide Level

| *Natriuretic peptide as marker* | *Optimal peptide level* | *Sensitivity* | *Specificity* |
|---|---|---|---|
| EF < 45% | | | |
| BNP | 18 | 0.83 | 0.81 |
| C-ANP | 20 | 0.79 | 0.67 |
| N-ANP | 222 | 0.67 | 0.63 |
| LV mass index > 120 g/m$^2$ | | | |
| BNP | 16 | 0.81 | 0.90 |
| N-ANP | 262 | 0.73 | 0.79 |
| C-ANP | 20 | 0.69 | 0.67 |
| $\tau$ > 55 ms[a] | | | |
| BNP | 16 | 0.85 | 0.74 |
| N-ANP | 253 | 0.70 | 0.61 |
| C-ANP | 21 | 0.60 | 0.65 |

[a]$\tau$ indicates the time constant of LV isovolumic pressure fall. Optimal peptide level was determined as the peptide level at the point closest to that of perfect separation (sensitivity = 1.0, specificity = 1.0) on each ROC curve *(48)*.

These investigators reported that ANP could identify patients at risk for CHF, suggesting that its use in the elderly could be utilized for prevention, early detection, and treatment in this population.

We and others have recently focused on the N-terminus of pro-ANP (N-ANP), which is the nonbiologically active fragment of the prohormone and is coreleased with the biologically active 28-amino acid C-terminal ANP (C-ANP) (Fig. 4). N-ANP is cleared more slowly, and thus circulates at higher concentrations than the biologically active C-ANP. Moreover, it is more stable in vitro than C-ANP. In recent studies, we examined its specificity and sensitivity as a diagnostic test in identifying subjects with early asymptomatic left ventricular dysfunction (ALVD), as documented prospectively with radionuclide angiography and clinical characterization *(44)*. These studies demonstrated that N-ANP was elevated consistently in NYHA class I patients with ALVD and was more sensitive and specific than C-terminal ANP, thus emerging as an important noninvasive serum marker in the identification of patients with ALVD. The diagnostic importance of N-ANP is also underscored by Hall and coworkers *(45)*, who have recently reported that N-ANP is a powerful prognostic indicator in patients with ALVD following acute myocardial infarction.

With the known elevation of plasma BNP in CHF, studies have focused on its diagnostic utility in CHF. Davis et al. reported that elevated BNP was an excellent discriminator of cardiac and noncardiac dyspnea *(46)*. Motwani et al. *(47)* reported that BNP was superior to ANP in correlating with improvement in ejection fraction with ACE inhibition following myocardial infarction. In studies by Yamamoto and coworkers *(48)*, BNP emerged as superior to either N-terminal or C-terminal ANP as a marker for ventricular systolic or diastolic dysfunction and ventricular hypertrophy in patients with, or at risk for, cardiac disease. Specifically, BNP demonstrated greater sensitivity and specificity than N-terminal or C-terminal ANP for reductions in ejection fraction, the time constant at ventricular relaxation, $\tau$, or increases in LV mass (Table 2).

## FUNCTIONAL ROLE OF THE NATRIURETIC PEPTIDE SYSTEM IN THE REGULATION OF CARDIORENAL HOMEOSTASIS AND INHIBITION OF THE RENIN–ANGIOTENSIN SYSTEM IN CHF

Several well-designed therapeutic trials have demonstrated drug efficacy in improving functional capacity and survival in humans with CHF *(49–51)*. The fact that patients classified as New York Heart Association (NYHA) functional class I or II experienced significant benefit, and that drug intervention was not completely effective in halting the progressive worsening process leading to death, has shifted interest to the early stages of heart failure. The NIH-sponsored "Studies of Left Ventricular Dysfunction" (SOLVD) *(52)* in patients with chronic left ventricular dysfunction but without signs of overt CHF (i.e., ALVD) demonstrated humoral activation that was characterized by increases in circulating ANP without activation of the circulating renin–angiotensin–aldosterone system in the absence of diuretic treatment *(53)*. Based on the known biology of the natriuretic peptide system, this humoral system may play a key role in preserving the compensated state of ALVD.

Evidence supports an important role for the natriuretic peptide system in ALVD to preserve cardiorenal homeostasis and contribute to the maintenance of sodium balance and to inhibit the RAAS despite ventricular dysfunction. In a low-ANP model of acute CHF produced by thoracic inferior vena caval constriction (TIVCC) *(54)*, which is characterized by decreased cardiac output without increases in atrial pressures or ANP, marked sodium retention, vasoconstriction, and activation of the RAAS resulted. These findings were not observed in a high-ANP model produced by rapid ventricular pacing in the same study despite similar reductions in cardiac output and mean arterial pressure. Exogenous administration of ANP in TIVCC to mimic circulating concentrations encountered in the high-ANP heart failure model prevented sodium retention, vasoconstriction, and activation of the RAAS.

Redfield et al. *(55)* have reported cardiorenal function in a conscious canine model of ALVD. This model mimics the humoral profile of patients with ALVD reported in SOLVD and was characterized by significant ventricular dysfunction but without sodium retention in association with elevated ANP and no activation of the RAAS. In response to acute intravascular volume expansion, a normal release of ANP and an intact renal natriuretic response were observed. More recently, Stevens et al. established that the transition from experimental ALVD to overt CHF could be accelerated utilizing a natriuretic peptide receptor antagonist that in ALVD lead to premature sodium retention, impaired renal natriuretic response to volume expansion, and activation of the RAAS *(56)*. In addition, these investigators suppressed activation of the natriuretic peptide system by surgically removing both the right and left atrial appendages, where ANP is richest in abundance. During the initial phase of ALVD, in the absence of the atrial appendages, the increase in plasma ANP and cGMP was blunted in association with activation of the RAAS (Fig. 5). In addition, there was the early onset of sodium retention secondary to enhanced tubular sodium reabsorption in the group with suppressed ANP (Fig. 6). These investigators concluded that increased ANP is activated with cardiac volume overload in early CHF, which contributes to the maintenance of sodium balance and inhibition of the RAAS. They speculated that therapeutic strategies that potentiate the biological actions of ANP as well as BNP may prolong the asymptomatic phase of ventricular dysfunction and delay progression to overt CHF. These studies are now

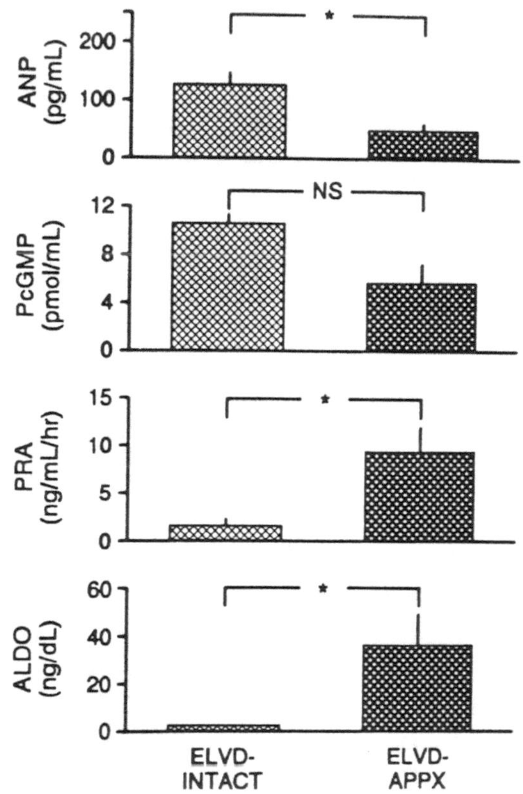


**Fig. 5.** Circulating neurohumoral activity in early left ventricular dysfunction (ELVD) in the presence (ELVD) and absence of atrial appendages (ELVD-APPX) *(56)*.

complemented by investigations in more advanced CHF in which natriuretic peptide receptor blockade augments the magnitude of sodium retention *(57,58)*.

Endogenous CNP may complement ANP and BNP in CHF but with an important role in vascular homeostasis. Supaporn et al. *(59)* recently reported that intracoronary administration of a natriuretic peptide receptor antagonist increased coronary vascular resistance in the normal dog. Preliminary studies have extended this to a model of CHF (unpublished data). Based on the known coronary vasorelaxing actions of CNP *(25)*, one can extend to the coronary vasculature a functional role for the natriuretic peptide system in CHF to preserve myocardial perfusion.

## RENAL HYPORESPONSIVENESS TO ANP AND BNP IN CHF AND RELATIONSHIP TO ANGIOTENSIN II

Overt CHF, unlike ALVD, is a syndrome characterized by sodium retention, activation of the RAAS, and elevation of both circulating ANP and BNP. Human and animal models of chronic CHF are characterized by an attenuated natriuretic response to endogenous and exogenous ANP *(60,61)*. It has been suggested that the diminished renal response to the cardiac natriuretic peptides plays an important role in the pathophysiology of sodium retention and systemic and renal vasoconstriction observed in overt heart failure. The mechanism(s) responsible for the renal hyporesponsiveness to cardiac natri-

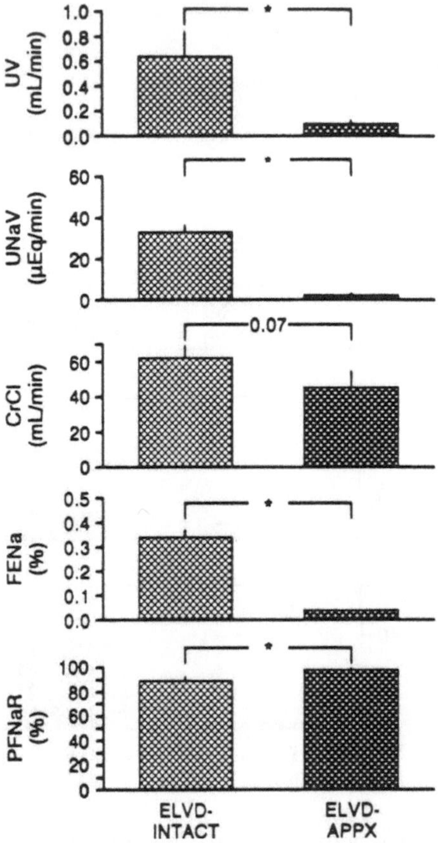


**Fig. 6.** Renal function in ELVD in the presence (ELVD) and absence of atrial appendages (ELVD-APPX) *(56)*.

uretic peptides in CHF are most likely multifactorial and include decreased renal perfusion pressure *(62)*, increased renal sympathetic nerve activity *(63)*, receptor downregulation *(64)*, and possibly enhanced enzymatic degradation of ANP and BNP by NEP *(65)*. A key role for ANG II in mediating the renal hyporesponsiveness to the natriuretic peptide system appears to be fundamental to CHF.

As can be seen in Fig. 7, the natriuretic peptides and ANG II have renal actions at the same vascular and tubular sites within the kidney. Thus, ANG II may oppose the renal effects of ANP at both the glomerulus and the renal tubule, preventing the full natriuretic activity of this peptide and thus contributing to sodium retention and edema formation of CHF.

Several studies support this concept that increased ANG II plays a key role in mediating the blunted renal response to the natriuretic peptides in overt CHF. The natriuretic response to exogenous ANP in salt-retaining rats with chronic arteriovenous fistula can be restored with chronic treatment with the ACE inhibitor enalapril *(66)*. This response occurred despite a decrease in mean arterial blood pressure and was unrelated to changes in endogenous levels of ANP. The mechanism of this enhancement secondary to decreased generation of ANG II probably includes reductions in renal vascular resistance and decrements in cGMP phosphodiesterase. Further, the opposing action of ANG II on ANP is supported by the work of Showalter et al. *(67)*, which observed a blunting of the

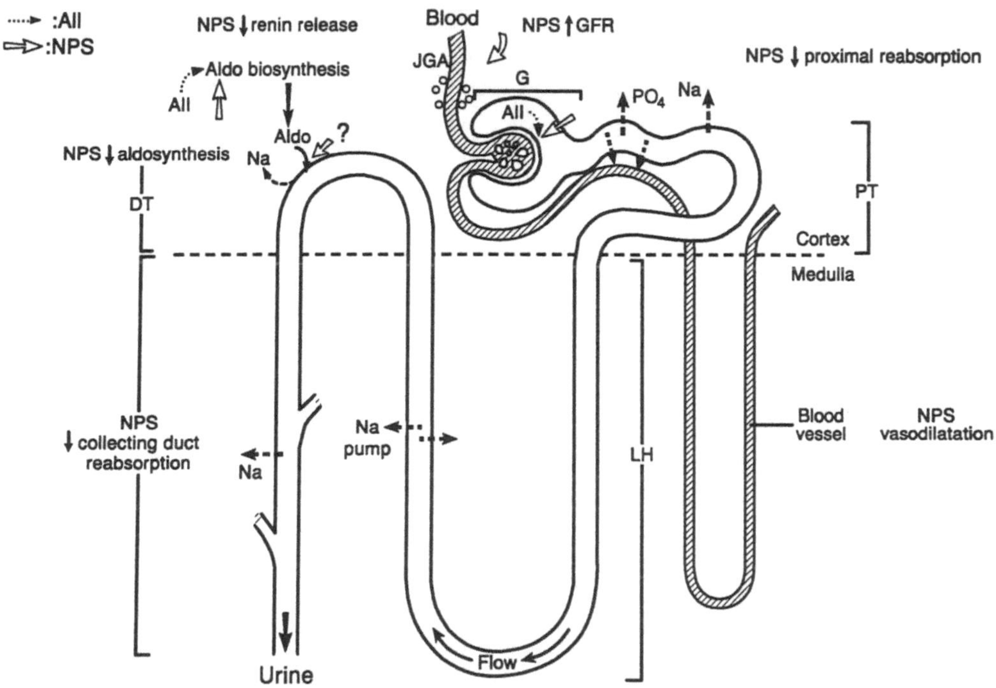


**Fig. 7.** Schematic diagram of the renal sites of action of the natriuretic peptides (NPS) and angiotensin II.

natriuretic effect to systematically administered ANP when ANG II was infused via the renal artery at a dose without systemic effects, indicating that at normal renal perfusion pressure, intrarenal ANG II can antagonize the natriuretic response to ANP.

Results from our laboratory indicate that sodium retention during the evolution of CHF occurs in association with a loss of the previously enhanced renal generation of cGMP and occurs in association with an increase in ANG II *(61)*. Smith and Lincoln *(68)* observed that ANG II decreased ANP-stimulated intracellular cGMP accumulation by stimulating cGMP hydrolysis in cultured vascular smooth-muscle cells. This augmented hydrolysis of cGMP appeared to be mediated via a $Ca^{2+}$-activated cGMP phosphodiesterase. In vivo, studies have demonstrated that cGMP phosphodiesterase inhibition markedly potentiates the renal response to acute volume expansion and low-dose ANP, an effect that was attenuated by administration of a monoclonal antibody (MAb) directed against ANP *(69)*. More recently, Supaporn et al. *(70)* have reported the increases in type I and type V cGMP phosphodiesterases in CHF that therefore could mediate the decrease in cGMP generation to ANP in CHF. In these studies in isolated glomeruli from kidneys of dogs with overt CHF, ANP cGMP generation was attenuated when compared to control. In an additional study, type V phosphodiesterase inhibition in experimental CHF enhanced sodium excretion and renal cGMP generation *(71)*. Thus, these studies support an important role for renal cGMP phosphodiesterase modulation of the renal response to endogenous and exogenous ANP. Moreover, since the renal glomerular cGMP response has been reported to be attenuated by ANG II, this peptide (ANG II) may be a key mechanism to explain the activation of cGMP phosphodiesterase activity in CHF and the renal hyporesponsiveness that characterizes this syndrome. Finally, the report that

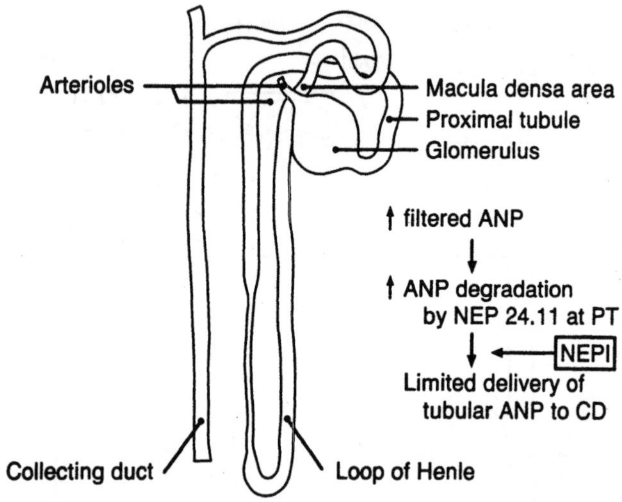


**Fig. 8.** The NEP-I hypothesis that illustrates the concept of NEP degradation of ANP at the proximal nephron, limiting delivery of biologically active peptide to receptor sites in the terminal nephron, and therapeutic role of enzyme inhibition.

intrarenal ANG II is natriuretic in ALVD but markedly increased in overt CHF also supports an important interaction between the natriuretic peptide and renin–angiotensin system in evolving CHF *(72)*.

## POTENTIATION OF ANP AND BNP IN THE THERAPEUTICS OF CHF

Since the discovery of ANP and subsequently BNP, strategies have emerged to utilize this peptide system in the treatment of disorders of cardiorenal function, such as CHF. Therapeutic strategies have included infusion of the peptides and potentiation their actions by inhibitors of cGMP phosphodiesterases. Recent therapeutic approaches have focused on inhibition of the enzyme that degrades the natriuretic peptides, neutral endopeptidase 24.11, which is colocalized with ACE (Fig. 1).

As has been discussed above, two pathways exist for natriuretic peptide metabolism; enzymatic degradation via the ectoenzyme-neutral endopeptidase 24.11 (NEP) and receptor-mediated endocytosis via the clearance receptor, NPR-C. With regard to the enzymatic pathway, endogenous natriuretic peptide degradation occurs through a membrane ectoenzyme system, neutral endopeptidase 24.11 (NEP), which cleaves peptides at the amino side of hydrophobic residues. The vasoactive peptide substrates of NEP include the natriuretic peptides, bradykinin, substance P, angiotensins, and endothelins *(73)*. Kenny and Stephenson *(74)* suggested that renal tubular NEP actually serves a physiologic function to rapidly and efficiently degrade filtered ANP to prevent biologically intact peptide from reaching the terminal nephron, which is a major site of action *(75)*.

In acute experimental CHF, we found a parallel increase of urinary cGMP excretion and urinary sodium excretion after neutral endopeptidase inhibition (NEPI) together with an increase in urinary ANP, suggesting that delivery of ANP to the terminal nephron indeed may activate cGMP and contribute to a natriuretic response *(76)*. This supports the hypothesis that NEP may limit the full natriuretic action of elevated endogenous ANP in CHF (Fig. 8). Such observations are also supported by studies demonstrating a luminal

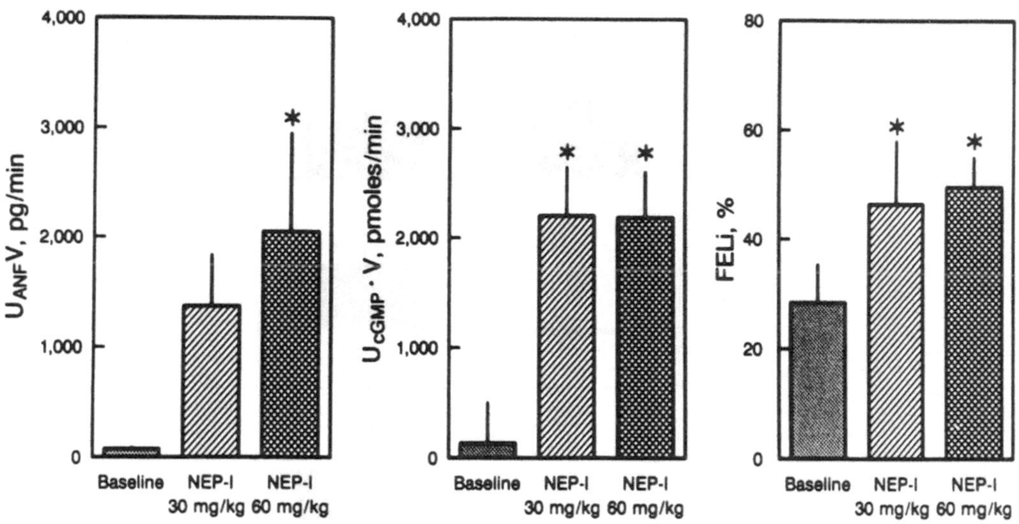


**Fig. 9.** Urinary ANP and cGMP excretion and fractional lithium excretion at baseline (solid bars) and during candroxatrilat infusion 30 (hatched bars) and 60 mg/kg (checked bars) *(65)*.

action of ANP in the inner medullary collecting duct to inhibit sodium transport, as discussed above *(75)*. In chronic experimental CHF, NEPI produced a decrease in atrial pressures and an initial maintenance of cardiac output. NEPI potentiated the natriuretic action of endogenous ANP by a mechanism independent of systemic or renal hemodynamics *(65)* (Fig. 9). More striking, this natriuretic action occurred in a model of severe CHF that was resistant to the natriuretic action of exogenous ANP, underscoring the potential efficacy of this form of therapy. The observed biologic responses in these studies in acute and chronic experimental CHF were also greater than that predicted from any increase in circulating ANP, indicating a local tissue potentiation of ANP or BNP. The unique natriuretic action of NEPI may also be associated with cardiovascular and humoral effects. Recently, Riegger and coworkers *(77)* have reported that chronic NEPI in humans with chronic CHF resulted in favorable hemodynamic responses together with suppression of vasoconstrictor humoral systems.

We discussed above the antagonism of ANP and BNP by ANG II. Such an antagonism is also observed in severe CHF to limit the full natriuretic action of NEPI. Chronic angiotensin antagonism with ACE inhibition potentiated both renal hemodynamic and excretory response to NEPI, an effect that was abolished by intrarenal infusion of low-dose ANG II *(78)* (Fig. 10). These latter studies support the concept that coinhibition of both NEP and ACE may emerge as a unique form of therapy in the treatment of CHF. Indeed, based on the knowledge of the delay in onset of symptoms with ACE inhibition in ALVD, one could speculate that coinhibition of both these ectoenzyme systems may indeed emerge as optimal therapy in delaying the transition to overt CHF.

## SUMMARY

A hallmark of the syndrome of CHF is the activation of local and circulating humoral systems. The natriuretic peptide system, specifically the cardiac hormones ANP and BNP, emerges as a humoral mechanism that may delay the progression of CHF, in part

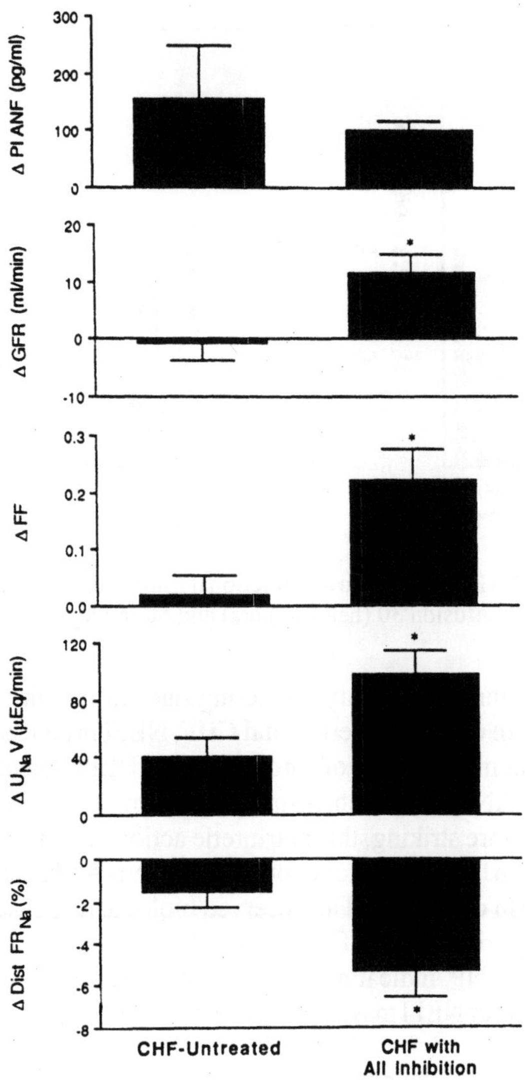


**Fig. 10.** Changes in renal hemodynamics and sodium handling during CHF in the presence (CHF untreated) and absence of converting enzyme inhibition CHF with ANG II inhibition *(78)*.

by its renal, vascular, and endocrine actions. The increase in circulating natriuretic peptides, which reflect activation of myocardial synthesis, may also serve important diagnostic and prognostic roles in the management of CHF. Finally pharmacological and perhaps even genetic potentiation of this system via inhibition of its degradation by NEP or enhancing the local expression of these peptides may complement current therapies for CHF. Clearly, the natriuretic peptides have become established in the pathophysiology, diagnostics, and therapeutics of this important cardiovascular disease state.

## REFERENCES

1. Kirsch B. Electron microscopy of the atrium of the heart. I. Guinea pig. Exp Med Surg 1956;14:99–112.
2. de Bold AJ. Heart atria granularity effects of changes in water–electrolyte balance. Proc Soc Exp Biol Med 1979;161:508–511.

3. de Bold AJ, Borenstein HB, Veress AT, Sonnenberg H. A rapid and potent natriuretic response to intravenous injection of atrial myocardial extract in rats. Life Sci 1981;28:89–94.
4. Bloch KD, Zisfein JB, Margolies MN, et al. A serum protease cleaves proANF into a 14-kilodalton peptide and ANF. Am J Physiol 1987;252:E147–151.
5. Burnett JC Jr, Granger JP, Opgenorth TJ. Effects of synthetic atrial natriuretic factor on renal function and renin release. Am J Physiol 1984;247:F863–866.
6. Cao L, Gardner DG. Natriuretic peptides inhibit DNA synthesis in cardiac fibroblasts. Hypertension 1995;88:81–86.
7. Sudoh T, Kangawa K, Minamino N, Matsuo H. A new natriuretic peptide in porcine brain. Nature 1988;332:78–81.
8. Ogawa Y, Nakao K, Mukoyama M, et al. Rat brain natriuretic peptide: tissue distribution and molecular form. Endocrinology 1990;126:2225–2227.
9. Mukoyama M, Nakao K, Hosoda K, et al. Brain natriuretic peptide as a novel cardiac hormone in humans. Evidence for an exquisite dual natriuretic peptide system, atrial natriuretic peptide and brain natriuretic peptide. J Clin Invest 1991;87:1402–1412.
10. Sudoh T, Minamino N, Kangawa K, Matsuo H. C-type natriuretic peptide (CNP): a new member of the natriuretic peptide family identified in porcine brain. Biochem Biophys Res Commun 1990;168:863–870.
11. Stingo AJ, Clavell AL, Heublein DM, Wei CM, Pittelkow MR, Burnett JC Jr. Presence of C-type natriuretic peptide in cultured human endothelial cells and plasma. Am J Physiol 1992;263:H1318–H1321.
12. Furuya M, Yoshida M, Hayashi Y, Ohnuma N, Minamino N, Kangawa K, Matsuo H. C-type natriuretic peptide is a growth inhibitor of rat vascular smooth muscle cells. Biochem Biophys Res Commun 1991;177:927–931.
13. Edwards BS, Zimmerman RS, Schwab TR, Heublein DM, Burnett JC Jr. Atrial stretch, not pressure, is the principal determinant controlling the acute release of atrial natriuretic factor. Circ Res 1988;62:191–195.
14. Schwab TR, Edwards BS, Heublein DM, Burnett JC Jr. Role of atrial natriuretic peptide in volume-expansion natriuresis. Am J Physiol 1986;251:R310–313.
15. Riegger GAJ, Elsner D, Forssman WG, Kromer EP. Effects of ANP-(95–126) in dogs before and after induction of heart failure. Am J Physiol 1990;259:H1643–1648.
16. Suzuki E, Hirata Y, Kohmoto O, et al. Cellular mechanisms for synthesis and secretion of atrial natriuretic peptide and brain natriuretic peptide in cultured rat atrial cells. Cir Res 1992;33:713–721.
17. Grantham JA, Borgeson DD, Burnett JC Jr. Brain natriuretic peptide in acute congestive heart failure: pathophysiologic and potential therapeutic implications. Am J Physiol 1997;R1077–1083.
18. Morita E, Yasue H, Yoshimura M, Ogawa H, Jougasaki M, Matsumura T, Mukoyama M, Nakao K. Increased plasma levels of brain natriuretic peptide in patients with acute myocardial infarction. Circulation 1993;88:82–91.
19. Suga S, Itoh H, Komatsu Y, Ogawa Y, Hama N, Yoshimasa Y, Nakao K. Cytokine-induced C-type natriuretic peptide (CNP) secretion from vascular endothelial cells. Endocrinology 1993;133:3038–3041.
20. Suga S, Nakao K, Itoh H, Komatsu Y, Ogawa Y, Hama N, Imura H. Endothelial production of C-type natriuretic peptide and its marked augmentation by transforming growth factor beta: possible existence of a vascular natriuretic peptide system. J Clin Inv 1992;90:1145–1149.
21. Maack T, Suzuki M, Almeida FA, Nussenzveig D, Scarborough RM, McEnroe GA, Lewicki FA. Physiological role of silent receptors of atrial natriuretic factor. Science 1987;238:675–678.
22. Kerr MA, Kenny AJ. The purification and specificity of a neutral endopeptidase from rabbit kidney brush border. Biochem J 1974;137:477–488.
23. Erdos EG, Skidgel RA. Neutral endopeptidase 24.11 (enkephalinase) and related regulators of peptide hormones. FASEB J 1989;3:145–151.
24. Skidgel RA, Engelbrecht S, Johnson AR, Erdos EG. Hydrolysis of substance P and neurotensin by converting enzyme and neutral endopeptidase. Peptides 1984;5:769–776.
25. Wei CM, Hu S, Miller VM, Burnett JC Jr. Vascular actions of C-type natriuretic peptide in isolated porcine coronary arteries and coronary vascular smooth muscle cells. Biochem Biophys Res Commun 1994;205:765 771.
26. Burnett JC Jr, Kao PC, Hu DC, et al. Atrial natriuretic peptide elevation in congestive heart failure in the human. Science 1986;231:1145–1147.
27. Yasue H, Yoshimura M, Sumida H, Kikuta K, Kugiyama K, Jougasaki M, Ogawa H, Okumura K, Mukoyama M, Nakao K. Localization and mechanism of secretion of B-type natriuretic peptide in

comparison with those of A-type natriuretic peptide in normal subjects and patients with heart failure. Circulation 1994;90:195–203.
28. Thibault G, Doubell AF, Garcia R, Lariviere R, Schiffrin EL. Endothelin-stimulated secretion of natriuretic peptides by rat atrial myocytes is mediated by endothelin A receptors. 1994;74:460–470.
29. Perrella MA, Schwab TR, O'Murchu B, Redfield MM, Wei CM, Edwards BS, Burnett JC Jr. Cardiac atrial natriuretic factor during evolution of congestive heart failure. Am J Physiol 1992;262:H1248–1255.
30. Edwards BS, Ackermann DM, Lee MU, Reeder GS, Wold LE, Burnett JC Jr. Identification of atrial natriuretic factor within ventricular tissue in hamsters and humans with congestive heart failure. J Clin Invest 1988;81:82–86.
31. Redfield MM, Edwards BS, McGoon MD, Heublein DM, Aarhus LL, Burnett JC Jr. Failure of atrial natriuretic factor to increase with volume expansion in acute and chronic congestive heart failure in the dog. Circulation 1989;80:651–657.
32. Volpe M, Tritto C, De Luca N, et al. Failure of atrial natriuretic factor to increase with saline load in patients with dilated cardiomyopathy and mild heart failure. J Clin Invest 1991;88:1481–1489.
33. Stevens TL, Borgeson DD, Luchner A, Wennberg PW, Redfield MM, Burnett JC Jr. A modified model of tachycardia-induced cardiomyopathy: insights into humoral and renal adaptations. In: Spinale F, ed. Tachycardia-Induced Cardiomyopathy. Futura, Armonk, NY; 1996:235.
34. Wei CM, Heublein DM, Perrella MA, Lerman A, Rodeheffer RJ, McGregor CGA, Edwards WD, Schaff HV, Burnett JC Jr. Natriuretic peptide system in human heart failure. Circulation 1993;88:1004–1009.
35. Wei C-M, Aarhus LL, Miller VM, Burnett JC Jr. Action of C-type natriuretic peptide in isolated canine arteries and veins. Am J Physiol 1993;264:H71–73.
36. Kullo I, Burnett JC Jr. C-type natriuretic peptide; the vascular component of the natriuretic peptide system. In: Sowers JR, ed. Contemporary Endocrinology: Endocrinology of the Vasculature. Humana, Totowa, NJ, 1996, pp. 79–93.
37. Takahashi T, Allen P, Izumo S. Expression of A-, B-, and C-type natriuretic peptide genes in failing and developing human ventricles. Circ Res 1992;71:9–17.
38. Izumo S, Nakal-Ginard B, Mahdavi V. Protooncogene induction and reprogramming of cardiac gene expression produced by pressure overload. Proc Natl Acad Sci USA 1988;85:339–343.
39. Thibault G, Nemer M, Drouin J, et al. Ventricles as a major site of atrial natriuretic factor synthesis and release in cardiomyopathic hamsters with heart failure. Circ Res 1989;65:71–82.
40. Wei C-M, Kao P, Lin JT, Heublein D, Schaff HV, Burnett JC Jr. Circulating β-atrial natriuretic factor in congestive heart failure. Circulation 1993;88:1016–1020.
41. Kambayashi Y, Nakajima S, Ueda M, Inouye K. A dicarba analog of beta-atrial natriuretic peptide (beta-ANP) inhibits guanosine 3',5'-cyclic monophosphate production induced by alpha-ANP in cultured rat vascular smooth muscle cells. FEBS Lett 1989;248:28–34.
42. Gottlieb SS, Kukin ML, Ahern D, Packer M. Prognostic importance of atrial natriuretic peptide in patients with chronic heart failure. J Am Coll Cardiol 1989;13:1534–1539.
43. Davis KM, Fish LC, Elahi D, Clark BA, Minaker KL. Atrial natriuretic peptide levels in the prediction of congestive heart failure risk in frail elderly. JAMA 1992;267:2625–2629.
44. Lerman A, Gibbons RJ, Rodeheffer RJ, et al. Circulating N-terminal atrial natriuretic peptide as a marker for symptomless left-ventricular dysfunction. Lancet 1993;341:1105–1109.
45. Hall C, Rouleau JL, Moye L, et al. N-terminal proatrial natriuretic factor. An independent predictor of long term prognosis after myocardial infarction. Circulation 1994;89:1934–1942.
46. Davis M, Espiner E, Richard G, Billings J, Town I, Neill A, Drennan C, Richards M, Turner J, Yandle T. Plasma brain natriuretic peptide in assessment of acute dyspnea. Lancet 1994;343:440–444.
47. Motwani JG, McAlpine H, Kennedy N, Struthers AD. Plasma brain natriuretic peptide as an indicator for angiotensin-converting-enzyme inhibition after myocardial infarction. Lancet 1993;341:1109–1113.
48. Yamamoto K, Burnett JC Jr, Jougasaki M, Nishimura RA, Bailey KR, Saito Y, Nakao K, Redfield MM. Superiority of brain natriuretic peptide as a hormonal marker of ventricular systolic and diastolic dysfunction and ventricular hypertrophy. Hypertension 1996;28:988–994.
49. Cohn JN, Archibald DG, Ziesche S, et al. Effect of vasodilator therapy on mortality in chronic congestive heart failure. Results of a Veterans Administration Cooperative Study. N Engl J Med 1986;314:1547–1552.
50. Cohn JN, Johnson G, Ziesche S, et al. A comparison of enalapril with hydralazine-isosorbide dinitrate in the treatment of chronic congestive heart failure. N Engl J Med 1991;325:303–310.

51. The CONSENSUS Trial Study Group. Effects of enalapril on mortality in severe congestive heart failure. Results of the Cooperative North Scandinavian Enalapril Survival Study (CONSENSUS). N Engl J Med 1987;316:1429–1435.
52. The SOLVD Investigators. Effect of enalapril on survival in patients with reduced left ventricular ejection fractions and congestive heart failure. N Engl J Med 1991;325:293–302.
53. Francis GS, Benedict C, Johnstone DE, et al. Comparison of neuroendocrine activation in patients with left-ventricular dysfunction with and without congestive heart failure. Circulation 1990;82: 1724–1729.
54. Lee ME, Miller WL, Edwards BS, Burnett JC Jr. Role of endogenous atrial natriuretic factor in acute congestive heart failure. J Clin Invest 1989;84:1962–1966.
55. Redfield MM, Aarhus LL, Wright RS, Burnett JC Jr. Cardiorenal and neurohumoral function in a canine model of early left ventricular dysfunction. Circulation 1993;87:2016–2022.
56. Stevens TL, Burnett JC Jr, Kinoshita M, Matsuda Y, Redfield MM. A functional role for endogenous atrial natriuretic peptide in a canine model of early left ventricular dysfunction. J Clin Invest 1995;95:1101–1108.
57. Zhang PL, Mackenzie HS, Totsune K, Troy JL, Brenner BM. Renal effects of high-dose natriuretic peptide receptor blockade in rats with congestive heart failure. Circ Res 1995;77:1240–1245.
58. Wada A, Tsutamoto T, Matsuda Y, Kinoshita M. Cardiorenal and neurohumoral effects of endogenous atrial natriuretic peptide in dogs with severe congestive heart failure using a specific antagonist for guanylate cyclase-coupled receptors. Circulation 1994.;89:2232–2240.
59. Supaporn T, Wennberg PW, Wei CM, Kinoshita M, Matsuda Y, Burnett JC Jr. Role for the endogenous natriuretic peptide system in the control of basal coronary vascular tone in dogs. Clin Sci 1996;90:357–362.
60. Scriven TA, Burnett JC Jr. Effects of synthetic atrial natriuretic peptide on renal function and renin release in acute experimental heart failure. Circulation 1985;72:892–897.
61. Margulies KB, Heublein DM, Perrella MA, Burnett JC Jr. ANF-mediated renal cGMP generation in congestive heart failure. Am J Physiol 1991;260:F562–F568.
62. Redfield MM, Edwards BS, Heublein DM, Burnett JC Jr. Restoration of renal response to atrial natriuretic factor in experimental low-output heart failure. Am J Physiol 1989;257:R917–923.
63. Koepke JP, Di Bona GF. Blunted natriuresis to atrial natriuretic peptide in chronic sodium retaining disorders. Am J Physiol 1987;252:854–871.
64. Schiffrin EL. Decreased density of binding sites for atrial natriuretic peptide on platelets of patients with severe congestive heart failure. Clin Sci 1988;74:213–218.
65. Cavero PG, Margulies KB, Winaver J, Seymour AA, Delaney NG, Burnett JC Jr. Cardiorenal actions of neutral endopeptidase inhibition in experimental congestive heart failure. Circulation 1990;82:196–201.
66. Abassi Z, Haramati A, Hoffman A, Burnett JC Jr, Winaver J. Effect of converting-enzyme inhibition on renal response to ANF in rats with experimental heart failure. Am J Physiol 1990;259:R84–89.
67. Showalter CJ, Zimmerman RS, Schwab TR, Edwards BS, Opgenorth TJ, Burnett JC Jr. Renal response to atrial natriuretic factor is modulated by intrarenal angiotensin II. Am J Physiol 1988;254: R453–456.
68. Smith JB, Lincoln TM. Angiotensin decreases cyclic GMP accumulation produced by atrial natriuretic factor. Am J Physiol 1987;253:C147–150.
69. Wilkins MR, Settle SL, Needleman P. Augmentation of the natriuretic activity of exogenous and endogenous atriopeptin in rats by inhibition of guanosine 3',5'-cyclic monophosphate degradation. J Clin Invest 1990;85:1274–1279.
70. Supaporn T, Sandberg SM, Borgeson DD, Heublein DM, Luchner A, Wei CM, Dousa TP, Burnett JC Jr. Blunted cGMP response to agonists and enhanced glomerular cyclic 3'5'-nucleotide phosphodiesterase activities in experimental congestive heart failure. Kidney Int 1996;50:1718–1725.
71. Margulies KB, Burnett JC Jr. Inhibition of cyclic GMP phosphodiesterases augments renal responses to atrial natriuretic factor in congestive heart failure. J Cardiac Failure 1994;1:71–80.
72. Luchner A, Stevens TL, Borgeson DD, Redfield MM, Bailey JE, Sandberg SM, Heublein DM, Burnett JC Jr. Angiotensin II in the evolution of experimental heart failure. Hypertension 1996;28:472–477.
73. Margulies KB, Barclay PL, Burnett JC Jr. The role of neutral endopeptidase in dogs with evolving congestive heart failure. Circulation 1995;91:2036–2042.
74. Kenny AJ, Stephenson SL. Role of endopeptidase-24.11 in the inactivation of atrial natriuretic peptide. FEBS Lett 1988;232:1–8.

75. Sonnenberg H. Mechanism of release and renal tubular action of ANF. Fed Proc Nat Acad Sci 1997;45:2106–2110.
76. Perrella MA, Margulies KB, Wei CM, Aahrus LL, Heublein DM, Burnett JC Jr. Pulmonary and urinary clearance of atrial natriuretic factor in acute congestive heart failure in dogs. J Clin Invest 1991;87: 1649–1655.
77. Elsner D, Muntze A, Kromer EP, Riegger GA. Effectiveness of endopeptidase inhibition (candoxatril) in congestive heart failure. Am J Cardiol 1992;70:494–498.
78. Margulies KB, Perrella MA, McKinley LJ, Burnett JC Jr. Angiotensin inhibition potentiates the renal responses to neutral endopeptidase inhibition in dogs with congestive heart failure. J Clin Invest 1991;88:1636–1642.

# PERSPECTIVE

Since the fundamental observations of DeBold and colleagues in 19881, the maturation of the natriuretic peptide field has been highlighted by the uncovering of multiple members of this protein family, the cloning of novel receptors, and the discovery of substantial unanticipated functions. A definitive role for the natriuretic peptides in regulating blood pressure, as well as salt and fluid homeostasis, has emerged from genetic inactivation models, and the use of antagonists to the guanylate cyclase receptors. As delineated in this text, newer reagents that interrupt signaling by specific receptors promise to further refine our understanding of the endocrine, paracrine, and autocrine functions of the natriuretic peptides.

A number of important controversies and questions remain in understanding the functions of these protein hormones. For instance, why do the ANP and GC-A gene in activation models not duplicate and thereby support the role of this system in defending against salt-induced hypertension? What are the additional roles, in vivo, of the natriuretic peptide clearance receptor, and what additional signaling mechanisms link these actions? Are the amino-terminal fragments of pro-ANP physiologically relevant? Are ANP and BNP redundant peptides, evolutionarily speaking, or do they each subserve unique, and as of yet undiscovered, functions? Is there a unique BNP receptor? What are the paracrine roles for CNP, amd under what circumstances are the actions of this peptide relevant? These and many other important questions can be addressed using the models and tools now available to disect the molecular, cellular, and whole organism physiology of the natriuretic peptides.

***Ellis R. Levin***

# INDEX

**D**



**E**



**F**

## R



## S



## T



## U



## V

MIX
Papier aus verantwortungsvollen Quellen
Paper from responsible sources
FSC® C105338

If you have any concerns about our products,
you can contact us on
**ProductSafety@springernature.com**

In case Publisher is established outside the EU,
the EU authorized representative is:
**Springer Nature Customer Service Center GmbH**
**Europaplatz 3, 69115 Heidelberg, Germany**

Printed by Libri Plureos GmbH
in Hamburg, Germany